T0199270

V. L. Chopra, PhD
K. V. Peter, PhD
Editors

Handbook
of Industrial Crops

Pre-publication
REVIEWS,
COMMENTARIES,
EVALUATIONS . . .

"**T**he *Handbook of Industrial Crops* has brought together a wealth of information on industrial crops that is not readily available in books previously written on each of the crops covered. The authors demonstrate their wealth of experience and expertise with the comprehensive and detailed way in which each crop is treated. This book will be of immense value to a wide audience, including academicians, industrialists, policymakers, extension workers, and students in tertiary institutions. The information provided in the book will bring about fresh ideas for research and development, as well as further expansion of the market for industrial crops."

Lanre Denton, PhD
Assistant Director,
National Horticultural
Research Institute,
Ibadan, Nigeria

"From the dawn of agriculture over 10,000 years ago, women and men have been screening native vegetation for plants of potential human and economic significance. Numerous plants, both annual and perennial, have been domesticated for use as food, medicine, and for the manufacture of a wide variety of industrial products during the course of the evolution of agriculture. It is a tribute to the power of observation of rural and tribal women and men that they have been able to identify from wild flora nearly all plants of significance to human life and economy. During the last couple of centuries we have been able to add to the list of domesticated flora only microorganisms and a few horticultural plants. However, modern biotechnology, particularly genetic engineering, is enabling us to go beyond the already domesticated plants for enlarging the gene pool for the isolation and incorporation of new genes of economic value across sexual barriers. Therefore, we are now conscious of the need for conserving genetic variability in macro- and microflora irrespective of their current economic value.

It is in the above context that the present *Handbook of Industrial Crops* is a timely contribution. It deals with major industrial crops such as arecanut, cardamom, cashew, cinchona, cocoa, coconut, coffee, oil palm, palmyra, rubber, tea, and wattle. The different crops have been dealt with by some of the most eminent authorities in this field. The chapters provide integrated information on conservation, cultivation, commerce, and consumption. This holistic approach makes the book very valuable. Recent scientific developments, including potential applications of biotechnology, have been dealt with in a lucid manner.

Several of the industrial crops covered in this book have made a profound impact on the economic well-being of the rural families where they have been introduced. For example, rubber introduced from Brazil first into Sri Lanka by Sir Henry Wickham nearly 130 years ago has helped to change the rural economy of many Asian countries. Similarly, oil palm has had a profound impact on the economics of land use in rural Malaysia. A valuable benefit conferred by industrial crops is the generation of opportunities for downstream employment. This is particularly significant since we now urgently need job-led economic growth and not job-less growth.

The *Handbook of Industrial Crops* will be of great value not only to scientists and scholars but also to policymakers, industrialists, and investors. We owe Chopra and Peter our deep sense of gratitude for compiling this very valuable publication. The Haworth Press is to be congratulated on its timely contribution to fostering an industrial crop-based economic revolution in many developing countries."

M. S. Swaminathan, PhD
Chairman,
National Commission on Farmers,
Government of India

Handbook
of Industrial Crops

FOOD PRODUCTS PRESS®
Crop Science
Amarjit S. Basra, PhD
Senior Editor

The Lowland Maya Area: Three Millennia at the Human-Wildland Interface edited by A. Gómez-Pompa, M. F. Allen, S. Fedick, and J. J. Jiménez-Osornio

Biodiversity and Pest Management in Agroecosystems, Second Edition by Miguel A. Altieri and Clara I. Nicholls

Plant-Derived Antimycotics: Current Trends and Future Prospects edited by Mahendra Rai and Donatella Mares

Concise Encyclopedia of Temperate Tree Fruit edited by Tara Auxt Baugher and Suman Singha

Landscape Agroecology by Paul A Wojkowski

Concise Encylcopedia of Plant Pathology by P. Vidhyaskdaran

Molecular Genetics and Breeding of Forest Trees edited by Sandeep Kumar and Matthias Fladung

Testing of Genetically Modified Organisms in Foods edited by Farid E. Ahmed

Fungal Disease Resistance in Plant: Biochemistry, Molecular Biology, and Genetic Engineering edited by Zamir K. Punja

Plant Functional Genomics edited by Dario Leister

Immunology in Plant Health and Its Impact on Food Safety by P. Narayanasamy

Abiotic Stresses: Plant Resistance Through Breeding and Molecular Approaches edited by M. Ashraf and P. J. C. Harris

Teaching in the Sciences: Learner-Centered Approaches edited by Catherine McLoughlin and Acram Taji

Handbook of Industrial Crops edited by V. L. Chopra and K. V. Peter

Durum Wheat Breeding: Current Approaches and Future Strategies edited by Conxita Royo, M. M. Nachit, N. Di Fonzo, J. L. Araus, W. H. Pfeiffer, and G. A. Slafer

Handbook of Statistics for Teaching and Research in Plant and Crop Science by Usha Rani Palaniswamy and Kodiveri Muniyappa Palaniswamy

Handbook of Microbial Fertilizers edited by M. K. Rai

Eating and Healing: Traditional Foods As Medicine edited by Andrea Pieroni and Lisa Leimar Price

Physiology of Crop Production by N. K. Fageria, V. C. Baligar, and R. B. Clark

Plant Conservation Genetics edited by Robert J. Henry

Introduction to Fruit Crops by Mark Rieger

Sourcebook for Intergenerational Therapeutic Horticulture: Bringing Elders and Children Together by Jean M. Larson and Mary Hockenberry Meyer

Agriculture Sustainability: Principles, Processes, and Prospects by Saroja Raman

Introduction to Agroecology: Principles and Practice by Paul A. Wojtkowski

Handbook of Molecular Technologies in Crop Disease Management by P. Vidhyasekaran

Handbook of Precision Agriculture: Principles and Applications edited by Ancha Srinivasan

Dictionary of Plant Tissue Culture by Alan C. Cassells and Peter B. Gahan

Handbook
of Industrial Crops

V. L. Chopra, PhD
K. V. Peter, PhD
Editors

Food Products Press®
The Haworth Reference Press™
Imprints of The Haworth Press, Inc.
New York • London • Oxford

For more information on this book or to order, visit
http://www.haworthpress.com/store/product.asp?sku=5306

or call 1-800-HAWORTH (800-429-6784) in the United States and Canada
or (607) 722-5857 outside the United States and Canada

or contact orders@HaworthPress.com

Published by

Food Products Press® and The Haworth Reference Press™, imprints of The Haworth Press, Inc.,
10 Alice Street, Binghamton, NY 13904-1580.

Cover design by Jennifer M. Gaska.

Library of Congress Cataloging-in-Publication Data

Handbook of industrial crops / V. L. Chopra, K. V. Peter, editors.
     p. cm.
Includes bibliographical references and index.
    ISBN 1-56022-282-4 (hard : alk. paper)—ISBN 1-56022-283-2 (soft : alk. paper)
    1. Crops—Handbooks, manuals, etc. 2. Botany, Economic—Handbooks, manuals, etc.
3. Produce trade—Handbooks, manuals, etc. I. Chopra, V. L. II. Peter, K. V.

SB91.H36 2005
633—dc22

                                                      2004012839

CONTENTS

Chapter 6. The Coconut Palm   235

M. A. Foale
G. R. Ashburner

Chapter 7. Coffee   295

J. M. Njoroge
C. O. Agwanda
P. N. Kingori
A. M. Karanja
M. P. H. Gathaara

ABOUT THE EDITORS

V. L. Chopra, PhD, is a geneticist of international repute. He is a former Professor of Genetics at the Indian Agricultural Research Institute, Director General of ICAR, and National Professor, B. P. Pal Chair, all in New Delhi. Professor Chopra is a former President of the International Genetics Congress and member, Board of Trustee of ICRISAT, CIMMYT, and IRRI. Professor Chopra is now President of the National Academy of Agricultural Sciences in New Delhi, India. He is also a member of the task force on Biotechnology, set up by Ministry of Agriculture, Government of India, Coffee Research Institute, and Chairman, Research Coordination Committee of Institutes under Central Silk Board, Government of India. He is a Fellow of the Indian National Science Academy, the Indian Academy of Sciences, and the Third World Academy of Sciences. The President of India conferred the prestigious award of Padmabhushan on Professor Chopra for his contribution to science.

K. V. Peter, PhD, is Vice Chancellor at Kerala Agricultural University in Trichur, India, and coordinator of the Department of Biotechnology (GOI) funded projects on plantation crops and spices. Professor Peter is a horticulturalist, a plant breeder, and an acknowledged research manager. He is the Founder Director for the Indian Institute of Spices Research and a former Professor and Director of Research for Kerala Agricultural University.

Professor Peter is the editor of seven books and author or co-author of thirteen books/bulletins and 132 research papers. The Indian Council of Agricultural Research conferred on Professor Peter the Rafi Ahmed Kidwai Prize, the National Academy of Agricultural Sciences awarded the Recognition Award 1999 for contributions to crop improvement, and the University of Agricultural Sciences conferred The M. H. Marygowda Award for being the Best Horticultural Scientist. Professor Peter is the Elected Fellow of the National Academy of Agricultural Sciences and Fellow of the National Academy of Sciences.

CONTRIBUTORS

Siti Nor Akmar Abdullah, Palm Oil Research Institute of Malaysia, Kuala Lumpur, Malaysia.

Saji T. Abraham, Rubber Research Institute of India, Kottayam, Kerala, India.

C. O. Agwanda, Coffee Research Foundation, Ruiru, Kenya.

S. Prasannakumari Amma, Cadbury–KAU Co-operative Coca Research Project, College of Horticulture, Kerala Agricultural University, Vellanikkara, Thrissur, Kerala, India.

G. R. Ashburner, Agriculture Victoria, Tatura, Victoria, Australia.

D. Balasimha, Central Plantation Crops Research Institute, Regional Station, Vittal, Karnataka, India.

M. Wahid Basri, Palm Oil Research Institute of Malaysia, Kuala Lumpur, Malaysia.

K. V. A. Bavappa, Karooth Village, Kumarenellur, Kerala, India.

M. A. Foale, CSIRO Sustainable Ecosystems, Indooroopilly, Brisbane, Australia.

M. P. H. Gathaara, Coffee Research Foundation, Ruiru, Kenya.

Rajendra Gupta, Zhandu Pharmaceutical Works, New Delhi, India.

Maizura Ithnin, Palm Oil Research Institute of Malaysia, Kuala Lumpur, Malaysia.

Norman Kamaruddin, Palm Oil Research Institute of Malaysia, Kuala Lumpur, Malaysia.

A. M. Karanja, Coffee Research Foundation, Ruiru, Kenya.

H. Hameed Khan, All India Coordinated Research Project on Palms, Central Plantation Crops Research Institute, Kasaragod, Kerala, India.

P. N. Kingori, Coffee Research Foundation, Ruiru, Kenya.

V. S. Korikanthimath, Cardamom Research Centre, Indian Institute of Spices Research, Appangala, Kodagu, Karnataka, India.

N. Kumar, Faculty of Horticulture, Tamil Nadu Agricultural University, Coimbatore, India.

V. K. Mallika, Cadbury–KAU Co-operative Coca Research Project, College of Horticulture, Kerala Agricultural University, Vellanikkara, Thrissur, Kerala, India.

W. W. D. Modder, The Tea Research Institute of Sri Lanka.

R. Vikraman Nair, Cadbury–KAU Co-operative Coca Research Project, College of Horticulture, Kerala Agricultural University, Vellanikkara, Thrissur, Kerala, India.

J. M. Njoroge, Coffee Research Foundation, Ruiru, Kenya.

E. V. V. Bhaskara Rao, National Research Centre for Cashew, Puttur, Dakshina Kannada, Karnataka, India.

A. Sankaralingam, All India Coordinated Research Project on Palms, Central Plantation Crops Research Institute, Kasaragod, Kerala, India.

M. R. Sudharshan, Indian Cardamom Research Institute, Spices Board Regional Station, Sakleshpur, Karnataka, India.

K. R. M. Swamy, National Research Centre for Cashew, Puttur, Dakshina Kannada, Karnataka, India.

Y. Annamma Varghese, Rubber Research Institute of India, Kottayam, Herala, India.

Introduction

Industrial crops as presented by arecanut, cardamom, cashew, cinchona, cocoa, coconut, coffee, oil palm, palmyra, rubber, tea, and wattle are important components of economy and export trade of many countries both developing and developed. Industrial crop-based products meet the requirements for edible oil (coconut, oil palm, masticatory [arecanut]); nuts (cashew, coconut); latex (rubber); beverage (coffee, tea); resins (wattle); antioxidants and flavor (cardamom); drugs (cinchona); chocolate (cocoa); and timber (arecanut, coconut, palmyra). Historically, tea, coffee, and rubber were raised as industrial crops in large estates. Now, a sizeable area under these crops is in smaller holdings and under diverse farming systems. Industrial crops are important in many respects. Coconut—the tree of heaven *(Kalpavriksha)*—provides materials for food, edible oil, and industrial lubricants. Tender coconut water is a healthy drink. The timber, leaf petiole, shell, husk, etc., are used for various purposes. Arecanut yields a masticatory used in association with betel leaf and also as pan masala and scented *"supari."* Oil palm yields palm oil rich in vitamins A and E. Cashew yields apple and nuts of commercial value. Cashew nutshell liquid (CNSL) is an industrial oil. Cocoa is grown for beans used for making cocoa butter and chocolates. Cardamom lends flavor to many food items. Rubber is grown for latex. Tea and coffee are beverage crops. Cinchona is a medicinal tree and wattle is grown for soft wood. The present handbook was launched to collate information on important crop groups from the viewpoints of evolutionary biology and commercial agriculture. For industrial crops, a more liberal view has been taken because for some crops basic information on genetic architecture and evolutionary processes are far from adequate. The treatment, therefore, includes all aspects of cultivation and trade in addition to available information on evolution and adaptation.

Arecanut, *Areca catechu* L., is one of the cash crops of South India grown for its masticatory nuts popularly known as betel nut or *supari.* Based on clustering pattern of cultivars from India, Vietnam, Singapore, and Indonesia, Bavappa (1974) reiterated that both *A. catechu* and *A. triandra* originated in the archipelago of Indonesia. He contended that these species moved west through Malaysia to India, Sri Lanka, and as far as Mauritius without losing their specific identity. The somatic chromosome number (2n) of *Areca catechu* is 32. High-yielding improved varieties (*mangala,*

1

sumangala, sreemangala, Mohit Nagar, Calicut-17, SAS-1) have been successfully evolved through introduction of indigenous and exotic types and refinement of selection procedures for mother palm, seed nuts, and seedlings. In recent years, hybridization and exploitation of genes for breeding dwarf and high yielding varieties have been initiated.

Lesser cardamom (*Elettaria cardamomum* (L.) Maton) ($2n = 48, 52$), green cardamom, Malabar cardamom or *chotta elaichi* is the true cardamom of commerce. Referred to as the "queen of spices," it is one of the most valued spices in the world. The Western Ghats of South India are the abode of cardamom from where it spread to Sri Lanka, Guatemala, and Thailand. It is now grown in Laos, Vietnam, Costa Rica, El Salvador, and Tanzania. Transition of cardamom from a forest product to the present-day plantation crop is a saga involving humans, animals, birds, rain forests, and even "witch doctors." In nature, it occurs in the canopy of evergreen forests, at latitudes ranging 600-1,500 m where warm humid atmosphere prevails, rainfall is evenly distributed, and soils are rich in humus. The cultivars Mysore, Malabar, and Vazhukka of cardamom arose through latitudinal effects associated with temperature and light intensity. In India, Kerala has the largest area of the three ecotypes mentioned eariler (about 62 percent), followed by Karnataka (31 percent) and Tamil Nadu (7 percent).

Cashew ($2n = 40, 42$) is a native of Brazil and is a discovery of the Portuguese who were responsible for spreading it to other parts of the world. Cashew cultivation in India dates back to the sixteenth century when it was introduced in the west coast region by the Portuguese. Cashew is a hardy plant that thrives even in poor soils and adverse climatic conditions. It was spread to various countries for soil conservation, afforestation, and wasteland development. Cashew is commercially cultivated for its edible kernels. In the global trade, cashew kernels are valued at U.S. $500 million in trade and U.S. $1 billion at the consumer level. India was the first country to exploit international trade of cashew kernels in the early part of the twentieth century and became the largest exporter of cashew kernels. India was perhaps the first country to initiate systematic research on cashew in the early 1950s. Now, research efforts are underway in Australia, Vietnam, Brazil, Tanzania, and China. Of the 35 cashew varieties released in India, 23 are selections from germplasm materials and 12 are hybrids involving 14 germplasm accessions as parents.

Cinchona ($2n = 34$), known for the antimalarial alkaloid quinine, is indigenous to the Andes. Initially, the entire world supply came from bark collected from wild trees in the Andes. It was introduced in Europe around 1640. The Dutch and British undertook extensive planting of the crop in the Far East from where more than 90 percent of the world supply of quinine was sourced. *Cinchona* is a relatively small genus with 15 species indigenous to the rain forests on the slopes of the Eastern Andes. The species *C. calisaya* and *C. ledgeriana* occur in Southern Peru and Bolivia at 1,200-1,700 m above MSL. *C. officinalis* occurs from Colombia to Northern Peru at higher altitudes, and *C. succirubra* occurs from Costa Rica to Bolivia. From an evolutionary point of view, nothing significant happened in the early phases of species development. The genetic base of cinchona is very narrow and many wild species are yet to be exploited.

The palmyra palm *(Borassus flabellifer)* (2n = 36) is a native of tropical Africa. It is also found in dry regions of India, Myanmar, and Sri Lanka.

Cocoa *(Theobroma cacao* L.) is the only cultivated species among twenty-two species of the genus. The primary center of diversity is the Upper Amazon basin. The tropical part of Central America is the secondary center of diversity. A diploid with 2n = 20, *T. cacao* is subdivided into *Theobroma cacao* L. ssp., *sphaerocarpum* consisting all the other populations, viz., *forastero* and *trinitario*. In little more than two centuries, commercial growing of cocoa has extended from its center of origin in South America to West Africa, the Far East, and Oceania. Cocoa is now an important plantation crop throughout the humid tropics. Cocoa beans originally used for making a drink—"chocolate"—assumed such value as to be used as currency; this practice continued in Yucatan markets as late as 1840.

Coconut belongs to the family Arecaceae under the class Monocotyledons. *Cocos* is a monotypic genus that has no known wild forms. It has a pantropic distribution mainly in coastal regions at 20° either side of equator. Grown in more than 80 countries in the tropics, the original home of coconut is still uncertain. Dr. H. C. Harries opines that Southeast Asia is the most probable center of diversity, and a Melanesian origin has also been advocated. Dispersal of coconut may have been effected through humans or through ocean currents. Recent studies, using restriction fragment length polymorphism (RFLP) confirm the movement of coconut from Southeast Asia to the Pacific and then to the West Coast of America. Coconut yields more products of use to mankind than any other tree. It has been variously glorified as *Kalpavriksha* (tree of heaven), the consoles of the East, Mankind's Greatest Provider in the tropics, Tree of Life, Tree of Abundance, and Tree of Plenty. In Sanskrit, coconut is called *Kalpavriksha,* the tree which provides all the necessities of life.

Genus *Coffea* consists of more than 80 species of which mostly two, *C. arabica* L. and *C. canephora* Pierre ex Froehner are in cultivation. The majority of *Coffea* species are natives of Africa. *Coffea arabica* is a native of Central Africa (Congo and Zaire). Robusta coffee was introduced to India from Indonesia and Sri Lanka in the late nineteenth century when *C. arabica* started showing severe incidence of leaf rust. Early selections were made from the estate plantings to improve the fruit size and bean size, which are generally small in robusta. A bold fruited selection S.274 was released for cultivation in 1950s. Another species, *C. congensis,* which has compact bush characteristics and better quality of seeds, was introduced to India in the 1930s. At present, the Central Coffee Research Institute, Chikmagalur, Karnataka (India), has a collection of 18 species belonging to *Coffea* and the closely related genus *Psilanthus.*

Oil palm, which originated in the Guinea Coast of West Africa, is regarded as the highest oil-yielding crop. Semiwild and wild oil palm groves still exist in the tropical West African countries. In the fifteenth century, oil palms were introduced to Brazil and to other tropical countries by the Portuguese. In 1848, the Dutch imported oil palm seeds from West Africa via Amsterdam and four seedlings were planted at Bogor, Indonesia. Commerical planting of oil palm started in Malaysia in 1917. The fruit in oil palm is a drupe, which varies in shape from heavy spherical to ovoid or elongate. Based on differences in fruit structure, three varieties of oil palm have been identified: dura, pisifera, and tenera. In India, oil palm was introduced toward the end of eighteenth century as a botanical collection at the National Botanical Gardens, Calcutta.

The Para rubber tree of commerce, *Hevea brasiliensis* Muell. Arg., produces about 99 percent of the world's natural rubber, an industrial raw material of strategic importance. The genus *Hevea* has its origins in the Amazon river basin in Brazil, covering parts of Brazil, Bolivia, Peru, Colombia, Ecuador, Venezuela, French Guiana, Suriname, and Guyana up to an altitude of about 800 m. The importance of rubber to the Aztec and Mayan civilizations of Mexico and the Yucatan was reported by numerous commentators from B. de Sahagun (1530) to Juan de Torquemada (1615). Rubber, *olli* or *ulli,* symbolizing blood or life force was burnt or paraded in the form of figurines or paintings on bark paper when human sacrifices were made to propitiate gods. As time passed, Spanish colonists in Mexico found another use for the magic substance: waterproofing textiles. This practice, using the latex, began in the seventeenth century and was reported in detail by Saavedra Miguel Cervantes (1794) who described the preparation of rubber-coated caps, boots, and even carriage hoods. *Hevea brasiliensis* and its congeners are diploids with $2n = 2x = 36$. The 36-chromosome genera are probably old tetraploids based on $x = 9$. Rubber is an outbreeder and shows both inbreeding depression and heterosis. Conventional methods of genetic improvement, viz., introduction, selection, and hybridization have been followed in *Hevea.* Molecular techniques such as RFLP, random amplified polymorphic DNA (RAPD), and microsatellites have also been deployed for the analysis of the extent and distribution of genetic diversity among promising clones.

The geological history of Eastern Yunnan and the distribution of *Camellia* species in that area indicate that the long, narrow region of Wenshan and Honghe located at 22°40' to 24°10' N and 103° 10' to 105° 20' E is the center of genesis of the tea plant. From the traditional centers of cultivation in South and Southeast Asia, tea was introduced into other domains of the world. It is now grown in environments ranging from Mediterranean climates to hot humid tropics. Controversy still persists as to the real place of origin of tea because of the nonavailability of true wild populations that exemplify the form from which the cultivated population of diverse morphological characters originated. Botanists distinguish three distinct commercial tea-producing taxa: *Camellia sinensis* (L.) O. Kuntze (China type), *C. assamica* (Masters) Wight (Assam type) and *C. assamica* ssp *lasiocalyx* (Planch ex. Watt) Wight (Cambod type). The non-tea-producing taxa are *C. irrawadiensis* Barua (Wilsons Camellia), *C. reticulata, C. sasanqua* Thumb; *C. lutescens* Dyer, *C. saluensis, C. taliensis* (W. W. Smith) Melchior, and other closely allied species. These species have contributed to the evolution of the present string of genetic pools of commercial tea populations. The commercial tea populations are polymorphic and represent a wide genetic diversity. The main taxa of tea are diploids ($2n = 30$). Natural triploids, tetraploids, pentaploids, and aneuploids have been detected in tea populations in Assam, but are present in very low frequency. Wu Tu, a mountain situated in the Szechwan province of China, is celebrated as the birthplace of the tea industry. The tea shrub is said to have been first cultivated here around 350 A.D. and was used to prepare a medicinal decoction. During 386-535 A.D., tea leaves were plucked and made into cakes which were roasted until reddish in color, pounded into tiny pieces, and placed in a chinaware pot. Boiling water was then poured over them. Onion, ginger, and orange were added to flavor the drink. China became the fountainhead from which tea culture spread to other countries. The first consignment of tea was brought to Europe by the Dutch in 1610. This was obtained from the Chinese in exchange for dried sage. Cultivation of tea started in Java in 1825 and in India in 1833. In India, merchants of the East India Company were primarily responsible for the development of the trade and planting, which together comprise the tea industry. In Africa, tea was grown at the Durban Botanic Gardens in 1850 and was the forerunner of modest local plantation industry in Natal. In Sri Lanka, serious cultivation of tea began in the 1870s. Russia is now also one of the chief areas where tea is grown. Tea is now widely dispersed and cultivated in Malaysia, Myanmar (Shan states), Thailand, Vietnam, Mauritius, Zaire, Rhodesia, Mozambique, Ethiopia, St. Helena, Cameroon, Brazil, Peru, Argentina, Paraguay, Colombia, Bolivia, Mexico, Fiji, Martinique, Iran, North and South Carolina, Australia (including New Guinea), Turkey, and Corsica.

Wattle *(Acacia mearnsii)* ($2n = 26$) grows as natural vegetation in arid regions. It is a native of eastern Australia and was introduced to South Africa in 1864 as a quick growing tree for shelter belts. Wattle bark, also known as mimosa bark, contains 35 to 39 percent tannin and is now the most widely used tanning material. Commercial plantations are raised from seed and the populations are heterogeneous. Main areas of production are in South Africa, Kenya, and Tanzania, far away from the center of origin.

It is clear from the foregoing that industrial crops have an interesting history of evolution, migration, adaptation, and fitment into the agricultural production scenario. Information about varied aspects of industrial crops, such as origin and distribution, botany, taxonomy, reproductive biology, cytogenetics, breeding, growing conditions, processing, industry, and commerce are scattered and not easily accessible without investment of considerable time and effort. This volume is an attempt to bring together the available relevant information at one place. Each chapter has been prepared by an authority who has long been associated with the crop dealt with and hence possesses in-depth knowledge about it. It is our hope that the book will meet, under one cover, the information needs of both the academic and commercial organizations.

FURTHER READING

Bala, Subramaniam. 1995. *Tea in India.* Wiley Eastern Ltd., New Delhi.

Banerjee, Barundeb. 1993. *Tea: Production and Processing.* Oxford and IBH Publishing Co. Pvt. Ltd., New Delhi.

Bavappa, K.V.A. 1974. Studies in the genus *Areca:* Cytological and genetic diversity of *A. catechu* L. and *A. triandra* Roxb. PhD thesis, University of Mysore.

Central Plantation Crops Research Institute (CPCRI) 1999. *Improvement of Plantation Crops.* Kasaragod 671124, Kerala.

Eden, T. 1976. *Tea.* Longman Group Ltd., London.

Hartley, C.W.S. 1979. *The Oil Palm.* Longman, New York.

Patel, J. S. 1938. *The Coconut: A monograph.* Superintendent, Government Press, Madras.

Webster, C.C. and Baulkwill, W.J. 1989. *The Rubber.* Longman Scientific and Technical, New York.

Wood, G.A.R. and Lass, R.A. 1985. *Cocoa.* Longman, New York.

Wright, Herbert. 1999. Cocoa: *Its Botany, Cultivation, Chemistry, and Diseases.* Biotech Books, New Delhi.

Wriglay, Gordon. 1988. *Coffee.* Longman Scientific and Technical, New York.

Chapter 1

Arecanut

K. V. A. Bavappa
D. Balasimha

INTRODUCTION

The arecanut palm (*Areca catechu* L.) is the source of a common masticatory nut, popularly known as arecanut, betelnut, or *supari*. It is used extensively in India by all sections of people as masticatory, and is an essential requisite for several religious and social ceremonies. Consequently, the arecanut palm occupies a prominent place among cultivated crops in the Indian states of Kerala, Karnataka, Assam, Meghalaya, Tamil Nadu, and West Bengal. The *Areca* palm is a monocot belonging to the family Palmae. The commonly cultivated species is *A. catechu* in most of the countries where it is used for chewing. In Sri Lanka, the fruits of *A. concinna* are occasionally chewed. India ranks first in the world in area of cultivation and production followed by China.

ORIGIN, HISTORY, AND DISTRIBUTION

The origin of arecanut is not exactly known. No fossil remains of the genus *Areca* exist, but the fossil records of closely related genera indicate its presence during the Tertiary period. The maximum species diversity (24 species) and other indicators suggest its original habitat in contiguous areas of Malaya, Borneo, and Celebes (Raghavan, 1957; Bavappa, 1963).

Innumerable references are found in several of the ancient Sanskrit texts to the arecanut palm, the arecanut and its uses. Bhat and Rao (1962) have produced evidence to prove the antiquity from a number of such works, among the most important one quoted being *Anjana Charitra* by Sisy Mayana (1300 B.C), where the reference had been made to groups of arecanut palms full of inflorescence and branches presenting a nice appearance.

Although it is not precisely known as to when arecanut found its way to the Indian subcontinent, several evidences about its antiquity exist (Mohan Rao, 1982). Arecanut is mentioned in various ancient Sanskrit scriptures (650-1300 B.C.); and its medicinal properties were known to the famous scholar Vagbatta (A.D. 500). One of the well-known Ajanta caves in central India (200 B.C. to A.D. 900) has one exquisitely painted arecanut palm providing a backdrop to the Padmapani Buddha. According to Furtado (1960), one of the earliest references to arecanut was in A.D. 1510. The abundant uses of arecanut in chewing and religious functions were indicated even during the times of the Aryans in India.

India has about 57 percent in area and 53 percent in production of the total world production or arecanut (Table 1.1). This is followed by China, Bangladesh, and Myanmar. Other countries where arecanut is grown are Malaysia, Indonesia, Vietnam, Thailand, and the Philippines. The highest productivity is recorded in China (3,752 kg/ha), followed by Malaysia (1,667 kg/ha), Thailand (1,611 kg/ha), and India (1,189 kg/ha).

Arecanut is cultivated largely in the plains and foothills of the Western Ghats and the northeastern regions of India. Area and production in different states are given in Table 1.2. Three states, Kerala, Karnataka, and Assam, account for 90 percent of area under cultivation and 95 percent of production. In India, productivity is highest in Maharashtra (3,947 kg/ha).

BOTANY AND TAXONOMY

Martius (1832-1850) was the first to attempt to restrict the limits of the genus, *Areca*. But the attempts were not satisfactory, as they were not based on real affinities. Later, various species grouped under *Areca* were separated into different genera and limited the genus to close relatives of the type of the genus, *Areca catechu* (Blume, 1836). Furtado (1933) described the limits of the genus Areca and its sections. A list of *Areca* species and their geographical distribution is given in Table 1.3.

Detailed morphology, floral biology, and embryology of arecanut have been described (Murthy and Pillai, 1982). The arecanut palm is a graceful, erect, and unbranched palm reaching heights up to 18-20 m. The stem has scars of fallen leaves in regular annulated forms. The girth of the stem depends on genetic variation, soil condition, and plant vigor. The arecanut palm has an adventitious root system. The crown of an adult palm contains 7 to 12 leaves. The apical bud produces leaves in succession every 40 days or so, and the age of leaf is about two years. The leaves are pinnatisect and consist of a sheath, a rachis, and leaflets. The leaf sheath completely encir-

TABLE 1.1. Area and production of arecanut in countries.

Country	Area (in thousand ha)					Production (in thousand million tonnes [MT])				
	1961	1971	1981	1991	1998	1961	1971	1981	1991	1998
Bangladesh	82.60	40.10	36.43	35.81	36.00	62.99	23.36	25.05	24.12	28.00
China	0.62	1.33	2.83	26.96	46.00	3.71	10.07	24.35	111.09	172.57
India	135.0	167.3	185.20	217.00	270.00	120.00	141.00	195.90	258.50	310.00
Indonesia	65.00	75.00	90.00	95.76	75.38	13.00	15.00	18.00	22.81	32.60
Malaysia	6.00	2.50	1.30	2.20	2.40	6.50	3.00	2.50	4.00	4.00
Maldives	0.003	0.003	0.006	0.030	0.030	0.001	0.001	0.005	0.016	0.016
Myanmar	11.33	24.68	26.47	28.93	29.50	8.00	19.20	25.80	92.27	31.50
Thailand	—	—	—	8.50	9.00	0.00	0.00	0.00	13.25	14.50
World	300.55	310.92	342.25	415.20	468.31	214.21	211.64	291.62	446.15	593.29

Source: Based on data from Food and Agriculture Organization of the United Nations.

TABLE 1.2. Statewise area and production of arecanut in India.

State	Area (in thousand ha)					Production (in thousand tonnes)				
	1966-1967	1971-1972	1981-1982	1991-1992	1997-1998	1966-1967	1971-1972	1981-1982	1991-1992	1997-1998
Andhra Pradesh	0.00	0.20	0.20	0.20	0.20	0.00	0.20	0.20	0.20	0.50
Andaman and Nicobar	0.00	0.00	0.00	0.00	3.60	0.00	0.00	0.00	0.00	5.20
Assam	26.20	25.90	47.20	66.00	74.10	25.20	29.00	48.10	50.50	64.00
Goa, Daman, and Diu	0.00	1.40	1.70	1.30	1.50	1.20	1.70	1.30	1.50	1.80
Karnataka	34.80	43.20	55.20	65.40	88.40	52.70	56.30	80.20	96.00	128.50
Kerala	71.20	86.80	61.20	63.51	76.10	44.30	53.40	66.00	65.14	94.00
Maharastra	0.00	2.30	2.10	1.90	1.90	0.00	2.80	2.40	2.60	7.50
Meghalaya	0.00	6.30	6.50	8.90	9.50	0.00	4.20	4.90	8.70	12.10
Mizoram	0.00	0.10	0.40	0.00	0.00	0.00	0.00	0.00	0.10	0.10
Pondicherry	0.00	0.00	0.00	0.10	0.10	0.00	0.00	0.00	0.10	0.20
Tamil Nadu	0.00	5.00	4.30	2.80	2.70	0.00	5.00	2.80	3.40	4.30
Tripura	0.00	0.00	0.70	1.20	1.80	0.00	0.00	1.00	2.30	3.50
West Bengal	0.00	3.10	3.10	5.90	8.10	0.00	0.80	0.80	7.90	12.40

Source: Directorate of Economics and Statistics, government of India.

TABLE 1.3. Geographical distribution of *Areca* species.

Country	Species
India	A. catechu, A. triandra
Andamans, India	A. catechu, A. laxa
Sumatra	A. catechu, A. triandra, A. latiloba
Sri Lanka	A. catechu, A. concinna
Malaysia	A. catechu, A. triandra, A. latiloba, A. montana, A. ridleyana
Borneo	A. catechu, A. borneensis, A. kinabaluensis, A. arundinacea, A. bongayensis, A. amojahi, A. mullettii, A. minuta, A. furcata
Java	A. catechu, A. latiloba
Celebes	A. celebica, A. oxicarpa, A. paniculata, A. henrici
Australia	A. catechu, A. alicae
Solomon Islands	A. niga-solu, A. rechingeriana, A. torulo, A. guppyana
New Guinea	A. congesta, A. jobiensis, A. ladermaniana, A. macrocalyx, A. nannospadix, A. warburgiana
Phillippines	A. catechu, A. hutchinsoniana, A. vidaliana, A. costulata, A. macrocarpa, A. parens, A. caliso, A. whitfordii, A. camariensis, A. ipot
Moluccas	A. glandiformis
Bismarck Islands	A. novo hibernica
Laos	A. laosensis
Cochin China	A. triandra
Lingga Islands	A. hewittii

cles the stem. It is about 54 cm in length and 15 cm in breadth. The average length of leaf is 1.65 m, bearing about 70 leaflets. The leaflets are 30.0 to 70.0 cm in length and 5.8 to 7.0 cm in breadth depending on the position of the leaf.

When grown under optimum conditions, flowering occurs in about five years, appearing at the tenth node in the South Kanara cultivar (Murthy and Bavappa, 1960b). In another "semi-tall" cultivar, Mangala, flowering occurs earlier, at about three years (Bavappa, 1977). The inflorescence is a spadix produced at the leaf axils. The number of spadices produced depends upon the number of leaves. The mean number of spadices produced are three to four depending on the age of the plant (Murthy and Bavappa, 1960a). The fruit of arecanut is a monocular one-seeded berry that is orange-red to scarlet when ripe encircled by a thick fibrous outer layer (husk). Usually 100 to 250 fruits are present in each bunch.

The distribution of arecanut roots under different spacing was reported (Shama Bhat and Leela, 1969). It was observed that 61 to 67 percent of all roots were concentrated within a 50 cm radius of the palm and 80 percent of all roots were within depths of an 85 cm radius. The cultural operations of arecanut base are also restricted within this radius. The 0-50 cm and 51-100 cm soil depth contained 66 to 67 percent and 18 to 24 percent of all roots respectively. The roots of closely planted palms penetrated deeper than those planted at wider spacing.

Arecanut is monoecious with both male and female flowers occurring on same spadix. It is cross-pollinated (Bavappa and Ramachander, 1967). The male phase lasts for twenty-five to forty-six days. Female flowers are cream colored turning green within a week. The flowers open between 2 h and 10 h. The female phase extends up to ten days. The stigma remains receptive up to six days (Shama Bhat et al., 1962b; Murthy and Bavappa, 1960a). Pollen is generally carried by the wind.

Fruit development takes place in three stages (Shama Bhat et al., 1962a). In the first phase, the size increase is rapid. The second stage is characterized by an increase in volume and dry matter accumulation in the kernel. During this period the embryo becomes macroscopic and develops rapidly. In the third stage swelling occurs and its green color fades. It takes 35 to 47 weeks for fruit development.

CYTOGENETICS

The chromosome number of *Areca catechu* L. was first determined and reported by Venkatasubban (1945) as $2n = 32$ (Figure 1.1). The chromosome number of the species was later confirmed by Sharma and Sarkar (1956), Raghavan and Baruah (1958), Abraham et al. (1961), and Bavappa and Raman (1965).

A chromosome number of $2n = 32$ reported by Darlington and Janaki Ammal (1945) for *A. triandra* Roxb. was later confirmed by Sharma and Sarkar (1956) and Bavappa and Raman (1965). Nair and Ratnambal (1978) determined the meiotic chromosome number of *A. macrocalyx* Becc. as $n = 16$. Meiotic abnormalities such as nondisjunction, lagging chromosomes, univalents, and pentads were reported in *A. catechu* by Sharma and Sarkar (1956). Bavappa and Raman (1965) observed in the meiosis of four ecotypes of *A. catechu,* abnormalities, such as univalents at diakinesis and metaphase I, nonsynchronization of orientation, clumpting, delayed disjunction, chromosome bridges and laggards at anaphase I and II, chromosome mosaics, and supernumerary spores.

FIGURE 1.1. Somatic chromosomes in *Areca catechu* cv. Mangala 2*n* = 32.

Sharma and Sarkar (1956) found the meiotic division quite normal in *A. triandra* except for the presence of 14 and 18 chromosomes occasionally at metaphase II. Bavappa and Raman (1965) also reported regular meiotic division in the types of *A. triandra* studied by them.

Intracultivar variation in meiotic behavior of *Areca* was reported by Bavappa (1974) and Bavappa and Nair (1978). Although normal bivalent formation was observed in some palms, others had maximum association of hexavalents, octovalents, and even decavalents. Abnormalities, such as bridges and laggards, and disorientation of chromosomes at anaphase I and anaphase II, were also reported in this species.

Intrapalm variation in chromosome numbers in the pollen mother cells of *A. catechu, A. triandra,* and their hybrids was reported by Bavappa and Nair (1978), and cytomixis to an extent of 39 percent seemed to have contributed to this abnormality. In spite of the high degree of multivalents in *A. catechu,* pollen fertility was very high. The possibility of the frequency of multivalent formation and disjunction being under genotypic control and being subjected to selection was suggested by Bavappa and Nair (1978).

Partial desynapsis of chromosomes at diakinesis was reported by Bavappa (1974) and Bavappa and Nair (1978) in *A. triandra* and *A. catechu* × *A. triandra* hybrids. Desynapsis observed at diakinesis was followed by an increase in pairing at metaphase I as reflected by the frequency of bivalents in *A. triandra* and *A. catechu* × *A. triandra* hybrids. This was attributed to distributive pairing, a mechanism that has been possibly adopted for ensuring their regular segregation (Bavappa and Nair, 1978). The extent of desynapsis was higher in the F1 hybrids of *A. catechu* and *A. triandra* as compared to *A. triandra*, suggesting that the gene controlling this character may be dominant. The large number of univalents observed in the hybrid as compared to *A. triandra* parent has been attributed to reduced homology of the parental chromosomes (Bavappa and Nair, 1978).

Two pairs of short satellite chromosomes in the somatic chromosome complement of *A. catechu* were observed (Venkatasubban, 1945). Three pairs of long chromosomes, six pairs of medium-sized chromosomes, and seven pairs of short chromosomes were observed by Sharma and Sarkar (1956) in *A. catechu*. They categorized the chromosomes into seven groups based on their morphology and relative length. Two pairs of long chromosomes next to the longest were found to have secondary constrictions. They also observed that the chromosomes of *A. triandra* were longer than those of *A. catechu*. Bavappa and Raman (1965) found the chromosomes of *A. catechu* and *A. triandra* differing in size, total chromatin length, position of primary and secondary constrictions, and number and position of satellites. Based on the assumption of Sharma and Sarkar (1956) that gradual reduction in chromatin matter had taken place in the evolution from primitive to advanced forms of different genera and tribe of Palmae, Bavappa, and Raman (1965) considered *A. catechu* as more advanced than *A. triandra*.

The pachytene chromosomes in *A. catechu* were found to be morphologically in close agreement with the somatic chromosomes, though the pachytene chromosomes were about ten times longer than the somatic chromosomes (Bavappa and Raman, 1965).

The chromosome morphology of a few cultivars of *A. catechu* from Assam was reported by Raghavan (1957). Minor variation in structure and length of individual chromosomes, total length of the complement, and position of constrictions among the types were noted. On the basis of morphology, he recognized nine groups in the somatic chromosomes of the cultivars.

Studies on the karyotypes of eight cultivars of *A. catechu* and four ecotypes of *A. triandra* (Bavappa, 1974; Bavappa et al., 1975) revealed considerable differences in their gross morphological characteristics. The karyotypes of the *A. triandra* ecotypes showed a higher frequency of submedian and median chromosomes as compared to *A. catechu*. A classification of the karyotype of the two species according to the degree of their asymmetry, which recognizes three grades of size differences and four grades of asymmetry in centromere position (Stebbins, 1958), showed that karyotypes, 1B, 2A, 2B, and 3B are represented in *A. catechu* cultivars and only 1A, 2A, and 2 are represented in the ecotypes of *A. triandra*. Even within the same cultivar of *A. catechu,* two different types of asymmetry in karyotypes were observed, whereas no such variation exists in *A. triandra* ecotypes. Evidently, *A. triandra* has a more symmetrical karyotype than *A. catechu*. It was concluded that delineating the cultivars of *A. catechu* on the basis of a standard karyotype seemed to be rather difficult. The fact that *A. catechu* has lesser chromatin matter and an asymmetrical karyotype compared to *A. triandra* shows that the latter is more primitive.

GERMPLASM COLLECTION

Several cultivars have been recognized in Karnataka (Rau, 1915) and the Philippines (Beccari, 1919) based on fruit and kernel. Based on stomatal characters, size of nuts, leaves shed, female flowers, and nut size, cultivars could be separated (Bavappa, 1966; Bavappa and Pillai, 1976).

The genetic resource program in arecanut has been undertaken at the Regional Station, Vittal, Karnataka. A collection of five species, viz., *A. catechu, A. triandra, A. macrocalyx, A. normanbyii,* and *A. concinna,* and two genera, *Actinorhytis* and *Pinanga dicksonii,* are available (Anuradha, 1999). The germplasm holding now consists of 113 accessions (Ananda, 1999b). Among these, 23 exotic accessions were introduced from different countries, viz., Fiji, Mauritius, China, Sri Lanka, Indonesia, Vietnam (Saigon), Singapore, British Solomon Islands, and Australia representing different species. About 90 collections have been included in the germplasm after exploration made in different arecanut growing states of India. About 39 accessions have been described based on descriptors.

A wide range of variations exist in fruit characters, stem height, internode length, and leaf size and shape. The nuts in the Malnad, parts of Shimoga, and Chikmagalur districts are small in size, whereas in North Kanara and Ratnagiri they are bigger (Murthy and Bavappa, 1962). In addition, wide variations exist in yield, earliness in bearing, fruits/bunch, quality, and dwarfness.

The exotic and indigenous collections have been under evaluation since 1957 for morphology, nut characters, and yield attributes (Bavappa and Nair, 1982; Ananda, 1999b). Yield evaluation resulted in the release of four high yielding cultivars, of which three are selections from the exotic collection. The characteristics of these varieties have been described (Ananda and Thampan, 1999; Ananda, 1999a). Among the exotic collection, cultivar VTL-3 introduced from China was released and named as 'Mangala' (Photo 1.1; Bavappa, 1977). This had earliness in bearing, greater number of fe-

PHOTO 1.1. Mangala cultivar.

male flowers, high yield, and shorter stem height compared to other accessions. Other varieties released for cultivation are 'Sumangala' (Photo 1.2) and 'Sreemangala' (Photo 1.3), which are introductions from Indonesia and Singapore respectively. High-yield potential was obtained in one of the indigenous collections from West Bengal and was released as the 'Mohitnagar' variety (Photo 1.4). The characteristics of these varieties are given in Table 1.4. Other promising cultivars are 'SAS-I,' 'Tirthahalli,' and 'Calicut-17.'

Bavappa (1974) recorded morphological, anatomical and yield characters for 13 cultivars of *A. catechu* and four ecotypes of *A. triandra* for the years 1963, 1966, and 1972. The analysis of variance of the results obtained in 1963 showed that the differences between cultivars are highly significant

PHOTO 1.2. Sumangala cultivar.

for all the six morphological characters. A combined analysis of the data for two years for 24 common characters recorded during 1967 and 1972 also revealed significant interaction between cultivars for all characters. A significant interaction between years and cultivars was seen in height, girth, internodal distance, number of bunches and inflorescences on the palm, length and breadth of leaf sheath, length and volume of nut and length, breadth, weight, and volume of kernel.

Bavappa (1974) also worked out 136 D^2 values between cultivars, the number of characters being unequal in different years. The magnitude of D^2 values indicated that considerable divergence exists between any of the cultivars in all the years. He grouped 13 cultivars and four ecotypes from

PHOTO 1.3. Sreemangala cultivar.

nine countries into six clusters for the independent years 1963, 1966, and 1972, and found that though the number of clusters were the same, constituents in the different clusters were slightly different in different years. The number of clusters and pattern of clustering were more or less similar for the years 1966 and 1972. In the pooled analysis, the number of clusters was reduced from six to five. However, the pattern of clustering was more or less in conformity with the groups obtained for the individual years. The spatial diagram showing the distribution of clusters in 1966 and 1972 (pooled) is given in Figure 1.2.

All of the four ecotypes of *A. triandra* were in one cluster in the pooled analysis, and this cluster continued to show maximum divergence from the

PHOTO 1.4. Mohitnagar cultivar.

rest. The divergence between clusters IV and V were due to the differences in nut and kernel characters, breadth of leaf sheath, breadth of leaflets, and number of leaflets. Based on the analysis, Bavappa (1974) concluded that detection of the genetic divergence in the early years of the productive phase is of considerable advantage in formulating breeding programs in a perennial crop such as arecanut.

The rankings obtained by the different characters during 1966 for their contribution toward overall genetic divergence showed that the mean volume of nut and breadth of kernel were the characters of primary importance. For divergence between *A. triandra* and *A. catechu,* mean length of fruit of characters for 1972 and the pooled data also revealed the importance of nut

TABLE 1.4. Yield and nut characteristics of arecanut varieties.

Variety	Growth habit	Shape and size of nut	Dry nut yield (kg/plant)	Recommended for
South Kanara	Tall	Round, Bold	2.00	Coastal Karataka, Kerala
Mangala	Semi-tall	Round, Small	3.00	Coastal Karataka, Kerala
Sumangala	Tall	Oval, Medium	3.20	Karnataka, Kerla
Sreemangala	Tall	Round, Bold	3.18	Karnataka, Kerla
Mohitnagar	Tall	Round, Medium	3.67	West Bengal, Kerala, Karnataka

Source: Ananda, 1999a.

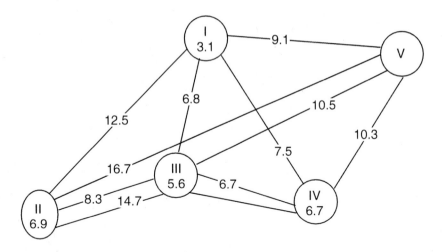

FIGURE 1.2. Spatial diagram showing distribution of clusters (Clusters I—Br.Sol.Is 1, Br.Sol.Is 2, Br.Sol.Is 3; Cluster II—Indonesia 1, Indonesia 2, Mauritius, Ceylon 3; Cluster III—Ceylon 1, Indonesia 6, Saigon 1, Saigon 2, Saigon 3, Singapore, South Kanara; Cluster IV—China, Fiji; Cluster V—Ceylon 2).

and kernel characters in differentiation within *A. catechu* cultivars and between *A. catechu* and *A. triandra* types. The results obtained from canonical analysis were also in broad agreement with the clustering pattern found from D² analysis. However, Bavappa (1974) concluded that the canonical analysis could be of only limited utility in view of the fact that the first two canonical roots accounted for only 85 percent of the variation or less.

The grouping obtained by D^2 analysis revealed that the three cultivars each from Saigon and the British Solomon Islands and the two ecotypes of *A. triandra* from Indonesia were invariably in one cluster each. As against this, close similarity between the cultivars from different countries has also been observed. The cultivar from Singapore was grouped with the three cultivars from Saigon in one cluster. A similar affinity between the two geographically distant cultivars was shown by 'Ceylon-1' and 'Indonesia-6', both always coming within the same cluster. The local cultivar was found to be invariably associated with the cultivar from Singapore in forming the cluster. Of the two cultivars of *A. catechu* from Ceylon, 'Ceylon-2' was always forming a separate cluster, indicating its distinct nature of divergence. The clustering pattern of cultivars and ecotypes revealed that geographic diversity need not always be related to genetic diversity (Bavappa, 1974).

CROP IMPROVEMENT

The hybridization program was initiated with the objectives of exploiting the existing variability in the *Areca* germplasm (Bavappa and Nair, 1982). The main concerns were to evolve high-yielding, regular-bearing, high-quality, and semi-tall ideotypes. Interspecific hybrids of *A. catechu* × *A. triandra* had only one stem, as in *A. catechu,* indicating dominance of this character (Bavappa, 1974). The hybrids mostly equaled the parents in internodal length. They also exhibited hybrid vigor for a number of characters such as number of male flowers, female flowers, spadix length, and stem girth.

The tall nature of the palm hinders various operations such as spraying and harvesting and is quite labor intensive. One of the major thrusts in the research on breeding in arecanut has been to induce dwarfness. A natural dwarf mutant was identified and was named 'Hirehalli Dwarf' (Naidu, 1963). This, however, has low yields coupled with very poor quality nuts that are not suitable for chewing. An attempt was made to cross the high-yielding varieties with 'Hirehalli Dwarf' to exploit the dwarfing nature (Ananda, 2000). Maximum dwarfs and intermediates were recovered in crosses 'Sumangala' × 'H.Dwarf', 'Mohitnagar' × 'H.Dwarf', and 'Mangala' × 'H.Dwarf' from among the twelve combinations made.

GROWING CONDITIONS

Soils and Climate

Arecanut is cultivated in a wide range of soils. The largest area under arecanut cultivation is in the gravelly laterite soils of red clay type of southern Kerala and coastal Karnataka (Nambiar, 1949). In the plains of Karnataka the soil is fertile clay loam (Naidu, 1962). The deep black fertile clay loams are highly suitable, whereas sticky clay, sandy, alluvial, brackish, and calcareous soils are not favorable for arecanut cultivation.

The arecanut palm is a crop of the subhumid tropics and thrives well in the regions 28°N and 28°S of the equator. The altitude at which arecanut grows depends on the latitude (Shama Bhat and Abdul Khader, 1982). In the northeastern region of India, it is grown mostly in the plains. At higher elevations lower temperature becomes the limiting factor. In general, however, the palm grows at altitudes up to 1,000 m above mean sea level (MSL) and beyond this the quality of nut produced is reduced (Nambiar, 1949). The palm requires a temperature range of 14°C to 36°C, though it is grown in places such as West Bengal where the minimum temperature may be as low as 4°C. Arecanut requires heavy rainfall ranging from 3,000 to 4,500 mm per annum. However, in the plains of Karnataka and Tamil Nadu, where the annual rainfall is only about 750 mm, the gardens are usually irrigated during summer months.

The climatic conditions are different in the arecanut growing areas of coastal India. Southern Kerala gets well distributed rainfall as compared to Northern Kerala and coastal Karnataka, which receive most of the rains during June to September followed by five to six months of drought. Hence, in these areas arecanut is grown principally as an irrigated crop.

Cultural Requirements

Arecanut is a seed-propagated crop. The age of mother palms and seed size are important in selecting the planting material. In general, seeds are collected from trees that have attained stabilized yield stages, i.e., more than ten years. Seeds that are mature and heavy are selected for obtaining good germination. Selected seeds are sown 5 cm apart with their stalk ends upward. Regular irrigation is essential. Three-month-old saplings can be transferred to a secondary nursery and planted at 30 cm distance. Seedlings can also be raised in polyethylene bags (25 × 15 cm size, 150 gauge) with potting mixture of topsoil, cattle manure, and sand in the ratio of 7:3:2. Twelve- to eighteen-month-old seedlings are selected for field planting.

The spacing of arecanut depends primarily on depth and fertility of soil. Studies on varying spacing conducted at several locations have shown the superiority of planting at a distance of 2.7 × 2.7 m (Table 1.5) (Shama Bhat, 1978; Shama Bhat and Abdul Khader, 1982). The seedlings are planted in pits 60 cm^3 in size after filling up to 50 cm with soil.

For efficient use of water in all methods of irrigation, frequency of irrigation should conform to soil water depletion requirements of the crop. Therefore, rather than using a fixed irrigation schedule, considerable flexibility should be given to accommodate varying crop evapotranspirational (ET) losses. Irrigation is done mostly by splashing and basin methods. Several researchers worked out the water requirement and irrigation scheduling for different locations. However, all these studies were done by taking into account frequency and quantity of water applied without considering the evaporative demand and temperature. Systematic study based on a climatological approach by Yadukumar et al (1985) and Abdul Khader and Havanagi (1991) revealed that irrigation at IW/CPE (irrigation water/cumulative pan evaporation) ratio of 1, i.e., 30 mm depth of water gave maximum yield. Based on a modified Penman method, a net irrigation requirement of 899 mm during post-monsoon season and an irrigation interval of six to seven days were found sufficient (Sandeep Nayak, 1996). Mahesha et al. (1990) estimated the evapotranspiration of arecanut using a modified Penman's method and stated that ET rates increased from 4.60 mm/day in December to 6.25 mm/day in April, and fell to 5.78 in May due to pre-monsoon showers. Drip irrigation significantly increased growth components, such as height, girth, and number of nodes; leaf area; and yield (Abdul Khader, 1988). Based on crop evaporative demand, Mahesha et al. (1990) found that optimum depth of water requirement per irrigation was 26 mm and that about 8-12 liters of water per palm per day was sufficient through

TABLE 1.5. Effect of spacing on cumulative yield of arecanut (kg/ha).

Spacing m x m	Vittal (Coastal Karnataka)	Hirehalli (Plain of Karnataka)	Kahikuchi (Assam)	Peechi (Kerala)
1.8 × 1.8	6290	3130	2.00	1749
1.8 × 2.7	8167	3705	3.00	1766
1.8 × 3.6	7829	4132	3.20	1710
2.7 × 2.7	10722	3867	3.18	1867
3.6 × 3.6	6169	2417	3.67	1448

Source: Sannamarappa, 1990.

drip irrigation. Balasimha et al. (1996) reported that the photosynthetic parameters and yield of arecanut increased with increases in drip irrigation levels from 10 liters/day to 30 liters/day in arecanut + cocoa mixed cropping system (Table 1.6). This saved water can be effectively used for irrigation of other crops/intercrops, and the depletion of ground water can be checked.

For effective irrigation water management, a combination of irrigation and mulching is desirable. Mulching is very important in arecanut because of the highly porous nature of the soil and greater seepage losses. Various plastic and organic mulching materials were found to be effective (Shama Bhat, 1978; Abdul Khader and Havanagi, 1991).

Nutrition

Availability of nutrients depends on the biological processes in the soil and organic and inorganic inputs to the system. Several fertilizer trials have revealed that each palm requires 100 g N, 40 g P_2O_5, and 140 g K_2O annually (Shama Bhat and Abdul Khader, 1982). It is also generally supplemented with about 12 kg of organic manure. Recent studies on fertilizer requirements of high-yielding varieties have shown that doubling this dosage results in increased yields (Sujatha et al., 1999). Economic analysis indicated an appreciable increase in net income and cost-benefit ratio with increase in fertilizer level to 200 g N, 80 g P_2O_5, and 280 g K_2O per palm. The fertilizer requirements and dynamics of nutrients have been standardized for an arecanut monocrop (Mohapatra and Bhat, 1982).

TABLE 1.6. Effect of irrigation levels on photosynthetic characteristics and dry nut yield (kg/ha) of arecanut.

Treatment	Photosynthesis ($\mu mol/m^2/s$)	Transpiration ($\mu mol/m^2/s$)	Stomatal conductance ($mol/m^2/s$)	Yield (kg/ha)
I_1 (10 l/day)	3.98	3.46	0.105	1,804
I_2 (20 l/day)	3.62	3.68	0.122	1,958
I_3 (30 l/day)	4.48	4.75	0.168	2,007
CD (5%)	NS	0.96	0.041	NS

Source: Balashimha et al., 1996.

Cropping Systems

Multiple cropping in arecanut as a productive land-use system through the use of interspaces has been practiced (Bavappa, 1961; Sannamarappa and Muralidharan, 1982). The long prebearing age of arecanut has prompted farmers to grow different annual or semiperennial crops for economic sustainability. The initial period of five to six years is ideal for growing short-duration crops. In later years, as the arecanut canopy increases in height, mixed cropping with other shade tolerant perennial crop species can be done.

A number of annual crops, such as paddy, sorghum, cowpea, vegetables, and yams are grown as intercrops of arecanut palms (Abdul Khader and Antony, 1968; Shama Bhat and Abdul Khader, 1970; Abraham, 1974; Shama Bhat, 1974; Thomas, 1978; Muralidharan, 1980). When these crops are cultivated the cultural and nutritional practices followed are used for the pure stands (Sannamarappa and Muralidharan, 1982). Leaving the 1.0 m radius around the arecanut palm, the interspaces are prepared for cultivation of the intercrops during the pre-monsoon period. Crops such as paddy, sorghum, corn, cowpea are sown in rows and groundnut and sweet potato sown in beds. Pits or trenches are taken for yam, elephant foot yam, banana, and pineapple. Except for banana and beans, the biological productivity of these intercrops was lower in comparison to sole crops. In most of the studies at different regions, no deleterious effects on main crops due to intercropping were noticed (Muralidharan and Nayar, 1979; Abraham, 1974; Sadanandan, 1974). Banana is the preferred intercrop in all arecanut-growing regions (Bavappa, 1961; Shama Bhat, 1974; Brahma, 1974). It also provides good shade during the early growth of arecanut plants.

Black pepper is an excellent crop for mixed cropping with arecanut, and high economic returns can be expected (Abraham, 1974; Nair, 1982). The arecanut stems are used as live standards for training black pepper. The pepper cultivars, Panniyur-I and Karimunda, performed better as mixed crop. The yield of arecanut was not affected by growing pepper.

The microclimate, especially shade, soil moisture, and temperature, in the arecanut gardens was found to be ideal for cocoa growth (Photo 1.5); and initial successful experiments were reported by Shama Bhat and Leela (1968) and Shama Bhat and Bavappa (1972). An experiment laid out in 1970 as a $6 \times 2 \times 4$ confounded asymmetrical factorial design having six different spacings and two fertilizer levels showed significant influence on the number and weight of cocoa pods in the normal spacing of arecanut. It was indicated that either an S1 (2.7×2.7 m) or S2 (2.7×5.4 m) combina-

PHOTO 1.5. Areca and cocoa mixed crops.

tion could be safely followed, although operational advantages are better in S2 spacing (Shama Bhat, 1983, 1988).

A high-density multispecies cropping system in arecanut typically comprises of arecanut black pepper, cocoa/clove pineapple/coffee and banana occupying different vertical airspace levels. A model was laid out in 1983 with six crop species in a 17-year-old arecanut garden and preliminary results were reported (Bavappa et al., 1986). A steady increase was observed in the yield of arecanut, and the intercrops started yielding after the third year of planting. The economic dry matter yield of intercrops accounted for about 27 percent of total economic yield. This system could accommodate 1,300 arecanut palms with pepper vines, cocoa, clove, banana, and pineapple plants numbering 3,700 in one hectare.

The study of rhizosphere microorganisms in high-density multiple cropping and arecanut monocrops showed that the population of bacteria, fungi, actinomycetes, N^2-fixers, and P-solubilizers was more in the rhizosphere of arecanut, cocoa, and pineapple compared to arecanut monocrops (Bopaiah, 1991). The spore count, vesicular-arbuscular mycorrhizal (VAM) root infection, and colonization were lowest in banana. The microbial biomass was also higher in multiple cropping systems. The asymbiotic N_2-fixers isolated from an arecanut-based high-density multiple cropping system had an N_2- fixing capacity in the range of 2.8 to 11.8 mg N/100 ml of medium (CPCRI, 1988a).

Physiology

Approximately 30 to 50 percent of photosynthetically active radiation (PAR) is transmitted through the arecanut canopy (Balasimha and Subramonian, 1984, Balasimha, 1986). This varies with season and time of day. Under the arecanut canopy, the light environment is highly dynamic due to variation of cloud cover, solar angle, and canopy. The pattern of light transmission varies with the spacing of arecanut palms also. The midday light profile revealed that spacings of 3.3 × 3.3 m and 1.8 × 5.4 m allowed maximum transmission. However, in the afternoon 3.9 × 3.9 m spacing showed maximum transmission. On average, the arecanut canopy intercepts 70 percent of incoming radiation. The interception of the remaining radiation depends on the nature of the intercrop canopy and leaf area index (LAI). For example, cocoa with a compact canopy and LAI of three to five intercepts nearly 90 percent of available light. Pepper, which is trained on arecanut stems, receives differential light depending upon directional and diurnal effects.

On average, eight to nine leaves are borne in adult arecanut palm. The canopy area and leaf area are about 11.2 m^2 and 22.0 m^2 respectively. The arecanut canopy covers a ground area of 9.1 m^2 and the LAI is 2.44. The photosynthesis in arecanut leaves ranged from 2.4 to 8.2 μmol CO$_2$ m^{-1}s^{-1} depending on the cultivar and leaf position. The first fully open and third leaves showed the highest photosynthesis. The total chlorophyll content also varied among the species of *Areca,* and the highest content was recorded in *A. trianda* (Yadava and Mathai, 1972).

DISEASES AND PESTS

Diseases

Arecanut palm is affected by several diseases, some of which, such as yellow leaf disease, *Phytophthora* fruit rot, Anabe caused by *Ganoderma,* and inflorescence dieback, cause considerable economic losses (Rawther et al., 1982; Sampath Kumar and Saraswathy, 1994).

Yellow leaf disease (YLD) is by far the most serious disease affecting arecanut, though it is restricted to certain areas such as Southern Kerala and parts of Karnataka. This is a debilitating disease, and an effective control measure is yet to be evolved. Phytoplasma is found constantly associated with the disease (Nayar and Selsikar, 1978). The plant hopper *Proutista moesta* acts as a vector in the spread of the disease (Ponnamma et al., 1997). The diseased palms treated with tetracycline showed remission of symptoms, providing additional supporting evidence for phytoplasmic etiology (CPCRI, 1994). The disease is characterized by yellowing of leaves followed by necrosis (Photo 1.6). A comparative physiology of healthy and diseased arecanut palms showed significantly higher stomatal resistance

PHOTO 1.6. Areca palm with yellow leaf disease.

with increased water potential and turgor pressure in the latter (Chowdappa et al., 1993, 1995), whereas photosynthesis and transpiration were lower. Chlorophyll content was also reduced (Chowdappa and Balasimha, 1992). The altered values of chlorophyll fluorescence indices reflected in normal arrangement of antennae pigments of photosystem II (PS-II) and reduction in photosynthetic quantum yield (Chowdappa and Balasimha, 1995). The nuts and roots are also affected. Various attempts were made to contain the disease by different management practices. One such method is to include multiple cropping or mixed farming systems in YLD-affected gardens.

Fruit rot of arecanut, which is called *Kolerog* or *Mahali* locally, is a major disease (Photo 1.7). This occurs in epidemic proportions during the rainy season, and losses are reported to be 10 to 15 percent (Coleman, 1910; Sampath Kumar and Saraswathy, 1994). The first symptom is the appearance of water-soaked lesions on the nut surface. The infected nuts lose their natural color and turn dark green. A white mycelial mass then covers the entire nut. The causal organism is identified as *Phytophthora mardi*. The effective control of the disease is proper periodic spraying with 1 percent bordeaux mixture on fruit bunches. Covering of bunches with polyethylene gives complete control of the disease (CPCRI, 1983; Sastry and Hegde, 1985; Chowdappa et al., 1999).

PHOTO 1.7. Fruit rot in areca bunches.

The fungus can also cause bud rot of the palm. The spindle leaf is affected, killing the plant. Early detection, removal of affected tissues, and treatment with 10 percent bordeaux paste can save affected plants.

Foot rot of Areca, commonly known as "Anabe disease," is caused by *Ganoderma lucidum,* which is a soilborne bracket-forming fungus (Sampath Kumar and Nambiar, 1990). Poor drainage and high water table are predisposing factors of the disease. The affected palms initially show yellowing of leaves in the outer whorl. Affected stems show dull brown patches. In later stages gummy exudation starts and fruiting bodies of the fungus appear. Proper management of the garden would involve checking the occurrence of the disease. Spread of the disease can be controlled through drenching of the root zone with 0.3 percent calixin (15-20 l) or root feeding with the same chemical (125 ml/palm).

Inflorescence dieback is another serious disease causing button shedding. The disease is characterized by yellowing and later drying of inflorescence rachillae. The disease is prevalent throughout the year, but severe during summer months (Saraswathy et al., 1977; Chandramohanan and Kaveriappa, 1985). The disease is caused by the fungus *Colletotrichum gloeosporioides.* The disease can be controlled by spraying Indofil M-45 at 3 g/l of water at the time female flowers open.

Stem bleeding is a minor disease occurring in some parts of India. The disease is caused by the fungus *Thielaviopsis paradoxa.* Disease symptoms appear in the basal portion of the stem as small discolored depressions. Later these patches coalesce and cracks appear, leading to the disintegration of fibrous tissues. Later, brown exudates ooze out of the cracks. The disease can be controlled by removing affected parts and applying coal tar or bordeaux paste.

Bacterial leaf stripe disease is largely confined to the Maidan Region of Karnataka. The disease is caused by *Xanthomonas campestris* (Rao and Mohan, 1970; Sampath Kumar, 1981). The infection causes partial to complete blighting of leaves. Spraying of streptocycline or tetracycline (500 ppm) can effectively control the disease.

Insect Pests

Among the insect pests causing economic losses are mites, spindle bugs, inflorescence caterpillars, and root grubs (Nair and Daniel, 1982). These are either seasonal or persistent on the crop. Two major foliage mites (*Oligonychus indicus* and *Raoiella indica*) occur abundantly during dry seasons. Spraying with dicofol, carbophenothion, or chlorobenzilate controls the pest.

The spindle bug *(Carvalhoia arecae)* infests the leaf axils of arecanut (Photo 1.8), and maximum populations are found in August and September (Nair and Das, 1962). Bugs suck the sap from spindles and young leaves resulting in linear necrotic lesions on leaflets. As a result of this the spindle dries up and fails to open. Spraying with insecticides or filling innermost leaf axils with systemics (Thimet 10G or Sevin 4G) at the rate of 10 g/palm can control the pest effectively.

Among the root grubs, *Leucopholis burmeisteri* and *L. lepidophora* are the most common species. The grub voraciously feed on the roots causing serious damage. The pest damage is more common in low-lying areas. The leaves turn yellow and stems taper at the top. Soil insecticides such as Rogor (dimethoate) and phorate control white grubs effectively (Prem

PHOTO 1.8. Spindle bugs on an arecanut leaf axil.

Kumar and Daniel, 1981). Collection and destruction of adult beetles on emergence from soil is an effective management practice.

The pentatomid bug, *Halyomorpha marmorea,* causes tendernut drop in arecanut (Photo 1.9; Vidyasagar and Shama Bhat, 1986). The infestation is prominent during March to August. The bugs pierce the nut and suck the kernel sap. The kernel dries up, and the nut abscises and falls. Spraying 0.05 percent endosulfan can effectively control the pest.

The inflorescence caterpillar *Tirathaba mundella* damages areca inflorescence. It feeds on tender rachillae. The pest can be controlled by the application of insecticides, such as malathion.

Scale insects *(Aonidiella orientalis),* which colonize on leaves, nuts, and leaf sheaths, suck the sap from tissues (Photo 1.10). Feeding on nuts results in premature yellowing and a loss in quality. Scale insects are present throughout the year, but are more abundant in October to February. Although scale insects were considered minor pests earlier, there have been widespread outbreaks of the insects in some of the districts in coastal India.

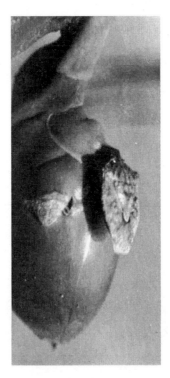

PHOTO 1.9. A pentatomid bug on an arecanut nut.

PHOTO 1.10. Scale insects on arecanut nuts.

Chemical control of scale insects is rather difficult, although spraying of the insecticides malathion (0.1 percent) and fenthion (0.1 percent) can be done with partial success. The best possible method is to use biocontrol agents. The coccinellid beetles, mainly the *Chilocorus* spp, are effective natural enemies of scale insects.

Nematodes

The burrowing nematode *Radopholus similis* is commonly associated with arecanut (Sundaraju and Koshy, 1988). The intercrops found to be infested with nematodes are banana, cardamom, and pepper. The interculti-vation of these crops with arecanut favors the increase in nematode popula-tions (Sundaraju and Koshy, 1988). Cocoa, pineapple, clove, cinnamon, and nutmeg were completely free of nematodes and thus are ideal species for mixed cropping with arecanuts (CPCRI, 1988b). However, the presence of nematodes does not adversely affect the crops to economic thresholds. Nematodes can be effectively kept under control by soil amendments with nematicides or neem cake.

HARVESTING AND PROCESSING

The stage of harvesting depends on the type of produce to be prepared for the markets. Harvesting of nuts at the correct stage is very important for obtaining quality products. For dry nut making, ripe nuts are harvested. Green nuts at six to seven months of age are harvested for tender nut processing.

The most popular type of arecanut is dried whole nuts known as *chali*. After harvesting, ripe nuts are sun dried for 40 to 45 days. It is essential to spread the nuts in a single layer for drying. Proper drying is essential to avoid fungal infections. Turning of nuts once a week helps in uniform drying. The dry nuts are dehusked manually or mechanically and marketed. Good quality *chali* is free from immature nuts, surface cracking, husk sticking, and fungal and insect infection.

If the requirement for the market is tender processed nuts, then the nuts are harvested at six to seven months maturity when they are green and soft. The processing consists of dehusking, cutting nuts into halves, and boiling with water or diluted extract from previous boiling. After boiling, arecanut pieces are coated with *kali* which is a concentrated extract after boiling three to four batches of arecanut, to get good quality processed nuts. These nuts are generally sun dried, although oven drying is also followed on occasion.

Dried arecanuts are broken into bits, blended with flavor, and packed for marketing. The flavoring of *supari* varies with region and market preferences. Tender processed nuts are widely used in making scented *supari*. Spices and synthetic flavors are added. Rose essence and menthol are common additives. The scented *supari* is packed in aluminium or butter paper pouches for marketing. Recently, a product blended with cashewnut bits called *kaju-supari* has been developed and is in the market.

FUTURE OUTLOOK

The future of arecanut in this country has become a matter of great concern to everyone connected with the industry. Based on an analysis of the then available production and demand data, Bavappa (1982) suggested that the internal capacity to utilize the increased production of arecanut in India is much more than the estimated figures. This has been found to be true even in the year 2000. The National Commission on Agriculture has projected the requirements of arecanut in 2000 as about 190,000 tonnes. Production in India was 330,000 tonnes during 1998-1999. In spite of this high production, the price that prevailed during 1999-2000 was Rs 13,181 per quintal, the highest ever recorded in the history of arecanut trade.

As a consequence of the implementation of the WTO agreement, 3,022 tonnes of arecanut were imported during the year 2000-2001 at a unit price of Rs 3,122/quintal, which was less than 25 percent of the price that prevailed during the year previous to this import. This was followed by a crash in price for the local nut during 2000-2001, the price falling to Rs 7,893/quintal.

Although the price of the imported nut was much lower than the prevailing local price, the reduction in the local rate was only about 60 percent. In spite of the import, the demand for the nut continued to be steady. It appears, therefore, that the quantity of the imported nut has played only a low role in the local price reduction. The import was mainly from Sri Lanka (2,426 tonnes) and Thailand (419 tonnes). India is the largest producer and consumer of arecanut in the world, accounting for 52 percent of the world production. No organized cultivation of arecanut exists in any country other than India. However, with the opening of the Indian market it is likely that the crop would receive greater attention in other countries as well. Keeping the cost of production of arecanut at the lowest level with sustainable high yields through innovative technology intervention appear to be the pragmatic approach to keeping the industry viable and competitive. All the same, it will be worthwhile to study the current scenario with an international outlook to develop appropriate development and marketing strategies.

REFERENCES

Abdul Khader, K.B. 1988. *Effect of drip irrigation on yield and yield attributes of arecanut.* Paper presented at the National Seminar on Drip Irrigation, Water Technology Centre, Tamil Nadu Agricultural University (TNAU), Coimbatore, India.

Abdul Khader, K.B. and Antony, K.J. 1968. Intercropping: A paying proposition for areca growers—What crops to grow? *Indian Farming* 18: 14-45.

Abdul Khader, K.B. and Havanagi, G.V. 1991. Consumptive use of water in relation to cumulative pan evaporation (CPE) with and without mulching in arecanut. *J. Plantation Crops* 18 (Suppl): 139-146.

Abraham, A., Mathew, P.M., and Ninan, C.A. 1961. Cytology of *Cocos nucifera* L. and *Areca catechu* L. *Cytologia* 26: 327-332.

Abraham, K.J. 1974. Intercropping in arecanut helps to build up farmer's economy. *Arecanut Spices Bulletin* 5: 73-75.

Ananda, K.S. 1999a. *Arecanut varieties*. Extension publication, CPCRI, Kasaragod, India.

Ananda, K.S. 1999b. Genetic improvements in arecanut. In Ranambal, M.J., Kumaran, P.M., Muralidharan, K., Niral, V., and Arunachalm, V., Eds., *Improvement of Plantation Crops*. Central Plantation Crops Research Institute (CPCRI), Kasaragod, India, pp. 52-57.

Ananda, K.S. 2000. Exploitation of a dwarf mutant in arecanut breeding. In Muraleedharan, N. and Raj Kumar, R., Eds., *Recent Advances in Plantation Crops Research*. Allied Publ. Ltd., New Delhi, pp. 69-72.

Ananda, K.S. and Thampan, C. 1999. Promising cultivars and improved varieties of arecanut (*Areca catechu* L.). *Indian J. Arecanut Spices and Medicinal Plants* 1: 24-28.

Anuradha, S. 1999. Genetic resources and conservation strategies in arecanut (*Areca catechu*). In Ranambal, M.J., Kumaran, P.M., Muralidharan, K., Niral, V., and Arunachalm, V., Eds., *Improvement of Plantation Crops*. CPCRI, Kasaragod, India, pp. 48-51.

Balasimha, D. 1986. Light penetration patterns through arecanut canopy and leaf physiological characteristics of intercrops. *J. Plantation Crops* 16 (Suppl): 61-67.

Balasimha, D., Abdul Khader, K.B., and Bhat, R. 1996. Gas exchange characteristics of cocoa grown under arecanut in response to drip irrigation. Paper presented at All India Seminars on Modern Irrigation Techniques, June 26-27, 1996, pp. 96-101.

Balasimha, D. and Subramonian, N. 1984. Nitrate reductase activity and specific leaf weight of cocoa and light profile in arecanut-cocoa mixed cropping. Proceedings of the Plantation Crops Symposium VI, Kottayam, ISPC, Kasaragod, India, pp. 83-88.

Bavappa, K.V.A. 1961. Some common intercrops in an arecanut garden. *ICAC Monthly Bulletin* 2: 16-17.

Bavappa, K.V.A. 1963. Morphological and cytological studies in *Areca catechu* Linn. and *Areca triandra* Roxb. MSc (Ag.) thesis, University of Madras, India.

Bavappa, K.V.A. 1966. Morphological and anatomical studies in *Areca catechu* Linn. and *Areca triandra* Roxb. *Phytomorphology* 16: 436-443.

Bavappa, K.V.A. 1974. Studies in the genus *Areca* L. (Cytological and genetic diversity of *A. catechu* L and *A. triandra* Roxb.). PhD thesis, University of Mysore.

Bavappa, K.V.A. 1977. Mangala—a superior arecanut variety. *Arecanut and Spices Bulletin* 9: 55-56.

Bavappa, K.V.A. 1982. Future of arecanut. In Bavappa, K.V.A., Nair, M.K., and Prem Kumar,T., Eds., *The Arecanut Palm*. CPCRI, Kaasragod, India, pp. 319-325.

Bavappa, K.V.A., Kailasam, C., Abdul Khader, K.B, Biddappa, C.C., Khan, H.H., Kasturi Bai, K.V., Ramadasan, A., Sundararaju, P., Bopaiah, B.M., Thomas,

G.V., et al. 1986. Coconut and arecanut based high density multispecies cropping systems. *J. Plantation Crops* 14: 74-87.

Bavappa, K.V.A. and Nair, M.K. 1978. Cytogenetics of *Areca catechu* L, *A. triandra* Roxb. and their F1 hybrids (Palmae). *Genetica* 49: 1-8.

Bavappa, K.V.A. and Nair, M.K. 1982. Cytogenetics and breeding. In: Bavappa, K.V.A., Nair, M.K., and Premkumar, T., Eds., *The Arecanut Palm*. CPCRI, Kasaragod, India, pp. 51-96.

Bavappa, K.V.A., Nair, M.K., and Ratnambal, M.J. 1975. Karyotype studies in *Areca catechu* L, *A. triandra* Roxb. and *A. catechu x A. triandra* hybrids. *Nucleus* 18: 146-151.

Bavappa, K.V.A. and Pillai, S.S. 1976. Yield and yield component analysis in different exotic cultivars and species of *Areca*. In Chadha, K.L., Ed., *Improvement of Horticulture, Plantation, and Medicinal Plants*, Volume 1. Today and Tomorrow Printers and Publ., New Delhi, pp. 243-246.

Bavappa, K.V.A. and Ramachander, P.R. 1967. Improvement of arecanut palm, *Areca catechu* L. *Indian J. Genet.* 27: 93-100.

Bavappa, K.V.A. and Raman, V.S. 1965. Cytological studies in *Areca catechu* Linn. and *Areca triandra* Roxb. *J. Indian Bot. Soc.* 44: 495-505.

Beccari, O. 1919. The palms of the Philippine Islands. *Philipp. J. Sci.* 14: 295-362.

Bhat, P.S.I. and Rao, K.S.N. 1962. On the antiquity of arecanut. *Arecanut J.* 13: 13-21.

Blume, C.L. 1836. *Rumhiasive Commntations Botanicae de Plantis Indiae Orientalis*. Leyden.

Bopaiah, B.M. 1991. Soil microflora and VA mycorhizae in areca based high density multispecies cropping and areca monocropping systems. *J. Plantation Crops* 18 (Suppl): 224-228.

Brahma, R.N. 1974. In North Bengal banana is a paying intercrop in areca gardens. *Arecanut Spices Bull.* 5: 80-81.

Central Plantation Crops Research Institute (CPCRI). 1983. Annual Report. Kasaragod, India.

Central Plantation Crops Research Institute (CPCRI). 1988a. Annual Report for 1986. Kasaragod, India.

Central Plantation Crops Research Institute (CPCRI). 1988b. Annual Report for 1987. Kasaragod, India.

Central Plantation Crops Research Institute (CPCRI). 1994. Annual Report. Kasaragod, India.

Chandramohanan, R. and Kaveriappa, K.M. 1985. Epidemiological studies on inflorescence dieback of arecanut caused by *Colletotrichum gloeosporioides*. *Proceedings of the Silver Jubilee Symposium on Arecanut Research and Development*. CPCRI, Vittal, India, 1982, pp. 104-106.

Chowdappa, P. and Balasimha, D. 1992. Non-stomatal inhibition of photosynthesis in arecanut palms affected with yellow leaf disease. *Indian Phytopath.* 45: 312-313.

Chowdappa, P. and Balasimha, D. 1995. Chlorophyll fluorescence characteristics of arecanut palms affected with yellow leaf disease. *Indian Phytopath.* 47: 87-88.

Chowdappa, P. Balasimha, D., and Daniel, E.V. 1993. Water relations and net photosynthesis of arecanut palms affected with yellow leaf disease. *Indian J. Microbial Ecol.* 3: 19-30.

Chowdappa, P., Balasimha, D., Rajagopal, V., and Ravindran, P.S. 1995. Stomatal responses in arecanut palms affected with yellow leaf disease. *J. Plantation Crops* 23: 116-121.

Chowdappa, P., Saraswathy, N., Vinaya Gopal, K. and Somala, M. 1999. Control of fruit rot of arecanut through polythene covering of bunches. *Proceedings of the Symposium on Plant Disease Management for Sustainable Agriculture, Indian Phytopathological Society.* CPCRI, Kayangulam, India, pp. 18-19.

Coleman, L.C. 1910. Diseases of the areca palm (*Areca catechu* L). 1. Koleroga or rot disease. *Ann. Mycol.* 8: 591-626.

Darlington, C.D. and Janaki Ammal, E.K. 1945. *Chromosome Atlas of Cultivated Plants*. George Allen and Unwin Ltd., London.

Furtado, C.X. 1933. The limits of *Areca* Linn. and its sections. *Fedde's Repertrium Specierum Novarum. Regnum Vegetabilis* 33: 217-239.

Furtado, C.X. 1960. The philological origin of *Areca catechu. Principes* 4: 26-31.

Mahesha, A., Abdul Khader, K.B., and Ranganna, G. 1990. Consumptive use and irrigation requirement of arecanut palm (*Areca catechu* L). *Indian J. Agric. Sci.* 60: 609-611.

Martius, C.F.P.Von. 1832-1850. *Historia Naturalis Palmarum.* Leipzig.

Mohan Rao, M. 1982. Introduction. In Bavappa, K.V.A., Nair, M.K., and Premkumar, T., Eds., *The Arecanut Palm.* CPCRI, Kasaragod, India, pp. 1-9.

Mohapatra, A.R. and Bhat, N.T. 1982. Crop management. A. Soils and manures. In Bavappa, K.V.A., Nair, M.K., and Premkumar, T., Eds., *The Arecanut Palm.* CPCRI, Kasaragod, India, pp. 97-104.

Muralidharan, A. 1980. Biomass productivity, plant interactions, and economics of intercropping in arecanut. PhD thesis, University of Agricultural Sciences, Bangalore, India.

Muralidharan, A. and Nayar, T.V.R. 1979. Intercropping in arecanut gardens. In Nelliat, E.V. and Shama Bhat, K., Eds., *Multiple Cropping in Coconut and Arecanut Gardens.* CPCRI, Kasaragod, India, pp. 24-27.

Murthy, K.N. and Bavappa, K.V.A. 1960a. Floral biology of areca (*Areca catechu* Linn.). *Arecanut J.* 11: 51-55.

Murthy, K.N. and Bavappa, K.V.A. 1960b. Morphology of arecanut palm—the shoot. *Arecanut J.* 11: 99-102.

Murthy, K.N. and Bavappa, K.V.A. 1962. Species and ecotypes (cultivars) of arecanut. *Arecanut J.* 13: 59-78.

Murthy, K.N. and Pillai, R.S.N. 1982. Botany. In Bavappa, K.V.A., Nair, M.K., and Premkumar, T., Eds., *The Arecanut Palm.* CPCRI, Kasaragod, India, pp. 11-49.

Naidu, G.V.B. 1962. *Arecanut in Mysore.* Department of Agriculture, Bangalore, India.

Naidu, G.V.B. 1963. Seen a dwarf areca palm? *Indian Farming* 12: 16-17.

Nair, C.P.R. and Daniel, M. 1982. Pests. In Bavappa, K.V.A., Nair, M.K., and Premkumar, T., Eds., *The Arecanut Palm.* CPCRI, Kasaragod, India, pp. 151-184.

Nair, M.G. 1982. Intercropping with pepper. *Indian Farming* 32: 17-19.

Nair, M.K. and Ratnambal, M.J. 1978. Cytology of *Areca macrocalyx* Becc. *Curr. Sci.* 47: 172-173.

Nair, M.R.G.K. and Das, N.M. 1962. On the biology of *Carvalhoia arecae* Miller and China, a pest of areca palms in Kerala. *Indian J. Ent.* 24: 86-93.

Nambiar, K.K. 1949. *A Survey of Arecanut Crop in Indian Union.* Indian Central Arecanut Committee, Calicut, India.

Nayar, R. and Selsikar, C.E. 1978. Mycoplasma-like organisms associated with yellow leaf disease of *Areca catechu. European J. Forest Pathol.* 8: 125-128.

Ponnamma, K.N., Solomon, J.J., Rajeev, K., Govindankutty, M.P., and Karnavar, G.K. 1997. Evidence for transmission of yellow leaf disease of areca palm, *Areca catechu* L. by *Proutista moesta* (West Wood) (Homoptera: Derbidae). *J. Plantation Crops* 25: 197-200.

Prem Kumar, T. and Daniel, M. 1981. Studies on the control of soil grubs of arecanut palm. *Pesticides* 15: 29-30.

Raghavan, V. 1957. On certain aspects of the biology of arecanut (*Areca catechu* Linn.) and utilization of its by-products in industry. DPhil thesis, Gauhati University, India.

Raghavan, V. and Baruah, H.K. 1958. Arecanut: India's popular masticatory—History, chemistry, and utilization. *Econ. Bot.* 12: 315-345.

Rao, Y.P. and Mohan, S.K. 1970. A new bacterial stripe disease of arecanut (*Areca catechu*) in Mysore State. *Indian Phytopath.* 23: 702-704.

Rau, M.K.T. 1915. The sweet arecanut, *Areca catechu* L. var. deliciosa. *J. Bombay Nat. Hist. Soc.* 23: 793.

Rawther, T.S.S., Nair, R.R., and Saraswathy, N. 1982. Diseases. In Bavappa, K.V.A., Nair, M.K., and Premkumar, T., Eds., *The Arecanut Palm.* CPCRI, Kasaragod, India, pp. 185-224.

Sadanandan, A.K. 1974. Raise intercrops in arecanut plantation for higher returns. *Arecanut Spices Bulletin* 9: 70-72.

Sampath Kumar, S.N. 1981. Bacterial leaf stripe disease of arecanut *(Areca catechu* L.) caused by *Xanthomonas campestris* pv. *arecae.* PhD thesis, Indian Institute of Science, Bangalore, India.

Sampath Kumar, S.N. and Nambiar, K.K. 1990. *Ganoderma* disease of arecanut palm—Isolation, pathogenicity, and control. *J. Plantation Crops* 18: 14-18.

Sampath Kumar, S.N. and Saraswathy, N. 1994. Diseases of arecanut. In Chadha, K.L. and Rethinam, P., Eds., *Advances in Horticulture.* Malhotra Pupl. House, New Delhi, Volume 10, pp. 939-967.

Sandeep Nayak, J. 1996. Crop water assessment for arecanut. Paper presented at the All India Seminar on Modern Irrigation Techniques, Bangalore, India.

Sannamarappa, M. 1990. Spacing trial in arecanut. *J. Plantation Crops* 17(Suppl): 51-53.

Sannamarappa, M. and Muralidharan, A. 1982. Multiple cropping. In Bavappa, K.V.A., Nair, M.K., and Premkumar, T., Eds., *The Arecanut Palm.* CPCRI, Kasaragod, India, pp. 133-149.

Saraswathy, N., Koti Reddy, M., and Nair, R.R. 1977. *Colletotrichum gloeosporioides* causing inflorescence dieback, button shedding, and nut rot of betelnut palm. *Plant Disease Reptr.* 61: 172-174.

Sastry, M.N.L. and Hegde, R.K. 1985. Control of koleroga of arecanut. *Proceedings of the Silver Jubilee Symposium on Arecanut Research and Development.* CPCRI, Vittal, India, pp. 86-91.

Shama Bhat, K. 1974. Intensified inter/mixed cropping in areca garden—The need of the day. *Arecanut Spices Bulletin* 5: 67-69.

Shama Bhat, K. 1978. Agronomic research in Arecanut—A review. *J. Plantation Crops* 6: 67-80.

Shama Bhat, K. 1983. Plant interactions in a mixed crop community of arecanut (*Areca catechu* L.) and cacao (*Theobroma cacao* L.). PhD thesis, University of Mysore, Mysore, India.

Shama Bhat, K. 1988. Growth and performance of cacao (*Theobroma cacao* L.) and arecanut (*Areca catechu* L.) under mixed cropping system. *Proceedings of the Tenth International Cocoa Research Conference.* Cocoa Producers' Alliance, Lagos, Nigeria, pp. 15-19.

Shama Bhat, K. and Abdul Khader, K.B. 1970. Inter and mixed cropping in arecanut gardens. *Indian Farming* 20: 35.

Shama Bhat, K. and Abdul Khader, K.B. 1982. Crop management—Agronomy. In Bavappa, K.V.A., Nair, M.K., and Premkumar, T., Eds., *The Arecanut Palm.* CPCRI, Kasaragod, India, pp. 105-131.

Shama Bhat, K. and Bavappa, K.V.A. 1972. Cacao under palms. In Wastie, R.L. and Earp, D.A., Eds., *Cacao and Coconuts in Malaysia.* Incorporated Soc. Planters, Kuala Lumpur, pp. 116-121.

Shama Bhat, K., Krishna Murthi, S., and Madhava Rao, V.N. 1962a. Studies in the development of the fruit of arecanut. *South Indian Hort.* 10: 1-17.

Shama Bhat, K., Krishna Murthi, S., and Madhava Rao, V.N. 1962b. Studies on certain aspects of floral biology of the arecanut. *South Indian Hort.* 10: 22-34.

Shama Bhat, K. and Leela, M. 1968. Cacao and arecanut are good companions for more cash. *Indian Farming* 18: 19-20.

Shama Bhat, K. and Leela, M. 1969. The effect of density of planting on the distribution of arecanut roots. *Trop. Agric.* 46: 55-61.

Sharma, A.K. and Sarkar, S.K. 1956. Cytology of different species of palms and its bearing on the solution of problems of phylogeny and speciation. *Genetics* 28: 361-488.

Stebbins, G.L. 1958. Longevity, habitat, and release of genetic variability in the higher plants. *Cold Spring Harb. Symp. Quant. Biol.* 23: 365-378.

Sujatha, S., Ravi Bhat, Reddy, V.M., and Abdul Haris, A. 1999. Response of high yielding varieties of arecanut to fertilizer levels in coastal Karnataka. *J. Plantation Crops* 27: 187-192.

Sundaraju, P. and Koshy, P.K. 1988. Effect of intercrops on occurrence of *Radopholus similis* in arecanut palms. *J. Plantation Crops* 18 (Suppl): 299-301.

Thomas, K.G. 1978. Crops diversification in arecanut gardens. *Indian Arecanut Spices and Cocoa J.* 1: 97-99.

Venkatasubban, K.R. 1945. Cytological studies in Palmae. Part I. Chromosome number in a few species of palms in British India and Ceylon. *Proc. Indian Acad. Sci.* 22: 193-207.

Vidyasagar, P.S.P.V. and Shama Bhat, K. 1986. A pentatomid bug causes tendernut drop in arecanut. *Curr. Sci.* 55: 1096-1097.

Yadava, R.B.R. and Mathai, C.K. 1972. Chlorophyll and organic acid contents of arecanut. *Madras Agric. J.* 59: 306-307.

Yadukumar, N., Abdul Khader, K.B., and Shama Bhat, K. 1985. Scheduling irrigation for arecanut with pan evaporation. In *Arecanut Research and Development: Proceedings of the Silver Jubilee Symposium on Arecanut Research and Development*. CPCRI, Vittal, 1982, pp. 33-37.

Chapter 2

Cardamom

M. R. Sudharshan
V. S. Korikanthimath

INTRODUCTION

Cardamom (*Elettaria cardamomum* (L.) Maton), a major oriental spice known as the "Queen of Spices," is the dried fruit of a tall perennial herbaceous plant, belonging to the family Zingiberaceae. India is the major producer and exporter of this spice and an estimated 7,900 tonnes of cardamom were produced during 1997-1998 from an area of 72,444 ha. Until the late 1970s, India enjoyed a near monopoly in the world trade of cardamom. But the situation changed drastically, and the share of Indian cardamom in the world trade declined rapidly. The main competitor is Guatemala. India's productivity is 149 kg per ha, whereas in Guatemala it is more than 250 kg per ha.

Cardamom seeds have a pleasant aroma and a characteristic warm, slightly pungent taste. This spice is used for flavoring curries, cakes, and other culinary preparations; it is also used as masticatory and for the preparation of herbal medicine. Substantial quantities are imported into the Middle East and Arab countries, particularly in Saudi Arabia where cardamom is by far the most popular spice. Cardamom coffee, commonly known as *gahwa,* is a symbol of Arab hospitality, as well as a habitual beverage. Other important markets are Sweden and Finland, where cardamom is widely used in confectionery. Cardamom is thought to have aphrodisiac properties. Cardamom is listed in the British and U.S. pharmacopoeias, and is used as an aromatic stimulant, carminative, and flavoring agent. The steam-distilled essential oil of cardamom is used for flavoring processed foods, in perfumery, and for flavoring liquors and other beverages. Cardamom substitutes, which appear in trade and are given the name cardamom, are obtained from *Aframomum* spp. in Africa and *Amomum* spp. in Asia (Purseglove et al., 1981).

From time immemorial, India has been known as the home of cardamom. Until 1800 the world's supply came from the evergreen monsoon forests of the Western Ghats in South India and Sri Lanka. The erstwhile Travancore government in A.D. 1823 actively took up cultivation of cardamom in India. The area under cardamom cultivation during 1997-1998 was 72,444 ha distributed in three southern states, Kerala (40,867 ha), Karnataka (25,686 ha), and Tamil Nadu (5,891 ha). Of late, Guatemala has emerged as a keen competitor to Indian cardamom in the international spice market. Tanzania, Sri Lanka, El Salvador, Vietnam, Laos, Cambodia, and Papua New Guinea are the other cardamom-growing countries.

In India, cardamom thrives in the tropical rain forests of the Western Ghats situated between 8°30' and 14°30' N latitude and longitude of 75-77°. From north to south, the area is an elongated tract extending more than 2,000 km, from Sirsi of Karnataka to Thirunelveli of Tamil Nadu. From east to west, it is a narrow belt of land distributed over the Western Ghats at an elevation ranging from 600 to 1,500 m above mean sea level (MSL).

Area, Production, and Productivity

The area under cardamom in India declined from 1.75 lakh hectares in the early 1960s to 0.72 lakh hectares in 1997-1998 (Table 2.1). The productivity has gone up from 44 kg (1975-1976) to 149 kg per hectare (1997-1998) (Table 2.2).

TABLE 2.1. Area, production, and productivity of cardamom in India.

Year	Area (ha)	Production (t)	Productivity (kg/ha)
1970-1971	91,480	3,170	46
1975-1976	91,480	3,000	44
1980-1981	93,950	4,400	62
1985-1986	100,000	4,700	77
1990-1991	81,554	4,750	78
1992-1993	82,392	4,250	70
1993-1994	82,860	6,600	108
1994-1995	83,651	7,000	113
1995-1996	83,902	7,900	128
1996-1997	73,592	6,625	125
1997-1998	72,444	7,900	149
1998-1999*	72,140	7,170	135
1999-2000*	72,430	9,330	174

Source: Spices Board of India, 1998, trade estimates.
*Provisional figures

TABLE 2.2. Production (t) in India by state (ha).

Year	Kerala		Karnataka		Tamil Nadu	
	Area	Production	Area	Production	Area	Production
1992-1993	43,388	2,570	32,881	1,200	6,123	480
1993-1994	43,459	4,430	33,414	1,510	6,087	660
1994-1995	44,237	4,720	33,518	1,580	5,896	700
1995-1996	44,248	5,380	33,743	1,745	5,811	775
1996-1997	41,268	4,550	25,996	1,360	6,329	715
1997-1998	40,867	5,290	25,686	1,860	5,891	750

Source: Based on data from Spices Board of India, trade statistics.

Growing Conditions

Cardamom is a shade-loving plant. Ideal conditions for its cultivation are altitudes of 600 to 1,200 m above MSL, an annual rainfall of 1,500 to 4,000 mm, and a temperature range of 10 to 35°C.

Cardamom is generally grown in evergreen forest areas. The soils here are clay loam, distinctly acidic, rich in organic matter, and low in available phosphorus and potassium.

WORLD TRADE

Domestic Demand

At present, India is the second largest consumer of small cardamom in the world after Saudi Arabia. The domestic consumption of small cardamom in India is expected to go up to 9,500 tonnes in 2000 and 12,500 tonnes in 2005.

Exports

Until 1986-1987, the Indian cardamom industry was largely export-oriented with 70 percent of the produce of 4,700 tonnes being exported. In 1970-1971, India exported 3,500 tonnes of cardamom but in 1987-1988 the export dipped to 270 tonnes (Table 2.3). In 1990, exports came down further to an all-time low of just 180 tonnes because of stiff competition from Guatemala.

The drought of 1981-1983 damaged about 60 percent of the cardamom plantations in India. India's export of small cardamom fell to 260 tonnes in 1983-1984 from 2,320 tonnes in 1981-1982. During this period, the global prices of cardamom saw a sharp upswing which helped Guatemala to increase its output and capture the international market held earlier by India.

The major consumers of cardamom are Saudi Arabia, other Arab countries, Europe, and Japan.

TABLE 2.3. India's exports of small cardamom.

Year	Quantity (t)	Total value (Rs.crores)	Unit value (Rs./kg)
1980-1981	2,340	34.75	149
1981-1982	2,320	30.20	130
1982-1983	1,030	16.37	159
1983-1984	260	5.44	209
1984-1985	2,380	64.80	272
1985-1986	3,270	53.46	163
1986-1987	1,447	18.49	128
1987-1988	270	3.40	126
1988-1989	780	10.37	133
1989-1990	180	3.07	171
1990-1991	400	10.87	272
1991-1992	553	16.07	278
1992-1993	189	5.93	314
1993-1994	342	13.86	405
1994-1995	255	7.37	288
1995-1996	500	12.3	246
1996-1997	226	8.7	384
1997-1998	297	10.64	358
1998-1999	355	22.49	634

Source: Based on data from Spices Board of India, trade statistics.

HISTORY

Rosengarten (1969) records that

> Cardamom was an article of Greek trade during the fourth century
> B.C. The inferior grades were known as *amomon;* the superior as
> *Kardamomon.* By the first century A.D. Rome was importing substan-
> tial quantities of cardamom from India. It was one of the most popular
> oriental spices in the Roman cuisine. Cardamom was listed among the
> Indian Spices liable to duty in Alexandria in A.D. 176.

Ridley (1912), noted that

> there was a spice known to the Greeks and Romans as *Cardamomum*
> and *Amomum,* but it appears to be certain that these spice plants,
> whatever they were, were not the cardamoms of present day; Al-
> though the name of this spice, as we know it, is evidently taken from
> these words.

Burkill (1956) also doubts whether the Greeks and Romans had the true
cardamom from *Elettaria,* the cardamoms have long been an article of trade
in India, certainly from the time of Ibn Sena (A.D. 980-1037). The writer
Edrisi described the spice as a product of Sri Lanka in A.D. 1154. Surprisingly Marco Polo did not mention it in his travels. Portuguese traveler Gar-
cia da Orta in 1563 differentiated between the smaller, aromatic form
(*Elettaria cardamomum* (L.) Maton var. *cardamomum*) from India and the
larger fruited form (*Elettaria cardamomum* (L.) Maton var. *major*) from Sri
Lanka (Ridley 1912).

The world's supply of cardamom up to 1900 came from the evergreen
monsoon forests of the Western Ghats in Southern India and Sri Lanka. The
cultivation of cardamom commenced much later. The earliest detailed account of the cultivation of cardamom to reach the West is contained in Isaac
Buchanan-Hamilton's description of his journey through the West Coast of
India published in 1807. This was the system he saw in North Kanara
(Karnataka). The collectors of cardamoms probably assisted nature by selective felling of trees to produce the desirable ecological conditions for
cardamom development. In Coorg, such a refinement of jungle collection
developed into a more or less regular system in the nineteenth century. This
system, "the Malai," existed until recently as one of the principal types of
cardamom production in that province (Mayne 1954).

Suresh (1980), tracing the history of cardamom plantations in Kerala, recorded that the cultivation of cardamom was actively encouraged by the

Travancore government in A.D. 1823, during which time special cardamom staff were attached to the forest department and the government had control over cardamom plantations and a monopoly over trade. 1903 was a disastrous year due to excess production in Ceylon (now known as Sri Lanka) resulting in low prices and abolition of the government monopoly on cardamom plantations and trade. Mayne (1954) recorded that as a plantation crop, cardamoms came into some prominence as a secondary crop in the wetter and more heavily shaded parts of Coffee estates in Mysore (Karnataka) quite early in the history of East Indian Coffee. But major development occurred at the end of the nineteenth century, when the Travancore government gave up its monopoly over the cardamom trade. From that time, the industry took its modern aspect.

SYSTEMATICS, EVOLUTION, AND ADAPTATION

The generic name *Elettaria* originated from Maton (Abraham and Thulsidas, 1958) who segregated the Malabar cardamom, then known as *Amomum cardamomum* L., from *Amomum* as a distinct genus *Elettaria*. Maton derived the scientific name *Elettaria* from the local (Malayalam) name *Elathari.*

Thwaites (1864, cited in Abraham and Thulsidas, 1958) named the larger type of cardamom produced in Sri Lanka as *E. cardamomum* (L.) Maton var. *major.* Schumann (1904, cited in Abraham and Thulsidas, 1958) recognized two species under *Elettaria, E. cardamomum* (L.) Maton and *E. major* Scn. corresponding to *E. cardamomum* (L.) Maton var. *major* Thw.

Molegode (1938) recognized three varieties of *Elettaria: E. cardamomum* Maton as found in Sri Lanka, one of which is indigenous to that island, and the cultivated varieties, *Malabar* and *Mysore,* which appeared to have been introduced from India.

Abraham and Thulsidas (1958) applied the experimental method of taxonomy to the South Indian cardamoms after an extensive study from 1944 to 1953. According to them, the cardamoms of southern India fall into two main groups:

1. A smaller type that rarely if ever exceeds about 10 feet (3 m) in height and is characterized by prostrate panicles with globular or oblong fruits
2. A larger type which under favorable situations may reach 15 feet (4.5 m) in height and is characterized by larger leaf lamina and by flexuous or erect panicles with larger longish fruits

As cardamom has become an agricultural crop in comparatively recent times, it can be considered to be only in the transition stage from the wild to the cultivated state. From the natural distribution of the two main groups of cardamom, it is interesting to note that the larger type with erect or flexuous panicles and large fruits is indigenous to the extreme south of the Western Ghats, i.e., south of the Thamraparni river in the Tirunelveli District (Tamil Nadu); whereas the smaller cardamom with prostrate panicles and smaller fruits is indigenous to the vast cardamom area extending from Thamraparni along the Western Ghats up to North Kanara (Karnataka) in the North. This appears to indicate the natural adaptability of the larger cardamom to the lower but well distributed rainfall of the extreme south of the peninsula, and the adaptability of the smaller "Malabar" cardamom to the wetter northern portions of the Western Ghats up to North Kanara.

From their studies, Abraham and Thulsidas (1958) concluded that the larger type of cardamom with erect or flexuous panicles found in the Singampatti hills of the Thirunelveli district was indeed an indigenous type of those tracts south of the Thamraparni River. Their studies on the distribution, habits, and morphology of different cardamom in their natural habitats and when grown side by side under identical conditions, revealed that the cardamom of each locality or habitat formed a freely interbreeding population; the individuals of which represented various recombinations of the genetic composition. Different gene types maintained in Singampatty Research Station bred true, hence they suggested that these ecogeographical types of cardamom were "clines."

It appears that the larger cardamom of the extreme south of the peninsula and Sri Lanka was the original cardamom from which the smaller type originated. The following observations support this theory:

1. Greater range of variability exists in respect of the morphology of the panicles and the fruits as well as the productivity of individual plants in the larger cardamom than in the smaller cardamom. In smaller cardamom, the variability on these features even among its forms is limited.
2. The habitat of the larger cardamom is always the evergreen jungles, and it is satisfactorily productive between elevations of 3,000-4,500 feet (900-1,400 m). The smaller cardamom is usually productive from an elevation 4,500 feet down to about 2,000 feet (1,400-600 m). The fact that a form of smaller cardamom peculiar to and cultivated in North Kanara is productive even at an elevation of 1,500 feet (450 m) is indicative of a tendency of the smaller cardamom to gradually come

down to lower elevations, which is undoubtedly a greater advance in evolution.

3. The local types of the smaller variety, especially *Pattiveeranpatty* (Lower Pulneys in Tamil Nadu), grow well and give good yields even with 50 or 60 inches of rainfall (1,270-1,530 mm) per year.

4. Prostrate panicled plants and plants with globular fruits as well as plants combining both these characters were observed to occur as mutations in the larger cardamom of the Singampatti hills situated south of the Thamraparni River. These two types of cardamoms are well distributed into two ecogeographically different regions. The hills immediately south of Thamraparni River in the Singampatti and Kalakad areas, the home of larger cardamom in India, are on an average 3,000-5,000 feet (900-1,525 m) above sea level. Those in the high mountain ranges are covered with thick evergreen forests. In this southern home of larger cardamom, grass-covered hills devoid of forests are few and far between. In general, the rainfall in the southern portion is much less, i.e., 100-150 inches (2,540-3,810 mm).

5. The amount of rainfall increases as we go north. The rainfall is much better distributed in the southern hills than in the north, where it is more seasonal and broken by long spells of dry weather. In the extreme south of the peninsula, the northeast monsoon extends up to February. This accounts for the much better distribution of rainfall in the south.

In the northern part of the Western Ghats, the rainfall distribution is very irregular and the northeast monsoon is scanty. The northern hills of the Western Ghats are characterized by the existence of vast grasslands covering hill tops; only in valleys and saddles (*shola* vegetation) in this undulating plateau is forest growth seen. These variations in ecogeographical locations are likely to have contributed to the evolution and adaptation of two distinct types of cardamom.

Two distinct forms or local types of smaller cardamom have also been noticed: *Pattiveeranpatty* type from the Lower Pulneys in Tamil Nadu and North Kanara type from Karnataka.

1. The Pattiveeranpatty type is the tallest among the smaller types, it has very short petioles and large laminae. The pubescence of leaves persists only up to the sixth or seventh leaf (in this respect it occupies a middle position between the smaller and larger cardamom), and it has oblong shaped fruits, characteristics that make the cardamom from the Lower Pulneys a distinct local type.

2. The fruit of smaller cardamom cultivated in North Kanara are distinctly three angled and ovoid in shape (Mayne 1954). One of the most distinguishing features of this type is its capacity to be productive at low elevations of 1,500 to 2,000 feet above MSL (450-600 m), while none of the other cardamoms are productive below 1,500 feet (450 m) elevation. This seems to be adapting itself gradually to lower elevations and shows a very important change in habit as a result of the peculiar ecological conditions to which it is subjected, resulting in "ecoclines."

Classical taxonomists have attempted to classify cardamom, viz., Thwaites (1861-1864), Baker (in Hooker 1894), and Schumann (1904, as cited in Abraham and Thulsidas, 1958). Their attempts led them all to the general conclusion that the cardamom, *Elettaria* Maton, fall into two distinct groups, the larger and smaller. The larger type found in Sri Lanka, the former two, named it var. *major,* a single variety of *Elettaria cardamomum* (L.) Maton, whereas Schumann raised these two groups to species rank as *E. cardamomum* (L.) Maton and *E . major* Scn. According to the modern experimental method of taxonomy, if two very closely allied groups of plants of the same genus, although morphologically different and ecologically and geographically segregated, freely interbreed and form stable hybrid populations, when they are grown together, they should be considered as belonging to one "ecospecies," which often corresponds to the taxonomic "species."

According to studies conducted by Abraham and Thulsidas (1958), the cardamom *Elettaria* comprises a freely interbreeding population and forms a single cenospecies completely separated from other closely allied genera such as *Amomum* and *Alpinia* by genetic barriers. The cenospecies *Elettaria* consists of only one ecospecies comprising all cardamom populations of South India and Sri Lanka and may be termed ecospecies *cardamomum.* The ecospecies *cardamomum* consists of two ecotypes corresponding to the two, ecogeographical populations, and the larger and smaller cardamoms. The larger ecotype may be named ecotype *major,* whereas the smaller may be named ecotype *minor.*

The ecotype *minor* may be divided into three main local populations or local types. The one commonly found in the cardamom hills of Travancore, Anamalai, Coorg, and Mysore might be named ecotype *minor,* local type *travancorica.* The distinct local population of the Lower Pulneys may be named ecotype *minor,* local type *oblongata,* as it is characterized by oblong fruits. The peculiar local population cultivated in Sirsi of North Kanara may be named ecotype *minor,* local type *Kanarensis.*

According to Holttum (1950), the genus *Elettaria* has been regarded as comprising only two species, i.e., *E. cardamomum* and *E. major* from Southern India and Sri Lanka, respectively. The Malayan *Elettariopsis longituba* Ridl. and Sumatran species *Elettariopsis multiflora* Ridl. have exactly the same inflorescence structure as *Elettaria cardamomum*. From the morphology of inflorescence and flower, it is suggested that *Elettaria* cannot be a derivative of *Amomum:* rather it is another and distinct offshoot from the *Alpinia* stock. The resemblance of coloring of the flowers in the two genera indicates a near origin in that stock; it is stated that *Elettaria* occurs in Sri Lanka, Southern India, Sumatra, Malaya, and Borneo. It is, however, possible that *Elettaria cardamomum* and the Malaysian species represent parallel developments from different points of origin in the *Alpinia* stock.

BOTANY

Cardamom (*Elettaria cardamomum* (L.) Maton) belongs to the natural order Scitaminae under the family Zingiberaceae (Photos 2.1 and 2.2). It is

PHOTO 2.1. Cardamom.

PHOTO 2.2 Cardamom.

a herbaceous perennial (2-5 m in height) having underground (subterrain) rhizomes with aerial pseudostems (tillers) made of leaf sheaths. Studies on vegetative growth indicated that suckers continue their growth for a period of about 18 months from time of emergence. The rate of linear growth was maximum during June and July when the suckers attained an age of about one year. The development of reproductive buds (panicles) was noticed in 89 percent of them, indicating that the suckers require about ten to 12 months to attain maturity (Kuruvilla et al. 1990).

Inflorescence is a long panicle with racemose clusters arising from the underground stem, but coming up above the soil. The linear growth of the panicles extends over a period of about seven months. The rate of growth was maximum during April; it was slower both during the earlier and later stages. The growth habit of the panicles and shape and size of the capsules vary in different cultivated varieties/types of cardamom.

Based on the size of fruits, two types are recognized: *Elettaria cardamomum* (L.) var. *major* and var. *minor*. Considering the nature of panicles, three cultivars of cardamom ('Malabar', 'Mysore', and 'Vazhukka') are recognized in general. Their major characteristics include:

	Malabar	Vazhukka	Mysore
Adaptability	Lower elevation (below 900 m MSL)	Higher elevation (900-1,200 m MSL)	Higher elevation (above 1,200 m MSL)
General area of cultivation	Karnataka	Kerala	Kerala and parts of Tamil Nadu
Tolerance to drought	Withstand long dry spell	Prefer well-distributed rain	Prefer well-distributed rain
Plant stature	Dwarf (2-3 m)	Tall (3-5 m)	Tall (3-5 m)
Leaf petiole	Short	Long	Long
Panicle	Prostrate	Semi-erect	Erect
Bearing	Early, short span of flowering	Late, long flowering span	Late, long flowering span
Capsule color at maturity	Pale/golden green	Green	Green

The flowers are bisexual; bracts linear, oblong, and persistent; racemose sepals three; petals three, unequal, lip longer with violet carpels three; style one; ovary—trilocular; axile placentation; ovules—numerous in each carpel. Normally, flowering in cardamom is seen throughout the year on panicles produced during the current season as well as on panicles produced during the previous year. The peak flowering is spread over a period of six months from May to October. The time required from flower/bud initiation to full bloom stage ranges from 26 to 34 days. Capsule development takes about 110 to 120 days from the full bloom stage.

The maximum number of flowers open during the early hours of the day. Anthesis follows immediately. In the Mudigere region of Karnataka, flowering commences at 3:30 a.m. and continues to 7:30 a.m. The dehiscence of anthers takes place immediately, followed by anthesis at 3:30 a.m., which continues up to 7:30 a.m. The maximum pollen bursting occurs between 5:30 a.m. and 6:30 a.m. The pollen grains are round, found mostly in singles, and measure 87.6 µm in diameter on average. By appearance, 85.2 percent of the pollen grains are fertile, but a maximum of 70.1 percent germinate on artificial medium containing 20 percent sucrose and 1 percent agar. Studies on the viability of pollen grains indicate that only 6.5 percent of pollen grains remain viable after two hours of storage. After six to eight hours of storage, the percentage of viability was practically zero. With respect to morphology, pollen of all three varieties of cardamom is round in shape and appears as a creamy powder. The pollen grains of the 'Mysore' type are the largest and that of the 'Vazhukka' type are the smallest.

Though cardamom has bisexual flowers and is self-compatible, cross-pollination is the rule. Self-pollination is hindered due to slight protrusion of the stigma above the stamens. In cardamom, cross-pollination is mediated by the activity of bees (*Apis cerana* and *Apis dorsata*) as pollinators. Cardamom flowers remain in bloom for 15-18 hours. Stigma receptivity and pollen viability are reported maximum during morning hours. The receptivity is maximum between 8 a.m. and 10 a.m., when 72 percent of the opened flowers set fruit. After 10 a.m., stigma receptivity decreases gradually and only 24 percent of the flowers opening at 4 p.m. set fruit. The active foraging of bees is seen in the morning hours and is instrumental in increasing the fruit set in cardamom. The extent of fruit set recorded in different months indicates that fruit set is high (50 to 59 percent) during June, July, August, and September because of high prevailing humidity. During the dry season from December to March, practically no fruit set occurred.

CYTOLOGY

Darlington and Wylie (1955) give the basic chromosome number of *Elettaria* as $x = 12$ and the somatic number of *E. cardamomum* (L.) Maton as $2n = 48$. Ramachandran (1968) and Sudharshan (1987) also recorded $2n = 48$ in *E. cardamomum* (L.) Maton. Only Chakravarty (1952, as cited in Purseglove et al., 1988) recorded $2n = 52$ in cardamom.

GERMPLASM COLLECTION, AVAILABILITY, AND EXCHANGE

In spite of its prominence in world trade, cardamom received research attention for genetic upgradation only in the second half of the twentieth century. Six research organizations, viz., Indian Institute of Spices Research (IISR); Cardamom Research Centre, Appangala, Kodagu, Karnataka; Indian Cardamom Research Institute (ICRI), Myladumpara, and its regional research stations at Sakleshpur, Karnataka, and Tadiyankudisai in Tamil Nadu; Regional Research Station, Mudigere (University of Agricultural Sciences, Bangalore); Cardamom Research Station, Pampadumpara (Kerala Agricultural University); Horticultural Research Station, Yercaud (Tamil Nadu Agricultural University); and United Planters Association of South India

(UPASI), Vandiperiyar, are at present engaged in research for the improvement of cardamom. Regular surveys are being undertaken by these institutions for gathering available variable germplasm and exploiting the desirable genes of these accessions through various crop improvement techniques. The collection of a large number of indigenous lines are being conserved in different centers. A descriptor for cardamom was put forward by the International Plant Genetic Resources Institute (IPGRI), Rome, Italy.

Morphological Variations in Cardamom

Plant growth	Robust moderate/dwarf
Pseudostem	Red pigmentation/green
Leaf shape	Oblong/lanceolate/ovate
Leaf pubescence	Pubescent/puberulent/glabrous
Leaf breadth	Narrow
Ligule color	Red with a green tint

Inflorescence
Place of origin	Basal/both basal and terminal (Alfred clones) basal with a few leafy bracts
Panicle type	Prostrate/erect/semierect
Panicle branching (distal/entire/proximal)	Multibranched/compound panicle
Panicle length	Long/short panicles
Internodal length	Short/long

Capsules
Capsules/raceme	Two to six
Shape	Round/oval/elongate
Size	Bold/medium/small
Color	Golden yellow/pale green/dark green at maturity

Germplasm Accessions Available in Different Organizations

No.	Conservatory	No. of accessions
1.	IISR Cardamom Research Centre Appangala, Madikeri, Coorg District, Karnataka	310

2. ICRI, Mylandumpara (Spices Board)   660
   Idukki, Kerala

3. Cardamom Research Station   87
   Pampadumpara, Idukki
   Kerala

4. Regional Research Station   243
   Mudigere, Karnataka

5. UPASI, Vandiperiyar   45
   Kerala

6. Horticultural Research Station   35
   Yercaud,
   Tamil Nadu

Systematic survey, collection, and conservation of germplasm from the reserve forests are being undertaken by the Spices Board in collaboration with the National Bureau of Plant Genetic Resources (NBPGR, ICAR) under the National Agricultural Technology Project (Biodiversity). Since cardamom is also grown in other countries, efforts are made to acquire the variability existing in other cardamom-growing countries. A collaborative approach between IPGRI and NBPGR is required to collect and exploit the exotic germplasm.

CROP IMPROVEMENT

Being a cross-pollinated crop, considerable variability occurs in the seedling progeny of cardamom. As cardamoms are amenable to vegetative multiplication, high-yielding clones could be developed by selection.

Systematic survey, collection, and selection work were initiated at Singampatti Hills, Tamil Nadu, as early as 1944. A few outstanding selections were developed by culling (Aiyadurai 1966). Crop improvement programs are designed to achieve:

1. High yield of capsules
2. Resistance to biotic stresses, viz., viral diseases such as *"katt"* and *"kokke kandu"* and fungal diseases such as rhizome rot, clump rot, and capsule rot to get cardamom capsules free from pesticide residues.
3. Tolerance to abiotic stresses, primarily drought

4. Selection of plants with bold capsules and more number of seeds/fruit
5. Higher percentage of dry capsules recovery (> 22 percent)
6. Higher seed:husk ratio
7. Product of higher quality, i.e., higher percentage of essential oils, α-terpinyl acetate responsible for the aroma and flavor
8. Developing location-specific varieties suited to different agroclimatic conditions

Methods of crop improvement in cardamom include:

Clonal Selection

The improved varieties identified so far are evolved by selecting superior plants with desirable characters from land race populations. The highly heterozygous nature of the crop and the ability to multiply selections by clonal propagation has contributed to this success. Selection is made in situ in planters' fields and forests; the selected plants are multiplied clonally, subjected to preliminary evaluation, and subsequently evaluated in comparative yield trials and multilocation trials to confirm their superiority and adaptability.

Selections of cardamom are highly location specific in their agroecological requirements. The improved selections are much superior to the local clones with regard to yield and capsule characters. Salient features of the varieties in vogue are summarized in Table 2.4.

Resistance to cardamom mosaic virus which causes *katte* disease was identified among 134 disease escapes collected from hot spots of virus infection in southern India. These were screened in green house, sick plots, and hot spots. Testing in four hot spots and also against natural infection confirmed the resistance. Preliminary screening of cardamom varieties and improved selections for tolerance to azhukal disease *(Phytophthora meadii)* revealed that varieties such as 'Malabar' and 'Vazhukka' were more susceptible than the 'Mysore' types.

Hybridization

Intervarietal Hybridization

Hybridization among improved cultivars/varieties provides avenue of combining desirable attributes of yield, *"katte"* resistance, and drought tolerance. Research at various institutes has led to the isolation of high-yielding recombinants and heterotic hybrids. On-farm trials of these varieties are

TABLE 2.4. Characteristics of high-yielding selections of cardamom.

No.	Variety and type	Source	Yield (kg/ha)	Capsule shape	Essential Oil (%)	Cineole	α-Terpinyl accetate	Area recommended for cultivation
1.	CCS-1 ('Malabar')	IISR (ICAR) CRC, Appangala Karnataka	409	Oblong	8.7	42	37	All cardamom-growing tracts of Karnataka and Wynad of Kerala
2.	PV-1 ('Malabar')	KAU, Pampa-dumpara	260	Long	6.8	33	46	All cardamom-growing tracts of Kerala and parts of Tamil Nadu
3.	Mudigere-1	UAS (Bangalore) (Mudigere)	275	Oval	8.0	36	42	In the traditional ('Malabar') cardamom-growing Malanad areas of Karnataka
4.	Mudigere-2	UAS (Bangalore)	476	Round	8.0	45	38	Traditional cardamom-growing tracts of hill zones of Karnataka

TABLE 2.4 *(continued)*

No.	Variety and type	Source	Yield (kg/ha)	Capsule shape	Essential Oil (%)	Cineole	α-Terpinyl accetate	Area recommended for cultivation
5.	ICRI-3 ('Malabar')	ICRI (Spices Board) Myladumpara	325	Round	8.3	29	38	South Idukki zone of Kerala
6.	ICRI-2	ICRI (Spices Board) Myladumpara	375	Oblong	9.0	29	36	Vandanmedu and (Mysore) Nelliampathy of Kerala, and Annamalai and Meghamalai of Tamil Nadu
7.	ICRI-3 ('Malabar')	ICRI (Spices Board) Myladumpara	439	Oblong	6.6	54	24	Cardamom-growing tracts of Karnataka
8.	ICRI-4 ('Malabar')	ICRI (Spices Board) Myladumpara	455	Globose	6.4	—	—	Lower Pulneys in Tamil Nadu

in progress. Cardamom F1 hybrids evolved at ICRI, Myladumpara, show the following yields:

Hybrids	Projected yield (kg/ha)
MCC 16 × MCC 40	610
MCC 61 × MCC 40	675
MCC 21 ×MCC 16	650
MCC 21 × MCC 40	870
MCC 16 × MCC 61	800

A large number of crosses have been made to combine yield, rhizome rot resistance, and cardamom mosaic resistance. The hybrids are currently under evaluation at the Indian Institute of Spices Research, Cardamom Research Centre, Appangala. Varying degrees of heterosis ranging from –26.22 to 171.81 for plant height and –42.82 to 177.30 for total number of tillers per plant were recorded in the pre-bearing stage of these crosses.

Polycross Breeding

Since cardamom is a cross-pollinated crop, the polycross method of breeding is ideal to evolve superior types. Elite clones having predominantly desirable characters are planted together in an isolated plot. Beehives are maintained in the plot for assured pollination so that maximum fruit set and high number of seeds per capsule can be obtained. Initial results from this method of cardamom improvement are encouraging.

Intergeneric Hybridization

Cardamom mosaic disease *(katte)* causes severe yield losses (up to 68 percent). Although a few improved high-yielding varieties of cardamom have been evolved, combining yield and cardamom mosaic resistance has not been possible. To achieve this objective, intergeneric crosses were made using *Amomum subulatum, Alpinia nutans, Hedychium flavescens,* and *Hedychium coronarium* as male parents. A few fruits have been obtained only in the cross involving *A. nutans;* in other crosses no seed was set. Investigations are in progress to locate the causes of cross incompatibility.

Mutation Breeding

For developing clones tolerant to cardamom mosaic virus *(katte)* and drought, seeds and rhizomes of cardamom have been subjected to X-rays, nitrosomethyl urea (NMU), diethyl sulphate (DES), and ethyl methane sulphate (EMS). No desirable mutant has so far been obtained.

Polyploidy Breeding

Polyploids were induced in cardamom by treating sprouting seeds with 0.5 percent aqueous solution of colchicine. The tetraploid lines exhibited an increase in layers of epidermal cells, thick cuticle, and thicker wax coating on the leaves. These characters are generally associated with drought tolerance (Sudharshan 1987).

CROP MANAGEMENT

Planting Material Production

An important reason for low productivity of cardamom is poor quality of planting materials. Cardamom is propagated both by seeds and rhizomes. Although suckers free from pests and diseases should be employed for clonal multiplication of high-yielding selections, most plantations are still raised from seeds that yield heterogenous stands of poor genetic yield potential.

Clonal multiplication that ensures genetically uniform planting material can be raised by (1) macropropagation, i.e., clonal multiplication through rhizome under intensive care in a clonal nursery or (2) through micropropagation by tissue culture under aseptic conditions. Also, cardamom clones are highly location specific and planting material should be appropriate to its agroecological requirements.

Rapid Clonal Multiplication

A rapid clonal multiplication technique was evolved at the Indian Institute of Spices Research, Cardamom Research Centre, Appangala. The steps involved follow:

1. The clonal nursery is established on a gentle eastern slope with adequate drainage.
2. Trenches of 45 cm width, 45 cm depth, and a convenient length are made across the slope 1.8 m apart. The top 15 cm depth of soil is excavated and heaped on the upper side of the trench, and the lower 30 cm depth of soil is heaped on the lower side of the trench. Soil from the

first 15 cm depth is mixed with equal proportions of humus-rich forest soil, sand, and coffee compost at 5 kg/plant. The mixture is filled in layers up to a depth of 2.5 cm and mixed thoroughly by leaving a depression of 5 cm at the top to facilitate mulching and retention of soil moisture.

3. A part of the clump (38-40 planting units), earmarked in the main plantation, is carefully uprooted without injuring rhizomes. After trimming roots, suckers are separated. The minimum planting unit consists of an adult tiller along with a growing young shoot. The planting units are placed at a spacing of 1.8 m × 0.6 m in trenches. The top of the longest tiller is trimmed. The rhizomes are treated for five minutes with Emisan (methoxyethyl mercury chloride) (0.2 percent).

4. An overhead pandal is erected at a height of 4 m and covered with locally available silver oak twigs to allow 50 percent filtered sunlight. Regular plant protection measures are taken to control thrips, borers, and fruit flies.

Within a span of ten months, the rate of multiplication was 1:20 and the cost per planting unit was Rs 1.11 only. This method of clonal multiplication of cardamom is simple, reliable, and economically feasible for the production of quality planting material.

Propagation by Seeds

Seed used as planting material should be obtained from a mother plant of high quality whose performance has been watched continuously. Cardamom seed loses viability quickly, and therefore only freshly extracted seeds should be used. The time of sowing of cardamom seeds varies according to location, but the best germination was obtained in September (79.8 percent) and the least in January (0.8 percent).

Cardamom seeds have a hard seed coat, which impedes germination. Treating freshly extracted seeds with 25 percent nitric acid for ten minutes enhances germination. The selected nursery site should be on gentle slope and in vicinity of perennial sources of irrigation water. Soaking the soil of the nursery bed to a depth of 15 cm with 1:15 formaldehyde solution provides effective control of damping-off disease of seedlings.

Paddy straw or *Phyllanthus emblica* leaves are effective mulch to promote better germination.

The nursery operations are in two stages, viz., primary and secondary nursery. The seedlings are transplanted into secondary nursery when they reach the four-to-five leaves stage. Transplanting of seedlings in the secondary nursery is done in December or January in Karnataka and May-June in Kerala. In Karnataka, ten-month-old seedlings are used for transplanting, whereas in Kerala, it is common to plant 18- to 22-month-old seedlings.

Field Planting

Before taking up planting, the field should be prepared systematically. For planting in a new area, the ground needs to be cleared. If the land is sloppy, it is advisable to start clearing from the top and work downward. Branches of the shade trees should be thinned to provide an evenly dense overhead canopy.

Spacing (Geometry of Planting)

Plant spacing depends upon the variety and the intended duration of the crop. When it is intended to grow cardamom on a limited short cycle, with regular replanting, it is desirable to plant as closely as possible. For 'Mysore' and 'Vazhukka' cultivars, plant-to-plant distance can be 3 × 3 m or 2.4 × 2.4 m when planted in high rainfall or irrigated areas. A spacing of 1.8 × 1.8 m or 1.2 × 1.8 m is suitable in Karnataka. A spacing of 2.0 × 1.0 m is recommended for hill slopes, along the contour, and 2.0 × 2.0 m is recommended in the flat lands.

Field Preparation and Planting

In sloppy areas, soil is protected from erosion due to rains by planting cardamom in terraces, which are made across the slope. During terrace construction, 8 to 15 cm depth of topsoil is kept aside and used for pit filling. Pits of 90 × 90 × 45 cm are prepared before commencement of the monsoon season. About one-third of the pit is filled with topsoil, and one-third is filled with a 1:3 mixture of organic manure and topsoil. In low rainfall areas, trenches of 75 cm width and 30 cm depth are made and plants are spaced 1 to 1.5 m apart. The trench system of planting is advantageous in areas with moderate slope and adequate drainage. The trench system retains the highest percentage of moisture followed by the pit system.

Planting can be commenced when the soil is moist. Cloudy days with light drizzling provide ideal planting conditions. Deep planting should be avoided as it suppresses the growth of new shoots. The plant is be supported by staking immediately after planting to prevent damage due to wind. The base of the plant is provided with mulch of dried fallen leaves of shade trees.

Season of Planting

Planting during the rainy season, commencing from June, is ideal for steep and moderate slopes, which are well drained. Early planting gives the benefit of regular and distributed rains of the southwest monsoons and results in better establishment and growth. The ideal time for planting in low-lying areas is after the cessation of heavy monsoon showers.

Aftercare and Upkeep

Mulching

Cardamom is grown largely as a rainfed crop in India, and has to face dry spells for four to five months. Mulching is the practical solution for conserving soil moisture, and has been acclaimed as the best cultural operation for the overall improvement of soil and yield of cardamom.

Weed Control

Cardamom is a surface feeder. In the first year of planting, frequent weeding is necessary to avoid competition between young seedlings and weeds. As many as 21 dicotyledonous weeds have been identified in cardamom estates. Depending on the density of weeds, two to three rounds of weeding in a year would be necessary. Since manual weeding is difficult in large plantations, spraying of parquet (625 ml in 500 liters of water for one ha) in the interspaces between rows is recommended.

Trashing

Removal of old and dry shoots of cardamom plants is called *trashing*. Trashing improves penetration of sunlight and aeration, reduces infestation by thrips and aphids, and promotes plant growth. It also promotes pollination by honeybees and formation of green capsules. Trashing is carried out two to three times in a year. The trashed material can also be used as mulch.

Raking/Digging

Toward the end of the monsoon rains, light raking or soil digging is carried out around the plants. Digging done in one-year-old plantations facilitates root development. However, deep digging should be avoided.

Light Earthing Up

The humus-rich topsoil around the plant to a distance of 75 cm is scraped and applied as a thin layer to the base of the clumps. It forms the soil mulch and covers the crop roots and rhizomes. This practice also effectively checks the walking habit (radial growth) of cardamom.

Shade Regulation

Sufficient shade has to be maintained in the plantations to protect the plant from the scorching sun. Too much shade, however, causes over tillering, lanky growth of tillers, and consequently poor yield. A medium to high light intensity (45 to 65 percent) is ideal for growth and yield.

For providing adequate light during the rainy season, when light intensity is low, shade regulation before the onset of monsoons is necessary. Trees having well-distributed branching habits and small leaves are ideal for cardamom. Balangi *(Artocarpus fraxinifolius)* and red cedar *(Cedrela toona)* belong to this category. They are also self-regulating; they shed leaves in the monsoon season and put forth new flushes of leaves from December to January.

Nutrient Management

Fertilizer Requirements and Recommendations

In the 1800s, cardamom was harvested from plants growing in the natural ecosystem of rich forest soils. Its organized cultivation was prompted by increased demand. Up to the mid-1950s, cardamom was cultivated with little, if any, application of organic manures. Although it was reported that cardamom responds to the application of manures and fertilizers, the latter did not become a practice with cardamom growers.

With the establishment of research centers in India in the early 1960s, several fertilizer trials were initiated to work out the response and requirement of balanced fertilization for increased yields. In general, fertilizers to supply 30 kg N, 30 kg P_2O_5, and 60 kg K_2O per ha, appeared necessary for healthy and vigorous growth. An earlier recommendation of the Regional Research Station, Mudigere, was 35:30:20 kg NPK per ha. Based on tissue analysis at Mudigere, and considering the absorption of nutrients and factors affecting the availability of nutrients in soils, a fertilizer dose of 75:75:150 kg NPK per ha was recommended for normal crop of 100 kg dry capsules per hectare. For higher yields, the fertilizer doses are increased proportionately. Additional doses of 0.65 Kg N, 0.65 Kg P, and 1.3 kg K per ha are recommended for every increase in yield of 2.5 kg of capsules over the normal yield. Study of the effect of lime and nitrogenous fertilizers on bacterial population and nitrification in red acid soils showed that urea is a better source of nitrogenous fertilizer compared to ammonium sulfate. Liming corrects soil acidity, enhances the rate of nitrification, and results in better growth. The recommendation of the Spices Board of India was 75:75:150 kg and 125:125:250 kg NPK per ha for rainfed and irrigated areas respectively. As cardamom is still cultivated under rainfed condition (75 to 80 percent), conservation of soil moisture for efficient absorption and utilization of applied nutrients is imperative. A field experiment was laid out at the Indian Institute of Spices Research, Cardamom Research Centre, Appangala, comprising two systems of planting and five levels of fertilizers using the Cl-37 clone and a plant population of 5,000 plants per ha. Pooled analysis of the yield for three crop seasons (1987-1988 to 1989-1990) revealed that the trench method of planting was significantly superior to the pit method of planting. Application of 120:120:240 kg NPK per ha, which resulted in dry cardamom yield of 367.49 kg per ha, and 160:160:320 kg NPK per ha, which resulted in a dry cardamom yield of 390.44 kg per ha was on par in trench system of planting. Hence, a fertilizer dose of 120:120:240 kg NPK per ha accommodating a high density of 5,000 plants per ha is recommended.

Micronutrients

A survey conducted on micronutrients showed that zinc deficiency is widespread in cardamom soils. Application of 500 or 750 ppm of zinc in the form of zinc sulfate as a foliar application enhanced yield and quality of cardamom. The survey results also showed that iron and copper are not deficient in soils. Boron, when applied together with zinc, may have an antagonistic effect.

Irrigation

Judicious irrigation during the summer increases yield by at least 50 percent. Irrigation is required generally from February to April but at times from January to May depending upon rainfall. In Tamil Nadu, where the southwest monsoon is not very effective, irrigation from March to August is advisable. Cardamom plants irrigated at 75 percent available soil moisture recorded better yield. As cardamom is cultivated mainly on an undulating topography of hills and hill slopes, overhead (sprinkler) irrigation is the most appropriate. For sprinkler irrigation, water equivalent to 35 to 45 mm rains at fortnightly intervals is recommended. In the case of drip irrigation, water at the rate of 4-6 liters per clump per day may be given.

Cropping Systems

Land committed to plantation and spice crops for several decades is common in the tropics and subtropics, which have equitable climates, plentiful precipitation, and abundant sunshine. These situations favor plant growth round the year. This calls for effective utilization of horizontal and vertical space and efficient capture of solar energy to get maximum returns per unit time. A practical way of increasing the farm-level income, to withstand sharp fluctuations in price and maintaining employment opportunities on small holdings, is to adopt mixed cropping by growing compatible high-value perennial crops in the interspaces.

Cardamom offers great scope as a mixed crop with coffee, arecanut, and coconut. The latter provide overhead shade, which is essential for the survival and productivity of cardamom in high ranges of the Western Ghats in India.

Three long-term field experiments were conducted in the Kodagu district of Karnataka on mixed cropping of robusta coffee with cardamom (double hedge), mixed cropping of robusta coffee with Coorg mandarin, black pepper, and cardamom (single hedge), and robusta coffee alone. Mixed cropping of arecanut with cardamom and coconut with cardamom were tried in Sirsi, Uttar Kannada district, Karnataka.

Mixed cropping of cardamom with coffee, black pepper, arecanut, and coconut enhanced the overall productivity of the crop combination. Integrated management resulted in efficient utilization of various cash inputs in these crop combinations. Combined cultural operations followed in the crop combinations reduced the total cost of cultivation. Mixed cropping of coffee, arecanut, and coconut is more remunerative than sole crop cultivation. Mixed cropping of cardamom with coffee, arecanut, and coconut generated additional gainful employment.

HARVESTING AND PROCESSING

Although every stage of production is important, quality of produce is critically influenced by stage of maturity, time of harvesting, and pre-drying treatments. In most areas, peak harvest is in October-November. Picking is carried out at intervals of 15 days and completed in seven to eight rounds. The stage of harvest has a direct relationship with recovery of cardamom. Percentage recovery was highest (29 percent) in the harvest of fully ripened capsules, followed by harvest at physiological maturity (24 percent) and harvest at immature stage (14 percent).

Quality

The quality of cardamom is assessed on the basis of its appearance (size, color) and the aroma/flavor character and strength (content) of volatile oil. Uniform olive-green color without surface blemish and good size commands the market. The stage of maturity of the fruits at harvest is the most important factor, which determines quality of the product.

The dried fruits contain a steam-volatile oil, fixed (fatty) oil, pigments, proteins, cellulose, pentosans, sugars, starch, silica, calcium oxalate, and minerals. The major constituent of the seeds is starch (up to 50 percent), whereas the husk has crude fiber (up to 31 percent). The volatile oil is located predominantly in the seeds, which comprise 65 to 70 percent of whole dried fruits. Cardamom oil is obtained by steam distillation, Oil is pale yellow or colorless which darkens on exposure to light.

The major components of cardamom oils are 1,8-cineole (20 to 60 percent) and alpha-terpinyl acetate (20 to 53 percent). The other important components are linalyl acetate, linalool, and borneol (each up to 8 percent); alpha-terpineol (4.3 percent); alpha-pinene, limonene, and myrcene (each up to 3 percent). The oils of two distinct types of cardamoms, 'Malabar' and 'Mysore', differ in proportions of components in their oils. The 'Malabar' cardamoms have more of 1,8-cineol (>50 percent) and less of alpha-terpinyl acetate (<30 percent), whereas the 'Mysore' cardamoms have less of 1,8-cineol (<30 percent) and more of alpha-terpinyl acetate (>50 percent) (Lewis et al., 1966).

CARDAMOM IN COUNTRIES OTHER THAN INDIA

Sri Lanka

Sri Lanka is the original home of the larger type of small cardamom, which grows wild in the island state. Abeysinghe (1980) records that cardamom is the second most important of the spice crops grown in Sri Lanka. The cardamom industry in Sri Lanka is concentrated in two administrative districts: Kandy with 54 percent and Matale with 18 percent of the acreage.

The rest of the area is distributed in Kegalle, Nuwara Eliya, Ratnapura, Kurunegala, and the Moneragale district in Sri Lanka.

The distribution of holdings according to size indicated that 83 percent of the holdings are less than five acres (2 ha) in extent and account for only 11 percent of the total area. On the other hand, only 5 percent of the holdings are over 50 acres of land (10 Ha) but they account for 78 percent of the total area. Only two varieties of cardamom are grown commercially in Sri Lanka, the Mysore and the Malabar cultivars. 'Mysore' probably accounts for over 75 percent of the acreage. According to Molegode (1938) three varieties are found in Sri Lanka. One of the three is in the higher elevations, chiefly in the Ratnapura and Lungula districts. The cultivated varieties Malabar and Mysore appear to have been introduced from India. The cardamom is cultivated in about 400-500 hectares with an annual export of about 150-200 tonnes.

Guatemala

Cardamom is reported to have been introduced to Guatemala in 1920 in the department of Alta Verapaz by a German planter-settler named Carlos Datez Villela. According to the planter's son, the seeds of cardamom, of Indian origin, were procured from New York. From this, cardamom cultivation expanded to other regions in the north and later to the southern part of Guatemala. The hill tracts of Guatemala receive 80-200 inches (2,000-5,000 mm) of rainfall, which is well distributed. December to March are comparatively dry, but the rest of the year receives good rainfall. The northern belt receives comparatively more rainfall than the southern belt. The southern belt has evenly distributed rainfall. Many Zingiberaceae members, such as *Amomum* spp. and *Alpinia* spp., grow wild in that region. Major plantations in Guatemala have been developed in the southwestern departments of Suchitepequez, Solola, and Quezaltenango.

Cardamom is cultivated in Guatemala at altitudes of 2,000 to 4,000 feet (600-1,200 m). The predominant variety cultivated is 'Mysore' with erect panicles and oblong capsules, propagated mainly through suckers. General spacing adopted is 8" × 8", and hedgerows of 10' × 5' are also seen. Leaf spot disease has been recorded. Of late, mosaic disease has also been reported from Guatemala. Among important insect pests recorded, shoot and capsule borers are prominent. The most persistent pest, thrips, has not been reported from Guatemala.

The following are striking differences in the cultivation of Guatemala cardamom: Cardamoms are cultivated comparatively under open conditions in Guatemala, and luxuriant growth of plants with high productivity is observed. During World War II, the German plantations were expropriated, and for a number of years thereafter, production declined. In the mid-1950s, coffee producers who were dissatisfied with low coffee prices diversified into cardamom. This coupled with prevailing congenial conditions helped Guatemala to expand the cardamom industry and become the leading producer and exporter of cardamom since the 1990s. Average productivity is reported to be over 250 kg per hectare.

A similar development took place in El Salvador in the 1950s, but cardamom production could not be sustained. Tanzania likewise introduced cardamom quite early, but it has been a small supplier to the world market. During the mid-1970s, Tanzania marketed 300-350 tonnes of cardamom per annum; the produce is mainly sun dried and hence pale straw in color. Among the other countries that took interest in cardamom cultivation early on, Papua New Guinea has limited production with inconsistent quality.

END PRODUCTS

    1. Dried cardamom/cured cardamom
    2. Bleached cardamom
    3. Decorticated seeds and seed powder
    4. Cardamom volatile oil
    5. Cardamom oleoresin

Uses

The major use of cardamom worldwide is for domestic culinary purposes in whole or ground form. In Arab countries, cardamom is used traditionally for flavoring coffee. In the Middle East, it is used for flavoring a variety of baked goods, including cakes, pastries, and bread. In European countries and in North America, cardamom is used mostly in ground form as an ingredient in curry powder, sausage products, soups, canned fish, and to a small extent in the flavoring of tobacco. Cardamom-flavored cola and instant *gahwa*, carbonated *gahwa*, biscuits, Danish pastries, and toffees are other new products developed using cardamom.

The main application of cardamom essential oil is to flavor processed foods, but it is also used in cordials, bitters, liqueurs, and occasionally in perfumery. Cardamom oleoresin finds application in the flavoring of processed foods but is used less extensively.

It is common to use tinctures of cardamom in medicines for gassiness or stomachics. Powdered cardamom seeds are mixed with ground ginger, clove, and caraway and used for combating digestive ailments. Cardamom is used as a powerful, pleasant, aromatic stimulant; carminative; stomachic; diuretic; and cardiac stimulant.

Large Cardamom

The large cardamom, *Amomum subulatum* Roxb. is also known as greater Indian or Nepal cardamom. It is a native of the Eastern Himalayan region. It is cultivated in the Darjeeling district of West Bengal, Sikkim, Bhutan, and Nepal. This is grown in areas with maximum temperature of 14-33°C, minimum temperature of 6-22°C and annual rainfall of 2,000-2,500 mm.

The plant, which is a perennial, grows to a height of 2 m, with oblong to lanceolate glabrous leaves. The inflorescence is a dense short spike borne near the ground. The flowers have a large, white, central strap-shaped lip, and open from the base of the spike upward. The fruit, about 2.5 cm long, is ovoid and triangular, deep red when ripe, and ribbed. It contains 40-50 hard, dark brown, round aromatic seeds embedded in a soft, slimy pulp, which is sweet to taste. The pollinating agents are reported to be bumblebees. The crop is largely propagated vegetatively using pieces of rhizome.

In Darjeeling, spikes emerge during June-July and harvesting starts from September and continues until January. In general, capsules are harvested with the spike. They are dried in the sun or on platforms under shade with artificial heat. Dried capsules are dark brown to black.

Four well-known cultivated varieties of large cardamom are grown in Sikkim. They are 'Ramsey', 'Sawaney', 'Golsey', and 'Ramnag' (Gyatso et al. 1980). Apart from these, there are some other local varieties cultivated that are presumed to be derivatives of the cultivars named previously.

1. *Ramsey:* It is a Bhutia (local tribe) word meaning yellow color. This is well adapted to higher altitudes and produces fruits, which are the smallest of cultivated varieties and hence inferior in quality.
2. *Sawaney:* It is a Nepali word meaning the fruits are harvested in the month of "Sawn" (August). The variety is widely adopted in the state. It is cultivated generally in the middle and lower elevations. The plants are tall (220-240 cm), green in color, and the fruits are bold and brown in color.
3. *Golsey:* The word is derived from Hindi and Bhutia meaning round and yellow. This is a predominant variety of lower elevations. Plants are medium statured and green. Fruits are bold and round, contain 60-

80 seeds, are yellowish brown in color, and fetch premium prices in the market.

4. *Ramnag:* It is a Bhutia word for black color. This is cultivated in certain areas in high altitude. The plant size is intermediary between 'Sawaney' and 'Golsey'. The fruits are bold and dark brown in color and fetch the highest price in the market.

Cardamom in Sikkim is planted as a pure crop under natural shade trees on hill slopes ranging from 600 m to 1,800 m. It is cultivated in an area of 26,358 hectares in the states of West Bengal and Sikkim. Annual production is 5,265 tonnes, out of which about 1,200-1,700 tonnes are exported annually (Spices Board of India, 1998).

Large cardamom is subjected to a number of diseases and pests. "Foorkey" and "Chirkey" are common virus diseases, whereas hairy caterpillars, thrips, and aphids are important insect pests of this crop.

"Foorkey," or yellows, causes heavy losses; the infected plant becomes stunted and new tillers do not grow fully, hence the plants become dwarf, sterile, and unproductive. It is a viral disease and the virus is transmitted by banana aphids. In "Chirkey," or mosaic streak disease, the affected plants have mosaic streaks on the leaves. In addition, the productivity is reduced and results in severe crop loss. This is reported to be sap transmissible and also through aphids.

The seeds of large cardamom are aromatic and pungent, and are used for seasoning and flavoring food and various fruit and vegetable preserves. They are also used as a spice or condiment and as a flavoring agent in cookery and confectionery. The essential oil obtained after steam distillation of the crushed seeds yields 2.8 percent oil having a characteristic smell of creosote with the following physical and chemical properties:

Specific gravity at 29°C	0.9142
Refractive index at 29°C	1.4600
optical rotations in chloroform	18°3'
Acid value	2.90 percent
Saponification valu	14.53
Saponification value after acetylation	40.20
Cineole	64.94 percent
Terpinene	10.7 percent
Sabinine	6.60 percent
Terpinyl acetate	5.10 percent
Bisebolene	3.60 percent
Polymerized oil	3.60 percent
Terpineol	7.15 percent

RESEARCH AND DEVELOPMENT ORGANIZATIONS
CONCERNED WITH CARDAMOM

India

Indian Institute of Spices Research
(Indian Council of Agricultural Research)
Cardamom Research Centre, Appangala
Madikeri 571 201, Kodagu, Karnataka, India

Indian Cardamom Research Institute
(Spices Board of India)
Myladumpara, Idukki District
Kerala, India

Kerala Agricultural University
Cardamom Research Station
Pampadumpara, Idukki District
Kerala—685 556, India

University of Agricultural Sciences
Regional Research Station
Mudigere, Karnataka, India

United Planters Association of South India (UPASI)
Vandiperiyar
Kerala, India

Tamil Nadu Agricultural University
Horticultural Research Station
Yercaud, Tamil Nadu, India

Other Countries

International Plant Genetic Resources Institute
Via delle Sette Chiese 142
Rome, Italy

University of Peradeniya
Department of Crop Science,
Faculty of Agriculture,
Peradeniya, Sri Lanka

REFERENCES

Abeysinghe, P.D. 1980. Cardamom plantation industry in Sri Lanka. *Cardamom* 12(2): 3-13.

Abraham, P. and Thulsidas, G. 1958. South Indian cardamoms: Their evolution and natural relationships. *ICAR Bulletin* 79: 1-27.

Aiyadurai, S.G. 1966. *A Review of Research on Spices and Cashew Nut in India.* Directorate of Cashew Development, Ernakulam.

Anonymous. 1977. Report of the Indian cardamom delegation to Guatemala. *Cardamom* 9(6): 3-10.

Burkill, I.H. 1955. Cited in South Indian Cardamoms . . . , Abraham, P. and Tulsidas, G. 1958.

Darlington, C.D. and Wylie, A.P. 1955. *Chromosome Atlas of Flowering Plants.* George Allen & Unwin, London.

Gyatso, K., Tshering, P., and Basnet, B.S. 1980. Large cardamom of Sikkim. *Krishi Samachar* 2(4): 90-95.

Holttum, R.E. 1950. The Zingiberaceae of the Malay peninsula. *Gardens Bull. Singapore* 13: 236-239.

Hooker, J.D. 1894. *Flora of British India.*

Kuruvilla, K.M., Sudharshan, M.R., Madhusoodanan, K.J., Priyadarshan, P.M., Radhakrishnan, V.V., and Naidu, R. 1990. Phenology of tiller and panicle in cardamom *E. cardamomum* Maton. *Journal of Plantation Crops* 20(1): 162-165.

Lewis, Y.S., Nambudri, E.S., and Philip, T. 1966. Composition of cardamom oils. *Perf. Essent. Oil Record.* 57: 623-628.

Mayne, W.W. 1954. Cardamom in South Western India. *World Crops* (6): 397-400.

Molegode, W. 1938. Cardamoms-1. *Tropical Agriculturist* 7: 325-329.

Purseglove, J.W., Brown, E.G., Green, C.L., and Robbins, S.R.J. 1981. *Spices,* Volume 2. Longman Inc., New York.

Ramachandran, K. 1968. Chromosome numbers in Zingiberaceae. *Cytologia* 34: 213-221.

Ridley, H.N. 1912. *Spices.* Macmillan, London.

Rosengarten, F. Jr. 1969. *The Book of Spices.* Livingston Publishing Co., Wynnewood.

Spices Board of India. 1998. *Spices Statistics,* Fourth Edition. Spices Board, Kochi, India.

Sudharshan, M.R. 1987. Colchicine-induced tetraploids in cardamom (*Elettaria cardamomum* Maton). *Current Science* 56(1): 36-37.

Suresh, K.A. 1980. The history of cardamom plantations in Kerala. *Cardamom* 12(7): 3-7.

Chapter 3

Cashew

E. V. V. Bhaskara Rao
K. R. M. Swamy

INTRODUCTION

Cashew (*Anacardium occidentale* L.), widely cultivated throughout the tropics for its nuts, is a native of Brazil. It was one of the first fruit trees from the New World to be widely distributed throughout the tropics by the early Portuguese and Spanish adventurers (Purseglove, 1988). The cashew has a long history as a useful plant, but only in the present century has it become an important tropical tree crop. Small-scale local exploitation of the cashew for its nuts and cashew apples appears to have been the pattern for more than 300 years in Asia and Africa. It was not until the early years of the twentieth century that international trade in cashew kernels began with the first exports from India. This was a very slow beginning, but recent decades have seen the cashew become an important commercial tree crop (Johnson, 1973). In India, use of cashew apples and nuts was adopted by local peoples, and accounts from Africa are similar; making cashew wine appears to have been a common practice in both Asia and Africa (Johnson, 1973). The Maconde tribe in Mozambique call it the Devil's Nut. It was offered at wedding banquets as a token of fertility, and research carried out at the University of Bologna has in fact indicated the presence in cashew kernel of numerous vitamins including vitamin E, considered by many to be an aphrodisiac (Massari, 1994). At the time of the first Portuguese colonization, the name used by local populations (Tupi natives of Brazil) for the cashew was *acajú* (nut), which turned into *cajú* in Portuguese, and "cashew" in English. Most of the names for cashew in Indian languages are also derived from the Portuguese name *cajú* (Johnson, 1973). Cashew is a versatile tree nut. It is, in fact, a precious gift of nature to humankind. The cashew kernels contain a unique combination of fats, proteins, carbohydrates, minerals, and vitamins. Cashews are 47 percent fat, but 82 percent of this fat is unsaturated fatty acids. The unsaturated fat content of cashew not only eliminates the possibility of the increase of cholesterol, but also balances or reduces the cholesterol level in the blood. Cashew also contains 21 percent proteins, 22 percent carbohydrates, and the right combination of amino acids, minerals, and vitamins, and therefore, nutritionally, it stands on a par with milk, eggs, and meat. As cashew has a very low content of carbohydrates, almost as low as 1 percent soluble sugar, the consumer of cashew is privileged to get a sweet taste without having to worry about excess calories. Cashew nuts do not lead to obesity and help to control diabetes. In short, it is a good appetizer, an excellent nerve tonic, a stimulant, and a body builder. Cashew is indigenous to Brazil, but India is the country that nourished this crop and made it a commodity of international trade and acclaim. Even today, India is the largest producer, processor, exporter, and second largest consumer of cashew kernels in the world (Nayar, 1998).

ORIGIN, EVOLUTIONARY HISTORY, AND DISTRIBUTION

Anacardium is distributed naturally from Honduras south to Parana, Brazil, and eastern Paraguay. It is not indigenous to South America west of the Andes except in Venezuela, Colombia, and Ecuador where *A. excelsum* occurs. *Anacardium occidentale* is cultivated or adventive throughout the New and Old World tropics. The genus has two centers of diversity—central Amazonia and the Planalto of Brazil. This is illustrated by the occurrence of four species in the vicinity of Manaus and by three species occupying the same habitat in the Distrito Federal, Brazil. The following five distribution patterns are found in *Anacardium*.

- *Anacardium excelsum* is isolated taxonomically and geographically from its congeners by the Andes. The uplift of the Andes was probably the driving force in the early differentiation of *A. excelsum* from the rest of the genus.
- *Anacardium giganteum* and *A. spruceanum* have Amazonian-Guyanan distributions.
- *Anacardium occidentale,* which is the most widespread species in the genus, has disjunct populations in the Planalto of Brazil, the *restingas* of eastern Brazil, the savannas of the Amazon basin, and the llanos of Colombia and Venezuela. It should be kept in mind, however, that the natural distribution of this species is obscured by its widespread cultivation in both the Old and New World.

- Three closely related species, *A. humile, A. nanum,* and *A. corymbosum,* are restricted to the Planalto of central Brazil.
- Two species of *Anacardium* are narrow endemics. *Anacardium corymbosum,* which is restricted to south-central Mato Grosso, is an allospecies of *A. nanum,* and *A. fruticosum* (a new species) is endemic to the upper Mazaruni River basin in Guyana. It is closely related to the Amazonian *A. parvifolium.*

The eastern portion of the Amazon River figures prominently in distributions of many plants and animals, many of which are found either exclusively to the north or south of the river. However, in the case of *Anacardium,* all Amazonian species are found on both sides of the Amazon River. The reason for this is probably the ease with which bats, large birds, and water (in the case of *A. microsepalum*) carry fruits across water barriers (Mitchell and Mori, 1987). *Anacardium occidentale* is cultivated and adventive throughout the Old and New World Tropics where the geographical limits of its cultivation are latitudes 27°N and 28°S, respectively (Nambiar, 1977). *Anacardium occidentale* is native to tropical America where its natural distribution is unclear because of its long and intimate association with man. The problem of its origin and distribution has been investigated by Johnson (1973) who suggested that it originated in the *restinga* (low vegetation found in sandy soil along the coast of eastern and northeastern Brazil). Johnson is probably correct in assuming that the cultivated form of *A. occidentale* came from eastern Brazil, because cashew trees cultivated in the Old and New Worlds are identical in appearance to native trees found in *restinga* vegetation. In particular, cultivated and wild populations of cashew from eastern Brazil share chartaceous leaf blades and long petioles. *Anacardium occidentale* is probably an indigenous element of the savannas of Colombia, Venezuela, and the Guyanas. It is clearly a native, and occasionally a dominant feature of the *cerrados* (savanna-like vegetation) of central and Amazonian Brazil. The *cerrado* populations of *A. occidentale* differ from the *restinga* populations by having undulate, thickly coriaceous leaves with short, stout petioles. The hypocarps (cashew apples) of *cerrado* trees are usually smaller and sometimes have a more acidic flavor than those of the *restinga.* The natural distribution of *A. occidentale* extends from northern South America south to Sao Paulo, Brazil. It is probably not native to Central America, the West Indies, or South America west of the Andes. It is believed that *A. occidentale* originally evolved in the *cerrados* of Central Brazil and later colonized the more recent *restingas* of the coast. Central Brazil is a center of diversity for *Anacardium* where the distribution of *A. occidentale* overlaps the ranges of *A. humile, A. nanum,* and *A. corymbosum. Anacardium humile,* the closest relative of the cultivated cashew, is closer morphologically to the *cerrado* ecotype than it is to the *restinga* and cultivated populations of *A. occidentale* (Mitchell and Mori, 1987). The earliest reports of cashew are from Brazil coming from French, Portuguese, and Dutch observers (Johnson, 1973). The French naturalist and monk A. Thevet was the first to describe, in 1558, a wild plant extremely common in Brazil: the cashew tree and its fruits. He recounted that cashew apple and their juice were consumed and that the nuts were roasted in fires and the kernels eaten. Thevet provided the first drawing of the cashew showing the local people harvesting fruits and squeezing juice from the cashew apples into a large jar (Johnson, 1973; NOMISMA, 1994). There are indications that the local Tupi Indians had used cashew fruits for centuries. They probably played a major role in the species dispersion in their temporary migrations toward the coast of northeastern Brazil, where a considerable intraspecific variation has been recorded (Ascenso, 1986). The entire cashew fruit, nut and peduncle, will float when mature. This could account in Brazil for coastward dispersal of the species by rivers draining north and east. Fruit bats may also have been involved in seed movement. Within the Amazon forests fruit bats are the most important agents of seed dispersal of tree species (Johnson, 1973). From its origin in northeastern Brazil, cashew spread into South and Central America (Van Eijnatten, 1991). The presence of cashew in other continents is to be attributed to human intervention (Johnson, 1973). The Portuguese discovered cashew in Brazil and spread it first to Mozambique (Africa) and later into India between the sixteenth and seventeenth centuries (De Castro, 1994). According to Agnoloni and Giuliani (1977), it arrived in Africa during the second half of the sixteenth century, first on the east coast, then on the west coast, and finally in the islands. Although it can be guessed that the cashew was introduced to Africa at an early period by the Portuguese, no records exist which provide specific dates. Dispersal of the cashew in eastern Africa may in part be due to the elephant, whose fondness for fruits is well known (Johnson, 1973). Attracted by the color of the false fruit, they swallowed this together with the nut which was too hard to be digested. This was then expelled with their droppings, a natural manure, and trodden far enough into the ground by the animals following along behind to root and grow, into a seedling first and then a tree. This is how the cashew was spread along the east coast of Africa facing the Indian Ocean (Massari, 1994). The spreading of the cashew within the South American continent was gradual and spontaneous (NOMISMA, 1994). It is believed that the Portuguese brought the cashew to India between 1563 and 1578. It was first described in gardens in Cochin on the Malabar coast. Following its introduction into southwestern India, the cashew probably diffused throughout the Indian subcontinent to some degree by means of birds, bats, and, most important, human elements. Cochin served as a dispersal point for the cashew in India, and perhaps for Southeast Asia as well (Johnson, 1973). According to Johnson (1973), the reason for the introduction is not documented, although the popular explanation is that it was for the purpose of checking soil erosion in the coastal areas of India. This interpretation, however, smacks of a twentieth-century concept being applied to a sixteenth-century event. The Portuguese learned of the reported medicinal properties of the cashew and also that the juice of the cashew apple could be fermented into a good wine. It seems plausible, therefore, that they visualized cashew as a crop of potential value to India. After India, it was introduced into Southeast Asia (NOMISMA, 1994). Dispersal in Southeast Asia appears to have been aided by monkeys. Whether the cashew reached the Philippines via India is uncertain. It may have come directly from the New World on the Manila galleons (Johnson, 1973). The cashew later spread to Australia and some parts of the North American continent, such as Florida. Finally, its present diffusion can be geographically located between 31°N latitude and 31°S latitude, both as a wild species and under cultivation (NOMISMA, 1994). At present, cashew is cultivated in many tropical countries, mainly in coastal areas (Van Eijnatten, 1991; Ascenso, 1986). In the nineteenth century, proper plantations were planted and the tree then spread to a number of other countries in Africa, Asia, and Latin America (Massari, 1994). Traditionally, cashew has been cultivated on a commercial scale in Brazil, India, Tanzania, Mozambique, Kenya, and Madagascar; in recent years plantations have also been raised in Southeast Asian countries, such as Vietnam, Myanmar, and Thailand, on commercial scale (Bhaskara Rao, 1996).

AREA AND PRODUCTION

World Scenario

Cashew is grown in India, Brazil, Vietnam, Tanzania, Mozambique, Indonesia, China, Sri Lanka, and other tropical Asian and African countries. Cashew-growing countries in the world are listed in Table 3.1. The world production of cashew is around 1.09 million tonnes during 2000 (Balasubramanian, 2000; Bhaskara Rao and Nagaraja, 2000). Between 1980 and 1995, world raw nut production has increased from 0.422 million tonnes to 0.878 million tonnes registering an increase of 108 percent. The growth rate between 1995 and 2000 is less (24 percent) compared to the growth rate between 1980 and 1995. Subsequent to 1995, the world raw nut production has been around 1.09 million tonnes. World production of cashew between

TABLE 3.1. Cashew-producing countries in the world.

Africa	Latin America	Southeast Asia	Indian subcontinent
Angola	Brazil*	Vietnam*	India*
Benin	Barbados	Thailand*	Bangladesh
Burkina Faso	Dominican Republic	Indonesia*	Sri Lanka
Guinea-Bissau	El Salvador	Malaysia*	
Madagascar	Guadeloupe	Philippines	
Mozambique*	Honduras	China	
Mali			
Nigeria*			
Kenya*			
Senegal			
Tanzania*			
Togo			

*Major Producers

1980 and 2000 is furnished in Table 3.2. Country production of raw cashew nuts is furnished in Table 3.3. India's share in the world raw nut production accounts to 47 percent. The share of Southeast Asian countries has ranged from 14 to 16 percent during the past five years. Raw nut production in Southeast Asia has registered an increase of 45.2 percent from 1980 to 2000. Similarly, Latin American countries have registered an increase of 114 percent during the past twenty years (1980-2000).

Indian Scenario

In India, cashew is grown mainly in Maharashtra, Goa, Karnataka, and Kerala along the west coast, and Tamil Nadu, Andhra Pradesh, Orissa, and West Bengal along the east coast. To a limited extent it is grown in Manipur, Meghalaya, Tripura, Andaman and Nicobar Islands, and Madhya Pradesh. India's raw nut production has increased from 0.079 million tonnes in 1955 to 0.52 million tonnes in 2000. Raw nut production since 1955 is furnished

TABLE 3.2. World production of raw nuts (million tonnes).

Year	Africa	Latin America	Southeast Asia	Indian subcontinent	Total
1980	0.612 (38.4)	0.0841(19.9)	0.0275 (6.5)	0.1483(35.1)	0.4219
1981	0.1856(40.0)	0.0835(17.9)	0.0296 (6.4)	0.1651(35.6)	0.4638
1985	0.1114(22.4)	0.1246(25.0)	0.0305 (6.1)	0.2309(46.4)	0.4974
1990	0.1119(18.3)	0.1200(19.6)	0.0815(13.3)	0.2957(48.2)	0.6127
1995	0.1397(15.9)	0.1993(22.7)	0.157 (17.9)	0.382 (44.4)	0.878
1996	0.3205(29.5)	0.1689(15.6)	0.1638(15.1)	0.4328(39.8)	1.0861
1997	0.2917(27.7)	0.1175(11.6)	0.1708(16.8)	0.445 (43.8)	1.0149
1998	0.369 (36.9)	0.0380 (3.8)	0.1485(14.8)	0.445 (44.5)	1.0005
2000	0.200 (18.3)	0.18 (16.5)	0.152 (13.9)	0.52 (47.7)	1.090*

Source: Cashew Export Promotion Council of India, 2003.
*Includes 0.38 million tonnes under others. Figures within parentheses indicate percent of total raw nut production.

TABLE 3.3 Cashew raw nut production by country (2000).

Country	Production (million tons)
Indonesia	0.030
Nigeria	0.040
Vietnam	0.122
India	0.520
Brazil	0.180
Tanzania	0.130
Mozambique	0.030
Others	0.038
Total	1.090

Source: Cashew Export Promotion Council of India, 2003.

in Table 3.4. Between 1990 and 2000, raw nut production has almost doubled. If this trend continues, it would be possible to achieve the production of 7 lakh tonnes by 2005, which would be sufficient to meet the requirement of processors. In 1955, cashew in India was grown in an area of 0.11 million ha. In the year 2000, cashew was being grown in an area of 0.683 million ha.

TABLE 3.4. Production of cashew raw nuts in India.

Year	Area Million ha	Area % Increase/ 5 years	Production Million tonnes	Production % Increase/ 5 years	Productivity (kg/ha)
1955	0.11	—	0.079	—	720
1960	0.176	60.0	0.11	39.2	630
1965	0.232	31.8	0.141	21.9	610
1970	0.281	21.1	0.176	24.8	630
1975	0.358	26.7	0.166	−5.7	460
1980	0.451	25.9	0.142	−14.4	310
1985	0.509	11.4	0.221	55.6	430
1990	0.531	4.1	0.286	29.4	540
1995	0.577	8.7	0.371	22.9	640
1996	0.635	—	0.418	—	720
1997	0.65	—	0.43	—	835
1998	0.70	—	0.36	—	740
1999	0.73	—	0.46	—	800
2000	0.683	18.4	0.52	40.2	865

Source: Cashew Export Promotion Council of India, 2003.

Over the last 45 years, area under cashew has registered an increase of 520 percent. Area under cashew has been steadily increasing. Between 1970 and 1980, although area under cashew increased, the percent increase in production was negative. During 1995-2000, the increase in both area and production is phenomenal. In order to sustain India's presence in the international market, productivity has to be increased. Up to 1970, the productivity of cashew was around 630 kg/ha. Between 1975 and 1985, productivity was low (430 kg/ha). Since 1985, productivity has been steadily increasing from 430 kg/ha to 865 kg/productive ha in 2000 (Balasubramanian, 2000; Bhaskara Rao and Nagaraja, 2000). This is mainly due to improved technologies available, replanting of large areas of old plantations, and the availability of necessary high-yielding planting material through government agencies and private nurseries. Research institutions and private nurseries are producing nearly 10 million grafts annually. Statewise area under cashew and production and productivity of cashew in India during the year 1999-2000 is given in Table 3.5.

TABLE 3.5. Area, production, and productivity of cashew in India (1999-2000).

State	Area (in thousand ha)	Productive Area (in thousand ha)	Production (in thousand MT)	Productivity (MT/ha)
Maharashtra	121.20	85.00	125.00	1.47
Andhra Pradesh	100.00	90.00	100.00	1.10
Kerala	122.20	118.00	100.00	0.85
Karnataka	90.50	86.00	60.00	0.70
Goa	54.40	49.00	30.00	0.61
Tamil Nadu	85.20	84.00	45.00	0.54
Orissa	84.10	65.00	40.00	0.62
West Bengal	9.10	9.00	8.00	0.90
Others*	16.70	15.00	12.00	0.80
Total	683.40	601.00	520.00	0.865

Source: Cashew Export Promotion Council of India, 2003.
* Madhya Pradesh, Manipur, Tripura, Meghalaya, and Andaman and Nicobar Islands

WORLD TRADE OF CASHEW

India has been exporting cashew kernels since the 1950s. Over the years, both the export earnings as well as quality of kernels has been increasing. The established processing capacity of raw nuts is around 7 lakh tonnes. However, domestic production is around 5.2 lakh tonnes. Thus, India has been importing raw nuts from African countries to meet the demand of cashew-processing industries. The growth of export and import of cashew since 1955 is presented in Table 3.6. Export earnings have been on the increase since 1955. India has earned an all-time high export earnings of Rs 2,500 crores during 2000. Between 1980 and 1985, although export earnings increased, the quantity of cashew kernels exported decreased. Since 1985, the quantity of cashew kernels exported has grown steadily. It is estimated that the processing industries can absorb up to 10 lakh tonnes of raw nuts for processing (Bhaskara Rao and Nagaraja, 2000).

TABLE 3.6. Import of cashew raw nuts and export of cashew kernels.

Year	Import of rawnuts (tonnes)	Export of kernels (tonnes)	Export earnings (Million Rs)
1955	63,000	31,000	12.9
1960	95,000	39,000	16.1
1965	191,000	56,000	29.0
1970	163,000	60,000	57.4
1975	160,000	65,000	118.1
1980	24,000	38,000	118.0
1985	33,000	32,000	180.0
1990	59,000	45,000	3,650.7
1995	222,000	77,000	12,458.0
1996	222,819	70,334	12,405.0
1997	192,285	68,663	12,855.0
1998	224,968	76,593	13,961.0
1999	181,009	75,026	16,099.0
2000	199,000	95,000	25,000.0

Source: Cashew Export Promotion Council of India, 2003.

ECONOMIC BOTANY

Anacardium is one of the most economically important genera in the Anacardiaceae. This is due to *Anacardium occidentale* (the cashew of commerce), which yields: roasted cashew nuts (seeds), which are a major third world export to industrialized nations; cashew apples (hypocarps), which are consumed locally or used to make a widely marketed juice in South America, especially Brazil; and cashew nutshell liquid, which has medical and industrial applications. Some of the other *Anacardium* species have economic potential but they are currently underutilized. *Anacardium excelsum* is used for construction and as a shade tree for coffee and cocoa plantations. *Anacardium giganteum* is a locally important timber in South America, and its hypocarps are relished by local people. The spectacular white leaves associated with the inflorescences of *A. spruceanum* make it a tree with excellent ornamental potential. *Anacardium humile*, a subshrub closely related to *A. occidentale*, possess edible hypocarps and seeds. Selective breeding for higher quality hypocarps and seeds, as well as hybridizations with *A. occidentale,* could yield subshrubs with fruits that could be harvested mechanically. The economic potential of the other two subshrubs, *A. nanum* and *A. corymbosum* also should be investigated (Mitchell and Mori, 1987).

TAXONOMY

Cashew belongs to the family Anacardiaceae, the genus *Anacardium,* and species *occidentale.* A taxonomic treatment of *Anacardium* (Anacardiaceae; tribe Anacardieae), a Latin American genus of trees, shrubs, and geoxylic subshrubs, is provided by Mitchell and Mori (1987). Anacardiaceae is a moderately large family consisting of 74 genera and 600 species. It is subdivided into five tribes, namely Anacardieae, Spondiadeae, Semecarpeae, Rhoeae, and Dobineae. The tribe Anacardieae consists of eight genera, namely, *Androtium, Buchanania, Bouea, Gluta, Swintonia, Mangifera, Fegimanra,* and *Anacardium* (Mitchell and Mori, 1987). According to Bailey (1958) *Anacardium* is a small genus of eight species indigenous to South America. However, Agnoloni and Giuliani (1977) and Johnson (1973) have recognized eleven and sixteen species, respectively. Valeriano (1972) names five different species, namely *Anacardium occidentale* L., *A. pumilum* St Hilaire, *A. giganteum* Hanca, *A. rhinocarpus,* and *A. spruceanum* Benth. He also suggests recognition of only two species, namely *A. nanum* and *A. giganteum,* which can further be subdivided based on the color (yellow or red) and shape (round, pear-shaped, or elongated) of the pseudofruit. Valeriano (1972) also considers the division into dwarf and giant species to be the only way to classify cashew in a rational and practical way. His arguments are based on the characteristics of pseudofruits. However the description provided by Peixoto (1960) separates recognition of more than two species. It appears from the published accounts that *A. occidentale* L. is the only species that has been introduced outside the New World. Within Central and South America as many as 20 species of *Anacardium* are known to exist (Table 3.7). Mitchell and Mori (1987) recognize ten species in the genus *Anacardium,* one of which, *A. fruticosum,* is described as new. The genus has a primary center of diversity in Amazonia and a secondary center in the Planalto of Brazil. All known species of the *Anacardium* genus can be found in the South American continent; only four of them *(A. coracoli, A. encardium, A. excelsum, A. rhinocarpus)* do not exist in Brazil. There,

TABLE 3.7. Species of *Anacardium* Linn.

Botanical name	Country
Anacardium brasiliense Barb. Rodr.	Brazil
A. curatellaefolium St. Hil (= *A. subcordatum* Presl.)	Brazil
A. encardium Noronha	Malayasia
A. giganteum Hancock ex Engl.	Brazil
A. humile St. Hil (= *A. subterraneum* Liais)	Brazil
A. mediterraneum Vell. Fl. Flum	Brazil
A. nanum St. Hil (= *A. humile* Engl., *A. pumilum* Walp)	Brazil
A. occidentale L. (cashewnut)	Brazil
A. rhinocarpus D. C. Prod.	Brazil
A. spruceanum Benth ex Engl.	Brazil
A. microsepalum Loes	Amazon region
A. corymbosum Barb. Rodr.	Brazil
A. excelsum Skeels (= *Rhinocarpus excelsa*)	Brazil
A. parvifolium Ducke	Amazon region
A. amilcarianum Machado	Brazil
A. Kuhlmannianum Machado	Brazil
A. negrense Pires and Fro'es	Brazil
A. rondonianum Machado	Brazil
A. tenuifolium Ducke	Brazil
A. microcarpum Ducke	Amazon region

Source: Index Kewensis, 1996, Royal Botanical Gardens, Kew.

the high number of wild species suggests that the northeast coast is the site of origin for *Anacardium* genus and namely for *Ancardium occidentale* L. In fact, here different forms of cashew can be found with a high variability for local populations, namely along the coast and dune areas. Nowadays, most species belonging to the *Anacardium* genus are found everywhere in Brazil (NOMISMA, 1994). Ascenso (1986) reported that cashew (*Anacardium occidentale* L.) is the only species in the genus that attained economic importance. The *Anacardium* genus appeared to have originated in the Amazon region of Brazil and hence speciation followed different geographic patterns.

CYTOGENETICS

The cytology of *Anacardium occidentale* L. has not been studied in detail. The chromosome number is reported only for *A. occidentale*. This morphologically polymorphic species also exhibits choromosome polymorphism (Mitchell and Mori, 1987). Chromosome numbers reported in the literature range from $2n = 24$ (Goldblatt, 1984; Khosla et al., 1973), $2n = 30$ (Machado, 1944), and $2n = 40$ (Goldblatt, 1984; Simmonds, 1954) to $2n = 42$ (Goldblatt, 1984; Khosla et al., 1973; Darlington and Janaki Ammal, 1945; Purseglove, 1988). Such chromosome polymorphism is well known in many domesticated trees (Khosla et al., 1973).

Genetic Resources

Collection, Conservation, and Cataloging

Even though no reliable records of the introductions are available, it is presumed that the initial introductions in the Malabar coast of Kerala were from only a few trees, and due to the hardy nature of the crop it has spread to all the coastal regions of India naturally. All these introductions originated from *Anacardium occidentale*. The initial emphasis was only on the establishment of plantations of seedling origin. As cashew is primarily a cross-pollinated crop, it is highly heterozygous. Considerable segregation has resulted in large variation in the populations (Bhaskara Rao and Bhat, 1996). Subsequent to the initiation of research under the Indian Council of Agricultural Research (ICAR), the germplasm collection for collecting these variants/segregants has been undertaken at ICAR and State Agricultural University (SAU) research centers. Until the establishment of the National Research Centre for Cashew (NRCC) in India, germplasm accessions were collected as seed samples only. Variability recorded for some of the important characters in seedling germplasm accessions is given in Table 3.8 (Bhaskara Rao and Swamy, 1994). Wide variation within the accessions due to the cross-pollinated nature of the crop was noticed in these germplasm accessions of seed origin (Photos 3.1 and 3.2).

The National Cashew Gene Bank (NCGB), which was established at the NRCC, Puttur, is concerned exclusively with the clonal accessions. In the current research efforts, the accessions to be collected are identified in the survey taken up during the fruiting season and the scions from the identified mother tree are collected during the propagation season (June to September). The grafts are produced and the clonal accessions are planted in the NCGB. Efforts are also underway to establish conservation blocks with

TABLE 3.8. Variability recorded for some of the characters in the germplasm accessions (seedling progenies).

Character	Range
Flowering season	October-January
Flowering duration	40-127 days
Harvesting duration	30-105 days
Number of fruits per panicle	1-8
Apple weight	30-150 g
Nut weight	2.4-18.0 g
Apple:nut ratio	4:1-12:1
Shelling percentage	19.0-35.0
Kernel weight	0.5-4.5 g
Kernel count/lb	100-900
Shell thickness	1.5-5.0 mm
Mean yield/plant/year (ten annual harvests)	0.50-11.75 kg

Source: Bhaskara Rao and Swamy, 1994.

the clonal accessions in coordinating centers of the All India Coordinated Research Project on Cashew. A total of 1,490 accessions of cashew have been conserved in India (Bhat et al., 1999). At the NRCC, Puttur, a total of 433 clonal accessions have been conserved in the NCGB. As per the International Plant Genetic Resources Institute (IPGRI) cashew descriptors, so far 255 accessions have been characterized and cataloged after six annual harvest (after ten years of planting), and the *Catalogue of Minimum Descriptors of Cashew (Anacardium occidentale* L.) *Germplasm Accessions—I, II,* and *III* have been published (Swamy et al., 1997, 1998, and 2000).

Genetic Improvement

Breeding Perspectives

Yield in cashew is a complex character involving an integrated set of attributes, namely, number of inflorescence per unit area, number of nuts per inflorescence, and mean nut weight. These variables either directly or through their interaction influence the total nut yield in cashew. Any attempt to improve the yield should be preceded by understanding the process governing these components. The process of differentiation of reproductive shoots from the vegetative shoots is an important aspect that needs to be in-

PHOTO 3.1. A young cashew plant in fruiting.

vestigated. Although the data available at present are not adequate to offer an explanation of the differentiation of vegetative shoots into reproductive shoots, the indications are that this could be governed by environmental variables, such as the nutritional factors, availability of moisture, and weather parameters. Understanding this process will be helpful in breeding the varieties that have higher yield-contributing factors such as number of inflorescence per unit area, number of nuts per inflorescence, and fruit-to-nut ratio (Bhaskara Rao et al., 1998). The study by Foltan and Ludders (1995) indicated that there are no significant differences in fruit set following selfing compared to cross-pollination, except in one case where selfing H-3-13 resulted in significantly lower fruit set while the same variety when crossed with Guntur accessions gave maximum fruit set. The reciprocal combination of these parents resulted in a lower fruit set, which indicates

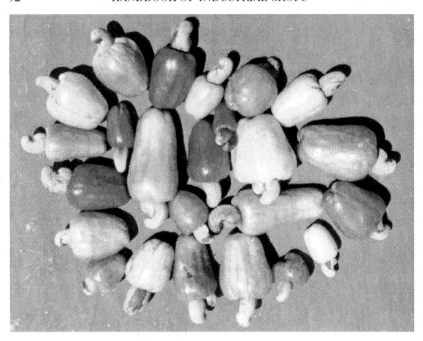

PHOTO 3.2. Variability for cashew apple shading, shape, and size.

the need to understand the cross-pollinated nature between preferential combinations of parents for realizing higher yields. Therefore, studies on the compatibility relationship of cashew varieties and designing the models for establishing orchards with polyclones to ensure highest compatibility and for realizing higher yields in cashew is a priority. One of the options available for increasing productivity of cashew per unit area is to resort to high-density plantings ranging from normal spacings of 8 × 8 m (156 plants/ha) or 7.5 × 7.5 m (175 plants/ha) up to 625 plants/ha depending upon the fertility of the soil and the canopy structure of the variety to be planted. High-density plantings ranging from 200 (10 × 5 m) to 625 plants/ha (4 × 4 m) will be possible only with the genotypes having dwarf stature, compact canopy, and intensive branching with high proportion of flowering laterals per unit area. Therefore the breeding strategy should also give priority for identifying dwarfing genotypes in cashew that can be either used as root stock or plant types by themselves that can be put through the hybridization program to include other yield attributes to structure a genotype which can fit into the high-density planting systems. This also should be supported through suitable canopy management techniques, such as pruning, either through conventional methods or application of chemicals such as Paclobutrazol. Among the various species reported in *Anacardium,* only *Anacardium microcarpum* is supposed to be a dwarf genotype that can be tried as the root stock for the multiplication of varieties that have compact canopies (Bhaskara Rao et al., 1998).

Tree nuts are generally considered as highly nutritive and have been placed in the base of the Medeterranian Diet Pyramid developed by the World Health Organization, which recommends their daily consumption. Nagaraja (1987a,b) reported differences with respect to neutral lipids and glycolipids, whereas the composition of phospholipids did not differ among the varieties. A quality index was developed (Anonymous, 1994) based on the protein, lysine, and sugar content of cashew kernels. However, the recent emphasis is also on low fat content so that the misapprehension that consumption of cashew kernels is deleterious to health is not propagated among consumers. Some of the varieties having > 35 percent protein, lysine > 50 µg/mg protein, and < 14 percent of sugar were identified. These can be used in the breeding program to develop varieties with better nutritive value for the diet-conscious consumer markets. One of the major production constraints in India as well as in other cashew-growing countries is the incidence of tea mosquito bug (TMB) in the flushing and flowering season. Studies made so far on screening for resistance indicated that among the germplasm accessions available in the country, resistance is unlikely to be encountered. However, in one of the accessions, Goa 11/6, a phenological evasion has been noticed which enables the accession to escape severe infestation of tea mosquito bug (Sundararaju, 1999). Similarly, reports have also been made that hybrids like H-3-17, H-1600, H-8-1, H-8-7, H-8-8, and H-15 show moderate tolerance to this pest. It will be worthwhile to look into the varieties whose flowering seasons do not coincide with the peak population of TMB and escape the infestation. The possibility of identifying tolerant types through screening of somaclonal variants is also a worthwhile proposition. Finally, it may be mentioned that the yield structure in cashew involves the integrated set of complex characters enumerated earlier. Even though each appears to be important on its own, specific partitioning of these variables needs to be understood through experimental approaches or through statistical models such as multiple linear regression, path analysis, principal component analysis, etc. Through such efforts, high heritability components contributing to yield need to be identified, and efforts should be made to integrate these characters through hybridization. These breeding efforts should be coupled with suitable management practices to achieve the desired result of achieving higher productivity of cashew (Bhaskara Rao et al., 1998).

Crop Improvement Through Selection

Of the 40 cashew varieties released in India, 25 are selections made from the germplasm materials available in different cashew research stations in the country (Abdul Salam and Bhaskara Rao, 2001). These 25 varieties were identified and released based on the germplasm evaluation carried out at different centers (Table 3.9). As the crop was propagated initially from "plus trees" for soil conservation and afforestation, not much emphasis has been given on the varietal concept of cashew. The concept of the varieties in cashew is of recent origin. The initial identification of varieties was based on total yield realized per tree only. This has resulted in the release of varieties with kernel grades of more than W 320. Important attributes such as kernel weight, shelling percentage, and recovery of whole kernels received little attention. In recent years, with increasing concern for quality, emphasis has been placed on the identification of varieties with kernel weights of over 2 g falling into the export grade of W 210 to W 240. To realize higher recovery of whole kernels, standards have been fixed for shelling percentage (not less than 30 percent). This calls for the identification of donor parents having these characters that can transmit them to their progeny (Bhaskara Rao et al., 1998) (Photos 3.3 and 3.4).

Crop Improvement Through Hybridization

Crop improvement through hybridization is receiving greater attention in almost all the cashew research centers of India. A crop improvement pro-

TABLE 3.9. Germplasm selections released as cashew varieties in India.

State	Center	Varieties
Andhra Pradesh	Bapatla	BPP-3, BPP-4, BPP-5, BPP-6
Goa	ICAR, RC, Goa	Goa-1
Karnataka	NRCC, Puttur	NRCC Selection-1, NRCC Selection-2
	Chintamani	Chintamani-1
	Ullal	Ullal-1, Ullal-2, Ullal-3, Ullal-4, UN-50
Kerala	Anakkayam	Anakkayam-1 (BLA 139-1)
	Madakkathara	Madakkathara-1(BLA 39-4), Madakkathara-2 (NDR 2-1), K-22-1, Sulabha
Maharashtra	Vengurla	Vengurla-1, Vengurla-2
Orissa	Bhubaneswar	Bhubaneswar-1
Tamil Nadu	Vridhachalam	VRI-1, VRI-2, VRI-3
West Bengal	Jhargram	Jhargram-1

PHOTO 3.3. NRCC Selection-1.

gram in Australia also is centered around the development of hybrids wherein thousands of hybrids are produced using parents of wide genetic diversity obtained from different countries, especially from India and Brazil (Chacko et al., 1990; Chacko, 1993). This has necessitated the standardization of a pollination technique in cashew that is reliable, foolproof, and simple. A simple technique of pollination in cashew has been developed at the National Research Center for Cashew, Puttur (Bhat et al., 1998). The new pollination technique developed at NRCC involves the use of butter paper rolls or pantographic paper rolls. The new pollination/crossing technique is as follows:

- Panicles having flower buds that will open the next day are selected both on male and female parental trees. All the opened flowers and nuts, if any, are removed from the selected panicles on the female parental tree.

- Every morning between 8 and 9:30 a.m., all the opened male flowers from the selected panicles on female parental trees are removed. Then, anthers are removed (emasculated) using ordinary pins before anther dehiscence from freshly opened hermaphrodite flowers of the panicles. The stigma along with the style is enclosed with a butter paper roll (pantographic paper roll), which is prepared using a small piece of butter paper sheet of 2.5 × 1.5 cm in size by rolling it with the help of fingers.
- Freshly opened male flowers with undehisced anthers are collected in a petri dish from selected male parents (between 8 and 9:30 a.m.) and the anthers are allowed to dehisce under partial shade.
- The butter paper roll from the emasculated flower is removed and the stigma is pollinated with pollen from freshly dehisced anthers of the male parent collected in a petri dish.
- The pollinated stigma along with style is re-enclosed with butter paper roll.
- Each panicle is labelled indicating the names of female and male parents of the cross as well as the panicle number. Each panicle is used for only one cross combination.
- The previous procedure is continued until eight to ten hermaphrodite flowers are pollinated in each selected panicle.
- All the opened hermaphrodite flowers that are not used for pollination are removed daily.
- All the remaining flower buds are removed from the panicle on the last day of pollination for that panicle.
- Each panicle with developing hybrid nuts is enclosed in a cloth bag in order to collect the nuts on maturity. The details of the crosses should also be written on the cloth bags.
- The hybrid seedlings are raised in polybags utilizing the hybrid nuts so obtained.

This pollination technique is easier to adopt and also gives higher percentage of hybrid nuts as compared to the existing procedure of pollination due to less injury to the delicate cashew flowers.

Fruit set and fruit retention. Studies on flowering, fruiting, and genotype compatibility were carried out at Darwin, Australia (Foltan and Ludders, 1995). Among the five cultivars used for the study, one cultivar, namely, H-3-13, behaved differently from the others. No significant differences were observed in the fruit set following selfing compared to cross pollination in all the combinations except in H-3-13, where selfing resulted in significantly lower fruit set. When H-3-13 was crossed with Guntur, the maximum

PHOTO 3.4. NRCC Selection-2.

fruit set of 51.7 percent was obtained. In the reciprocal combination of Guntur × H-3-13, the fruit set was only 38 percent. This indicates that careful selection of parental combination is necessary in cashew for obtaining a genotype with good fruit set (Bhaskara Rao, 1996). Wunnachit et al. (1992) recorded a general reduction in the yield in selfed progenies when compared to crossed progenies, and the authors attributed postzygotic mechanism to be responsible for observed self-sterility. Premature preferential shedding of selfed fruits is also noticed in avocado (Degani et al., 1989). These studies emphasize the need for better understanding of the compatibility relationships of cashew genotypes, since premature fruit drop is one of the major problems in cashew production (Bhaskara Rao, 1996).

Performance of hybrids. The review of performance of the 34 cashew varieties indicated that in the states of India where both selections and hybrids were released for cultivation, the performance of hybrids has been better than the selections. Hybrid vigor can easily be exploited in cashew because of the amenability of this crop for vegetative propagation. The technique of softwood grafting has been standardized and is the best method available for commercial multiplication of cashew varieties/clones. Hybridization in cashew was started in Kerala at Kottarakkara in 1963 and later continued at the Cashew Research Station, Anakkayam; it is currently being pursued at the Cashew Research Station, Madakkathara. In the initial breeding programs, three parents with prolific bearing (T.No.12A, 30, and 30A) and three bold nut type parents (T.No. 27, 8A, and Brazil-18) were used in hybridization at the Cashew Research Station, Anakkayam (Damodaran, 1977). The reports on the evaluation of these hybrids indicated marked variation in the progenies derived from the same parental combinations. It was also reported that wherever Brazil-18, an exotic bold nut accession, was used in the hybridization, the percentage of progenies with high yield (greater than 8 kg of raw nuts/tree) was higher (35 percent) than those involving the accessions that are collected within the country (9.1 percent). Of the 28 parental combinations evaluated at the Cashew Research Station, Anakkayam (191 hybrid progenies), and Vellanikkara (114 progenies), two hybrids, namely, H-3-17 and H-4-7, were found to be superior than all other combinations tried (Damodaran et al., 1978). Further, it may be mentioned that both these hybrids had exotic accession Brazil-18 as the male parent. Results reported from other research stations that have undertaken hybridization, namely, Vengurla (Maharashtra) and Bapatla (Andhra Pradesh), also indicated that whenever a prolific bearer is crossed with bold nut type, the chances of realizing the hybrid with better nut weight are greater (Nagabhushanam, et al., 1977; Salvi, 1979). Based on these results, varieties with smaller nut size but high nut yield were crossed with bold nut types, namely Vetore-56 and Brazil-711, in the Cashew Research Stations at Maharashtra, Andhra Pradesh, and Kerala. Nawale and Salvi (1990) found that Vetore-56 exhibited higher ability of transmitting bold nut character to the progenies. Seven hybrids recently released from Kerala Agricultural University, namely Dhana, Kanaka, Priyanka, Dharashree, Amrutha, Akshaya, and Anagha (Table 3.10) have at least one parent with bold nut character. Dhana is a cross between ALGD-1 and K 30/1, and Priyanka is a cross between BLA 139-1 and K 30/1, where K 30/1 has a good nut weight (over 8 g). Kanaka is a cross between BLA 139-1 and H 3-13. H 3-13 is itself a hybrid having one of its parents with bold nut size, namely, Brazil-18. Among the 15 cashew hybrids released in India (Abdul Salam and Bhaskara Rao, 2001) (Table 3.10), three hybrids, namely, BPP-1, BPP-2, and Vengurla-5 have small nuts (4.0 to 5.0 g) with kernel grades of W 400 to W 450, whereas the remaining 12 have kernel grades of W 180 to W 240. These 12 hybrids have at least one of the parents with bold nut character, and thus confirm the advantage of selecting the parent with bold nut type for realizing the hybrid with good kernel weight—an important factor in the international cashew trade. Coupled with this, it is desirable that at least one of the

TABLE 3.10. Cashew hybrids released in India and their salient features.

Center	Hybrid	Parentage	Yield potential (kg/tree)	Nut weight (g)	Kernel weight (g)	Shelling percentage	Kernel grade
Bapatla	BPP-1	T.No.1 × T.No.273	10.0	5.0	1.3	27.5	W-400
	BPP-2	T.No.1 × T.No.273	11.0	4.0	1.0	25.7	W-450
	BPP-8	T.No.1 × T.No.39	14.5	8.2	2.3	29.0	W-210
Madakkathara	Dhana (H1608)	ALGD-1 × K 30-1	17.5	9.5	2.2	28.0	W-210
	Kanaka (H1598)	BLA 139-1 × H 3-13	19.0	6.8	2.1	31.0	W-210
	Priyanka (H1591)	BLA 139-1 × K-30-1	16.9	10.8	2.8	26.5	W-180
	Dharashree (H 3-17)	T30 × Brazil	15.0	7.8	2.4	30.5	W-240
	Amrutha (H 1597)	BLA 139-1 × H3-13	18.3	7.1	2.2	31.5	W-210
	Akshaya (H 7-6)	H4-7 × k 30-1	11.7	11.0	3.1	28.3	W-180
	Anagha (H 8-1)	T20 × k 30-1	13.7	10.0	2.9	29.0	W-180
Vengurla	Vengurla-3	Ansur-1 × Vetore-56	14.4	9.1	2.4	27.0	W-210
	Vengurla-4	Midnapore Red × Vetore-56	17.2	7.7	2.4	31.0	W-210
	Vengurla-5	Ansur Early × Mysore Kotekar 1/61	16.6	4.5	1.3	30.0	W-400
	Vengurla-6	Vetore-56 × Ansur-1	13.8	8.0	2.2	28.0	W-210
	Vengurla-7	Vengurla-3 × M 10/4	18.5	10.0	2.9	30.5	W-180

Source: Abdul Salam and Bhaskara Rao, 2001.

parents of the hybrid has better shelling percentage so that a hybrid with higher kernel output can be realized. Among the hybrids released so far, Kanaka and Priyanka, both of which have BLA 139-1 as a parent, have very short flowering phases (Bhaskara Rao and Bhat, 1996). The future thrust in cashew improvement should be to test large numbers of hybrids that not only have high yielding potential, but also have other important attributes, such as export-grade kernels, higher shelling percentages, and also high nutritive value of their kernels. A desirable hybrid can be multiplied by softwood grafting for commercial cultivation. Current research strategy is to plant a large number of hybrid progeny at closer spacing for preliminary evaluation and multiply the identified hybrids with the desirable characters (which takes six to seven years for evaluation) through softwood grafting for final testing, preferably in several locations. However, this method of evaluation could be modified to reduce the time lag between production of hybrid combinations and final testing (Bhaskara Rao et al., 1998).

Biotechnology Interventions

Micropropagation will be a useful tool for faster multiplication of cashew elite lines, useful selections, and hybrids. Micropropagation is also a tool for producing clonal rootstocks. Micropropagation using seedling explants has been standardized (Lievens et al., 1989; D'Silva and D'Souza, 1992; Thimmappaiah and Shirly, 1996, 1999). Regeneration from mature tree explants has been difficult due to the high rate of contamination, browning, slow growth, and poor rooting of microshoots. However, micrografting in cashew is being attempted to rejuvenate mature cashew tree explant material. Mantell et al. (1997) earlier demonstrated micrografting as a means for germplasm exchange. Ramanayake and Kovoor (1999) reported micrografting success in cashew using a scion of seedling origin on in vitro rootstock. Somatic embryogenesis from maternal tissue such as the nucellus and leaf is an alternative for micropropagation and development of synthetic seeds. Somatic embryos can be used as target organs for transformation studies and also as organs for conserving germplasm. Hegde et al. (1993) reported somatic embryos from cotyledon leaves. Thimmappaiah (1997) observed embryogenesis from both cotyledons and nucelli. However, improper germination of the embryoids occurred in all cases. Somaclonal variation induced through culture processes can be exploited for the selection of useful variants, but regeneration from callus (direct organogenesis) is still to be demonstrated to exploit this tool for breeding. However, Bessa and Sardinha (1994) showed in vitro multiplication of cashew by culturing callus induced at the base of the microcuttings repeatedly over a period of time. Since immature embryos could be regenerated into a complete plant (Das et al., 1996), embryo rescue techniques can be used for retrieval and regeneration of inviable hybrids. Anther culture in cashew can be useful for producing haploids and dihaploids, which in turn can be used for genetic studies and for producing homozygous lines (inbreds). Cell cultures or protoplast cultures are useful for making somatic hybridization for the transfer of characters from alien sources. Protoplast isolation in cashew has been reported by Thimmappaiah (1997). However, protoplast cultures and regeneration are yet to be reported. Protoplasts can be used as target organs for transformation provided they are made regenerative to a complete plantlet.

Clonal propagation of elite lines, in vitro conservation, and international germplasm exchange are possible using micropropagation techniques. Molecular markers, such as DNA markers (random amplified polymorphic DNA [RAPD], random fragment length polymorphism [RFLP], amplified fragment length polymorphism [AFLP]), and biochemical markers (isozyme, protein) can be used for characterization of germplasm and somaclonal variants. DNA fingerprinting of varieties using RAPD markers is being done at the NRC for DNA fingerprinting, New Delhi, and Department of Horticulture, the Universtiy of Agricultural Sciences (UAS), Bangalore, in collaboration with the NRCC, Puttur. Mneney et al. (1997) reported RAPD profiles of 20 Tanzanian accessions of cashew. Similarly, RAPD profiles of 19 accessions were completed at NRCDNAF, New Delhi, and DNA fingerprints of 34 released varieties and one TMB-resistant accession were collected by Murali Raghavendra Rao (1999). Such techniques can be used for correlating markers with economically important characters, which will aid in marker-assisted selection. Genetic transformation techniques, such as *Agrobacterium*-mediated gene transfers, can be used in cashew for transfer of genes for biotic (TMB/CSRB [cashew stem and root borer]-resistant genes) and abiotic stress (drought).

CROP MANAGEMENT

Growing Conditions

Cashew is essentially a crop of the tropics with its distribution extending in nature both as wild species and under cultivation between 31°N and 31°S latitude (NOMISMA, 1994). However, the major production areas appear to be confined to 27°N (South Florida) up to 28°S latitude (South Africa) (Joubert and Des Thomas, 1965). Cashew is one of the tropical plants adapted to the most varied and hot climatic conditions. The optimum conditions for vegetative development are found in the tropical climates with sufficient rainfall and a pronounced dry season. It is highly sensitive to cold, and hence higher altitudes with lower temperatures are not conducive for production. The coastal belts between sea level and up to 700 m, where the temperature does not fall below 20°C for prolonged periods, are normally selected in most of the cashew-producing countries (Nair et al., 1979). In India the major producing areas are the coastal districts. However, in recent years the crop has also been cultivated in the interior of central India in the state of Madhya Pradesh (presently Chhattisgarh). The main factor that limits the distribution of cashew within the tropical belt is its inability to tolerate prolonged periods of cold and frost. Younger plantations are more sensitive to cold. With age they can withstand light frost or cold temperatures for brief spells. Cashew is grown in areas with rainfall ranging from 500 to 4,000 mm. For proper vegetative development and regular fruit setting, the ideal conditions appear to be annual rainfall between 800-1,600 mm spread over five to seven months with a clear, definite dry season. Cashew also cannot withstand stagnant water and hence in high rainfall regions (with rainfall over 2,000 mm) it is imperative to have proper drainage. Total rainfall is not the only factor for obtaining a good crop, as can be seen from the data available from some of the Asian countries where the crop is grown in high rainfall areas. Availability of sufficient moisture during flowering and fruit setting was found to be advantageous in the reduction of fruit drop as well as proper development of nuts with good filling. Although it may once again be stressed that even though cashew is adapted to rainfall ranging as low as 500 mm up to 4,000 mm, well-distributed rainfall between 1,000 mm and 2,000 mm is ideal for commercial cultivation of cashew (Ohler, 1979). Cashew is a sun-loving tree and does not tolerate excessive shade. Therefore, it is necessary to clear the site selected for cashew of all forest growth to avoid excessive shade. Average sunshine of nine hours per day was found to be optimum for the best production. As mentioned earlier, the crop is mainly confined to hot, humid coastal areas in most of the producing countries. Even though the crop can be cultivated in the interior, where the relative humidity is comparatively less (between 40 and 60 percent), production and nut size are generally low. Areas with low rainfall and relative humidity over 80 percent during the dry season enable the plant to balance its water requirement and get over the critical period much better than the areas where relative humidity is low. In the majority of cashew-growing countries, the coastal areas are exposed to wind. The strong winds may cause flower drop and fruit fall. In the low rainfall areas, which are away from the coast, strong winds are detrimental to crop production. In the plantations that are extremely close to the sea, high wind speed may lead to scorching of the buds in new flushes. In such areas, it will be necessary to raise fast-growing perennials as windbreaks around the plantations. This will not only protect the cashew plants from unfavorable hot wind, but also will provide sufficient relative humidity through transpiration.

Soil, Water, and Nutrient Requirements

Soils. One of the myths for cashew cultivation in many of the countries is the concept that cashew is an ideal candidate crop for soil conservation, wasteland development, and afforestation programs. In most of the countries, cashew is relegated to poor soils leading to the present crisis of low productivity in most of the cashew-growing countries in Asia. However, cashew can thrive well in a wide variety of soils, namely hard laterite-degraded soils, red sandy loam soils, sandy loam soils, and coastal sands. Mahopatra and Bhujan (1974) suggested a rating chart for land selection for cashew (Table 3.11). They have advocated that, instead of considering the type of the soil alone, the class of a soil with a grading from Class-I to Class-V should be adopted while selecting the site for raising cashew orchards. Class-I to Class-III types of soil with medium acidic range to nearer neutral (6.3 to 7.3 pH), with a slope of 0° to 15°, and with water tables up to 10 m, were recommended as the best soils suited for higher production of cashew. However, in many countries, other plantation crops such as coconut and rubber compete for similar types of lands, and hence only Class-IV and Class-V soils are at present committed to cashew. It is suggested that Class-V soils that are unsuitable for good production of cashew should be avoided, whereas Class-IV soils require strong soil conservation measures as well as other soil amelioration technologies.

Water. Cashew is cultivated mainly as a rainfed crop, and the areas in which cashew is planted are usually devoid of surface water sources. However, new plantations are being raised where supplementary irrigation during the dry months is possible by tapping underground water sources. Experiments conducted in India indicated that supplementary irrigation at 200 liters per tree from November to March can enhance fruit retention and also double the yield over the plantations that do not receive supplementary irrigation. The yield increase was attributed primarily to the higher retention of the fruit set with supplementary irrigation as compared to plots that did not receive any supplementary irrigation. Irrigation at the rate of 200 liters per tree from November to March at fortnightly intervals (ten irrigations during the period) resulted in retention of 44 percent of the fruits; in comparison, 30 percent of the fruits are retained in other plots. Even the fruit set is doubled in the irrigated plots as compared to the control plots (Yadukumar and Mandal, 1994) (Table 3.12). This package can be adopted in the homestead

TABLE 3.11. Guidelines for selection of land for cashew.

	Very good Class I	Good Class II	Fair Class III	Poor Class IV	Unsuitable Class V
Soil characteristics					
Depth of soil	>1.5 m	90 cm-1 m	45 cm-90 cm	23 cm-45 cm	<23 cm
Texture	Loam Sandy loam	Loamy sand Silty loam Coastal sand	Clay loam Silty clay loam Sandy clay loam Loamy skeletal	Gravelly clay loam Gravelly silty loam Gravelly sandy loam	Gravelly clay Sandy clay Silty clay Clay
Reaction	Very slightly acidic to neutral (pH 6.3 to 7.3)	Slightly acidic (pH 6 to 6.3)	Medium acidic (pH 5.6 to 5.9)	Strongly acidic (pH 5.1 to 5.5) or mildly alkaline (pH 7.4 to 7.8)	Very strongly acidic (pH less than 5) or alkaline (pH more than 7.8)
Land features					
Slope (%)	<3	3-5	5-15	15-25	>25
Water table (m)	2-5	1.5-2 (coastal belt)	8-10	10-13	>13
Erosion condition	None to slight (e_0)	Slight (e_1) (sheet erosion)	Moderate (e_2) (rill and sheet erosion)	Severe (e_3) (gully erosion)	Very severe (e_4) (gully and ravine erosion)
Drainage	Well drained	Well drained to somewhat excessively drained	Moderately well drained	Excessively and imperfectly drained	Poorly drained
Physiography	Coastal plains	Alluvial plain	Plateaus	Denuded hill slopes with shallow soils	Swamps
	Delta reaches	Natural levees	Hills	Ridges	Valley bottoms

	Shield plains	Upland plains	domes, mounds	Steeply undulating terrain with severe erosion	Escarpments
	Inland lateritic region adjoining coastal plain	Coastal ridges			Steeply sloping mountains Creek plain
Climate and environmental factors					
Altitude (m)	<20	20-120	120-450	450-750	750
Rainfall (cm/yr)	150-250	130-150	110-130	90-110	<250
Proximity to sea (miles)	<50	50-100	100-150	150-200	<200
Temperature (°F)					
Max. in summer	90-100	100-103	103-106	106-110	<110
Min. in winter	60	57-60	53-56	48-52	<48
Humidity (%)	70-80	65-70	60-65	50-60	<50 or >80
Occurrence of frost	None (once in 20 years)	None (once in 15 years)	Very rare (once in 10 years)	Occasional (once in 5 years)	Very often to frequent (every year)

Source: Mahopatra and Bhujan, 1974.

TABLE 3.12. Fruit retention as affected by irrigation treatments.

Treatments	Fruit set (average of five panicles)	Number of fruits harvested	Fruit retention (%)	Yield (kg/tree)
Irrigation once in 15 days @ 200 liters/tree from				
November-January	27	9	33.0	4.53
January-March	16	6	37.5	4.93
November-March	25	11	44.0	7.32
Control	31	4	30.4	3.54
CD				1.47

Source: Yadukumar and Mandal, 1994.

gardens, especially in rural areas, where it should be possible to provide supplementary irrigation without incurring large additional expenditures by the farmer. Similar effects of supplementary irrigation coupled with black polyethylene mulching were reported by Nawale et al. (1985). In China, supplementary irrigation is provided only during the establishment of the orchards. The monocrop orchards or the adult orchards rarely receive supplementary irrigation, whereas the practice is seen in some gardens where intercrops are cultivated. In other Asian countries, supplementary irrigation is rarely contemplated. Trials were also conducted at the National Research Center for Cashew in India on drip irrigation coupled with graded doses of nitrogen ranging from 250 to 750 g N and 62.5 g to 187.5 g P_2O_5 and K_2O respectively. It was found that irrigation alone at 60 to 80 liters without fertilizers increased the yield by 60 to 70 percent when compared to trees receiving no irrigation and no fertilizers. When the same level of irrigation was given once in four days during dry months along with the highest dose of fertilizers (750 g N and 187.5 g each of P_2O_5 and K_2O) the yield increased up to 114 to 117 percent over the plots which received no irrigation and fertilizers (NRCC, 1998) (Table 3.13).

Manures and Fertilizers. Fertilizer recommendations for cashew vary in different countries and even within the country in India. Mohapatra et al. (1973) reported that a cashew tree bearing 24 kg nuts and 155 kg apples removes annually 2.85 kg of N, 0.35 kg of $P_{22}O_5$, and 1.26 kg of K_2O through root stem, nut, and apple. Beena et al. (1995) have estimated that every kg of nut harvested along with apple requires 64.1 g N, 2.05 g P, 25.7 g K, 4.19 g Ca, and 1.57 g S. Almost all the reports available on fertilizer application of cashew from different countries indicate that a marked response to nitrogen application is seen in all trials. However, for balanced application of

TABLE 3.13. Effect of drip irrigation and NPK doses on cumulative nut yield (kg/tree) up to eight years after planting.

Treatments	Fertilizer (g/tree)				
	M1	M2	M3	M4	Mean
Irrigation (1/tree)					
0	7.6	10.2	9.5	9.9	9.3
20	10.4	12.4	14.4	14.5	12.9
40	10.8	12.4	13.3	14.5	12.8
60	12.9	12.9	15.0	16.5	14.3
80	12.3	14.0	16.4	16.6	14.8
Mean	10.8	12.3	13.7	14.4	
M1 = 0:0:0 M2 = 250:62.5:62.5		M3 = 500:125:125		M4 = 750:187.5:187.5	

Source: NRCC, 1998.

fertilizers, it is necessary to incorporate P_2O_5 and K_2O also. In China, trees are fertilized twice a year in July-September. In the second year when the plants are about 40-50 cm tall, 0.25 kg of urea is applied; this is increased to 0.5 kg and 1 kg in the third and fourth years respectively. 0.5 kg calcium phosphate, 0.3 kg of muriate of potash, or 20-30 kg of organic manure is also added in every application. The current recommendations for fertilizer application in India is 500 g N, 125 g each of P_2O_5 and K_2O per tree per year. In high-yielding varieties, the response of nitrogen was noticed up to 750 g. Since fertilizers available in India do not have the proportion of NPK mentioned earlier, application of straight fertilizers is recommended. Fertilizers are applied only at the end of the rainy season into a shallow trench at the drip line of the tree. It is also recommended that fertilizers be applied in split doses during the premonsoon season (May-June) and postmonsoon season (September-October). However, if single application is preferred, the postmonsoon period is recommended when there is adequate soil moisture. In the first year, one-third of the dose is recommended, which is gradually increased to two-thirds and full dose in the second and third years respectively. The rainfall pattern in the east-coast region and west-coast region of India is quite different. Whereas the east-coast states are characterized by low rainfall, states on the west coast receive high rainfall. In high rainfall areas, fertilizer application in circular trenches of about 25 cm width and 15 cm depth at a distance of 1.5 m away from the trunk is recommended. In the low rainfall areas, fertilizer is applied in the top and raked into the soil. In Indonesia, the practice is to give graded doses from the first year to the third year (Abdullah, 1994) (Table 3.14). In Myanmar, cashew-

TABLE 3.14. General recommendations for fertilizer application in cashew.

Age (Years)	Nitrogen (N)	Phosphorus (P_2O_5)	Potash (K_2O)
1	100	80	—
2	200	80	60
3	400	120	120
4	500	130	130
5	700	250	420
6	900	250	420
7	1000	500	300

Source: Abdulla, 1994.

growing farmers seldom apply fertilizers. The availability of chemical fertilizers is also low, and hence application of green manures and organic manures is recommended. Growing of *subabul* in the interspaces and cutting the stem of *subabul* and incorporating into the soil are the ideal practice under Myanmar conditions. This will also help in improving cashew cultivation (Bhaskara Rao, 1994). In Sri Lanka, adoption of fertilizer recommendations is very low, and it is estimated that only 3.8 percent of the cashew grown in the country is being fertilized. Locally available NPK mix (3:2:1) at 2.5 kg per tree is used.

PLANTING TECHNOLOGY

Planting of soft wood grafts is usually done during the monsoon season (July-August) both in the west and east coast of India. Therefore, land preparation such as clearing of bushes and other wild growth and digging of pits for planting, should be done during the premonsoon season (May-June). A spacing of 7.5 × 7.5 m or 8 × 8 m is recommended for cashew (156-175 plants per ha). A closer spacing of 4 × 4 m in the beginning and thinning out in stages and thereby maintaining a spacing of 8 × 8 m by the tenth year can also be followed. This enables higher returns during the initial years, and as the trees grow in volume, the final thinning is done. However, in level lands it will be advantageous to plant the grafts at 10 × 5 m spacing, which will accommodate about 200 plants per ha and at the same time leave adequate interspace for growing intercrops in the initial years of orchard establishment. Normally, cashew grafts are planted in pits 60 × 60 × 60 cm in size.

The size of the pits can be 1 × 1 × 1 m if hard laterite substratum occurs in the subsoil. It is preferable to dig the pits at least 15 to 20 days before planting and expose them to sun. The pits should be completely filled with a mixture of top soil, compost (5 kg) or poultry manure (2 kg), and rock phosphate (200 g). This will provide a good organic medium for obtaining better growth of plants. Planting of grafts is done preferably during July-August. Usually five- to 12-month-old grafts are supplied by the research stations and private nurseries in polyethylene bags. Cashew is commonly grown in sloppy lands both on the west coast and east coast of India. Soil erosion and leaching of plant nutrients are generally expected under such situations. To overcome this problem, preparing terraces around the plant trunk and opening of catch pits are highly essential. Therefore, before the onset of the southwest monsoon (May-June), terraces of 2.0 m radius should be made before the opening of pits. This helps in soil and moisture conservation, resulting in very good growth of plants in the first year of planting. Terraces are prepared by removing the soil from the elevated portion of the slope. The soil is then spread to the lower side, and a flat basin of 2.0 m radius is made. Terraces may be crescent shaped with the slope of the terrace toward the elevated side of the land, so that the topsoil that is washed off from the upper side due to rainwater is deposited in the basin of the plant. A catch pit (200 cm long, 30 cm wide, and 45 cm deep) across the slope at the peripheral end of the terrace is made for withholding water during the premonsoon and postmonsoon showers in sloppy areas. A small channel connecting the catch pit sideways is made to drain out the excess water during the rainy season (Bhaskara Rao and Swamy, 2000).

High-Density Planting

Conventionally, cashew is planted in a square or triangle system, at a spacing of 7.5 or 8 m. Trials with varying plant densities ranging from 156 plants to 2,500 plants per ha were taken up at the Central Plantation Crop Research Institute (CPCRI)/NRCC. Plant density of 625 plants at a spacing of 4 × 4 m for the first 11 years and thinning them after that to reduce the population to 312 plants (at a spacing of 8 × 5.7 × 5.7 m) gave the maximum cumulative yield of nuts. Results available from West Bengal (Jhargram center) also indicated higher production from the high-density plots. High-density plantings conserve soil more effectively and also suppress weed growth, especially in forestlands. For sustainable high-density models, judicious training and pruning of trees from the initial years will be required (Bhaskara Rao and Swamy, 2000).

Cover Cropping

Leguminous cover crops, such as *Pueraria javanica, Calapagonium muconoides,* and *Centrosema pubescens,* enrich soil with plant nutrients and organic matter, prevent soil erosion, and also help in conserving soil moisture. When sown at the beginning of the rainy season, seeds at the rate of 7 kg per ha will establish these cover crops in cashew orchards. The seeds should be soaked in water for six hours before sowing, and are sown in 30 × 30 cm beds that are prepared in the interspaces of the main crop, cashew. These cover crops will also help in checking soil erosion. As cashew responds to nitrogen, cultivation of the leguminous cover crops will also enrich the soils with nitrogen, which will be beneficial to the growth of cashew. However, it should be ensured that the basins of cashew trees are cleared before the harvesting season, so that the growth of cover crops does not interfere with gathering fallen fruits. However, in totally degraded laterite soils, it is difficult to establish cover crops. In China, natural grass and leguminous cover crops are usually maintained at the time of land clearing in order to conserve soil. Green manure crops are also cultivated during the initial years. Creeping cover crops, such as *Pueraria phaseoloides* and *Centrosema pubescens,* bush cover crops, such as *Gliricidia maculata* and *Leucaena leucocephala,* and nitrogen-fixing trees, such as *Acacia mangium,* are the principal cover crops used in cashew in Sri Lanka.

Intercropping

Intercropping received very little attention when there was no systematic planting of cashew on a large scale. The main objective of raising intercrops is to obtain some returns from the land during the initial years of cashew orchards. Once the canopy of the cashew tree covers the area, it leaves very little scope because of the dense nature of its canopy and shading of interspaces. Further, heavy leaf fall in cashew is not conducive for any field crops. Trials were conducted in India with fruit species, such as pineapple and sapota, forest species such as casuarina and acacia, and green manure crops/cover crops, such as *subabul* and mucuna. Best profit was realized from the cashew and pineapple combination. Cultivation of pineapple is also very advantageous as a soil conservation measure when pineapple is grown in trenches across slopes in hilly terrain. Tree species such as casuarina and acacia were grown for the first five years and removed due to their adverse effect on the main cashew crop. Subsequently, pineapple was planted in the same plots. The cumulative yield of cashew after six years of planting was 61.41 kg in the cashew and pineapple combination plots, as compared to 37.74 kg in plots in which cashew was grown as a monocrop. Cashew and acacia plots and cashew and casuarina plots showed the least yield of cashew before the removal of intercrops; after removing intercrops and planting pineapple, the yield in these plots also increased considerably (Table 3.15; Photo 3.5).

Since the primary concern of a cashew farmer is to increase the production of cashew, it is advisable to resort to intercropping pineapple, which also helps in soil conservation from the beginning, instead of cultivating tree species. An alternative option is to plant cashew itself in high density at the spacing of 5 × 5 m in the square system or 5 × 4 m in the hedgerow system and adopting judicious pruning for realizing higher yields in the initial years (NRCC, 1998). In India, popular intercrops are horsegram, cowpea, and groundnut. Casuarina is also planted as an intercrop at a spacing of 1.5 × 1.5 m in cashew orchards in Andhra Pradesh and Orissa. Medicinal and aromatic plants can also be cultivated wherever supplementary irrigation is being given to cashew. In Indonesia intercrops such as peanut, sweet potato, etc., are popular. In recent years, cashew has been intercropped in some areas with melon (sweet melon and watermelon) and vegetables, such as hot pepper. Using vegetables as intercrops is possible wherever supplementary irrigation is given to cashew. By cultivating melons, a large quantity of green manure will also become available for incorporation into the soil. In

TABLE 3.15. Yield of cashew two years after the removal of tree species and cumulative yield (kg/plot of 384 m^2 area).

Cropping system	5 years after planting before removal of intercrops (tree crops)	6 years after planting and 1 year after removal of tree species	7 years after planting and 2 years after removal of tree species	Cumulative yield for the past 6 years
Cashew monocrop	5.60	7.78	14.42	34.74
Cashew + pineapple	8.80	14.37	28.34	61.41
Cashew + casuarina	4.06	6.75	12.12	26.16
Cashew + acacia	2.03	2.15	10.32	16.23
Cashew + subabul	4.12	5.44	13.15	26.65
Cashew + mucuna	4.46	8.43	15.32	35.40
Cashew + guava	5.30	5.94	13.55	31.07
	CD for treatments	3.27	CD (5%)	7.07
			Sem	2.29

Source: NRCC, 1995.

PHOTO 3.5. Pineapple as intercrop in cashew orchard.

Myanmar, several intercrops, predominantly annuals, such as sweet potato, sesame, peanut, maize, cassava, and pigeon pea, are popular with the cashew farmers. Banana is a popular intercrop in many cashew plantations in Sri Lanka. Pineapple, papaya, pomegranate, and coconut are also cultivated as semiperennial and perennial intercrops in some areas. In Sri Lanka, the common annuals grown in cashew plantations are legumes (cowpea, black gram, green gram), oil crops (sesame, groundnut), and condiments, such as hot pepper and onion. Trials conducted at the Si Sa Ket Horticultural Research Center in Thailand have indicated that sweet corn, groundnut, and vegetables can be grown profitably in the interspaces of cashew orchards in the initial years. As mentioned at the beginning of this section, lack of moisture in the cashew orchards is one of the serious limitations for crop diversification. However, it is very important to include cultivation of intercrops in the management of cashew orchards in the initial years to get early returns from the land committed to cashew. This will also enable the farmer to adopt improved packages recommended for cashew cultivation. In small holdings this becomes essential to overcome the resource constraints for adoption of production technologies recommended for cashew in the initial years.

PEST AND DISEASE MANAGEMENT

Pests

One of the important constraints in many parts of Asia is the insect pest attack, especially on the foliage and inflorescences of crops. Many farmers are not aware of the initial symptoms of different pest attacks, and thereby neglect remedial action in the initial stages. Subsequently, even if they adopt curative measures, the damage that is already done cannot be reversed. More than 194 species of insects and mites have been listed as pests occurring in different cashew-growing countries in the world (Nair et al., 1979). In China, more than 40 insect pests have been observed attacking stems, branches, leaves, tender shoots, flowers, and fruits in the Hainan plantations (Liu Kangde et al., 1998). In India, more than 84 species have been reported by Pillai (1979), of which 79 are insects and 5 are mites. In a subsequent compilation by Rai (1984), another 26 species of pests (17 insects and 9 vertebrate species) were added to the list of pests that damage cashew crops. Sundararaju (1993) compiled a list of 70 species, in addition to the previous reports, which cause damage to cashew. In total, 151 insects, eight mites, and 21 vertebrate species damage cashew crops in India. In Indonesia, low yields in cashew have been attributed to pest attacks by *Cricula* and *Helopeltis* spp. In Myanmar, the major pest infesting cashew is stem borer, mainly due to the lack of phytosanitation measures. Sporadic incidence of shoot-tip caterpillar and leaf webber were also noticed (Maung Maung Lay, 1998). Hence the present level of infestation in Myanmar is reported not to cause any economic injury by their pests. Control of stem and root borers is an essential operation to save the high-yielding trees in the plantations. The most common insect pests in the Philippines are termites, leaf miners, shoot and root borers, and the tea mosquito bug. Chemical control measures are rarely adopted in pest control programs in the country. In Sri Lanka, stem and root borers and tea mosquito bugs are the major problems, whereas leaf miners and leaf and blossom webbers are noticed sporadically and endemic in their infestation. In Thailand, the major problem is the tea mosquito bug (*Helopeltis antonii* Sign.) which causes heavy economic loss to cashew. Thrips (*Haplothrips* species) are also noticed to cause damage to shoots and inflorescences by causing dieback. Although a large number of pests are reported to attack cashew trees (Nair et al., 1979; Rai, 1984; and Sundararaju, 1993), the most important ones which limit the production are the cashew stem and root borer and tea mosquito bug in many countries. Leaf miners *(Acrocercops syngramma)* and leaf and blossom webbers *(Lamida moncusalis)* are also the major pests in certain areas. In addition to these, there are some pests of minor importance in general, but in certain endemic areas, they become very serious. Such pests are defoliating caterpillars, leaf beetles, shoot-tip caterpillars, foliage thrips, flower thrips, and apple and nut borers.

In most Asian countries, recommendations are available for pests of local importance. In China the recommendations include pesticidal spray of 20 percent fenvalerate with a dilution of 1 in 200 and a mixture of 40 percent dimethoate and 80 percent diarotophos (1:2 proportion) with a dilution 1 in 200 given as a low-volume spray. For fruit borers, the recommendation is spraying 20 percent fenvalerate or 2.5 percent deltamethrin (1 in 200 dilution). In India, three sprays are recommended for the control of foliage and inflorescence pests (with monocrotophos and carbaryl) during flushing, flowering, and fruit setting. Endosulphan or monocrotophos at 0.05 percent for the first and second spray, and carbaryl at 0.15 percent for the third spray are recommended. In Myanmar, plant protection practices are rarely used. However, considering the level of infestation at present, it is not necessary to have a general recommendation for all the cashew-growing areas. For the control of thrips (*Haplothrips* species) in Thailand, application of carbosulfan at 30 ml per 20 liters of water or carbaryl at 50 g per 20 liters of water is recommended. For the control of tea mosquito bug, application of either carbaryl at 20 g per 20 liters of water or cyhalothrin at 10 ml per 20 liters of water is recommended. The recommendations made in different Asian countries can control not only tea mosquito bug, but also almost all the foliage and inflorescence pests attacking cashew in the respective countries. However adoption of these recommendations in many of the countries is minimal, mainly due to a lack of awareness of the pest incidence among farmers.

Foliage and Inflorescence Pests

Tea mosquito bug (Helopeltis *sp.*). This is the most serious pest in many of the cashew-growing countries. The adult and immature stages of this mirid bug suck sap from tender shoots, leaves, floral branches, developing nuts, and apples. The injury made by the suctorial mouth parts of the insect causes the tender shoots to exude resinous gummy substances. Tissues around the point of entry of stylets become necrotic and form brown or black scabs, presumably due to the action of the phytotoxin present in the saliva of the insect injected to the plant tissue at the time of feeding. Finally, the adjoining lesions coalesce and the affected portion of shoot/panicle dries up. Severe infestations on the floral branches may also attract fungal infestations, which result in inflorescence blight. Immature nuts infested by this pest develop characteristic eruptive spots, and finally shrivel and fall off. As mentioned earlier, prophylactic spray at the time of flushing, flowering, and fruiting can minimize the losses (Photos 3.6, 3.7, and 3.8).

Leaf miner (Acrocercops syngramma *M.*). The spray schedule recommended for tea mosquito bug normally acts as a prophylactic measure. However, if a serious leaf miner outbreak is noticed, spraying of phosphomidon, fenitrothion, or monocrotophos at 0.05 percent concentrations is effective.

Leaf and blossom webber (Lamida moncusalis *Walker, and* Orthaga exvinacea *Hamps*). Application of carbaryl at 0.15 percent is effective against this pest.

Shoot-tip caterpillars (Hypatima haligramma *M.*). The tiny yellowish or greenish-brown caterpillars of the moth *Hypatima haligramma* M. damage shoot tips and inflorescence. Use of systemic insecticides, such as monocrotophos (0.05 percent), will be more suitable if the outbreak of this pest is noticed.

Foliage thrips (Selenothrips rubrocinctus *Giard,* Rhipiphorothrips cruentatus *Hood, and* Retithrips syriacus *M.*) *and flower thrips* (Rhynchothrips raoensis

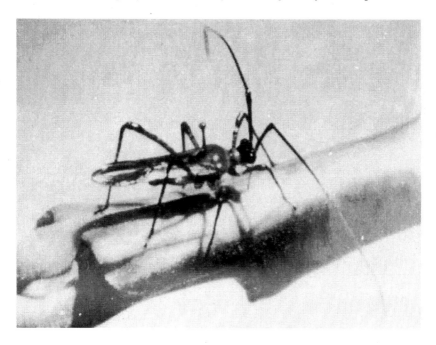

PHOTO 3.6. Tea mosquito bug (TMB).

PHOTO 3.7. TMB-infested shoot.

G., Scirtothrips dorsalis *H.*, Haplothrips ganglabaueri *[Schmutz]*, Thrips hawaiiensis *[Morgan]*, H. ceylonicus *Schmutz, and* Frankliniella schultzei *[Tryborn]*). Cashew plantations, especially those raised with grafts that flush continuously, are more prone to the damage by foliage thrips. There are three species, namely, *Selenothrips rubrocinctus* Giard, *Rhipiphorothrips cruentatus* Hood, and *Retithrips syriacus* M.

Flower thrips cause premature shedding of flowers and scabs on floral branches, apples and nuts. Infestation on developing nuts results in formation of corky layers on the affected parts. Malformation of nuts and even immature fruit drop and nuts also noticed. Both the foliage and flower thrips can be controlled by spraying of endosulfan, monocrotophos, or quinalphos at 0.05 percent concentration.

Apple and nut borer (Thylocoptila panrosema *M. and* Nephopterix *spp.*). Apple and nut borers are responsible for heavy crop losses (Dharmaraju et al., 1974). The caterpillars attack the fruits at all stages and cause shrivelling and premature fall of nuts. Spraying of carbaryl at 0.1 percent or endosulfan at 0.05 percent is effective in controlling apple and nut borers.

PHOTO 3.8. TMB-infested panicles, nuts, and apples.

Stem and Root Borer

Cashew trees infested by stem and root borers, if not treated at appropriate time, will succumb to the attack and die. One of the main reasons is the lack of phytosanitation in the cashew orchards. The extent of attacks on cashew plantations, especially those raised by forest departments, could be as high as 35 percent, and the loss of trees could be anywhere between 20 to 30 percent. The primary species infesting cashew is *Plocaederus ferrugineus* L. Two other species, namely *Plocaederus obesus* Gahan and *Batocera rufomaculata* De G., infest cashew. The symptoms of infestation include the presence of small holes in the collar region, gummosis, extrusion of frass through holes, yellowing of leaves, drying of twigs, and, finally, death of the tree (Pillai, 1975; Pillai et al., 1976). The adult is a medium-sized, reddish-brown, longicorn beetle, the head and thorax of which are dark brown or almost black. The beetle lays eggs in the crevices of the bark as well as on the exposed portion of the roots of the tree. When the eggs hatch, the grubs make irregular tunnels into the fresh tissue and bark and feed on the subepidermal tissues and sapwood. This results in injury to the cells, and a resinous material oozes out when the vascular tissues are damaged.

The ascent of plant sap is arrested; leaves become yellow and later fall (Photos 3.9 and 3.10).

Several pest management techniques incorporating mechanical, chemical, cultural, and biological methods were tried against this pest. Mechanical removal of the eggs, grubs, and pupae from the infested trees, and swabbing on the trunk after extraction of the grubs from the infested tree with carbaryl, lindane (0.2 percent), or painting a mixture of coal tar and kerosene at 1:2 proportion will revive the tree. Whenever an infested tree is noticed in the plantation, it should be treated in the early stages of infestation, and all the adjacent trees also should be given prophylactic treatment of swabbing the trunk with coal tar and kerosene.

Biological Control of Pests

A number of natural enemies are reported on *Helopeltis* from West Africa and a few from Asian countries. The recorded natural enemies of *Helopeltis* are listed by Simmonds (1970) from both Africa and Asia. In India, *Telenomus* spp. and *Chaetostricha minor* were reported to infest *Helopeltis antonii* eggs (Sundararaju, 1993). *Erythmelus helopeltidis* was also reported to parasitize on the eggs of *H. antonii* (Devasahayam and Radhakrishnan Nair, 1986). However, efforts to multiply the egg parasitoids met with little

PHOTO 3.9. Adult beetles of cashew stem and root borer (male–left; female–right).

PHOTO 3.10. CSRB-infested cashew tree.

success as these are specialized parasitoids. *Crematogaster wroughtonii* Forel (Formicidae) has been recorded as a predator of nymphs of the pest (Ambika and Abraham, 1979). Spiders, *Hyllus* spp. (Salticidae); *Oxyopes schirato, P. hidippes* Patch, and *Matidia* sp. have been observed as predators of *H. antonii* (Sundararaju, 1984; Devasahyam and Radhakrishnan Nair, 1986). Three species of reduviid bugs (*Sycanus collaris* [Fab.], *S. phadanolastas signatus* Dist., and *Endochus inornatus* Stil.) have also been noticed as predators of the tea mosquito bug (Sundararaju, 1984). Recently, Rickson and Rickson (1998) observed that a number of ants regularly visit cashew, and indicated the possibility of ant defense in controlling *Helopeltis*. They have surveyed the plantations in Sri Lanka, India, and Malaysia, and indicated the possibility that *Oceophyllas maragdina* could be a promising ant species. Among those recorded by Rickson and Rickson (1988), three small ants, *Crematogaster* spp., *Monomorium latinode,* and *Tapinoma indicum,* entered the open flowers and were able to prey on flower thrips or mites. These ants did not appear to interfere with other pollinators. Rickson and Rickson (1988) have suggested that *Helopeltis* kills about 80 percent of a tree's current vegetative growth. The extent of outbreak is normally patchy, and a pest scouting and tree-by-tree spraying program could be implemented for checking the *Helopeltis* outbreak. This approach demands an ideal program of attention to small outbreaks by staff trained to spot trouble situations, followed by pesticide use for outbreak areas only (Pimentel, 1986; Vandermer and Andow, 1986; Greathead, 1995). For the control of cashew stem and root borers, the possibility of using bacteria, such as *Bacillus thuringiensis, B. popillae,* and fungi, such as *Metarhizium anisopliae* and *Beauveria bassiana,* were indicated by Pillai et al. (1976). The presence of nymphal and adult endoparasitoid, and mermithid parasite nematodes, was detected in adult populations of tea mosquito bugs for the first time in the cashew ecosystem (NRCC, 1998).

Diseases

Compared to the number of pests that attack cashew and their intensity, the magnitude of the disease problem is rather negligible. More than four dozen fungi are reported on cashew, but most of them are of very little economic significance. The disease problem appears mainly in the nursery. In China, root rot, stem rot, and dieback are reported to occur in the nursery stage, whereas gummosis, defoliation, and root rot have been observed in the adult orchards. In India, dieback or pink disease caused by *Corticium salmonicolor,* damping-off of seedlings (Kumararaj and Bhide 1962), and anthracnose disease (Singh et al., 1967) were found to be of importance. The other diseases reported are shoot rot and leaf fall (Thankamma, 1974), decline in cashew nut (Ramakrishnan, 1955), and yellow leaf spot (Subbaiah et al., 1986). Powdery mildew caused by *Oidium* spp. was also reported on the west coast of India in Maharashtra by Phadnis and Elijah (1968). Gummosis is also a problem in certain endemic areas, but none of the diseases from any of the Asian countries are reported to be causing economic losses. Some of the important diseases are as follows.

Anthracnose

The fungus *Colletotrichum gloeosporioides* is the causal agent identified causing anthracnose in many fruit plants, such as mango, citrus, avocado, and papaya. According to Singh et al. (1967), this pathogen continues its growth on the dead parts of the host tissues and perpetuates in unfavorable periods. Effective control of anthracnose can be obtained by spraying bordeaux mixture or copper oxychlorate. The infection of the fungus is usually preceded by infestation of *Helopeltis* (Nambiar, 1974). However, instead of resorting to fungicidal spray in the small holdings, it is better to remove the infested parts and burn removed stems and leaves, which will effectively control this disease. However, in large plantations, spraying of bordeaux mixture or copper oxychlorate will be essential.

Inflorescence Blight

As the very name indicates, this is characterized by drying of floral branches, and gummy exudation can be seen at the lesion site caused primarily by *Helopeltis antonii*. The fungi identified were *Gloeosporium mangiferae* and *Phomopsis anacardii* (Nambiar et al., 1973). The fungi were found to be secondary saprophytic colonizers, and are not the pathogens.

Dieback

Several fungi cause drying of the terminal twigs in cashew. These diseases are also called pink disease caused by *Pellicularia salmonicolor* or *Corticium salmonicolor* (Anonymous, 1950). Bordeaux mixture 1 percent spray during the post-monsoon periods is an effective prophylactic control measure against this disease (Nambiar, 1974).

Leaf Spots

Many types of leaf spots have been reported on cashew. Grey blight is caused by *Pestalotia microspora,* red leaf spot by *Phyllosticta* spp., brown leaf spot by *Colletotrichum gloeosporioides,* etc. Spraying of 1 percent bordeaux mixture or Benlate 0.3 percent was recommended for the control of these leaf spots. For a long time, yellow leaf spot was reported as a disease of unknown etiology. However, Subbaiah et al. (1986) indicated that yellow leaf spot is associated with low soil pH of 4.5 to 5; the affected leaves were deficient in molybdenum and had excessive amounts of magnesium. This disorder could be corrected by foliar sprays of moleybdenum salt.

Powdery Mildew

Phadnis and Elijah (1968) reported that cashew blossoms were affected by *Oidium* species in Maharashtra during cloudy days, but the infestation by

Oidium was reported to be very rare in Asian countries. However, it was reported to be very severe in African countries. Dusting of sulfur was recommended as the control measure against this disease. As mentioned earlier, economic losses caused by diseases are much less in cashew. Only diseases such as those that affect the nursery plants are of economic importance, and necessary prophylactic/curative measures have to be adopted for protecting grafts.

Adverse Weather Conditions

The vegetative phase of cashew extended due to untimely rains during November and December 1995 and 1996 which resulted in late flowering. A high number of bright sunshine hours also induced early bud break as was evident in 1995. A rise in night temperature (about 20°C) together with fewer dewy nights, which coincide with the flowering phase, was detrimental for cashew flower production. Heavy cloudiness during the flowering phase also appeared to be detrimental for the opening of hermaphrodite flowers. However, the mechanism of flower withering under heavy cloudiness and less dew is yet to be understood as it was seen even when the flowering phase was free from tea mosquito menace. This situation led to the poorest ever cashew output in the Kannur and Kasaragod districts of Kerala during 1996-1997. The impact of adverse weather could be mitigated to some extent if early flowering cultivars or a mix of cultivars having different flowering periods are chosen for the replanting program (Rao et al., 1999).

END PRODUCTS USE

Cashew Kernels

Cashew kernel is the most popular nut used by the confectionary industry. As per Indian specifications, there are 33 different grades of cashew kernels, of which 26 are commercially available and exported. These can be broadly classified into white wholes, scorched wholes, dessert wholes, white pieces, scorched pieces, and dessert pieces. At present, cashew kernels are used mainly as a snack food in the roasted and salted forms. The broken cashew kernels are used mainly in the confectionary, bakery, and chocolate industries. Lately, many cashew recipes have been developed and are gaining popularity among cooks all over the world. Cashew nut blends easily with many of the food preparations used in Southeast and western Asia, or for that matter, in any country. Quality has become the vital criterion for all items seeking entry in the international markets. In the emerging food markets, price no longer commands the governing edge in trade. The emphasis has shifted to quality encompassing the aspects of safety, reliability, and acceptability of the product to the customer/consumer. To ensure a place in the emerging new markets, especially in a competitive environment, every product has to achieve internationally accepted quality standards, and cashew being a predominantly export-oriented food product, it is all the more relevant and important. Also, of late, a growing level of environmental concern has been seen in all the developed countries leading to new and stringent regulations on quality of food products and on the type of packing materials used. These regulations relate to the use of lead-free solder in tin containers, avoiding use of toxic/carcinogenic chemicals in preservation and storage, use of environmentally friendly and recyclable materials for packaging, storage, etc. (Nayar, 1998).

Cashew Kernel Peel

Kernel peels are a rich source of tannin (25 percent), which is in great demand by the leather industry (Nayudamma and Koteswara Rao, 1967; Nair et al., 1979). Kernel peels with adhering pieces of cashew kernel are an excellent poultry feed (Nair et al., 1979).

Cashew Apple Products

Whereas the cashew kernel is invariably marketed either in the domestic market or in the international market by all the cashew-producing countries, the apple is wasted except in very few countries. The fruit is very juicy and rich in vitamin C, up to five times that of citrus juice. It also contains 10 to 30 percent sugar. The apple is eaten as such by sucking the juice and discarding the residual fibrous mass. The astringent and/or active acid is due to 0.35 percent tannins and other substances present in the apple (Jain et al., 1954). Steaming the cashew fruit was found to be most efficient in removing the astringent and acrid principles. Pressure of steam and time of exposure vary from 2-6 kg and 5-15 min respectively, according to the quality of fruit and the products that are to be made. The astringent principle can also be removed by boiling in common salt solution of 2 percent for four to five minutes. Alternatively, the fruit juice can be treated with gelatin (0.25-0.4

percent) and pectin (above 0.35 percent) or lime juice (25 percent). A number of cashew apple beverages, such as clarified cashew apple juice, cloudy juice, cashew apple syrup, or cashew apple juice concentrate, can be made from the juice in which the astringent content is removed as explained earlier. Other popular products are cashew apple vinegar, cashew apple candy and jam, canning of cashew apple, cashew apple chutney, and pickles. As cashew is a seasonal fruit, and many of these cashew apple products have low shelf life, they have not become popular in many countries. However, the alcoholic beverage known as *fenni,* which is made in Goa in India, is quite popular and contributes substantially to the state revenue. Popularizing cashew apple processing in cashew-growing countries will also enhance economic returns to the cashew farmers.

Cashew apple residue left after juice extraction, which accounts for 30 to 40 percent of the apple, was found to be nutritious. It contains protein (9 percent), fat (4 percent), crude fiber (8 percent), pectin (10 percent), etc. The residue can be utilized for making various products such as drink, jam, chutney, and preserve (Joshi et al., 1993). The residue can also be used as cattle feed after drying or can be utilized for the recovery of low methoxy pectin (Nanjundaswamy, 1984).

Cashew Nutshell Liquid (CNSL)

CNSL is a by-product of cashew nut processing. The shell oil is extracted by means of motor-driven expellers. Following extraction, the shell liquid is heated and filtered, then sealed into metal drums for export. Percentage of shell liquid ranges from 33 to 38 percent by weight of the shell (Johnson, 1982).

CNSL offers itself as a valuable raw material for unsaturated phenols. Described often as a versatile industrial raw material, CNSL has innumerable applications in polymer based industries such as friction lining, paints and varnishes, laminating resins, rubber compounding resins, cashew cements, polyurethane based polymers, surfactants, epoxtresins, foundary chemicals, and intermediates for the chemical industry (Aggarwal, 1954, 1973; Anonymous, 1993).

Cashew Shell Cake

The residual shell cake left over after extraction of the shell liquid is currently used only as fuel in the processing factories and CNSL extraction plants. This oil cake could also serve as a raw material in the manufacture of plastics and container boards (Johnson, 1982).

Value-Added Products

Several products are made from raw cashew kernels. Product development leads to diversified uses and thus adds value to raw kernels.

Cashew Kernel Flour

Lower-grade kernels are processed into cashew flour, which has a high protein content and is easily digested (Johnson, 1982).

Cashew Kernels Oil (Caribbean Oil)

Lower grade kernels are processed into kernel oil, which is a high-quality edible oil that has been favorably compared to olive oil (Johnson, 1982). The kernel contain 35 to 40 percent oil (Van Eijnatten, 1991).

Cashew Kernel Butter

Kernel residue after extraction of kernel oil is used to produce cashew kernel butter, which is similar to peanut butter (Nair et al., 1979). The cake remaining after extraction of oil serves as animal feed (Van Eijnatten, 1991). Central Food Technological Research Institute (CFTRI) at Mysore has extracted cashew butter for Cashew Export Promotion Council (CEPC), Cochin, and technology for the extraction of cashew butter is available.

COATED CASHEW KERNELS

Results obtained so far at NRCC on value addition have indicated the possibility of developing sugar, honey, and salt-coated baby bits, which are organoleptically acceptable. Baby bits are the lowest-grade kernels marketed commercially (Bhaskara Rao and Swamy, 2000).

Cashew Kernel Milk

It has been shown at NRCC that sweetened and flavored milk could be prepared from cashew kernel baby bits (Bhaskara Rao and Swamy, 2000).

Cashew Spread

Cashew spread has been prepared from kernel baby bits at NRCC (Bhaskara Rao and Swamy, 2000). Sweetened and vanillin-flavored cashew spread is most preferred to salted spread (Bhaskara Rao and Swamy, 2000).

Scope for Generating Value-Added Products

- Commercial exploitation of cashew butter/oil in the cosmetic industry for production of cold creams.
- As 2 to 3 percent kernel rejects are obtained during processing, and cashew kernels are rich in fat, efforts are to be made to extract and refine the kernel oil, which could be used for developing various oil-based products as cashew kernel oil contains considerable quantities of vitamin E.
- Cashew apple is rich in fiber, which needs to be extracted, and attempts are to be made to blend this with others to develop fiber rich foods. In this regard, studies at NRCC have been initiated to isolate and characterize the fiber from cashew apple after extracting cashew apple juice. As cashew apple is rich in fiber, attempts are to be made to look into antidiabetic factors that could be present in cashew apple.
- Cashew kernels are rich in proteins, and these proteins are known to contain all the essential amino acids. At NRCC, attempts have been made to study and compare the functional properties of defatted cashew kernel flour with soybean and almond. Functional properties of cashew kernel flour are comparable with other flours. It has been shown that stable foam could be produced from cashew kernel flour over a wide range of pH levels. Attempts are to be made to blend cereal/pulses flour to develop nutritionally rich blends of flours. Fortification of lower-grade flour with cashew kernel flour needs to be attempted.
- Cashew kernel testa after the removal of tannin has been shown to contain a considerable quantity of protein and carbohydrates. Efforts are to be made to develop food or feed blends from tannin-free testa.

RESEARCH AND DEVELOPMENT ORGANIZATIONS

India was the first country to initiate systematic research on cashew in the early 1950s, which was further strengthened in the 1970s with the establishment of the Central Plantation Crops Research Institute, Kasaragod

(Kerala), and later in 1986 with the establishment of an independent National Research Center for Cashew at Puttur (Karnataka). In the recent years, considerable research efforts have also been made in Australia (Commonwealth Scientific and Industrial Research Organisation, Darwin), Vietnam (Cashew Training Research Center, Binh Duang), Brazil (National Research Center on Cashew, Fortaleza), Tanzania (Tanzanian Agricultural Research Organization-Research Institute, Naliendele, Mtwara), and China (Hainan Cashew High Yield Research center, Hainan) (NOMISMA, 1994; Bhaskara Rao, 1996). A present, there are nine research centers in eight cashew-growing states of India under the All India Co-ordinated Research Project (AICRP) on cashew (Table 3.16).

TABLE 3.16. List of research centers of AICRP on cashew.

Institution	Location	State	Year established
CRS, Andhra Pradesh Agricultural University	Bapatla	Andhra Pradesh	1971
ARS, University of Agriuclture Sciences, Bangalore	Chintamani	Karnataka	1980
CRS, Kerala Agricultural University	Madakkathara	Kerala	1972
RARS, Kerala Agricultural University	Pilicode	Kerala	1993
ZARS, Indira Gandhi Krishi Vishwa Vidyalaya	Jagadalpur	Madhya Pradesh	1993
RFRS, Konkan Krishi Vidyapeeth	Vengurla	Maharashtra	1970
CRS, Orissa University for Agriculture and Technology	Bhubaneswar	Orissa	1975
RRS, Tamil Nadu Agricultural University	Vridhachalam	Tamil Nadu	1970
RRS, Bidhan Chandra Krishi Vishwa Vidyalaya	Jhargram	West Bengal	1982

Note: CRS = Cashew Research Station; ARS = Agricultural Research Station; RARS = Regional Agricultural Research Station; ZARS = Zonal Agricultural Research Station; RFRS = Regional Fruit Research Station; RRS = Regional Research Station.

FUTURE OUTLOOK

Considerable progress has been made during the past fifty years in terms of developing high-yielding varieties, fixing nutrient requirements, standardizing vegetative propagation techniques, chemical control of major pests of cashew, cashew apple utilization, and production of a large number of planting material required for fresh planting and replanting programs. Keeping in view the changing global scenario, and to meet the demand and challenge in the international market, research programs will need to be intensified in the following lines for increasing the production and productivity of cashew:

Genetic Resources

- Consolidation of cashew germplasm material so far conserved in NCGB at NRCC, Puttur, and preparation of a districtwise collection map of the country
- Collection of germplasm material from other cashew research centers such as ARS, Ullal; RRS, Brahmavar; CRS, Anakkayam; and CRS, Kavali; which do not function under the ICAR setup
- Collection of germplasm from nontraditional areas, such as hills of Garo (Meghalaya) and Bastar (Chhattisgarh)
- Characterization of accessions in NCGB to support the breeding program, with respect to processing quality of raw nuts, apples for better fiber quality, and tolerance/resistance to major pests of cashew

Varietal Improvement

- Genetic studies on dwarfing, cluster bearing, and bold nut characters through molecular markers/isozyme banding pattern in different cashew genotypes
- Studies on reciprocal differences in the hybrids, polyclonal interactions, pollen viability, incompatibility, fertility, and postzygotic abortions

Biotechnology

- Characterization of released varieties and germplasm accessions of cashew by DNA fingerprinting in collaboration with UAS, Bangalore, and National Research Centre for DNA Finger Printing (NRC DNAFP), New Delhi

- Development of DNA fingerprinting facilities for cashew at NRCC
- Evaluation of biochemical and physiological basis for variation observed in response from mature tree explants
- Comparison of micrografts with normal grafts

Crop Management

- Working out absolutely essential or minimum requirements for achieving targeted yield
- Working out optional (irrigation, pruning) requirements for further enhancing the yield and its economics
- Considering data on soil characters, nutrient reserve in the soil, weather parameters, fertilizer/manure sources, and economics; determining basic land-use planning requirements of cashew
- Evaluation of less vigorous (Ullal-1, H-2-16, BLA 39-4, etc) varieties for high-density planting for maximizing yield per unit area
- Reorienting the drip irrigation approach to determine optimum water requirements of cashew in different types of soil
- Monitoring of micronutrient status of soil, leaf, and kernel, with emphasis on nutrients, such as iron, calcium, magnesium, and potassium, which are important in human nutrition
- Field evaluation of (1) performance of a scion variety on its own root stock and on different root stock varieties and (2) performance of a root stock variety with its own scion and scions from different varieties
- Basic studies on the use of hormones, including growth promoters, retardants, and inhibitors, in promoting cashew yields

Crop Protection

- Standardization of revival techniques in cashew stem and root borer (CSRB)-infested trees through induction of bark regrowth and root development
- Design and development of traps for CSRB based on kairomones or utilizing infested trees as a bait
- Approaches for reducing latent or residual CSRB inoculum
- Developing models for identifying the most vulnerable trees for CSRB attack in the garden
- Investigations on the bioecological aspects of CSRB incidence
- Possibilities of CSRB management with entomophilic nematodes
- Management of flower/fruit pests with least pesticidal approaches, including pheromones

- Biology and bionomics of flower pests (thrips, apple and nut borers, and shoot-tip caterpillars)
- Monitoring of fauna in the cashew ecosystem
- Studies on the pest complex of nuts in postharvest and preprocessing stages
- Developing a forecasting system for inflorescence pests
- Studies on pesticide residues in soil, water, and weeds in cashew plantations and nearby areas

Postharvest Technology

- Development of protocols for the extension of cashew apple shelf life for fresh consumption through appropriate storage and packing studies
- Development of products from cashew apple that are cost-effective and could generate employment among rural women in all cashew-producting countries

Technology Transfer

- Assessment of the impact of technologies recommended to the cashew farmers of Karnataka and other cashew-growing states
- Refinement of various recommended practices on crop management, plant protection, postharvest management through field assessment, training programs, and demonstrations
- Recommendation of varieties depending upon ecological conditions
- Production of quality planting materials

REFERENCES

Abdul Salam, M. and Bhaskara Rao, E.V.V. (Eds.) 2001. *Cashew Varietal Wealth of India.* Directorate of Cashew and Cocoa Development, Government of India, Ministry of Agriculture, Cochin.

Abdullah, A. 1994. *Technological Packages for Cashew Development.* Upland Farmer Development Project, Directorate-General Estate, Ministry of Agriculture, Indonesia.

Aggarwal, J.S. 1954. The cashewnut and CSNL. *Oleagineux* 8(8-9): 559-564.

Aggarwal, J.S. 1973. Resins, varnishes, and surface coatings from cashewnut shell liquid. *Paint Manuf.* 43(8): 29-31.

Agnoloni, M. and Giuliani, F. 1977. *Cashew Cultivation.* Library of Tropical Agriculture, Ministry of Foreign Affairs, Instituto Agronomico Per L'oltremare, Florence.

Ambika, B. and Abraham, C.C. 1979. Bioecology of *Helopeltis antonii* Sign. (Miridae: Hemiptera) infesting cashew trees. *Entomon.* 4: 335-342.

Anonymous. 1950. List of common names of Indian plant diseases. *Indian J. Agric. Sci.* 20(1): 107-114.

Anonymous. 1993. Indian cashewnut shell liquid—A versatile industrial raw material. *Indian Cashew J.* 21(1): 8-9.

Anonymous. 1994. *Annual Report 1993-94.* National Research Centre for Cashew, Puttur, Karnataka, India.

Ascenso, J.C. 1986. Potential of the cashew crop-1. *Agric. Intl.* 38(11): 324-327.

Bailey, L.H. 1958. *The Standard Encyclopedia of Horticulture,* Volume 1, Seventeenth Edition. Macmillan, New York.

Balasubramanian, P.P. 2000. Cashew—The premier crop of Indian commerce. In *Selected Articles on Cashew—State Level Seminar on Cashew* (compiled by K.K. Mohanty, P.C. Lenka, and A.K. Patnaik, eds.). OSCDC, Bhubaneswar, pp. 3-11.

Beena, B., Abdul Salam, M., and Wahid, P.A. 1995. Nutrient uptake in cashew (*Anacardium occidentale* L.). *The Cashew* 9(3): 9-16.

Bessa, A.M.S. and Sardinha, R.M.A. 1994. In vitro multiplication of cashew (*Anacardiuam occidentale* L.) by callus culture. *Proceedings of the Eighth International Congress on Plant Tissue Cell Culture,* Firenze, Italy, June 12-17, pp. 2-37.

Bhaskara Rao, E.V.V. 1994. *Cashew Cultivation in Myanmar.* Deputation Report—Project UNDP/MTA/86/018, ARCPC (Plantation Crops Division, Yangon, Myanmar).

Bhaskara Rao, E.V.V. 1996. Emerging trends in cashew improvement. In Chopra, V.L., Singh, R.B., and Varma, A., Eds., *Crop Productivity and Sustainability—Shaping the Future.* Second International Crop Science Congress, Oxford and IBH Publishing Co. Pvt. Ltd., New Delhi, pp. 675-682.

Bhaskara Rao, E.V.V. and Bhat, M.G. 1996. Cashew breeding—Achievements and priorities. *Proceedings of the Seminar on Crop Breeding in Kerala.* Department of Botony, University of Kerala, India, pp. 47-53.

Bhaskara Rao, E.V.V and Nagaraja, K.V. 2000. Status report on cashew. In Rethinam, P., Ed., *Souvenir International Conference on Plantation Crops: 12-15 December 2000.* National Research Center for Oilpalm, Pedavegi, India, pp. 1-7.

Bhaskara Rao, E.V.V. and Swamy, K.R.M. 1994. Genetic resources of cashew. In Chadha, K.L. and Rethinam, P., Eds., *Advances in Horticulure,* Volume 9, *Plantation and Spice Crops,* Part 1. Malhotra Publishing House, New Delhi, pp. 79-97.

Bhaskara Rao, E.V.V. and Swamy, K.R.M. 2000. Cashew research scenario in India. In Balasubramanian, P.P., Ed., *The Cashew Industry in India.* DCCD, Cochin. (In press).

Bhaskara Rao, E.V.V., Swamy, K.R.M., and Bhat, M.G. 1998. Status of cashew breeding and future priorities. *J. Plant. Crops* 9(4): 18-24.

Bhat, M.G., Bhaskara Rao, E.V.V., and Swamy, K.R.M. 1999. Genetic resources of cashew and their utilization in crop improvement. In Ratnambal, M.J., Kumaran, P.M., Muralidharan, K., Niral, N., and Arunachalam, V., Eds., *Improvement of Plantation Crops.* Central Plantation Crops Research Institute, Kasaragod, Kerala, India, pp. 91-98.

Bhat, M.G., Kumaran, P.M., Bhaskara Rao, E.V.V., Mohan, K.V.J., and Thimmappaiah. 1998. Pollination technique in cashew. *The Cashew* 12(4): 21-26.

Cashew Export Promotion Council of India. 2003. Cashew statistics. CEPCI, Cochin.

Chacko, E.K. 1993. *Genetic Improvement of Cashew Through Hybridisation and Evaluation of Hybrid Progenies.* A final report prepared for the Rural Industries R&D Corporation (mimeographed), CSIRO Division of Horticulture, Winnellie, Australia.

Chacko, E.K., Baker, I., and Downton, J. 1990. Toward a sustainable cashew industry for Australia. *J. Australian Inst. Agric. Sci.* 3(5): 39-43.

Damodaran, V.K. 1977. F_1 population variability in cashew. *J. Plant. Crops* 5(2): 89-91.

Damodaran, V.K., Veera Raghaavn, P.G., and Vasavan, M.G. 1978. Cashew breeding. *Agric. Res. J. Kerala* 15(1): 9-13.

Darlington, C.D. and Janaki Ammal, E.K. 1945. *Chromosome Atlas of Cultivated Plants.* Allen & Unwin, London.

Das, S., Jha, T.B., and Jha, S. 1996. In vitro propagation of cashew nut. *Plant Cell Rep.* 15: 615-619.

De Castro, P. 1994. Summary of the study. In Delogu, A.M. and Haeuster, G. *The World Cashew Economy.* NOMISMA, L'Inchiostroblu, Bologna, Italy, pp. 11-12.

Degani, C., Goldring, A., and Gazit, S. 1989. Pollen parent effect on out crossing rate in "Hay" and "Fuerte" Avocado plots during fruit development. *J. Amer. Soc. Hort. Sci.* 114: 106-111.

Devasahayam, S. and Radhakrishnan Nair, C.P. 1986. The tea mosquito bug *Helopeltis antonii* Signoret on cashew in India. *J. Plant. Crops* 14: 1-10.

Dharmaraju, E., Rao, P.A., and Ayyanna, T. 1974. A new record of *Nephopteryx* sp. as an apple and nut borer on cashew in Andhra Pradesh. *J. Res. Andhra Pradesh Agricultural University* 1(4 and 5): 198.

D'Silva, I. and D'Souza, L. 1992. In vitro bud proliferation of *Anacardium occidentale* L. *Plant Cell Tissue and Organ Culture* 29: 1-6.

Foltan, H. and Ludders, P. 1995. Flowering, fruit set, and genotype compatiblity in cashew. *Angew. Bot.* 69: 215-220.

Goldblatt, P (Ed.) 1984. *Index to Plant Chromosome Numbers, 1979-1981.* Missouri Botanical Garden.

Greathead, D.J. 1995. Natural enemies in combination with pesticides for integrated pest management. In Reuveni, R., Ed., *Novel Approaches to Integrated Pest Management.* Lewis Publishers, Boca Raton, FL, pp. 183-197.

Hegde, M., Kulasekaran, M., Jayasankar, S., and Shanmugavelu, K.G. 1993. In Vitro embryogenesis in cashew (*Anacardium occidentale* L.) *Indian Cashew J.* 21(4): 17-25.

Jain, N.L., Das, D.P., and Lal, G. 1954. *Procceedings of the Symposium on Fruits and Vegetables Preservation Industry.* CFTRI, Mysore.

Johnson, D. 1973. The botany, origin, and spread of the cashew *Anacardium occidentale* L. *J. Plant. Crops* 1: 1-7.

Johnson, D. 1982. Cashewnut processing in Brazil. *Indian Cashew J.* 12(2): 11-13.

Joshi, G.D., Prab, B.P., Relekar, P.P., and Magdum, M.B. 1993. Studies on utilization of cashew apple waste. *Abstract, Golden Jubilee Symposium on Horticultural Research—Changing Scenario,* Bangalore, India, pp. 372.

Joubert, A.S. and Des Thomas, S. 1965. The cashew nut. *Fmg. S. Afr.* 40: 6-7.

Khosla, P.K., Sareen, T.S., and Mehra, P.N. 1973. Cytological studies on Himalayan Anacardiaceae. *Nucleus* 16: 205-209.

Kumararaj, K. and Bhide, V.P. 1962. Damping off of cashewnut (*Anacardium occidentale* L.) seedlings caused *Phytophthora palmivora* Butler in Maharashtra State. *Curr. Sci.* 31: 23.

Lievens, C., Pylyser, M., and Boxus, P.H. 1989. First results about micropropagation of *Anacardium occidentale* by tissue culture. *Fruits* 44: 553-557.

Liu Kangde, Liang Shibang, and Deng Suisheng. 1998. Integrated production practices of cashew in China. In Papademetriou, M.K. and Herath, E.M., Eds., *Integrated Production Practices of Cashew in Asia.* Food and Agricultural Organization of the United Nations, Regional Office for Asia and the Pacific, Bangkok, Thailand, pp. 6-14.

Machado, O. 1944. Estudos novos sobre uma planta velha cajueiro (*Anacardium occidentale* L.). *Rodriguesia* 8: 19-48.

Mahopatra, G. and Bhujan. 1974. Land selection for cashew plantation—A survey report. *Cashew Bull.* 11(8): 8-15.

Mantell, S.H., Boggetti, B., Bessa, A.M.S., Lemos, E.P., Abdelhadi, A., and Mneney, E.E. 1997. Micropropagation and micrografting methods suitable for international transfers of cashew. *Proceedings of the International Cashew and Coconut Conference.* Dar es Salaam, Tanzania, p. 24.

Massari, F. 1994. Introduction. In Delogu, A.M. and Haeuster, G., Eds., *The World Cashew Economy.* NOMISMA, L'Inchiostroblu, Bologna, Italy, pp. 3-4.

Maung Maung Lay. 1998. Integrated production practices of cashew in Myanmar. In Papademetriou, M.K. and Herath, E.M., Eds., *Integrated Production Practices of Cashew in Asia.* Food and Agricultural Organization of the United Nations, Regional Office for Asia and the Pacific, Bangkok, Thailand, pp. 33-46.

Mitchell, J.D. and Mori, S.A. 1987. The cashew and its relatives (*Anacardium:* Anacardiaceae). *Memoirs of the New York Botanical Garden* 42: 1-76.

Mneney, E.E., Mantell, S.H., Tsoktouridis, G., Amin, S., Bessa, A.M.S., and Thangavelu, M. 1997. RAPD-profiling of Tanzanian cashew (*Anacardium occidentale* L.). *Proceedings of the International Cashew and Coconut Conference,* Dar es Salaam, Tanzania, p. 24.

Mohapatra, A.R., Vijaya Kumar, K., and Bhat, N.T. 1973. A study on nutrient removal by the cashew tree. *Indian Cashew J.* 9(2): 19-20.

Murali Raghavendra Rao, M.K. 1999. Estimation of genetic diversity in cashew (*Anacardium occidentale* L.) cultivars by using molecular markers. MSc thesis, University of Agriculture Sciences, Bangalore.

Nagabhushanam, S., Dasaradhi, T.B., and Venkata Rao, P. 1977. Some promising cashew selections and hybrids in Andhra Pradesh. *Cashew Bull.* 14(7): 1-4.

Nagaraja, K.V. 1987a. Lipids of high yielding varieties of cashew. *Plant Foods for Human Nutrition* 37: 307-311.

Nagaraja, K.V. 1987b. Protein of high yielding varieties of cashew. *Plant Foods for Human Nutrition* 37: 69-75.

Nair, M.K., Bhaskara Rao, E.V.V., Nambiar, K.K.N., and Nambiar, M.C. 1979. *Cashew (Anacardium occidentale* L.). *Monograph on Plantation Crops-1*, pp. 169. Central Plantation Crops Research Institute, Kasaragos.

Nambiar, K.K.N., Sarma, Y.R., and Pillai, G.B. 1973. Inflorescence blight of cashew (*Anacardium occidentale* L.) *J. Plant. Crops* 1: 44-46.

Nambiar, M.C. 1974. Recent trends in cashew research. *Indian Cashew J.* 9(3): 20-22.

Nambiar, M.C. 1977. Cashew. In Paulo de T. Alvin and Koslowski, T.T., Eds., *Ecophysiology of Tropical Crops*. Academic Press, Inc., San Francisco, pp. 461-478.

Nanjundaswamy, A.M. 1984. Economic utilization of cashew apple. *Cashew Causerie* 6(2): 2-7.

National Research Center for Cashew (NRCC). 1995. *Annual Report 1994-95*. NRCC, Puttur, Karnataka.

National Research Center for Cashew (NRCC). 1998. *Annual Report 1997-98*. NRCC, Puttur, Karnataka.

Nawale, R.N. and Salvi, M.J. 1990. The inheritance of certain characters in F-1 hybrid progenies of cashewnut. *The Cashew* 4(1): 11-14.

Nawale, R.N., Sawke, D.P., Deshmukh, M.T., and Salvi, M.J. 1985. Effect of black polythene mulch and supplementary irrigation on fruit retention in cashewnut. *Cashew Causerie*, (July-September): 89.

Nayar, K.G. 1998. Cashew: A versatile nut for health. In Topper, C.T., Caligari, P.D.S., Kulaya, A.K., Shomari, S.H., Kasuga, L.H., Masawe, P.A.L., and Mpunami, A.A., Eds., *Proceedings of International Cashew and Coconut Conference. Trees for Life—the Key to Development*. BioHybrids International Ltd., Reading, United Kingdom, pp. 195-199.

Nayudamma, Y. and Koteswara Rao, C. 1967. Cashew testa—Its use in leather industry. *Indian Cashew J.* 4(2): 12-13.

NOMISMA. 1994. *The World Cashew Economy* (Delogu, A.M. and Haeuster, G., Eds.). L'Inchiostroblu, Bologna, Italy.

Ohler, J.G. 1979. *Cashew*. Communication 71, Department of Agricultural Research, Koninklijk Instituut voor de Tropen, Amsterdam.

Peixoto, A. 1960. *Caju*. Servico de Informacao Agricola, Ministerio de Agricultura, Rio de Janerio (Brazil).

Phadnis, A. and Elijah, S.N. 1968. Development on cashewnut in Maharashtra state. *Cashew News Teller* 1(8,9, and 10): 7-12.

Pillai, G.B. 1975. Pests of cashewnut and how to combat them. *Cashew News Teller* (October-December): 31-33.

Pillai, G.B. 1979. Pests. In Bhaskara Rao, E.V.V. and Hameed Khan, H., Eds., *Cashew Research and Development*. Indian Society for Plantation Crops, CPCRI, Kasaragod, Kerala, India, pp. 55-72.

Pillai, G.B., Dubey, O.P., and Singh, V. 1976. Pests of cashew and their control in India: A review of current status. *J. Plant. Crops* 4: 37-50.

Pimentel, D. 1986. Acroceology (Sic) and economics. In Kogan, M., Ed., *Ecological Theory and Integrated Pest Management Practice.* John Wiley & Sons, New York, pp. 299-319.

Purseglove, J.W. 1988. Anacardiaceae. In *Tropical Crops—Dicotyledons.* English Language Book Society, Longman, London, pp. 18-32.

Rai, P.S. 1984. *Hand Book on Cashew Pests.* Research Publications, Delhi, India.

Ramakrishnan, T.S. 1955. Decline in cashewnut. *Indian Phytopath.* 8: 58-63.

Ramanayake, S.M.S.D. and Kovoor, A. 1999. In vitro micrografting of cashew (*Anacardium occidenatle* L.). *J. Hortl. Sci. and Biotech.* 74(2): 265-268.

Rao, G.S.L.H.V.P., Giridharan, M.P., Naik, B.J., and Gopakumar, C.S. 1999. Weather inflicted damage on cashew production—A remedy. *The Cashew* 13(4): 36-43.

Rickson F.R. and Rickson, M.M. 1998. The cashewnut, *Anacardium occidentale* (Anacardiaceae), and its perennial association with ant: Extrafloral nectary location and the potential for ant defense. *American J. Bot.* 85(6): 835-849.

Salvi, P.V. 1979. Cashew hybrids for increased production. *Indian Fmg.* 28(12): 11-12.

Simmonds F.J. 1970. A memorandum on the possibilities of biological control of cocoa-infesting mirids. *Commonwealth Institute of Biological Control, Bangalore, Report* (unpublished).

Simmonds, N.W. 1954. Chromosome behaviour in some tropical plants. *Heredity* 8: 139.

Singh, S., Sehgal, H.S., Pandey, P.C., and Bakshi, B.K. 1967. Anthracnose disease of cashew (*Anacardium occidentale* L.), its cause, epidemiology and control. *Indian Forester* 93: 374-376.

Subbaiah, C.C., Manikandan, P., and Joshi, Y. 1986. Yellow leaf spot of cashew: A case of molybdenum deficiency. *Plant and Soil* (Netherlands) 94(1): 35-42.

Sundararaju, D. 1984. Studies on cashew pests and their natural enemies in Goa. *J. Plant. Crops* 12: 38-46.

Sundararaju, D. 1993. Compilation of recently recorded and some new pests of cashew in India. *The Cashew* 8(1): 15-19.

Sundararaju, D. 1999. Screening of cashew accessions to tea mosqutio bug (*Helopeltis antonii* L. Sign.). *The Cashew* 13(4): 20-26.

Swamy, K.R.M, Bhaskara Rao, E.V.V., and Bhat, M.G. 1997. *Catalogue of Minimum Descriptors of Cashew (*Anacardium occidentale *L.) Germplasm Accessions—I.* National Research Center for Cashew, Puttur, Karnataka, India.

Swamy, K.R.M., Bhaskara Rao, E.V.V., and Bhat, M.G. 1998. *Catalogue of Minimum Descriptors of Cashew (*Anacardium occidentale *L.) Germplasm Accession—II.* National Research Center for Cashew, Puttur, Karnataka, India.

Swamy, K.R.M, Bhaskara Rao, E.V.V., and Bhat, M.G. 2000. *Catalogue of Minimum Descriptors of Cashew (*Anacardium occidentale *L.) Germplasm Accessions—III.* National Research Center for Cashew, Puttur, Karnataka, India.

Thankamma, L. 1974. *Phytophthora nicotianae* var. *nicotianae* on *Anacardium occidentale* in South India. *Plant Dis. Reptr.* 58: 767-768.

Thimmappaiah 1997. In vitro studies in cashew (*Anacardium occidentale* L.). PhD thesis, Mangalore University, India, p. 239.

Thimmappaiah and Shirly, Raichal Samuel. 1996. Micropropagation studies in cashew (*Anacardium occidentale* L.). Natl. Symp. on Hort. Biotech., (Souvenir), Bangalore, p. 65.

Thimmappaiah and Shirly, Raichal Samuel. 1999. In vitro regeneration of cashew (*Anacardium occidentale* L.). *Indian J. Exp. Biol.* 37: 384-390.

Valeriano, C. 1972. O cajueiro. *Boletim do Instituto Biologico de Bahia (Brazil)* 11(1): 19-58.

Van Eijnatten, C.L.M. 1991. *Anacardium occidentale* L. In Verheij, E.W.M. and Coronel, R.E., Eds., *Plant Resources of South-East Asia, No. 2. Edible Fruits and Nuts.* Pudoc-DLO, Wageningen, The Netherlands, p. 446.

Vandermer, J. and Andow, D.A. 1986. Prophylactic and responsive components of an integrated pest management program. *J. Econ. Entom.* 79: 299-302.

Wunnachit, W., Pattison, S.J., Giles, L., Millington, A.J., and Sedgley, M. 1992. Pollen tube growth and genotype compatibility in cashew in relation to yield. *J. Hortic. Sci.* 67(1): 67-75.

Yadukumar, N. and Mandal, R.C. 1994. Effect of supplementary irrigation on cashew nut yield. In Satheesan, K.V., Ed., *Water Management for Plantation Crops: Problems and Prospects.* CWRDM, Calicut, Kerala, India, pp. 79-84.

Chapter 4

Cinchona

Rajendra Gupta

ORIGIN AND USES

The bark of the cinchona tree is the exclusive source of quinine and related alkaloids, which have come to be known as remedies against malarial parasites. Quinine is one of the oldest naturally occurring phytochemicals known in medicine. Dr. Jaime Jaramillo Arango, a former rector of the National Faculty of Medicine at Bogotá, has established decisively in his book (1950), *The Conquest of Malaria,* that natives in Peru knew of the curative properties of this bark, then called *quina-quina,* against periodic fevers long before the Spaniards arrived in the Andes. It is believed that around the early sixteenth century, Father Burtolome Tafur brought some bark with him to a religious conclave in Rome (Italy), considered, then, the most malarial locality in Europe. It proved effective in the treatment of malarial fever and soon gained popularity. Another Jesuit priest stationed in the Andes then began to supply the bark regularly to the Holy City and distributed it as "Jesuitus bark" or "fever bark" for treatment of malarial fever. The bark received increasingly greater use and gained popularity in the next 200 years until two French chemists isolated and purified quinine alkaloid, as the most potent compound to fight against malaria. This period, therefore, witnessed vast natural populations of cinchona trees continuously uprooted and exploited in the Andes and tonnes of the bark were shipped to different parts of the globe annually. It brought the high-rainfall-receiving mountain ranges of the Andes, mainly in Peru, Bolivia, Colombia, and Ecuador, in focus of explorers, herbalists, and botanists, as it was known to hold rich populations, not only of many taxa of cinchona, but also other valuable sources of herbal products and pharmaceuticals. In fact, ipecac root was another source of a new phytopharmaceutical discovered from the region for the treatment of hill diarrhea. Malaria has been the single most dreadful scourge of human sufferings, killing a large part of the human population annually; thus, the likely scarcity of the source (of antimalarial drugs) on one hand and periodic restrictions on its collections and export imposed by local governments on the other, brought large investments for raising plantations of cinchona in far-off places during the latter part of the eighteenth century. This led to the establishment of a lucrative plantation industry for pharmaceutical purposes in Indonesia and India, and, later, manufacturing capacity increased in Western Europe.

Quinine is the most important alkaloid of cinchona bark, and in the form of its salts, such as sulfate, bisulfate, hydrochloride, and dihydrochloride, it is used for the prevention and treatment of malaria. In cases of malignant malaria, in which the disease was (later) found resistant to synthetic antimalarials, quinine hydrochloride was given as an intravenous injection slowly for rapid cure. In combination with other compounds, quinine hydrochloride is also used as a sclerosing agent in the treatment of varicose veins and internal hemorrhoids. Quinine sulfate has long been used in preparations made in Europe for the treatment of night cramps, but has largely been given up now due to occasional complications causing thrombocytopenia (Pin, 1998). Quinine water (100 ppm) is used as a gargle to cure sore throats and foul smells.

Cinchona bark has astringent and bitter tonic properties. Thus, quinine hydrochloride and quinine sulfate have been used extensively as bitter substances in soft drinks, alcoholic bitters, and liquors. In France, ground cinchona bark was used chiefly as a raw material in the manufacture of aperitifs and restorative liquors. Quinine has also been used as an abortifacient (Morton, 1977). Quinine is bacteriostatic and a highly active in vitro compound against protozoa. It also inhibits the fermentation of yeast.

Another chemical from cinchona bark, quinidine, has been found of utility as a cardiac depressant, and its sulfate form is used for the treatment of cardiac arrhythmias. Quinine can be catalytically converted into quinidine, and it was thus used as a major source for quinidine production.

Evolutionary History

A widely circulated legend of the discovery of cinchona bark by European settlers is attributed to the first countess of Cinchon, the wife of the viceroy of Peru, who fell sick in 1638. When the governor of Loja, Ecuador, heard of it, he sent a packet of "quina-quina" bark that cured her completely. On her return to Spain, the countess is said to have liberally distributed the cure, and for this reason the bark powder during early times was also called "countessa powder." Haggis (1941), through painstaking investigations, discovered that the first countess of Cinchon died in Spain three years before her husband was appointed as viceroy of Peru. The second countess remained quite healthy until she died in Colombia, without ever returning to Spain. Keeble (1997) further discredited this legendary theory. He stated that the first authentic use of cinchona bark against malaria in Europe was made around the year 1630; it arrived in trade in Europe around 1643, and came to England in 1655 (Holmes, 1930). It entered the pharmacopoeia in 1677 and consequently received official recognition. Yet the name of the countess of Cinchon lingers on because the famous botanist Carolus Linnaeus in 1747 named the fever bark tree as *Cinchona officinalis*.

Twenty-three distinct and validated taxa now belong to the genus *Cinchona*, ranging from large shrubs to small trees that are distributed from an elevation of 800 m to 3,000 m above mean sea level (MSL) all over the slopes of the Andes rain forests. Chemical studies in these taxa, including their innumerable natural hybrids, have shown that only a few species, such as *C. calisaya, C. ledgeriana, C. officinalis,* and *C. pubescens (C. succirubra),* are rich in quinine and cinchonine alkaloids. The range of quinine in bark (of twigs, trunk, and root) is 2 to 8 percent. Before World War II, almost 90 percent of the world supply came from Java (Indonesia), which produced and exported cinchona bark and its alkaloids for a value ranging from 16 to 22 million pounds (UK) annually. India was then the second most important source with annual production of 25 tonnes quinine salts, though its exports were relatively meagre; the average Indian export of quinine salts was worth around Rs 10 million annually. The Belgians, meanwhile, developed huge plantations in Zaire (now known as the Democratic Republic of the Congo) in Africa, which surpassed Indian production; being richer in quinine, Zaire established its place next to Indonesia in trade and reaped huge profits during World War II, when supplies of quinine were short. It may be recalled that Bernardino Gomes, of Portugal, produced a crude crystalline mixture of alkaloids from the bark, and later the French chemists P. J. Pettetier and J. V. Caventon isolated and purified quinine crystals from the bark in 1820. Quinidine was discovered in 1833 and cinchonidine in 1847.

Cinchona Travels Eastward to Indonesia

As soon as cinchona bark was established as the source of quinine alkaloids that were most useful in the control of malaria, the home governments in Peru, Bolivia, Columbia, and Ecuador placed embargoes on the export of seed, seedlings, and other living plant materials. However, the Dutch were successful in sending plants to Justus Hasskarl, superintendent of the Botanical Garden in Java. It is said that Miller, a Dutch tourist (according to his passport) went sightseeing through cinchona territory during 1852, quietly collecting seed as souvenirs. When he could not freely move into provinces, where cinchona grew abundantly, he approached the governor of the province with certain proposals, which were seemingly rejected. Undaunted, he pursued his objective through shady characters in the administration and succeeded in exchanging hundreds of seedlings and a bag full of live seed for a fabulous sum of money. Arriving back in Java, Miller (Hasskarl, in all probability) was honored with a knighthood. Interestingly, when a German newspaper uncovered the unscrupulous event, the Dutch government quickly denied it. Burkill (1935) stated that the first individual plant to reach Europe, however, was of *Cinchona calisaya* from seed collected during James Weddell's first journey to Bolivia in 1846. It was raised in Paris (France) and handed over to botonist Hugo De Vries in 1851. It was planted at Buitenzoirg, which is only at 280 m elevation in the tropics, and died there, but not until a cutting was taken from it. This cutting thrived in the mountain garden of Tjibodas, Indonsesia (1,300 m), where it was maintained. More plants followed from Holland, raised from part of the seed deposited from Hasskarl's collection there. By 1860, a million small healthy cinchona trees were reported growing in Java.

Saha (1972) recorded an interesting instance of an English cinchona bark trader, Charles Ledger, living near Lake Titicaca. In 1865, Ledger managed to collect the seed of a particular tree (possibly yielding potent bark) belonging to *Cinchona ledgeriana* from the highlands of Coroico, Bolivia. He sent the seed to his brother in London. The British government refused to buy the seed, but two Dutch consular officers bought them and sent them to Java. In Java, where cinchona cultivation had already begun, these new seeds produced 20,000 plants. It proved to be a richer source of quinine and laid the foundation of the Dutch monopoly on cinchona trade, which the Netherlands has retained ever since. Burkill (1935) inferred that the best species of cinchona thus reached Europe by accident, for Ledger's interest in the subject was secondary; he had not expected that the seeds which he sold were to those who were keen to invest in raising commercial plantations. By 1918, the production of cinchona bark based on the choice taxa of *C. ledgeriana* in Java had reached its zenith. The bark was brought to the Amsterdam factory for extraction and production of quinine salts. This facilitated the Dutch to reap huge profits annually. It displaced from market countries such as Bolivia, Peru, and Colombia. These countries maintained varying genetic material in natural reserves.

Introduction into India and Sri Lanka

In 1856, British botanist John Forbes Royle suggested to the (then) East India Company that introduction of cinchona into India should be taken up without delay. The company assigned an expedition to the Andes and chose Clements R. Markham for the job. Markham organized the expedition on a bigger and more systematic scale (Watts, 1893). He divided the work to three parties: one led by himself with the purpose of getting yellow bark in the forests of Bolivia and southern Peru. He traveled to the interiors, making tedious journeys and hiring local hands. He collected a few hundred seedlings of yellow bark and red bark tree *(C. pubescens),* which he brought to India in October 1859. The second party was entrusted to Richard Spruce and Cross to search for red bark and allies from eastern slopes of Ecuador. The third under Woolcock Pritchett was to look for grey bark from central Peru. All were to send collections to the Royal Botanic Gardens in Kew, United Kingdom. Markham entrusted his collection to W. G. McIvor, the then superintendent of Government Botanical Garden at Ootacamund (Tamil Nadu). Swamy (1953) discovered from records that all of the plants brought by Markham, when transferred to field at Dodabetta (Nilgiri hills) failed to survive, possibly because they could not stand the long sea journey through Panama. However, the seed gave encouraging results. Markham's attempt to raise seedlings in cool climate of Nilgiri hills proved successful. However, Spruce and Cross collected plants and seed *(C. pubescens)* which reached India safely. McIvor, soon after, received a fresh consignment of seed from Pritchett, surprisingly by mail, which arrived safely in viable condition. The seedlings established well in the field. It goes to the credit of McIvor for laying the foundation of commercial plantation in India. McIvor sent a small number of plants to Thomas Anderson, superintendent of the Royal Botanic Garden, Calcutta, with a view to establish cinchona in the state of Sikkim. He also sent a small quantity of seed to the Royal Botanic Garden at Peradeniya in Sri Lanka for trials at Hakgala Botanical Garden. This was the start of cinchona plantations in Sri Lanka, then under British rule. By 1881, 138 tonnes of bark was exported to England, and the quantity rose to 690 tonnes by 1885 (Watts, 1893).

Anderson raised seedlings from McIvor's seed near Sikkim during 1861, but his attempts to establish these seedlings over the Darjeeling hills in 1863 were unsuccessful due to the heavy mist over these hills. The following year, the same trial was repeated at Mungpoo, India, where it proved successful; thus, Mungpoo tract was chosen for its commercial cultivation. Later in 1864, the government of India sent Anderson to Java to familarize himself with the Dutch mode of cultivation and to carry plants to India, which the Dutch government had generously offered. He brought 50 plants of *C. calisaya,* 284 of *C. pahudiana,* and only four of *C. lancifolia,* and handed them over to McIvor at Ootacamund, from where he took 193 plants of *C. pubescens* to cultivate in Bengal (Nandi, 1993). After Anderson's death, the charge was taken over by C. B. Clarke in 1870 and then by

George King, both distinguished botanists of their time who contributed enormously to the establishment and expansion of cinchona plantations in Bengal. By 1871, it was clearly established that cinchona could be grown as an industrial crop over the high-rainfall-receiving slopes of the Eastern Himalayan ranges in Bengal. Cinchona needed rich soil for which recently cleared forest lands were preferred. The growing trees also needed partly shady locations. Thus, a second plantation was established at Munsung near Sikkim in 1890, a third plantation at Rango near the Indo-Bhutan border in 1938, and a fourth one at Latpanchor in 1943, adjacent to Mungpoo plantation. By 1954, the total area under cinchona was around 3,000 ha, but this was brought down to 2,100 ha in 1955 because of a slump in quinine market. It was brought down further to 1,820 ha in 1970 (Saha, 1972; Nandi, 1993).

About 230 ha were under cinchona in the hills of South India (Tamil Nadu) by 1866, viz, 70 ha at Dodabetta and 160 ha at Naduvattam (Swamy, 1953). By 1931, this had expanded to over 860 ha, and rose to 2,316 ha in 1970. The Annamalais held plantations in 1,875 ha, and a smaller area of 441 ha were in Nilgiri; 80 percent of these trees were of *C. ledgeriana* (Rao, 1974). The Annamalais plantations are located 80 km from the town of Pollachi in Tamil Nadu, along the border of Kerala State, and are divided into eight divisions for the purpose of management. The Nilgiri plantations have two divisions, viz., Dodabetta and Devashola, located at 14 and 22 km from Ootacamund town respectively.

In 1867, a cinchona plantation was also opened at Nunklow, Assam, on the eastern slopes of the Khasi hills. The trees thrived but scarcity of farm labor in this tract forced abandonment of this site. Cultivation was also attempted at Mahabaleshwar, Maharashtra, but it failed due to excessive atmospheric moisture and greater variation between day and night temperatures. Cinchona's cultivation was also given patient trials in the northwestern hills of India, but the young trees perished due to frost and low winter temperatures. Attempts to encourage private planters to take up cultivation picked up in South India, but the enthusiasm did not survive long because of a slump in the market (1920) and competition with tea and coffee plantations, which gave consistent income and had expanding markets. Cinchona plantations in India were promoted entirely as a state enterprise, i.e., as part of a public welfare measure, in spite of the periodic dips in demand and price of end products in the world market.

Other Major Producing Countries

In the Philippines, cinchona plantations were introduced in 1927, and covered 1,874 ha on Mount Kitanglad, Bukidnon. A census of the plantations during the late 1980s revealed a holding of 2.5 million trees, comprising seven species. Burundi was another Asian country producing bark in commercial quantity for export, but it has since ceased production. China established plantations in its southern provinces and Formosa (Taiwan) mainly to meet domestic demand. The plantations in Zaire (Congo) and Guatemala were raised by Belgium and the United States respectively during this period. The Japanese occupation of Java in 1942 disrupted supplies, causing scarcity. This led to the establishment of new plantations in several African countries, such as Tanzania, Kenya, Cameroon, and Rwanda. It also revived the interest of South American governments such as those in Peru, Columbia, Bolivia, Costa Rica, and Ecuador, where cinchona was brought under commercial plantations. Among these, the Guatemalan venture was the most modern because it employed advanced plantation and production technology. It was a joint venture between the United States government, Guatemalan coffee planters, and a pharmaceutical company (Merck). They mass produced vegetative propagules from selected clones and used improved cultural practices for high (bark) yield and superior content (quinine) to compete in the world market. The plantation covered 400 hectares with around 1.75 million trees by mid 1948. It continued production despite the excessive cyclic nature of the market (Popenoe, 1949).

CHARACTERISTICS AND CHEMISTRY OF PLANTATION SPECIES

The genus *Cinchona* is comprised of evergreen shrubs and small trees growing up to 20 m tall and between 15 and 30 cm in diameter, mainly over the eastern and northern slopes of the Andes from 19°S to 10°N, thus extending on either side of the equator. *Cinchona* occur from 800 m in the premountain tract and rise to 3,000 m above MSL in Bolivia, Columbia, Ecuador, and Peru. One species *(C. pubescens)* extends further into northern Venezuela. Taxonomists in the past have placed various numbers of species (30, 40, to 65) in this genus, possibly due to the existence of interspecific forms in the natural population. However, after a through study of Andean Rubiaceae, Anderson (1998) recognized only 23 valid species. Anderson (1995b) based his inference on cladistic analysis of structural data comprising 48 characters. He transferred a few species to allied genera, created a new genera under the name *Cinchonopsis,* and created a new taxa, *C. amazonica,* making the genus monophyletic in origin. He came to believe that although generic endemism is low in the Andes, cinchona falls among five subendemic genera because of large species endemism (recorded around 59 percent in the tropical Andes). The ancestors probably occupied low mountain forests of the Andes, and later (evolved) species may have arisen through adaptation to mountain conditions (Anderson, 1995a). These plants bear opposite elliptical to ovate lanceolate leaves with the entire margin. The flowers are small, pink, cream to brown in color, with small lobed calyx united at the base; the corolla is tubular, made up of five spreading lobes. The flowers emit a delightful fragrance. Fruit is an oblong capsule, containing 40 to 50 winged seeds. The Andes has given some of the most famous industrially important plants of the tropical rainforests of South America. The basic chromosome number (x) is 17 in all cultivated species, including the four species grown in India (Mathews and Philips, 1979). Several species are polyploids ($2n = 68$). Some hybrids, such as *"Hybrida"* and *"Robusta,"* have proved valuable in the plantations because they are hardier and invariably produce higher quinine content than either of the parents. Although a large number of cinchona species have been tested for their yielding ability in the plantation sector, commercial supplies of bark are obtained from the four species listed earlier. Of these, *C. ledgeriana* and *C. officinalis* are species of choice for the plantation industry because of their rich quinine content.

Taxonomy: Diversity Analysis of Source Species

C. calisaya *Wedd (Yellow Bark)*

It is a large bushy tree with a straight trunk, occurring in the lower reaches of Bolivia and southeastern Peru. It prefers lower elevations (400-1,000 m). Its leaves are thick, oblong to lanceolate with smooth surfaces. Inflorescence is a large panicle of numerous pale pink flowers. The capsule is 8 to 17 mm long and oblong in shape. The bark is thick (2-5 mm), has a greyish outer layer with a broad longitudinal, and a few transverse fissures peeling off in places. The trunk bark contains 3.89 to 7.24 percent of the total alkaloids, of which quinine content varies from 0.78 to 5.57 percent (Anonymous, 1992). The species show variable quinine content under plantation. It was initially introduced commercially in Sikkim and the Moyar valley of Nilgiri, but was replaced by the more consistent quinine-yielding *C. ledgeriana.*

C. ledgeriana *Moons (Yellow Bark)*

It is a weak, highly branched, fast-growing tree, attaining a height of 6 to 16 m at maturity. The species thrive between 1,000 and 1,900 m elevation in the hills of Darjeeling. Leaves are light green, elliptical, acuminate, with a small curve in the axil. Flowers are small, pale yellow, and appear from May to October. The capsule is oval-lanceolate and 15 to 19 mm long. The bark is similar to *C. calisaya;* it is thick (2-5 mm), but cracks are more numerous and less deep. The average total alkaloid content of the root and trunk (10-12 years age) is 7.21 and 6.01 percent, of which quinine is 5.4 and 1.98 percent respectively (Council of Scientific and Industrial Research, 1992). It is the richest source of quinine, and individual trees containing as high as 14 percent of quinine have been located in plantations (Chatterjee, 1982). The species is very exacting in its climatic requirement, and is a parent stock of major plantations in India. Under less favorable conditions, a hybrid, *C. ledgeriana* × *C. pubescens* called *'Hybrida'* is preferred. The hybrid has lower quinine content but produces robust growth (Gupta, 1980). Bark yields of 6 to 6.25 tonnes per ha are obtained, which contain 2 to 2.5 percent quinine (Chatterjee, 1982).

C. officinalis *Linn. (Pale or Crown Bark)*

It is a weak, struggling, slender, slow-growing tree, 6 to 10 m tall with dark green leaves, and prefers a cooler climate of 1,200 to 2,000 m elevation in the Nilgiri. Leaves are small, smooth, ovate-lanceolate, 4 × 10 cm in dimension with reddish petioles. The flowers are deep pink to rosy, 1.4-1.6 cm long, borne in small terminal panicles. The capsule is ovoid-oblong, 1.5-2.0 cm long. Bark is rough and brown (yellow within), around 1.5 mm thick; the outer surface has numerous transverse cracks with recurved edges. On average, the trunk bark yields 4 to 6 percent total alkaloids, up to half of which is quinine; the average in India however is low, around 1.5-2.0 percent (Chatterjee et al., 1988). Thus, a hybrid between *C. officinalis* × *C. pubescens* called 'Robusta' is preferred in plantations. This is a hardy tree, adapted to a wide range of elevation (1,200-3,000 m) and temperature regimes (Gupta, 1980); its average quinine content is 2.0 percent in the trunk bark (Chatterjee et al., 1988). Another hybrid (*C. officinalis* × *C. ledgeriana*) is also grown in Mungpoo (Directorate of Cinchona and Other Medicinal Plants, 1994).

C. pubescens *Vahl. (syn. C. succirubra Pav. ex klotzsch) (Red Bark)*

It is a hardy tree, native to Peru and Ecuador, producing a straight trunk and vigorous growth. It attains a height of 18-20 m and grows between 1,200 and 2,000 m elevations in the cooler climate of Annamalai's hills (Tamil Nadu). The species possesses the remarkable ability to withstand

both high humidity and drought conditions. Leaves are very large (40-50 ×
30-40 cm) in dimension, thin, light-green, and elliptical. Flowers are rosy
pink, and are produced almost round the year; 1-1.2 cm long, the upper surface of the corolla is white with pink stripes. The capsule is oblong, 2-3 cm
long. Bark is dull brown, 2-6 mm thick, with well-marked longitudinal
wrinkles and a few transverse cracks. It is rich in cinchonine but poor in quinine content (Chatterjee, 1993; Nandi, 1993). The bark of the root, trunk,
and twigs has 7.21, 6.09, and 4.00 total alkaloids, respectively, of which
quinine is 0.76-1.42, 1.1-1.74, and 0.8-1.16 percent respectively (Council
of Scientific and Industrial Research, 1992). A new alkaloid, quinamine, is
found in the bark. The species is used as a root stock for the grafting of *C.
ledgeriana* in Indonesia (Gupta, 1980) because of its hardy and vigorously
growing nature at lower elevations in Indonesia (Council of Scientific and
Industrial Research, 1992).

Chemistry

Cinchona bark contains 6 to 10 percent of the total quinoline alkaloids.
The alkaloid consists mainly of quinine, quinidine, cinchonine, and cinchonidine in addition to 30 other minor bases related to quinine. These alkaloids contain quinoline and quinuclidine rings with a vinyl group attached
to it. In addition to alkaloids, the bark has coloring matter (up to 10 percent), flavonoids, an essential oil, and polyphenols. The alkaloids exist
chiefly as salts of quininic and cinchotanic acids, and their relative concentrations vary in different species. The alkaloids are formed during descent
of the sap and therefore their percentage content is lowest in the twigs, high
in trunk (bark), and maximum in the root (bark). Of these, the collar portion
(30 to 45 cm in length, near the base) is the richest. The alkaloid content in
trees increases with age (eight-12 years), depending upon species. Among
these plantation species, quinine is the major alkaloid of *C. ledgeriana,
C. calisaya,* and *C. officinalis. Cinchona pubescens* has a larger proportion
of cinchonine (3.3 to 3.4 percent) out of higher (6.0 to 8.2 percent) total alkaloids. The cinchonidine content is the highest (3.7 to 4.9 percent) in the
bark of *C. kartamanahs* at 12 plus years and is also high (2.1-2.2 percent) in
the bark of *Hybrida* at around six to eight years of age (Dayrit, 1994). The
bark has high (3 to 4 percent) ash content and its quantitative estimation can
identify the source species. Doraiswami and Venkatraman (1982) reported
maximum ash content in crown bark grown at 2,000-2,500 m elevation,
lesser in red bark grown at 1,500-2,000 m, and the least in yellow bark
grown at 1,000-1,500 m elevation in India.

Quinine (m.p. 174.4-175°C) and quinidine (m.p. 173.5°C) contain a methoxy group and are stereoisomers. Thus, quinidine is dextrorotatory isomer of quinine. They show blue fluorescence with oxygenated acids such as sulfuric acid in filtered ultra-violet light. Cinchonine (m.p. 264°C) and cinchonidine (m.p. 204.5°C) do not possess a methoxy group and therefore do not show fluorescence. Quinine is isolated from the total alkaloids as quinine sulfate. It is a white, crystalline, odorless compound that is highly bitter in taste. It is sparingly soluble in water, but is highly soluble in organic solvents. Besides quinine salts, cinchona febrifuge (powder), which contains a mixture of alkaloids leftover after the extraction of quinine, is marketed. Similarly, a standard mixture of cinchonine, cinchonidine, and quinine is sold as "Totaquinine," and is also prescribed for treatment of malaria. Quinidine is present in small quantities (0.2 percent) in most cinchona barks, but is relatively higher in *C. calisaya* and is maximum in *C. tayansis*. It is commercially produced by the chemical conversion of quinine through oxidation, and its percentage of commercial conversion depends upon factors such as the oxidation agent, catalysts, and reducing agents employed. A rapid, accurate, and inexpensive quantitative high performance thin layer chromatography (HPTLC)/densitometry method is available for the simultaneous determination of four alkaloids in the bark (Dayrit, 1994).

The leaves contain up to 1 percent of total alkaloids; younger leaves have higher content. Analysis of *C. ledgeriana* leaves revealed five monomeric indole alkaloids (including quinomine, aricine, and 3-epi-quinine) and seven quasidimeric indole alkaloids. Analysis of *C. ledgeriana* leaves grown in Kenya and Zaire, of *C. pubescens* from Thailand, and *Hybrida* has shown that the alkaloid content was very low (c.f., bark) and no quinoline group of alkaloids was present to merit commercial extraction (Keene, et al., 1983).

Toxicology

It should be noted that quinine as well as ground cinchona bark have been reported (in rare cases) to cause urticaria, contact dermatitis, and other hypersensitivity reactions in human beings. The ingestion of these alkaloids can result in the chemical disorder known as "cinchonism." The disorder is characterized by severe headache, abdominal pain, convulsions, visual disturbances, blindness, and auditory disturbances (such as ringing in the ears), paralysis, and even collapse (Leung, 1980). Quinidine and related alkaloids are reported to be absorbed from gastro-intestinal tract and a single 2 to 8 g oral dose of quinidine may be fatal for an adult (Anonymous, 1993). The use of quinine is discouraged during pregnancy due to fetal and abortifacient effects.

EXTRACTION, PRODUCTION, AND TRADE

In India, commercial extraction of quinine alkaloids began in 1871 when quinine factories were established at Naduvattam (Annamalai district, Tamil Nadu) and Mungpoo (Darjeeling district, West Bengal). These factories produced quinine sulfate in 1887 for the first time; later they produced quinine hydrochloride and quinine ethyl carbonate (tasteless quinine). Their capacities were enlarged in 1905 to 27 tonnes (Mungpoo) and 20 tonnes (Naduvattam). The Naduvattam factory was closed due to fire, and a new factory was established in 1965 with enlarged capacity. However, these factories never ran on full capacity because of a shortage of bark.

Extraction

The Indian units have used century-old "open type" extraction plants, using mineral oil as solvent. For this purpose, moistened bark is ground to a fine powder (60 mesh) in disintegrators and mixed with slaked lime containing more than 60 percent calcium hydrochloride. Mixing is done thoroughly in mechanical mixers and kept as such for 24 hours. It is charged into extractors, where sodium hydrochloride solution is added and stirred well. This mixture is kept in extractors overnight and extracted with a mineral solvent at 88°C using live steam. The hot extracted oil, containing the alkaloids, is shaken thoroughly with diluted sulfuric acid in lead-lined tanks for transferring the alkaloids present in the oil to the acid. The oil from the mother liquor is separated and recycled, whereas the sulfuric acid layer containing the alkaloids is boiled and filtered. The hot and clear acidic liquor is neutralized with a hot sodium hydroxide solution at 5.5 pH and cooled. On cooling, crude quinine sulfate (gray in color) crystallizes out and is centrifuged. It is refined by boiling with water and activated carbon to obtain pure quinine sulfate (Doraiswami and Venkatraman, 1982). The remaining alkaloids are precipitated by making the mother liquor alkaline. Individual alkaloids, such as quinidine, are crystallized through selective solvent extraction. The production units in India were revamped in the mid-1970s by introducing mechanization and replacing mineral oil with a close circuit solvent extraction that uses toluene. This has improved efficiency and recovery of alkaloids. In European countries, further advance has taken place and they now use counter-current containers for efficiency and speed.

Manufacturing Capacities

European countries produce the bulk of the world's supply of quinine salts and have global marketing network. They control market price. The

Boehringer Mannhein Plant in Germany is controlled by a multinational company and has 125 tonnes capacity per annum. The Amsterdam Chemi Farma of Holland used to control the Java (Indonesian) plantation and has a 100 tonne capacity; after nationalization, the company has a collaboration with the government of Indonesia to produce the end products. It has attempted to regain its preeminent position. The Lake and Cruickshank of England, Societe Chimie Pointel Girard of France, and Bucher and Company of Germany have capacities of 35, 30, and 50 tonnes respectively per annum, and depend mainly upon purchases of bark from African and South American countries. India and Guatemala have their own extraction plants. In additon, Bolivia, Ecuador, and Columbia have erected smaller extraction plants with annual capacities of 2, 5, and 1.5 tonnes respectively (Saha, 1972; International Trade Centre, 1982).

Global Trade

The world production of cinchona bark fluctuates between 5,000 and 10,000 tonnes annually. This estimate is based on a consumption of 300 to 500 tonnes of cinchona alkaloids. Production of quinine in 1981 by the major manufacturing units was estimated at 265 tonnes (International Trade Centre, 1982).

Quality Standards

There is no grading system for cinchona bark, although classification is made into industrial bark which is in small pieces and "druggist quills" which are up to 30 cm long, 1 to 8 cm wide, and 2 to 6 mm thick. Industrial bark is used for alkaloid extraction and represents the bulk of demand for cinchona, whereas "druggist quills" are used in gelatine preparation (International Trade Centre, 1982).

In trade, commercial specifications include the country of origin, color of the inner side of the bark, moisture content, total alkaloid content, quinine content, quinidine content, and quinine sulfate content. Of these, quinine content is given as the percentage content of quinine as anhydrous alkaloid (QAA) and quinine sulfate content as SQ-7 (quinine sulfate with 7 water molecules) or SQ-2 (quinine sulfate with 2 water molecules). In practice, the QAA and SQ are not usually provided by suppliers in developing countries, but are determined by potential buyers from samples submitted by suppliers. Most pharmacopoeias prescribe a minimum total alkaloid of 5 to 7 percent in the bark.

Cinchona bark in pieces or chips is packed in jute bags of 50-60 kg, whereas ground bark is packed in polyethylene-lined jute bags or polyethylene sacks of 100 kg capacity. Before chipping or grinding, the bark must be dried to a minimum of 10 percent moisture content.

Both red and yellow barks have been approved for use in beverages, with a restriction that the total cinchona alkaloids should not exceed 83 parts per million (0.0083 percent) in the finished beverage offered for sale.

The bark of *Remijia pedunculata* (Rubiaceae) from Colombia also contains quinine and quinidine alkaloids, and is used as a source of quinidine. This can be distinguished from cinchona bark by a simple test; the powder of the latter, when heated in a dry test tube, preferably with a little glacial acetic acid, produces reddish fumes, which are condensed in the upper part of the test tube.

Periodic Market Upheavals

The market during the nineteenth century was driven by the European colonial powers, which were expanding their empires into humid tropical regions where malaria was endemic. Cinchona trees require ten years growth for commercial viability. Raising of plantations or their expansion appeared risky because of possible sudden surges or dips in demand. This, in turn, was the cause of anxiety among planters, and radically affected the fortunes of manufacturers and traders. The cyclic nature of demand in cinchona products have periodically landed violent fluctuations in global price, affecting plantations in the following years. The first major slump was recorded in 1920 due to the large production of quinine salts in Indonesia. India too recorded increasing production, from 12 tonnes in 1915 to 20 tonnes in 1920. This resulted in a drastic cut in imports. Regulated release of supplies from Indonesia later led to an increase in demand. In addition, supplies from natural sources had decreased significantly. Soon after, new plantations appeared in the Philippines, Thailand, Taiwan, Zaire, and Guatemala. But the occupation of Java (Indonesia) during World War II cut off supplies and prices suddenly shot up, forcing many European countries to invest in new plantations in their African colonies. During the early part of 1950s, most governments in tropical countries launched public health measures to eradicate mosquito breeding grounds through liberal use of DDT, as mosquitoes were identified as vectors of malarial parasites. These mosquito control measures met success and consequently reduced demand for antimalarial drugs. By this time, synthetic drugs began appearing in the market but did not make impact on demand immediately. However, a slump in demand occurred again from 1952 to 1962, and the international price of quinine fell to half of the sale price in India, which lost its export market. This led Sri Lanka, Burundi, Rwanda, and some other Africans countries to abandon plantations and even immature trees were cut and removed. The government of India lowered their plantation targets. A pharmaceutical inquiry committee, set up in 1953 by the Indian government, recommended the replacement of existing trees with selected clones of high quinine content. Investments were also made on research in plantation methods for lowering the cost of raising and managing the plantations and improving processing. Systematic research efforts in cinchona plantation technology in India thus began only after this period. Political turmoil occurred in Indonesia in 1963. Reports started appearing in 1964 of the emergence of a malignant malarial strain in several countries that was resistant to synthetics but amenable to natural quinine. The alkaloid quinidine also found use in cardiac ailments. These developments led to a revival of the market and fueled demand. The war in Vietnam also increased demand for natural quinine. Indian quinine was again sought after in trade, and its export rose significantly from 1966-1967 to 1971-1972. The price of quinine increased from Rs 110 to Rs 525 per kg by the close of 1965. This led to the rehabilitation of old plantations. Expansions were planned both in West Bengal and Tamil Nadu in the following years. The demand for quinine sulfate in soft drinks grew fast in Western countries. In general, the price of quinine rose to Rs 1,088 per kg in 1986 and reached Rs 4,067 per kg three years later. However, synthetic antimalarials had a much greater impact on the market during this period and lowered the demand of quinine sulfate as drug. In fact, synthetics such as choloroquine and primaquine showed the advantage of killing both gametes and the diploid stage of plasmodium (whereas quinine kills only the asexual form of parasite), and the consequent effectiveness of reducing the chances of a recurrence of malaria in the patient. Bruncton (1995) estimated that late in the 1980s, nearly half of cinchona bark deliveries were diverted to the food industry for the production of tonic bitters and additives, and between 30 to 50 percent were converted into quinidine, and (therefore) only a small proportion was channelized for making antimalarial drugs. A new drug, lidocaine (procainamide hydrochloride), came to market to compete with quinidine for its antiarrhymric property. The liquor industry too got a new source of bitter in quassia bark (stem wood of *Picrasma excelsa* from Jamaica [West Indies]) which largely replaced quinine from the soft drink industry. The quassia bark contains quassin and allied quassinoids, which are 50 to 60 times more bitter than quinine sulfate. The decline of the market began around 1982, and it gradually reached a level of collapse by the early 1990s. During this period, the Chinese bought huge quantities for use as abortifacients in their population control program, sustaining the market for a few years, but the program failed and was given up by 1989.

Soon, the Amsterdam-based factory collapsed and liquidated its assets. In India, the Nilgiri and Annamalai plantations were given back to state forest department, whereas the plantation program in Darjeeling was cut to a minimal output. The plantation land in Indonesia also lost ground to other crops due to continuing losses. However, quinidine still continues to find a market in North America.

RESEARCH: DEVELOPMENT OF IMPROVED CULTIVATION PRACTICES

The English botonist Sir William Jackson Hooker wrote his dissertation on cinchona in 1839. He brought out that the cinchona tree stands to coppicing (from base) well. The tree generated new growth and this could be harvested again after six to ten years. It was also discovered that cut and regrown cinchona has higher levels of effective alkaloids in its bark. This method of harvesting was adopted by plantations and has remained in practice ever since (Brockway, 1979). Moens, a Dutch biochemist working in Java (Indonesia), identified chemical traces in the progeny of *C. ledgeriana.* These were named cv. *chinidonifera* and cv. *cinchonidifera* based on the presence of cinchonine and cinchonidine alkaloids in their barks, respectively. This was a landmark development in recognizing intraspecific chemical taxa in medicinal plants research (Teteny, 1970). The Dutch in Indonesia devised a program on selection and clonal propagation of vigorously growing high assay trees which resulted in uniformly high alkaloid extraction. The work was done in the corporate sector and was never published. It was known that cinchona growing over moist substratum generated substantial hydrostatic pressure in the root system, and this was dependent on a continuing supply from the soil of magnesium and orthophosphate. These roots could not absorb water that is bound to soil colloids. The young leaves and buds require daily exposure to rain or mist to soften or remove the natural varnishlike coating that inhibits or even deforms new growth, if allowed to dry (Dawson, 1991). In Java, the richest trees (parent stocks) were never allowed to seed but were grafted to other seedlings of *C. pubescens.* The high-yielding trees were thus conserved and nurtured on plantation scale. However, when these trees of high-yielding stock were planted in India, the quinine yield was lower (half or even less) than in Java. It was suggested that volcanoic soils may influence productivity. Although *C. ledgeriana* has been the preferred choice in India and Indonesia, it was found to suffer from low winter hardiness and poor rooting ability.

Breeding and Selection

Systematic breeding has not been reported, although hybrids such as *"Hybrida"* and *"Robusta"* have performed better than either parents. The *C. ledgeriana*-based hybrids have given significantly higher yield over others. Individual trees of *C. ledgeriana* have been identified in Annamalai's plantation (and also in Darjeeling) as having 14 percent and more of quinine content. These selected trees, as in Java, were used as parental material in new plantations to improve their yielding ability (Rao, 1974). Clonal selection provided an easy method of raising the output remarkably in Indonesia, Zaire (now the Democratic Republic of the Congo), and Guatemala. Work in Balai Penelitiam (Indonesia) on correlations between morphological characters and production capacity of cinchona clones has shown that the quinine percentage in the bark and total (alkaloid) content per ring of six-year-old tree were closely correlated with the yield of quinine. It was suggested that this be used as the main criterion for selecting cinchona mother trees (Sukarja, 1997). Among the selected clones planted in field at a site 1,450 meters above MSL, based on the rate of growth (girth) and yield of trees, "clone cib-5" produced maximum yield (Sukarja and Munir Supralo, 1977). Polyploids have been evolved in three commercial species, *C. ledgeriana, C. calisaya,* and *C. pubescens,* through colchine treatment in Guatemala. The poyploids showed little visible differences and gave marginally higher quinine content over the control and were given up.

Soil and Climate

Cinchona needs well-drained, deep, fertile organic soils over open gravelly substratum, rich in humus, and of pH ranging from 5 to 6.5. It thrives over gentle slopes, and can grow also on steep slopes, but protection against soil erosion is needed. The Darjeeling hills have clay-loam acidic soil rich in organic matter (organic carbon 1.5 to 3.34) but deficient in phosphorous and potash. The available levels of N, P, and K of the land chosen for planting were 315-414, 1.01-21.00, and 0.8-13.7 mg per g of soil, respectively (Datta et al., 1990). Cinchona prefers a cooler climate that has a well distributed high annual rainfall of 190 to 500 cm; vigorous growth occurs when showers alternate with sunshine frequently. Chatterjee (1993) opined that high-quinine-yielding *C. ledgeriana* gave relatively low content over the hills of South India in comparison to Darjeeling (Eastern Himalayas). This may possibly be because of occasional high temperature and drought in the southern hills where soil is semi-alluvial red loam containing medium high organic carbon. A temperature of 21 to 24°C is ideal for luxuriant growth; it should not go below 8°C in winter and above 30°C in summer. The range of variation in day and night temperature should be minimal during the growing period. All species are susceptible to frost as well as subsoil water logging.

Raising Plantation: Nursery Practices

Cinchona trees produce a large quantity of fruits over clustered panicles. These fruits are collected from selected plus trees during November-December. The seeds are separated, cleaned, and stored in a dry place. A gram contains 300 to 400 seeds. The seeds begin to lose viability after six to eight weeks and lose it completely in one year. For raising seedlings, raised beds of 3.6 × 1.2 m size are laid out. Leaf compost and manure are thoroughly mixed in the soil. About 25 to 50 g seed is thickly broadcast during February-March (Darjeeling) on the surface of the bed, covered with a thin layer (0.2 to 0.5 cm) of soil under partial shade provided by thatched roofing. The sown beds are kept moist by sprinkling water through a fine nozzle. The seeds germinate in 20 to 40 days, with germination varying from 50 to 85 percent. It is estimated that 400 g seed can produce more than 1.0 lakh seedlings in the nursery. In three months, the seedlings grow a pair of leaves when they are transferred to another nursery in rows at 5 × 5 cm spacing. In five months, the seedlings grow 10 cm tall, bearing three to four pairs of leaves. They are transferred to another nursery at 10-12 cm spacing either way; this is mainly to avoid mortality due to closer spacing. Partial shade (roofing) is removed after three months and the seedlings are allowed to harden in the open for the next four months. It thus takes 16 months to attain 20-25 cm height, at which point they are ready for field planting during the wet season (Nandi, 1993). Although several reports suggested allelopathic inibition to seed germination where cinchona is planted, Aerts et al. (1991) rejected this notion after detailed field study. They found that the actual concentration of alkaloids in plantation soil is very low, and this level plays no role in inhibiting seed germination. However, propagation through seeds is often unsatisfactory as the viability period is short and progenies of alkaloid-rich clones often produce very low alkaloid content, suggesting a high percentage of heterogeneity in the seed.

For raising pure clonal blocks, vegetative propagation of high-yielding trees is resorted to during the wet season when root initiation occurs in 40-60 days; however, this propagation method adds to the cost considerably. Further, rooting is poor (around 10 percent) in high yielding trees. Top-worked cuttings have responded to rooting better, and the rate of success is 70 to 80 percent. Patch budding is practiced on a plantation scale in Java.

Similarly, cleft grafting of *C. ledgeriana* is done over stock of *C. pubescens* or even over *C.* × *robusta* and has given up to 90 percent success. *Gootee* planting is easier and is practiced in South India (Rao and Veeraghavan, 1954). It is made on two to three-year-old branches using 2 percent indole butyric acid (IBA) in lanolin paste for 24 hours before dressing with balls of soil (Thakurta and Dutt, 1944). It produces roots in four months with 88 percent success. Nandi (1993) reported studies on top-worked primary lateral stem cuttings of *C. ledgeriana* and *C.* × *hybrida,* which were planted during April-May in an experimental nursery. The cut ends were treated with growth hormones, viz., IBA, indole acetic acid (IAA), and kitinase (KN) at different concentrations (100, 200, 300, 500, and 1,000 mg/1). The results indicated that *C.* × *hybrida* is more responsive to treatment compared to *C. ledgeriana.* Wargadipura and Sutandjona (1977) found mixing equal proportions of top soil and composted cow manure to give the best result for survival and growth of plants in a nursery in Java. Thus, patch and slit budding, cincturing, air layering, veneer grafting, and mount layering were studied and recommended for raising new plantations (Mukerjee and Chatterjee, 1960; Directorate of Cinchona and Other Medicinal Plants, 1994). Mass propagation through tissue-culture now offers a better alternative. Koblitz et al. (1983) developed a protocol of micropropagation for *C. ledgeriana* and *C. pubescens,* but they found that the transfer of rooted shoots into soil was complicated because of inadequate development of the root system. More work will be needed to exploit this avenue.

Planting, Management, and Yield

Seedlings with four to six pairs of leaves are half defoliated for reducing transpiration and planted during the early monsoon season (June in Darjeeling). This helps in improving the survival rate in the field. Pits are dug out at 1.2 × 1.2 or 2.0 × 2.0 meter spacing, depending upon the slope, and half defoliated seedlings together with soil adhering to their roots are planted in the pits and staked properly to obtain erect clear boles. In general, 3,000 plants are stocked per hectare, saving room for the planting of shade trees. The trees come to flowering after the fourth year. Debudding of trees increases laminar growth and alkaloid content (Nandi, 1993). In the first year, two to three weedings are given until November. Light to deep forking of land is done from the second year onward yearly for soil aeration. Usually, 100 kg each of N and P in the form of inorganic fertilizers per hectare are given in rows and mixed in the soil during the second and third year. Application of 10 mg/1 of Mn and Mg is reported to augment growth and alkaloid contents (Nandi and Chatterjee, 1983). The use of micronutrients, such as Mn, Fe, Mo, B, and Zn, in producing higher alkaloid content and quinine has been encouraging (Chatterjee et al., 1988). Shade is essential to the growing plants, and planting both temporary and permanent shade trees is recommended. Seed of *Crotalaria anagyroides* and *Tephrosia candida* are sown at 3 × 3 or 4 × 4 m spacing in the first year itself and these fast growing leguminous plants also fix atmospheric nitrogen and protect soil from erosion during heavy rains. The temporary shade trees are thinned out after third year to 12 × 12 m spacing and finally removed when cinchona trees attain good height. In addition, permanent shade trees like *Alnus nepalensis, Aleurites montana, Aleurites fordii,* and *Mallotus philippinensis* are planted at 12 × 12 m spacing and maintained in the plantation. A plantation should have 80 percent mature trees of cinchona and 20 percent of shade trees (Nandi and Chatterjee, 1991).

The first crop is obtained from thinning operations after the third or fourth year, when half of the growing trees are removed by uprooting and their bark is collected. A second harvest is taken after the seventh or eighth year, when only one strong shoot is allowed to grow and others are cut at base. The final harvest is taken at 12 years of age, when the remaining trees are uprooted. It may be noted that the bark of the trunk and branches has maximum alkaloid contents from the fourth to seventh year but declines thereafter. The root at final harvest has a much higher alkaloid percentage. The bark is removed after ringing the stem at 45 to 60 cm intervals from the base. The ringed material is kept for one to two days in partial shade to loosen the bark. The bark is easily stripped by knife, leaving the woody central portion intact. Extraction of bark is done during dry weather to facilitate drying of the produce. The initial drying is carried in subdued sunlight for three to four days, when 70 to 76 percent weight loss occurs, and then it is dried in the open for the next three to four days. The dry produce for storage should contain not more than 10 percent of moisture.

In Java, the bark is shaved off as near the cambium layer as possible without injuring it. In this way, the bark is removed quickly. A modified method involves stripping and mossing (covering by moss) which is said to facilitate early growth. It is found that the renewed bark is never as thick as the original, but it has high alkaloid contents.

Root bark constitutes 33 to 40 percent of the total produce in *C. ledgeriana.* It is superior in quinine content. Bark yield from a well managed plantation at Darjeeling at 1,000-1,200 m elevation ranges from 3,900 to 5,000 kg, and of *C. × robusta* grown at 1,300-1,500 m elevation range from 2,900 to 3,250 kg per hectare (Chatterjee et al., 1988). Growth of *C. × hybrida* was better than the parent species *(C. ledgeriana)* and this is reflected in its high bark yield and quinine content (Nandi, 1993). In Guatemala, plantations yield 9 to 16 tonnes of bark per hectare (Popeno, 1949).

DISEASES AND PESTS

Damping-off is common and the most noticeable disease in nursery beds particularly at lower elevations; *Rhizoctonia soloni, Macrophomina phaseoli, Phytophthora* spp., *Fomes lamoensis,* and *Pythium vexans* are responsible for this condition. These fungi penetrate the seedlings through their roots and cause symptoms of sudden wilting and rotting of young succulant seedlings. Good drainage of beds, spraying with bordeaux mixture, and fumigation with chloropicrin are recommended for control (Sarma and Chatterjee, 1987). Soil application of thiram or pentachloronitrobenzene (PCNB) at 20 kg/ha and spraying seedlings with 0.2 percent zineb or other copper fundicide (0.2) reduces infection. Blight caused by *Phytophthora palmivora* is occasionally reported in India. It begins with necrosis of the terminal leaves and branches in the seedlings and results in death. This fungus also causes girdle canker in older trees. Dusting with bordeaux powder at regular intervals is recommended. Stem bark disease due to *Phytophthora cinnamomi* is a major disease in central African countries (Rwanda and the Democratic Republic of the Congo). Studies on infected material showed that anthraquinones were present in infected bark, but not in healthy bark, suggesting that these compounds in cinchona act as phytoalexins. The alkaloid content of the infected bark is lower (2.0 percent) than healthy bark (Wijnsma, 1986).

Stripe canker *(Sclerotium rolfsii),* stem blight (*Sporotrichum* spp.), top blight and girdle canker *(Phytophthora parasitica),* and collar rot *(Fomes noxious)* have been reported from Annamalai. Of these, *Phytophthora parasitica* is the most destructive because the organism spreads from infected plants and causes wilt on older trees as well. Application of copper-based fungicide over the soil is effective. It is recommended that diseased plants are uprooted and burnt. The damage due to pests is less because young *C. ledgeriana* plants contain cinhophyllene type of indole alkaloids with small amount of 5-methoxy-tryptamine. These compounds provide a chemical defense to the growing plants against herbivorous insects (Aerts, 1992). However, caterpillars of *Helopeltis* spp. and *Catalpa sphinx* cause damage to the nursery plants. Spraying of dimethyl phosphoric ester controls the infestation. The leaf bug *Disphinctus humeralis* occasionally attacks tender foliage, but can be controlled by spraying 0.2 percent malathion.

BIOTECHNOLOGICAL INTERVENTIONS

Many investigators have attempted to produce cinchona alkaloids in high concentration in cell-suspension cultures by using stress, precursors, elicitors, auxins, enzymes, and use of a hairy root system. Encouraging results have been reported. A literature search on this topic provides evidence of an increasingly larger number of investigations carried in European countries in the past ten years, although commercial success still appears a distant goal. Chung and Staba (1987) established leaf shoot and organ cultures of *C. ledgeriana* on MS media containing benzaldenine and studied the effect of age and growth regulators on alkaloid production. They found an increase in the content of alkaloids with an increase in the age of the cultured leaf shoot. The 32-week-old (tissue) cultures contained the same amount of alkaloids as one-year-old plants. Quinine production was favored by the presence of benzaldenine (5 mg/1) and gibberellic acid (5 mg/1). Quinidine production was higher in the presence of indole 3 acetic acid (5 mg/1). High concentrations of abscisic acid and mefluidide inhibited growth and alkaloid production. Feeding various precursors to eight-week-old leaf shoot cultures increased the total alkaloid content by approximately 66 percent with tryptophan, 42 percent with secologanin, and 5 percent with strictosidine type. A decrease of 10 percent was registered with methoxy-strictosidine-type alkaloid intermediates. Earlier, Hay et al. (1986) reported a fivefold increase in yield of quinine and quinidine on feeding unlabelled 1-tryptophan (500 mg/1) to root suspension cultures in this species. Harke (1985) found that higher auxin levels improved production of indole alkaloids (cinchonamine) in the callus, although many times growth may not occur in the light. The importance of light and dark periods was highlighted by Payne et al. (1987) in cultures transformed with *Agrobacterium tumefaciens* in a medium free of exogenous phytohormones. Growing transformed cultures in the dark resulted in 50 times more alkaloid accumulation than in cultures grown in light. This dark-dependent accumulation was not confined to a particular time in the growth cycle, although the extent of stimulatory effect increased when cultures were kept longer in darkness. Blue light was detrimental to alkaloid accumulation, but in red and green light the level of accumulation was equivalent to that in dark. Alternating cultures between light and dark conditions for up to a 28-day period resulted in alternate periods of low and high alkaloids production.

In an interesting study having bearing on biosynthetic pathways of indole alkaloids in plants, Isaac et al. (1987) isolated two nicotinamide adenine denucleotide phosphate [NADP(H)]-dependent enzymes from suspension cultures of *C. ledgeriana*. These enzymes catalyze the reversible reduction of cinchonidine to form both cinchonine and cinchonidine, and one of them shows reversible activity with its 6-methoxy derivative quinidine to form quinine and quinindine. Geerlings (1999) successfully integrated TDC (tryptophan decarboxylase) and STR (strictostidine synthase) enzymes from periwinkle *(Catharanthus roseus)* in terpenoid indole and quinoline alkaloid biosynthesis. The products of TDC and STR, tryptamine and strictosidine, were found in high amount (1,200 and 1,950 mg per g dry weight, respectively). Quinine and quinidine levels rose to 500 and 1,000 mg per g dry weight, respectively.

FUTURE OUTLOOK

The natural quinine salts produced in cinchona bark have played a great role both in prevention and treatment of malaria for nearly 400 years. Before quinine was discovered, malaria was treated by the release of humors, i.e., bleeding, purging, and use of emetics. Robert Talbor, who used it first in England, was regarded as a quack by English physicians. He won recognition only after successfully treating Charles II of France from malaria (Keeble, 1997). Even the synthetics that have replaced it are formulated on the novel model of the quinine molecule, which is a great tribute to its unique nature. The natural drug is produced in a tree, which takes a long growth cycle in plantation for optimum synthesis and accumulation of the target alkaloid, whereas the production of synthetics can be regulated in a laboratory and is therefore cheaper. Still, laboratories in several European countries are working to economically produce quinine and cinchonine through cell cultures, which hopefully stands a fair chance of success in the long run. Cinchonine has been used in population control programs because of its abortifacient nature but with uncertain results. It has not been given up. In fact, safer and cheaper birth control drugs are in great demand in developing countries, and if cinchona alkaloids or their natural analogues succeed in clinical trials, it will open the way for their comeback in pharmacy. Quinidine, on the other hand, has a limited demand, and is a drug of choice in North American countries. Cinchonidine has a weaker action than quinine but has been found useful in rheumatism, neuralgia, and sciatica. It is also used as an antispasmodic in whooping cough. Currently, demand for natural products is growing in Western societies. Cinchona bark is prescribed as a gargle for sore throat. In homeopathic medicine, it is given for nervous ergotamints, anemia, and convalescence. In European herbal medicine, cinchona bark is a top antiprotozoal medicament, given in all kinds of fevers. It provides relief in enlarged spleen, liver, and gall bladder disorders. It is a stimulant in hair growth products. The bark has made a comeback in "bark tea" for the management of malaria. This remedy calls for a cup of boiling water to be poured over approximately 1 g of ground natural bark and allowed to steep for ten minutes. One cupful of infusion is recommended to be taken half an hour before meals to stimulate appetite or after meals to treat indigestion (Witch, 1994). The popularity of natural products is gaining ground and it is believed that cinchona bark will remain in the medical chest among herbal medicines for a long time to come.

REFERENCES

Aerts, R.J. 1992. Detrimental effect of cinchona leaf alkaloids on larvae of the polyphagus insect *Spodoptera*. *Chemical Ecology* 18(11): 1955-1964.

Aerts, R.J., Snoeiger, W., Van der Moijden, E., and Veerpoorte, R. 1991. Allelopathic inhibition of seed germination by cinchona alkaloids. *Phytochemistry* 30(9): 2947-2951.

Anderson, L. 1995a. Diversity and origin of Andean Rubiaceae. In Churchill, S.P., Balsleu, H., Farero, E. and Luteyn, L.L., Eds., *Biodivesity and Conservation of Neotropical Mountain Forests: Proceedings of the New York Botanical Garden*, pp. 441-450.

Anderson, L. 1995b. Tribes and genera of the cinchoneae complex (Rubiacaeae): First international conference on the Rubiaceae. *Annals of the Missouri Botanical Garden* 82(3): 409-427.

Anderson, L. 1998. A revision of genus *Cinchona* (Rubiaceae—Cinchonae). *Memoirs of the New York Botanical Garden* 80: 1-75.

Anonymous 1993. The review of natural products. *Quinine: Facts and Comparisons*. Michael R Riley Publishers, St. Louis, pp. 1-2.

Brockway, L. 1979. *Science and Colonial Expansion: The Role of the British Botanical Garden*. New York Academy Press, New York.

Bruncton, J. 1995. *Pharmacology and Phytochemistry of Medicinal Plants*. Intrcept Ltd., Hampshire, England.

Burkill, I.H. 1935. Cinchona. In *Dictionary of the Economic Plants of Malaya Peninsula*, Volume I. Oxford, UK, pp. 538-543.

Chatterjee, S.K. 1982. Cultivation of cinchona in Darjeeling hills, West Bengal. In Atal, C.K. and Kapur, B.M., Eds., *Cultivation and Utilization of Medicinal Plants*. Regional Research Laboratory, Jammu, pp. 222-229.

Chatterjee, S.K. 1993. Domestication studies of some medicinally important exotic plants growing in India. *Acta Horticulture* 331: 151-160.

Chatterjee, S.K., Bhararti, P., Chhetri, K.B., and Ramsong, A.F. 1988. Agrotechnology of cultivating cinchona and ipecac in Darjeeling hills. *Proceedings of Fifth ISHS Symposium of Medicinal, Aromatic, and Spice Plants* 188-A: Spl issue, 25-27.

Chung, C.T.A. and Staba, E.J. 1987. Effect of age and growth regulators on growth and alkaloid production in *Cinchona ledgeriana* leaf shoot organ cultures. *Planta Med.* 53(2): 206-210.

Council of Scientific and Industrial Research. 1992. Cinchona. *Wealth of India: Raw Materials,* Revised Edition, Volume 3, pp. 563-572.

Datta, M., Sah, K.D., Gupta, S., and Banerjee, S.K. 1990. Characteristic of soils supporting cinchona plants in Darjeeling hills. *Indian Agriculturist* 34(2): 73-77.

Dawson, R.F. 1991. Quinine and quindine production in the Americas: A brief history. *Horticultural Technology* l(l): 17-21.

Dayrit, F.M. 1994. Determination of quinine content in the bark of cinchona trees grown in the Mt. Kilanglad, Bukidnon. *Philippine J. Science* 123(3): 215-227.

Directorate of Cinchona and Other Medicinal Plants. 1994. *Research Tech. Bull.* Directorate of Cinchona and Other Medicinal Plants, Darjeeling (West Bengal), India, 18: 3.

Doraiswami, K. and Vekatraman, K.P. 1982. Chemistry, production, and marketing of cinchona alkaloids. In Atal, C.K. and Kapur, B.M., Eds., *Cultivation and Utilization of Medicinal Plants*. Regional Research Laboratory, Jammu, pp. 230-243.

Geerlings, A. 1999. Alkaloid production by *Cinchona officinalis* Ledgeriana hairy root culture containing constructive expression constructs of tryptophan decarboxylase and strictosidine synthase DNA from *Catharanthus roseus. Plant Cell Reports* 19: 191-196.

Gupta, R. 1980. Cinchona. In ICAR *Aromatic and Medicinal Plants: Handbook of Agriculture*. Indian Council of Agricultural Research, New Delhi, pp. 1190-1192.

Haggis, A.W. 1941. Fundamental errors in early history of cinchona. *Bull. Hist. Med.*, Pt. 10: 417-459; Pt. 10: 568-592.

Harke, P.D.A. 1985. Influence of various media constituents on the growth of *Cinchona ledgeriana* tissue cultures and production of alkaloids and anthraquinones therein. *Plant Cell Tissue and Organ Culture* 4(3): 199-214.

Hay, C.A., Anderson, C.A., Robert, M.F., and Phillipson, J.D. 1986. In vitro culture of cinchona species: Precusor feeding of *C. ledgeriana* root organ suspension cultures with l-tryptophan. *Plant Cell Reports* 1: 1-4.

Holmes, E.M. 1930. Three hundred years of cinchona. *Chemist Drugg.* 112: 827-832.

International Trade Centre. 1982. Markets for selected medicinal plants and their derivatives. International Trade Centre, Geneva, pp. 95-102.

Issac, J.E., Robins, R.J., and Rhodes, M.J.C. 1987. Cinchonione: NADPH oxidoreductase I and II novel enzymes in the biosynthesis of quinoline alkaloids in *Cinchona ledgeriana. Phytochemistry* 26(2): 393-399.

Keeble, T.W. 1977. A cure for the ch: The contribution of Robert Talbor (1942-1981). *J. Royal Society of Medicine* 90(5): 285-290.

Keene, A.T., Anderson, L.A., and Philipson, J.D. 1983. Investigation of cinchona leaf alkaloids by high performance liquid chromatography. *J. Chromatography* 260(11): 123-128.

Koblitz, H., Koblitz, D., Schmauder, H.P., and Groger, D. 1983. Studies on tissue culture of the genus *Cinchona* L. *Plant Cell Reports* 2: 95-97.

Leung, A.V. 1980. *Encyclopaedia of Common Natural Ingredients Used in Food, Drugs and Cosmetics*. J. Wiley and Sons, New York.

Mathews, P.M. and Philips, K.O. 1979. Studies on South Indian Rubiaceae. Pt II. Cytology of *Cinchona* Linn. *Nucleus (India)* 22(2): 125-128.

Morton, J.F. 1977. *Major Medicinal Plants*. Charles C Thomas Publisher, Springfield, IL.

Mukerjee, S. and Chatterjee, S.K. 1960. A new method for propagation of *Cinchona*. *Sci. Cult.* 25(11): 648-650.

Nandi, R.P. 1993. *Cultivation of Cinchona and Production of Its Alkaloids in India.* Publ. Dr (Mrs.) S. Nandi, Sevoke Road, Siliguri, India, pp. 1-67.

Nandi, R.P. and Chatterjee, S.K. 1983. Effect of nitrogen, phosphorous, and potassium on growth, development, and alkaloid formulation in *Cinchona ledgeriana*. *Indian J. Forestry* 6(3): 230-232.

Nandi, R.P. and Chatterjee, S.K. 1991. Effect of shade on growth and alkaloid formation in *Cinchona ledgeriana* grown in Himalayas of Darjeeling. In Trivedi, R.N., Saima, P.K.S., and Singh, M.P., Eds., *Recent Researches in Ecology, Environment, and Pollution.* Today and Tomorrow Printers and Publishers, New Delhi, Series 5: 181-184.

Payne, J., Rhodes, M.J.C., and Robins, R.J. 1987. Quinole alkaloids production by transformed cultures of *Cinchona*. *Planta Med.* 53(4): 367-372.

Pin, G. 1998. Quinine for cramps. *Aust. Fam. Pharmacopea* 27(10): 922-923.

Popenoe, W. 1949. Cinchona cultivation in Guatemala. *Econ. Bot.* 3(13): 150-157.

Rao, M.M. 1974. A note on the Tamil Nadu Government Cinchona Department, pp. 1-21.

Rao, M.M. and Veeraghavan, R. 1954. Vegetative propagation of *Cinchona ledgeriana* by Gootees (Marcotte). *South Indian Horticulture* 2: 71-84.

Saha, J.C. 1972. A report on cinchona and quinine industry in India; Present and future prospects. ICAR Scientific Panel on Medicinal Plants and Minor Crops, pp. 1-14.

Sarma, P. and Chatterjee, S.K. 1987. Diseases of cinchona and other medicinal plants growing in Darjeeling hills of West Bengal. In Raychaudhari, S.P. and Verma, J.P., Eds., *Review of Tropical Plant Pathology.* Today and Tomorrow Printers and Publishers, New Delhi, Volume 4, pp. 331-340.

Sukarja, D. 1977. Correlation between some morphological characters and propagation capacity of cinchona trees. *Menara Perkebunan* 45(6): 275-278.

Sukarja, D. and Munir Supralo, A. 1977. Annual Report on Cinchona plantation. *Menara Perkebunan* 45(5): 233-236.

Swamy, A.Y. 1953. Cinchona industry in Madras State with particular reference to Nilgiries. In Krishnamurti, S., Ed., *Horticulture and Economic Plants of Nilgiries*, pp. 19-29.

Teteny, P. 1970. *Intra-Specific Chemical Taxa of Medicinal Plants.* Akademia Kiado, Budapest.

Thakurta, A.G. and Dutt, B.K. 1944. Vegetative propagation of *Cinchona ledgeriana* from Gootes (Marcotte) and cuttings by treatment with auxins. *Sci. and Cult.* 9(9): 401-402.

Wargadipura, R. and Sutandjona, S. 1977. Nursery media for Cinchona: Their influence on seedling growth. *Menara Perkebunan* 45(4): 167-173.

Watts, George. 1893. *Dictionary of Economic Products of India*, Volume 2.

Wijnsma, R.Q. 1986. Anthraquinones in *Cinchona ledgeriana* bark infected with *Phytophthora cinnamomi*. *Planta Med.* 3: 211-212.

Witch, M. 1994. *Herbal Drugs and Phytopharmaceuticals*. CRS Press, Boca Raton, FL.

Chapter 5

Cocoa

R. Vikraman Nair
S. Prasannakumari Amma
V. K. Mallika

Cocoa (*Theobroma cacao* L.) is the third important beverage crop next to coffee and tea. It is grown mainly in the tropical countries of the world. It is primarily an important item of confectionery industries and is the only source of chocolate. The cocoa of commerce is the cured dry beans, which contain 57.0 percent fat, 7.0 percent protein, 7.0 percent carbohydrate, and 1.7 percent theobromine (in additon to 6.0 percent moisture, 2.7 percent total ash, 1.1 percent minerals, 4.1 percent pectin, 2.1 percent fiber, 1.9 percent cellulose, 1.2 percent pentosans, 1.6 percent mucilage and gums, 6.2 percent tannins, and 0.4 percent acids). The stimulating effect is due to the presence of the alkaloid, theobromine. It is a crop cultivated in underdeveloped countries of the world. Major production is consumed by the people in affluent countries. The demand for cocoa-based products is registering a steady increase of about 12 percent every year, and there is immense potential for taking up cultivation of this crop in the coming years.

ORIGIN AND SPREAD

Cocoa (*Theobroma cacao* L.) is a native species of tropical humid forests on the lower eastern equatorial slopes of the Andes in South America (Cheesman, 1944). It was domesticated by the natives of Central America and was considered to be of divine origin. The generic name *Theobroma* literally means "food of the gods." Cocoa was domesticated and the produce used for consumption for the first time by the Maya and Aztecs. The first Europeans to drink cocoa were the Spanish who invaded and conquered the Aztec empire in Mexico in the sixteenth century. The Spanish learned from the Aztecs the technique of making *xocoatl*, a drink made from cocoa beans after roasting and grinding. The word *chocolate* is considered to have originated from *xocoatl.* The word *cacao* also was used by the Spanish and it probably originated from *cacahuatl,* a word that Aztecs used for cocoa beans. Even before the Spanish conquest, cocoa was taken to different regions by the Maya, Aztec, and Pipil-Nicarao peoples (Young, 1994; Coe and Coe, 1996).

From the center of origin, the species spread out creating two main groups: the *criòllos,* which resulted from dissemination through the Andes toward the lowlands of Venezuela, Colombia, and Ecuador and toward the north to central America and Mexico, and the *forasteros,* which resulted from dissemination toward the Amazon valley in Northern Brazil and the Guayanas (Alvim, 1987). *Criollo* types spread to Central America and to a large number of Caribbean Islands, including Trinidad in 1525, and thereafter to Jamaica. The dissemination to Venezuela and Costa Rica was made by the Spanish (Pittier, 1933). Introduction to Martinique and Haiti was by the French. Planting in Belem and Bahia in 1750 was attempted by the Portuguese.

Trinitario arose out of natural hybridization between *criollo* and *forastero.* It has been recorded that the *criollo* population from Venezuela and the Amelonado-type *forastero* from Guayana could have been involved in hybridization leading to the production of *trinitario.*

During 1822, cocoa seeds were taken from the Portuguese colonies of South America to the island of Sao Tome off the West Coast of West Africa. It also spread to the neighboring island, Principe. Cocoa cultivation was started in Fernando Po in 1840. The most successful introduction into African mainland was made by the Ghanian Tetteh Quashie in 1879. He brought a pod from Fernando Po, and the early population of cocoa in Ghana is considered to have originated from this pod. From Ghana, it spread to other African countries, the most important of which are Ivory Coast, Nigeria, and Cameroon. In these countries, there was immediate increase in area and they eventually turned out to be the largest producers of cocoa in the world. As it stands today, about 68 percent of the total world production of cocoa beans comes from these African countries where this crop was introduced relatively very late. A characteristic of African cocoa used to be, especially up to the 1950s, the homogeneity of cocoa populations with pods of "melon" shape (Amelonados).

Cocoa was introduced in the sixteenth century into Asia and the Pacific. Venezuelan *criollo* was introduced in Celebes by the Dutch in 1560. They also introduced the crop into Java. The Spanish took *criollo* types from Mexico to the Philippines in 1614. It was introduced into Sri Lanka from Trinidad about 1798. From Sri Lanka, cocoa was taken to Singapore and Fiji in 1880, Samoa in 1883, Queensland in 1886, and Bombay and Zanzibar in 1887. Cocoa was introduced into Malaysia in 1778, and in Hawaii in 1831. Cocoa was introduced in India in the early twentieth century, but its cultivation began in a big way only in the 1960s.

AREA AND PRODUCTION

As stated earlier, about 68 percent of the total world production of cocoa beans come from African countries, where the crop was introduced relatively very late. The world production of dry cocoa beans had been around 1.5 million tonnes in the 1970s and in the range from 1.5 to 2.5 million tonnes in the 1980s. It remained at around 2.5 million tonnes during the 1990s up to 1993-1994. The production (2000-2001) is 2.819 million tonnes (ED and F Man, 2002). Based on 2000-2001 statistics, the major producing countries are Ivory Coast, Ghana, Indonesia, Brazil, and Nigeria, their contribution being 82.0 percent of the world total (Table 5.1). The Af-

TABLE 5.1. Production of cocoa in different countries.

Country	Production (in thousand tonnes)	Percentage of total
Africa		
Ivory Coast	1,175	41.92
Ghana	398	14.2
Nigeria	202	7.21
Cameroon	121	4.32
Other Africa	40	1.43
Total Africa	1,936	69.08
Central and South America		
Brazil	130	4.64
Other America	177	6.31
Total America	307	10.95
West Indies	33	1.17
Asia and Oceania		
Indonesia	393	14.02
Malaysia	79	2.82
Papua New Guinea	35	1.25
Other Asia and Oceania	20	0.71
Total, Asia and Oceania	527	18.8
World total	2,803	100

Source: Cocoa Market Report, ED & F Man (April 2002), No 370.

rican countries produced 68 percent and the Central and South American countries produced 14 percent of the total. The Asian countries produced the remaining 18 percent. The contribution of India is negligible (0.21 percent).

Botany

Cocoa is one of the 22 species assigned to the genus *Theobroma,* a member of the family Sterculiaceae. *Theobroma cacao* is the only species of economic importance. *Theobroma bicolor* Humb. & Bonpl. is cultivated for the edible pulp around the beans, and the beans are used like those of cocoa. The beans of *T. angustifolium* Moc. & Sesse. are mixed with cocoa in Mexico and Cost Rica and sweet pulp around the beans of *T. grandiflorum* (Willd. ex Spreng.) Schumann is used for making a drink in parts of Brazil and is also eaten.

Roots

The tap root of cocoa grows predominantly downward with only a few branches. Under suitable growing conditions when the soil is deep, they grow to a depth of about 150 cm. The primary function of these roots is considered to be anchorage. The main feeding roots are those that arise from the tap root and grow laterally. Most of these roots are concentrated just below the soil surface up to a depth of 15 to 20 cm (Wahid et al., 1989). The lateral spread of such roots will be up to about 120 to 150 cm around an adult cocoa plant. As the bulk of feeding roots of cocoa is concentrated near the surface, any form of digging of soil around the cocoa plant will be harmful. The rooting pattern can be modified to an extent by the environment.

Stem

Cocoa grows in tiers. The shoot of a seedling that grows upward is called "chupon." After growing to a height of 1-1.5 meters, the growth of the chupon ceases and three to five lateral branches arise. These lateral branches are called "fans" or "fan branches." The point at which fans arise is called "jorquette," and the process of the formation of fans from jorquette is called "jorquetting." A layer of fans may be called a "tier." If allowed to grow unrestricted, new chupon buds arise on the main chupon stem below the first jorquette and grow up to jorquette again. Chupons may be distinguished from fans by their nature of growth and leaf arrangement. Fan growth will be predominantly to the sides, whereas chupons grow vertically up. Leaves of the fans are arranged in one plane and are alternate. Leaf arrangement of chupons will be spiral with a phyllotaxy of 3/8. Chupon leaves will have longer petioles than those of fans and will also have a more pronounced pulvinus.

Normally, buds arising on a chupon give rise to chupons, except at the jorquette when jorquetting occurs. Similarly, fans produce only fan branches. Occasionally, fans produce chupons.

Inflorescence

Cocoa flowers are borne on thickened leaf axils on stems called "cushions." The number of flowers per cushion season is up to 50. The cocoa flower is a compressed cyme and has five sepals, five petals, ten stamens in two whorls, and a superior ovary of five united carpels. Among the ten stamens, five of the outer whorl are sterile, and the five of the inner whorl, which occur opposite the petals, are fertile. These fertile stamens occur concealed in the pouched portions of petals. The ovary is simple and five lobed. The number of ovules per flower ranges from 40 to 60. The style has five stigmatic lobes.

Cocoa flowers are produced in large numbers but only a few of them develop into fruits. Those that are not fertilized fall off within 24 hours. The flowers are ill-adapted for pollination by natural methods as well as for self-pollination as the fertile stamen is concealed in the pouched portion of the petal and the stigma is surrounded by a ring of staminodes. The flowers are devoid of scent or nectar and the pollen grains are sticky. Natural pollination occurs only with the help of small crawling insects. The most important of the pollinating insects are the (ceratopogonid midges of the genus *Forcipomyia*). The insects are small and barely visible to the naked eye. The midges are attracted by the pigmented tissues of the staminodes and the guidelines of the petals. The midges moving on the guidelines near the anther pick up the pollen grains, and when they crawl to the staminodes, some of the pollen get transferred from their body to the stigma. Though midges are the most important pollinating agents, other insects, such as ants, are also implicated as probable pollinating agents. There is a probability, though slight, of wind pollination also. Flowers start to open late in the afternoon and are fully open by the forenoon the next day. As such, most pollination occurs in the early hours of the day.

Self-Incompatibility in Cocoa

A special feature of cocoa is self-incompatibility shown by some cocoa types. This was first reported by Harland in 1925. Upper Amazon and Ecuador types introduced in Trinidad were self-incompatible; but most of the self-incompatible plants were cross-incompatible also. Many of the homozygous types, such as West African Amelonados, are self-compatible. Even though the self-compatible types may have the advantage of better fruit set under varied situations, self-incompatibility is important in commercial hybrid seed production. Incompatibility in cocoa is unique in that the site of incompatibility is the embryo sac (Cope, 1962). After incompatible pollination, the pollen tube grows faster and delivers the gametes into the embryo sac in a normal fashion. The embryo sac is in no way abnormal and the rejection is due to the failure of male nuclei to unite with the egg. This incompatibility is referred to as "prefertilization inhibition in the ovule" and it is genetically controlled. Fusion or nonfusion is controlled by a series of alleles operating at a single locus (S), showing dominance or independence relationships (Purseglove, 1968). In incompatible matings, the flowers drop two to four days after pollination. If a population of cocoa is examined for self-incompatiblity reactions, it could be observed that the majority of the plants belong to the self-incompatible group. Cross-incompatible types frequently occur between two individuals with different genotypes, and it occurs only in diploid gametophytic systems when individuals share the same S genotype (Richards, 1986). Mallika, Amma, Nair, and Namboothiri (2002) studied compatibility relations among sixteen selected parents. Out of 128 crosses attempted, 23 were cross-incompatible. Cross-incompatibility is an indirect measure of the degree of closeness between the genotypes. When the parents used in crossing happen to be genetically similar, the incompatibility mechanism operates.

Fruits

The cocoa fruit is botanically a drupe, often called a "pod." The mature fruit consists of a thick husk containing 30 to 50 seeds. The seeds are covered with a sugary, mucilaginous coating called the "pulp." The seeds are held in position with the help of a placenta. The pods can be of green or reddish color when immature. Green pods change to yellow when mature and reddish pods to orange or yellow in color. The pericarp (husk) is fleshy and thick.

Seeds

Seeds called "beans" constitute the economic part of this crop. The pulp covering the seed contains about 10 to 15 percent sugars. The size of beans is of practical significance, and a minimum average bean size of 1 g or a bean count of not more than 100 is usually taken as the standard. The number of beans per pod ranges from 30 to 60. Each seed contains two convoluted cotyledons, a small embryo and a thin membrane, the remains of the endosperm, and a leathery testa (shell).

Classification

Cheesman (1944) classified the cultivated and wild cocoas into three groups based on Venezuelan trade names:

1. *Criollo:* Pods yellow or red when ripe, deeply 10 furrowed, markedly warty, conspicuously pointed, pod wall too thin, seeds large, plump and almost round, cotyledons white or pale violet which are less astringent. The beans ferment quickly, but yield is poor. It produces the highest quality cocoa. It is susceptible to stress and not adaptable to all situations. It can be subdivided into
   • Central American *criollo* and
   • Venezuelan *criollo.*
2. *Forastero:* Unripe pods green, turns to yellow on ripening, inconspicuously ridged and furrowed, surface smooth, ends rounded or bluntly pointed, pod wall thick, seeds flattened, fresh cotyledons deeply pigmented and dark violet giving an astringent product. The trees give high yields and are hardy. Quality is not comparable with *criollo.* The beans take five to six days for fermentation.
3. *Trinitario:* These originated in Trinidad from a genetic mixing of *criollos* and *forasteros.* These are very hetetrogenous and exhibit a wide range of morphological and physiological characters. It is difficult to specify the characters of *trinitarios* as they may have pod and bean characters ranging from those of typical *criollos* to those of *forasteros.*

In addition, several other subgroups fall under *forasteros.* The best known *forasteros* are the "Amelonados" of the African region, which were the predominant types traditionally cultivated in the West African countries since the nineteenth century. Amelonados are typically self-compatible and the pods have a "melon" shape with nearly smooth pod surface. The Amazonians showed wide genetic variability and are highly useful for breeding work in the major producing countries.

GERMPLASM COLLECTION

The distribution of cocoa germplasm must be through internationally recognized agencies. The International Cacoa Genebank, Trinidad, and the collection at Centro de Ensenanza y Investigacion de IICA, Turrialba, Costa Rica (CATIE) are designated as "universal collection depositories." The core of Trinidad collection is Pound's Ecuadorian and Peruvian collection which forms 70 percent of it, the 1952 Anglo Colombian collection, representatives of Chalmers' and Allen's material, and selections from cultivated cocoa in Trinidad and other Caribbean Islands. The core of the Turrialba collection is selections from cultivated cocoa, especially the United Fruit Company clones and their derivatives from Costa Rica, similar material from other American countries, and *criollo*. Large collections of primary material are also maintained in Colombia, Ecuador, French Guiana, Venezuela, and Brazil. Field collections are maintained in Puerto Rico, Ivory Coast, Jamaica, Malaysia, Grenada, Nigeria, Papua New Guinea, Ghana, and India. In Kerala Agricultural University, India, 544 diverse types of cocoa are being maintained (Photo 5.1).

The germplasm has been distributed from Trinidad and Costa Rica. Large quantities of seed were distributed from Trinidad to Ghana in 1944 and to Nigeria and Papua New Guinea in the 1960s. Long-distance distribution is done using intermediate quarantine facilities at the Royal Botanic Gardens, Kew (University of Reading, from 1983), and the United States Department of Agriculture in Miami, Florida. These transfers are carried out by authorized organizations such as the International Board for Plant Genetic Resources (IBPGR).

CROP IMPROVEMENT AND MANAGEMENT

Priorities in Breeding

In all cocoa-growing countries, yield improvement was the primary objective. With the spread of diseases, such as witches'-broom, cocoa swollen shoot virus (CSSV), vascular streak dieback, black pod, etc., which are difficult to be managed with chemicals, more emphasis is being placed on evolving disease-tolerant types. Emphasis is also being placed on retention of traditional flavor, adaptation to local environment, early and sustained bearing, tree shape, pod size, and bean characters.

PHOTO 5.1. Cocoa grove at Kerala Agricultural University.

Methods of Breeding

Introduction

A major portion of cocoa in the world is derived from countries situated away from its center of origin, and thus the introduction of diverse types plays an important role in cocoa improvement. The genetic base of cocoa is very narrow, and as such, crop improvement programs are not yielding spectacular progress. The introduced types are evaluated in the field for yield, pod and bean characters, incompatibility reaction, reaction to pests and diseases, and adaptation to the environment. The superior types emerging from this can be utilized for commercial planting or may be included in the breeding program.

Since cocoa is cross-pollinated, it is not advisable to import germplasm as pods or seeds. Vegetative materials such as bud wood can be imported, which is then budded to stock plants to get true-to-type plants. The introduced materials are maintained in isolation in quarantine houses until they are certified as free from pests and diseases. After proper evaluation, the introduced material may be directly released as improved clone(s). Some of the introductions can be used as sources of desirable genes for disease, pest, and drought resistance, quality, or other valuable characteristics which may then be incorporated into adapted varieties through hybridization procedures.

Selection

There is ample scope for selection in cocoa because of the highly heterozygous nature of the crop. Immense variability exists in the seedling populations. The variability is so high that, in a seedling population, about 75 percent yield is obtained from 25 percent of the trees. The remainder of the trees will be of low productivity. The yield is influenced by changes in environmental conditions. Yield per tree varies with spacing, shade, soil conditions, nutrient supply, etc. Longworth and Freeman (1963) suggested to consider tree yield along with trunk diameter for better efficiency in selection. In addition to pod production, the weight of cured beans per pod also may also be considered for selection. Number of beans per pod is a trait which is not influenced by environment but the bean weight is influenced by the environment to a considerable extent (Pound, 1931).

An easy approach to yield improvement in cocoa is to select plants superior in yield and their subsequent development into clones. For selecting individuals from the populations, certain criteria have been fixed: plants yielding not less than 100 pods per tree per year, each pod weighing 350-400 g or more with a pod value of not more than ten, and with 35-40 beans having a fermented dry weight of 1.0 g are selected as parents. In general, cocoa is well adapted to vegetative propagation by grafting, budding, or cuttings.

A number of superior clones have been selected throughout the world, and these are getting very high acceptability among the growers. Kerala Agricultural University has initially selected 70 clones, out of which seven (Cadbury Cocoa Research Project [CCRP] 1 to 7) were released for cultivation as clonal blends (Photos 5.2 and 5.3).

PHOTO 5.2. CCRP 2 clone at Kerala Agricultural University.

Hybridization

Hybrid vigor between parents showing good combining ability can be readily exploited in cocoa. A large number of crosses have been made in countries such as Trinidad, and the potentials of the parents have been assessed. Posnette (1951) demonstrated interpopulation heterosis in cocoa. The initial crosses involving Pound's seedling collection showed exceptional vigor, precocity, and high yield in Ghana. These observations and similar ones in Trinidad were attributed to hybrid vigor (Bell and Rogers, 1956; Montserrin et al., 1957). In trials with diverse crosses in Brazil, Costa Rica, Ghana, Ivory Coast, Nigeria, and Papua New Guinea, there were significant additive components for yield. A number of hybrids with high yield and other desirable characters have been evolved in different countries. Hybridization work in Kerala Agricultural University, India, led to the release

PHOTO 5.3. CCRP 7 clone at Kerala Agricultural University.

of three hybrids (CCRP 8 to 10) with high yield and tolerance to vascular streak dieback disease (Photo 5.4).

Heritability estimates ranging from moderate to high and large additive components of variation indicate easy progress toward high yields, at least in the early years of the program. However, after an initial boom in the first phase of hybridization, yield improvement in hybrids during the second phase of breeding was not as spectacular as expected. This is due to the poor genetic base. Most of the hybrids and clones are derived from a relatively lower number of types, and lack of yield improvement in intrapopulation crosses is due to inbreeding.

Method of hand pollination. The method is described by Mallika et al. (2000). In artificial pollination, a flower bud that will open the following day, recognized by its whitish color and swollen appearance, is selected.

PHOTO 5.4. Cocoa hybrid at Kerala Agricultural University with high yield and tolerance to vascular streak dieback disease.

The bud is covered with a hood of plastic tube/hose pipe piece 5 × 1.5-2 cm, which is sealed to the bark using materials such as Plasticine/glaze putty. The tube is covered with muslin cloth at the top, kept in place with a rubber band. This ensures circulation of air and exclusion of insects. Opened flowers are collected from the desired male parent and stamens are carefully taken out by pushing the corresponding petal. One entire anther with a part of the filament is deposited on the stigma. One or two staminodes may be pinched off to give access to the stigma. Emasculation is not necessary due to the presence of self-incompatibility. For selfing also, hand pollination is done using stamens from the same flower. The pollinated flowers are labeled using tin foil pieces fixed in the cushion using ball pins. The hoods are removed 24 hours after pollination, and in three to five days fertilization is confirmed by the visual swelling of the ovary. In order to prevent undue shedding and wilting of fruits from hand pollinations, it is usual to remove all the developing fruits on the tree produced by open pollination. Developing pods are covered with wire mesh after six to eight weeks to protect them from mammalian pests. The pods are collected at maturity, and beans are extracted and sown in the nursery.

Preselection method. In cocoa, the relationship between vegetative characters and yield was positive (Glendinning, 1966; Ngatchou and Lotode, 1971; Enriquez, 1981; Paulin et al., 1993). However, Francies (1998), Sridevi (1999), Verghese (1999), and Amma et al. (2002) recorded contradictory results, and they concluded that, after bearing, vegetative growth slows down and the correlation between growth reduction and yield became positive. This points out the need for evolving a viable preselection method in cocoa.

Selection of superior hybrids. The seedlings selected based on vigor/disease tolerance are field planted. On attainment of a steady yield, the hybrids are evaluated for their performance. The highest-yielding hybrids with other desirable attributes are multiplied and released as new clones.

The parents selected in hybridization programs are tested for both their general combining ability (GCA) and specific combining ability (SCA). To test the GCA, all the selected clones are crossed with a standard variety and the progenies are evaluated both in the nursery and in the field. A few best combiners are then selected and crossed in all possible combinations to assess their SCA. Parents of promising hybrids are identified as best combiners. The best combiners are multiplied and used as parents in seed gardens for the production of quality hybrid seeds.

Clonal seed gardens. The parents used in the seed gardens are selected based on the results of progeny trials. The search for the best combiners involves the screening and selection of a large number of crosses, both at the seedling and adult stages. Having selected the parents, they are propagated vegetatively. The female parent should be self-incompatible. The desired crosses can be ensured either by hand pollination or by the proper design of the seed garden where natural pollination is relied upon. With two self-incompatible parents, all the pods resulting from cross-pollination can be used for seed. Where one parent is self-incompatible, seed is collected from the self-incompatible parent only, and in such cases, the pollen parent is planted in a ratio of one to five female parent trees. The seed garden must be isolated to some extent from other cocoa; a distance of 200 m is considered sufficient to prevent unwanted cross-pollination.

Problems in hybridization. Cocoa is a perennial crop with an outbreeding nature. Most cocoa types are self-incompatible. Selection of self-incompatible parents in breeding programs makes hand pollination easy, as emasculation is not necessary. The existence of cross-incompatibility between some of the parents often poses problems to the breeder. Hand pollination often leads to no pod sets due to cross-incompatibility, and certain proportions of the developed pods wilt due to delayed incompatibility also. Each successful pollination gives rise to a pod, which contains about 35-60 beans. When a large number of crosses are made, the number of hybrid seedlings produced will be too high to be planted in the field on account of limitations in space and cost. This necessitates the development of selection criteria based on early growth parameters of the hybrid seedlings in the nursery itself, which bears positive correlation with the final yield. This must, however, ensure that a valuable hybrid produced out of crossing is not lost while screening the seedlings.

Inbreeding

Inbreeding often forms a part of the breeding activities not only to breed parents with some degree of homozygosity for the production of hybrids but also to breed materials homozygous for such desirable traits as disease resistance. Often, the incidence of self-incompatibility tenders inbreeding difficult or impossible. In cocoa, certain self-compatible trees are encountered in a population, and in these plants selfing is possible. The selfing needs to be continued up to six or seven generations to attain homozygosity, and thereafter these plants can be utilized for crossing to exploit hybrid vigor. An inbreeding program has been in progress in Kerala Agricultural University (Mallika, Amma, and Nair, 2002a) since 1987. Inbreds exhibit morphological abnormalities (Photo 5.5).

PHOTO 5.5. Inbred cocoa plant showing morphological abnormalities.

Breeding for Resistance to Diseases

Five major diseases, viz., witches'-broom (WB), black pod (phytoph-thora pod rot [PP], BP), moniliasis pod rot (MO), cocoa swollen shoot virus (CSSV), and vascular streak dieback (VSD), affect the crop, causing about 40 percent yield loss per year. Selection for disease resistance under field conditions is time consuming, and environmental factors plus genotype × environment interaction may affect the genotypic variation in host resistance. Screening tests on seeds (CSSV, Ghana), young seedlings (WB, Brazil), and seedlings (VSD, India) are of practical use in selection programs. Selection for host resistance requires standardization of the environment and inoculation methods to reveal maximum genotypic expression of major components of host resistance. A close correlation of the results of the preselection test with mature plant resistance should exist. The important diseases and their extent of damage and progress in breeding are outlined in the following sections.

Black Pod (BP) (Phytophthora palmivora *and* P. megakarya)

Worldwide, the most important disease of cocoa is black pod or pod rot. *Phytophthora palmivora* occurs in the center of origin of cocoa and causes 44 percent global crop loss. *Phytophthora megakarya* is restricted to Cameroon, Nigeria, Togo, and Ghana, causing 10 percent global crop loss. *P. capsici* parasitizes cocoa in Central and South America. It is the predominant cause of pod rot in Brazil where it is less aggressive than *P. palmivora*. *Phytophthora citrophthora* also attacks cocoa in Brazil. Much progress has been made recently in the study of nature of variation in host resistance as evidenced by field scores and artificial inoculation tests on *Phytophthora palmivora* in Trinidad (Iwaro et al., 1997), Costa Rica (Phillips-Mora, 1996), Papua New Guinea (Tan and Tan, 1990), and Ivory Coast (N'Goran et al., 1996). Similar studies have been conducted on *Phytophthora megakarya* in Cameroon (Nyasse et al., 1996).

Witches'-Broom (WB) (Crinipellis perniciosa)

This is endemic to wild cocoa and is restricted to the Western Hemisphere, causing 21 percent crop loss. It is prevalent in the center of diversification of cocoa in the Amazon and Orinoco River basins, Ecuador, Bolivia, Peru, Venezuela, Guyana, Suriname, Brazil, Trinidad, Tobago, and Grenada. The disease caused a dramatic decline in cocoa production in these countries up to the extent of 30 percent (Rudgard and Lass, 1985). The disease occurs in all species of *Theobroma* and the closely related genus *Herrania*. The fungus spreads through seeds as well, and hence quarantine measures should be strictly enforced.

The search for resistance started with F. J. Pound's expedition to the Amazon basin and was continued in Trinidad. Studies are in progress in Brazil (Gramacho et al., 1996), Trinidad (Laker et al., 1987), Ecuador (Aragudi et al., 1987), and in the United Kingdom (Wheeler and Mepstead, 1988). No genotypes are completely immune to WB infection and the resistance is generally of a quantitative and incomplete nature. Some indications of the role of a few major genes are also available. The level of resistance of SCA 6 to WB has remained the same for more than 60 years. However, under Ecuadorian conditions, SCA 6 shows a lower a level of resistance.

Cocoa Swollen Shoot Virus (CSSV)

Several viruses are found in cocoa such as the cocoa necrosis virus in Ghana and the cocoa swollen shoot virus in Sierra Leone. CSSV is the most serious disease in Ghana, Nigeria, and Togo. It occurs also in Ivory Coast and Sri Lanka. In Ghana, CSSV caused severe economic and political problems. In other West African countries, the virus strains have been less aggressive, and it has been possible to live with the disease. Mild strains of CSSV have been found in Trinidad, Sri Lanka, and some Trinitario clones in Indonesia. Work on CSSV has been reported from Tafo (Adu-Ampomah et al., 1996; Sackey, 2000), Nigeria (Williams and Akinwale, 1994), and Togo (Djekpor et al., 1994). Results indicate that there is no immunity or high level of field resistance, but in some Upper Amazon genotypes with Iquitos mixed calabacillo (IMC), Parinary (PA), or Nanay (NA) parentage, the rate of spread was only one-quarter of that of Amelonados. Amelonado cocoa is generally more susceptible to African CSSV than Upper Amazon and *trintario* types. Some resistance sources have been reported in Upper Amazon types. So far, only a limited number of CSSV-resistant genotypes have been utilized. It is therefore necessary to increase efforts to detect other resistant progenitors.

Vascular Streak Dieback (VSD) (Oncobasidium theobromae)

VSD is the most important disease in Indonesia, Malaysia, and Papua New Guinea, causing 9 percent crop loss. More recently, it has spread to all Southeast Asian countries. It is now reported to be severe in Kerala in India, Malaysia, Philippines, Indonesia, and Papua New Guinea. Inadequate plant quarantine measures appear to have played an important role in the rapid spread over these countries. The threat of the disease to the cocoa industry in Southeast Asia is very much reduced with the detection of partial levels of resistance in several Upper Amazon and *trinitario* genotypes.

The results published by workers in Papua New Guinea (Blaha, 1996), Malaysia (Lamin et al., 1996), and India (Mallika et al., 2000; Mallika, Amma, and Nair, 2002a) indicate that the nature of resistance is of a quantitative and incomplete nature. A high level of resistance exists in SCA 6 and 12, NA 33, and KA 2-106. The resistance is inherited in an additive manner, and heritability is high. Resistance has been reported in *trinitario* cocoa grown in Papua New Guinea and Upper Amazon selections. The resistance has been stable over 25 years and its use controlled the spread of VSD in Papua New Guinea. Thousands of hybrids tolerant to VSD have been field planted in Kerala Agricultural University, India, and some of these are precocious with very high yield (Photo 5.6).

Moniliasis or Frosty Pod Rot (MO) (Moniliophthora roreri)

The fungus is endemic on wild *Theobroma* and *Herrania* species. This disease is becoming increasingly serious in Ecuador, Colombia, and Central America causing 5 percent crop loss. It is also prevalent in Peru, Vene-

PHOTO 5.6. Hybrid cocoa plant at Kerala Agricultural University.

zuela, Panama, and Costa Rica. The number of genotypes with resistance under natural field conditions is low in Ecuador and Costa Rica. It appears that some relation exists between WB and MO resistance. Clone EET 233 was consistently resistant in Ecuador. Other sources of resistance to this disease are Scavina 6, PA 169, UF 296, IMC 67, SPA 9, and EET 59.

Ceratocystis Wilt (Ceratocystis fimbriata)

This has been very destructive in the center of diversity of cocoa. It has spread to Ecuador, Venezuela, Colombia, Trinidad, and Southeast Asian countries. Upper Amazon selections IMC 67 and Pound 12 are resistant (Lass and Wood, 1985). *Criollo* types are highly susceptible.

Climate

The important climatic factors that affect the growth of cocoa are temperature and rainfall. Other factors such as altitude and latitude influence the growth of this crop mainly through their effect on these two primary factors.

Temperature

The range in the mean monthly temperature of the majority of cocoa-growing regions is found to be from 15 to 32°C and this range is considered to be the optimum for the growth of this crop. The absolute minimum for any reasonable period is taken to be 10°C, below which frost injury is likely. These temperature limits set the latitude limits for the best growth of cocoa to within 8° north and south of the equator. Temperatures of a location are modified by altitude; therefore, when deciding the elevation limits up to which cocoa can be grown, the temperature regime in the plains can be used as a guide. The extent of decrease in temperature with increasing elevation over land will be around 4.5 to 5°C per kilometer rise in elevation.

Rainfall

The two parameters that are related to rainfall are the total amount and its distribution. In most of the cocoa-growing areas of the world, the total annual rainfall is in the range of 1,500 to 3,000 mm. Values lower than 1,500 mm may mean that the supply of water through rain may not meet the evapotranspiration demand, and regular supplementation from underground sources may be necessary to support unrestricted growth of the crop. Values beyond

3,000 mm may result in excessive and continuous rain during reasonably long periods of the year, which will favor incidence of diseases such as black pod *(Phytophthora palmivora)* and vascular streak dieback. In fact, the reason for the severity of VSD in Papua New Guinea and the rapid spread of the disease in India, the like of which is not noted in any other country, is attributed to the heavy rainfall. The reason for its rapid spread and severity in India also can be attributed to this factor, especially the monsoonal nature that provides for concentration of most of the rain to two or three months of the year. For ensuring best growth, proper distribution of rainfall is considered more important than the total amount. In most of the major South American, African, and Southeast Asian cocoa-producing countries, distribution is more or less even with minor peaks. It is so well distributed that around 10 cm of rain is received almost every month. Brazil, Ghana, and Malaysia can be taken as examples of countries with such well-distributed rain.

Soil

Cocoa is grown in a wide range of soils. Soils of high rainfall areas are relatively coarse textured and acidic to neutral in reaction. Very coarse, sandy soils are not usually put to cultivation of this crop in the countries where such choices are affordable. Also, virgin, freshly cleared forest soils are used for cultivation of this crop. These make soils of most of the cocoa-growing regions of the world rich in organic matter and nitrogen, well-drained, and acidic to neutral in pH. One basic soil requirement of this crop is a depth of up to 1.5 meters, which is the depth of penetration of roots.

PROPAGATION

Cocoa can be propagated both by seed and vegetative means.

Seed Propagation

Seed propagation is the cheap and easy method. However, the seedlings will be highly variable genetically. The chances for the recovery of better progeny are high from elite parents, with the following selection criteria.

- Trees of *forastero* type having medium or large pods of not less than 350 g weight or 400 cc volume; yield to be not less than 100 pods per tree per year
- Husk thickness of pods to be not more than 1 cm
- Pod value (number of pods required to give 1 kg of wet beans) to be not more than 12
- Number of beans per pod to be not less than 35
- Bean weight (average weight of fermented and dried beans) to be not less than 1 g

A more desirable procedure for ensuring quality will be to collect seeds from biclonal or polyclonal seed gardens.

The seeds of cocoa are nondormant and lose viability quickly within seven days of extraction. If the seeds are to be stored for more than seven days, seeds may be kept in moist charcoal and packed in polyethylene bags. The beans may be extracted, testa with pulp removed, and the beans stored in polyethylene bags. This can save space and weight of material for transport. The best method is to store as pods. The potting medium of farmyard manure, sand, and soil in equal proportion is good enough for raising cocoa seedlings. Though the seeds will germinate at any time of the year, nurseries may most conveniently be sown by December-January so that four- to six-month-old seedlings will become available by the onset of monsoons for field planting. Keshavachandran (1979) recorded 94.5 percent germination when fresh seeds were sown in February-March. Seeds are to be sown with the hilum end facing downward or sown flat. The depth of placing the seeds is to be such as to just cover the seeds with soil. Removal of pulp by abrasion with suitable materials has been found advantageous in enhancing the percentage of germination. The extent of advantage from this practice is likely to be marginal. However, peeling was suggested to have the advantage of rejecting non-*forastero* types with white or pale purple cotyledons. The seeds germinate in about a week's time and germination will continue for another week. The percentage of germination will be around 90.

Seedlings of four to six months age are suitable for planting. Experience has generally shown that when field conditions are suitable for seedling growth, there will be little difference in subsequent field growth of cocoa when seedlings of varying nursery age are used. When seedlings are raised under dense shade, and if shade intensity is likely to be substantially lower in the field, hardening the seedlings by exposing them to higher illumination levels will be necessary. The period of hardening may be about ten days.

Vegetative Propagation

Vegetative propagation as a means of large-scale production of superior planting material had been practiced in almost all the producing countries, especially prior to the development of hybrid seed production programs. Vegetative propagation ensures uniformity. As most of the plants in a population are self-incompatible, the use of vegetatively propagated material from a single plant can result in no yield. Hence when budded plants/ grafts/rooted cuttings are used, it is to be ensured that multiclonal blends are planted. The different methods, viz., budding, rooting of cuttings, and grafting, are successful.

Budding

Budding on rootstocks of about 6 to 12 months growth is most often resorted to, though green budding on seedlings of two to four months is also possible. The procedure for budding on older rootstocks has been described by Nair et al. (1995). Precuring the bud wood by cutting off the laminae of all the leaves of the selected branch to a distance of about 30 cm from the tip ten days before budding increases bud take. The bud wood, just hardened showing brown bark and just hardened green leaves, is to be selected. When bud sticks are to be transported long distances before budding, they are dipped in benzyl chloride and washed in water. The cut ends are sealed using molten wax and wrapped in wet cotton wool, wet tissue paper, or blotting paper. The bundle is then packed in a box using wet packing material. The packet is then covered with polyethylene sheets. This helps to extend viability about ten days from the date of collection.

Bud wood from fan branches or chupons can be used for budding. However, when bud wood is taken from fan shoots, the budded plant will finally have a bushy appearance. Plants growing from chupon buds will grow like seedlings. Even with this advantage of chupon buds, fan branches are often selected for budding just because the availability of bud wood from chupons will be less than that from fans. Any of the common methods of budding are suitable for cocoa. In Ghana, T-budding is considered to give the highest percentage of success, whereas the experimental results at Kerala Agricultural University favor the adoption of the patch method. These results are probably arising out of the differences in the skill of the workers. About three weeks after budding, the grafting tape is cut off, and if bud union has occurred, a vertical cut is made halfway through the stem above the bud and the stem is snapped back. Continued connection of root stock with the terminal leaves through the intact side is essential for bud sprouting and further growth. The stock portion is cut back after the bud has grown to a shoot, and at least two leaves have hardened. It is then allowed to grow for a further period of three to six months, after which it may be transplanted. Under normal situations, success can be around 70 to 90 percent.

Rooting of Cuttings

Cuttings are more frequently taken from fan branches, though chupons also can be used. Fairly young branches of 10-15 cm length, green at the top and brown below, are selected. The number of leaves on the stem should be at least four; the stem cutting should have a length of 20-30 cm. The apical half of the laminar portions of the upper leaves are cut off, and all the leaves of the lower portion are removed, leaving four to five leaves at the tip. The extra length of 2 cm of the stem is then cut, and the freshly cut end is dipped in a mixed solution containing 5 g 1-naphthalene acetic acid and 5 g of indole-3-butyric acid dissolved in 240 ml each of water and ethyl alcohol. The treated cuttings can be planted in polyethylene bags filled with potting mixture. A small hole is made in the center of the pot and the hole is filled with sawdust. Cuttings are then planted on the sawdust portion. It takes about a month for the cuttings to produce roots. Until this time, the cuttings are to be kept in a completely humid atmosphere and the soil is to be kept moist. This can be achieved by keeping them in a chamber that has provisions for spraying water at frequent intervals, or, preferably, they may be kept covered with a polyethylene sheet and watered every three days. The sides of the polyethylene sheet are to be kept pressed on to the ground by weighing them down on the sides (McKelvie, 1957). When this method is followed, the cuttings are to be kept in a heavily shaded area allowing only about 10 percent light to avoid excessive heat development.

The cuttings strike roots in about a month. The rooted cuttings are to be hardened by exposing them up to about 10 a.m. for the first week, up to 11 a.m. in the second week, and up to noon during the third week. From the fourth week onward, they are kept exposed throughout the day under shade. The cuttings will be ready for planting in about six months.

Grafting

Seedlings of about four months growth are used as root stock. Scion may consist of a shoot of comparable thickness that has just turned brown and has at least two hardened leaves. These leaves are cut back to stumps of about 5 cm length just before grafting. The stock stem is cut at a convenient height (about 5 cm from ground level) and a longitudinal slit of about 2 cm length is made. The stem portion of the scion shoot is also given a slanting cut on two sides to a similar length and it is made into a wedge. The scion portion is then inserted into the stock portion and tied round with budding tape to keep it in position. The stock is then kept covered with a poly-ethylene bag and tied round the stem to keep it in position. This bag is to be removed after about a week. It takes about three weeks for the graft union to be completed before the tape can be removed. The approximate percentage of success will be around 90. The advantage with this method is that the skill required for operation is less than that for budding. The fact that it needs a longer scion for each plant to be established and that the unsuccessful stock cannot be reused are the main disadvantages.

Shaping of Clonal Cocoa Plants Derived from Fan Shoots

As indicated elsewhere, plants derived from vegetative propagation using fan branches have diffused branching systems and are asymmetrical in growth habit. If a better shape of the plant is desired, which may be beneficial in the long run, appropriate formation pruning may be necessary. This involves the identification of a chupon arising on a fan shoot, allowing it to grow, and removing the original, lower fanlike shoots in stages. This, however, has to be done slowly, as an early drastic pruning will inhibit the growth.

FIELD MAINTENANCE

Planting

Cocoa seeds can be sown directly or seedlings planted at any time of the year if soil moisture conditions are suitable. Under Indian conditions, the best time for field planting of seedlings would be by the onset of the pre-monsoon showers in May-June. Although early planting would necessitate watering before the onset of monsoons, this will be beneficial for better establishment as in the case of other plants.

Each country has its own favored method of planting, as in the case of Sao Tome where big pits of up to 200 cm are dug, filled with soil and manure, and seedlings are planted. On the other extreme, in Ghana, seeds are pushed into the soil or seedlings are planted in pits that are just big enough to contain the ball of earth of the polybag seedling. In general, experimental results indicate a lack of any significant advantage out of making planting pits if soils are naturally deep enough and are fertile. However, if soils are gravelly or if hard pans occur within the depth of penetration of roots, such a practice may be advantageous. Again, if the soils are naturally of low fertility, especially on the surface, there may be an advantage arising out of the incorporation of manures that usually accompany the filling of planting pits. When soils are of low fertility, and where gravelly laterite zones occur at varying depths, it is better to dig pits of 50 cm length, width, and depth, fill them with a mixture of surface soil and organic manures, and plant the seedlings at the surface level. A point to be noted is that cocoa seedlings are to be planted on the soil surface as the feeding roots of cocoa get concentrated on the surface irrespective of the zone at which seedlings are initially planted. Barring the exceptions of India, Malaysia, and the Philippines, cocoa is planted as a monocrop under natural or planted shade trees.

Spacing

The spacing adopted for cocoa in the major producing countries is highly variable. It was generally very low in the in situ system of planting in Africa. Experimental work at the Cocoa Research Institute of Ghana has indicated that for the Amelonado cocoa, a close spacing in the range from 1.7 × 1.7 m to 2.7 × 2.7 m was the optimum. Within this optimum range, closer spacing was advantageous in the early years, especially for the unshaded cocoa. For the Amazonian types, a wider spacing in the range from 2.7 × 2.7 m to 3.3 × 3.3 m is recommended in Ghana. On the other extreme, a relatively wide spacing of 5 × 5 m is adopted in Sri Lanka. The advantage of closer planting in earlier years is generally noted in the spacing experiments. From a consideration of experimental results on spacing and the practices followed in producing countries, Wood and Lass (1985) concluded that spacing in the range from 2.3 × 2.3 m to 3 × 3 m would suit cocoa.

The ultimate spacing and population level depends upon the extent of canopy development, the variety used, and the type of management. For the African situation, where the less vigorous Amelonado is predominantly cultivated, and where practically no costly input is used, a closer planting may be beneficial. This will also mean a better crop in the early bearing period. Among the other cocoa-producing countries, such as the Philippines, Papua New Guinea, and Malaysia, where cocoa is cultivated along with coconut, the spacing followed in Malaysia is a rather close one, there being two rows of cocoa in between the rows of coconut at a plant-to-plant distance of around 2 m. The coconuts are spaced 8 to 10 m, and the cocoa population in a hectare would be about 1,000. Cocoa is also planted in the interspaces of arecanut. Arecanut is usually planted at a spacing of 2.7 ×

2.7 m and the usually adopted spacing of cocoa also is the same. The general experience is that such a spacing results in crowding of cocoa canopy. As a tentative recommendation for the Indian situation, a row of cocoa may be planted in between two rows of arecanut. For the space-planted coconut plantation, a row of cocoa at a plant-to-plant distance of 3-4.5 m is recommended for Indian cocoa.

Shade for Cocoa

Cocoa is a plant that originated under shade and has been traditionally cultivated under shade. The shade levels at which this crop was cultivated had been, however, highly variable. The results of a number of shade trials taken up since the 1950s in cocoa-producing countries have, however, shown that the shade requirement of this crop varies widely depending on the stage of growth, with it requiring as much as 75 percent shade in the early stages. This is gradually brought down to about 25 percent when cocoa comes to production. Such variations in shade levels are provided in the sole crop situations in Costa Rica, Ghana, Colombia, and Brazil by providing different types of shade plants and by their selective thinning. The studies conducted by Nair et al. (1996) at the Kerala Agricultural University on the response of cocoa to shade indicated that the girth of stem and yield increased with increases in illumination levels. The results suggested that it is possible to cultivate cocoa without shade under Kerala conditions and that the productivity will be the highest under shade-free situations. However, shading may be necessary in the early years using temporary shade plants.

In pure crop situations, permanent shade trees are planted or left without removal at a wide spacing of about 13 to 15 meters, and temporary shade plants are planted at the same spacing as cocoa, alternating with it. The common temporary shade plants in African countries are banana, tree cassava or cera rubber *(Manihot glaziovii),* and cocoyams *(Colocasia esculenta).* These shade plants are gradually removed as cocoa grows and the canopy develops. The permanent shade tree commonly used for planting in Ghana is *Terminalia ivorensis.* The other shade trees that may be used are *Gliricidia maculata* planted at about 3 × 5 m, dadap *(Erythrina lithosperma)* and *Leucaena leucocephala.* With both permanent and temporary shade plants, the shade level will be high resulting in the best vegetative growth of young cocoa. When temporary shade plants are removed as cocoa comes to its bearing stage, illumination will increase, stimulating production.

Manures and Fertilizers

Application of manures and fertilizer is not done on a routine basis in any of the major cocoa-producing countries. The primary reason for this is the high natural fertility of cocoa soils of these countries, these being mostly freshly cleared forest lands. The presence of shade trees, the dense canopy development of cocoa, and the large turnover to the soil of cocoa litter prevent any substantial loss of soil by erosion and depletion of nutrients. In a study at the Central Plantation Crops Research Institute at Kasaragod, India, the amount of organic material returned to the soil as cocoa litter was estimated at 818 and 1,985 kg/ha per year (on dry weight basis) in the single- and double-hedge systems of planting cocoa. The quantities of fertilizer nutrients contained in organic material from double-hedge planting were estimated as 50 kg N, 11 kg P_2O_5, and 35 kg K_2O per hectare. The quantities of N, P, and K removed by cocoa pods per kg of dry beans will work out to 43.80, 8.04, and 64.29 g, respectively. For a crop yielding about 2 kg of dry beans per plant (about 60 pods) per year, the average crop removal by pods would be around 85, 37, and 154 g each of N, P_2O_5, and K_2O. The fertilizer recommendation for cocoa under average management is 100:40:140 g of N, P_2O_5, and K_2O per plant for a year, which tallies with the crop removal figures. For cocoa under better management where the average annual yield is more than 60 pods, double this dose is tentatively recommended.

Method and Time of Application of Fertilizers

The feeding roots of cocoa are concentrated on the soil surface, occurring within a depth of 15 cm from the surface. Laterally, they are concentrated in a radius of 120 to 150 cm in established, adult cocoa. As such, fertilizers may preferably be applied in shallow basins of 120 to 150 cm radius and raked in without serious damage to the roots. The general recommendation in most cocoa-producing countries is to broadcast fertilizers in the entire field without any soil tillage. Although this may be suitable when the soil surface is wet and there is chance of immediate dissolution of fertilizers, a risk of volatile loss of fertilizer nitrogen exists when conditions are adverse. An immediate mixing of fertilizers with soil will reduce the chances of a volatile loss of nitrogen especially when urea is used as the nitrogenous source.

For timing the application of fertilizers, the stages of crop activity and the seasons of moisture availability may have to be considered. For unirrigated Indian cocoa, flushing occurs mostly in the rainy season starting from

May-June and continues up to November-December. Though essentially the same trend persists in irrigated cocoa, there will be some flushing in the summer season also. For rainfed cocoa, availability of soil moisture will impose another restriction. Taking these into account, fertilizer application of rainfed cocoa may be done in two splits, the first coinciding with the premonsoonal rains during May-June and the second by the close of the monsoons in September-October. For the irrigated cocoa, fertilizers may be applied in four equal splits during May-June, September-October, December, and February. Such four-split application has been found beneficial for coconut also.

For young cocoa in the field, the dose of fertilizers may be one-third the annual dose for adult plants for the first year and two-third for the second year. As cocoa under good management will start giving reasonable yield from the third year, it may be logical to supply full doses of fertilizers from this point on.

Application of organic manures for adult cocoa may not be essential, as there is a large return of organic debris to the soil by cocoa plants and a consequent substantial enhancement of soil organic matter content. However, for young cocoa, organic manures will be useful. These may be applied in the planting pits when seedlings are field planted and to the shallow basins afterward.

Pruning

Pruning is intended to restrict the growth of the trees to a convenient height, to have the first tier developed at the desired height, and to remove the excessive and inconvenient development of branches. Nair, Virakthmath, and Mallika (1994) observed significantly higher yield in unpruned control plants during the fourth and fifth years, which ceased to be statistically significant during the subsequent years though the trend of superiority of control continued with decreasing magnitude with advancing age. It is, thus, only for convenience that cocoa is pruned. Among all the major cocoa-producing countries, the only country in which cocoa is regularly pruned is Brazil. Pruning is not a part of cocoa management in any of the African countries.

The advantages of pruning are the convenience in harvesting, plant protection operations, and cultivation. If unrestricted growth of the tree is allowed, harvesting would require workers to use knives attached to poles. Climbing the cocoa plant for harvesting would not be feasible lest damage occurs to the flower cushions. It will be more convenient to undertake spraying if plant growth is restricted. Having the first tier developed at heights lower than 1 to 1.5 m will make cultivation operations difficult.

With these considerations in mind, the following pruning operations may be suggested:

- The first tier should develop at heights not less than 1-1.5 m. If plants jorquette at lower heights, the stem with the developing fans may be nipped off just below the jorquette. New chupons will arise on the main stem. One healthy chupon shoot may be allowed to grow up and jorquette. All the other chupon shoots may be removed. This process of nipping chupon shoots may be continued until the desired height is attained. The height at which jorquettes form is decided by the nature of the plant, but environmental factors also play a role. In general, jorquetting is higher for plants under heavy shade. Nonavailability of mineral nutrients and water tend to lower jorquette height. Some plants tend to grow very tall before jorquetting. No known method exists by which jorquetting height can be lowered in such cases. However, such plants are rare in a population.
- There should be only one main chupon stem. At times, additional chupons arise from the main chupon, which are to be removed periodically.
- Vertical growth is to be limited to a single tier. However, a second tier may be allowed to develop if the first tier is damaged. Arresting further vertical growth would require continuous removal of chupons that develop from below the jorquette. This will have to be a continuous process as the normal tendency is for the plants to put out new chupons. Normally, chupons arise from chupon stems only, and fan laterals arise only from fans. Rarely, chupons arise from fans, and should be removed at the early stages.
- Drooping fan branches may be cut off at a suitable distance from the jorquette once a year, preferably when the crop load is low. December-January and July-August may be convenient under the Indian situation, as cocoa will be nearly pod free during this period. Removal of part of the foliage may also help to reduce transpiration in the summer season.
- "Centering" as a part of pruning was recommended previously. This involves removal of all lateral fan branches that arise on the fans to a certain distance around the jorquette. The distance from the jorquette that is often suggested is in the range of 30 to 50 cm. This allows sunlight to fall on the main stem, which is beneficial for enhancing flowering. This speculation, however, lacks experimental support, and,

further, even if it enhances flowering, it may not mean enhanced cropping as flower production is rarely a limiting factor in cocoa. The general experience is that flower production is profuse even in plants with
dense canopies shading the main stem. A more probable advantage
from centering may accrue from better aeration of the fruit-bearing
main stem. This may reduce the incidence of black pod in the rainy
season.

- Although the previous recommendations may apply in the case of
  young cocoa, mature plants may be pruned gradually without much
  shock. The suitable period for this operation for Indian cocoa may be
  in December-January. Similarly, if the second tier that is already developed is to be removed, it may also be done in phases, removing the
  fans one by one.

Weed Control

The usual weed control operation common in major cocoa-producing
countries, where cocoa is grown as a sole crop, is slashing done twice a year
at a height of 5 to 15 cm. Experiments on near-complete removal of weeds
in comparison to slashing done in Ghana have indicated no additional advantage to complete removal. In the long run, it may also have the disadvan-
tage of enhanced erosion of bare soil. Slashing will have the advantage over
clean cultivation of being less labor consuming. Weeds compete with the
crop for water, nutrients, and light. Under conditions in which cocoa is
grown in the major producing countries where the rain-free period is short
and fertility superoptimal, weeds may not seriously compete with the crop.

Irrigation

As a general management practice, irrigation is not done in any of the
major cocoa-producing countries except for container-grown seedlings.
Experimental results also generally indicate the lack of response to irrigation. Obviously, the reason for such a lack of response must be attributable
to the well-distributed rainfall in these regions, the water stored in the root
zone being enough for normal functioning of the plants. In countries where
the rain-free period extends from four to six months, irrigation in the
summer months will be beneficial.

Top Working

Top working is useful to rejuvenate old and unproductive cocoa plants
and also to convert genetically poor yielders to high yielders. This tech-

nique was standardized at the Cadbury–KAU Co-operative Cocoa Research Project, Vellanikkara (Nair, Mallika, and Swapna, 1994). This technique consists of snapping back the desired trees below the jorquette after cutting half way. The snapped canopy continues to have contact with the trunk. A number of chupons arise below the point of snapping and this is triggered by the breakage of apical dominance. Patch budding, as practiced in the nursery, is done on three to five vigorous and healthy shoots using scions from high-yielding, disease-resistant clones, and the remaining chupons are removed. The polyethylene tape is removed three weeks after budding and the stock portion above the bud union is snapped back. The snapped portion is removed after the development of at least two hardened leaves from the bud. When sufficient shoots are hardened, the canopy of the mother tree can be removed completely. Because of the presence of an established root system and a trunk with reserve food, top-worked trees grow much faster and give prolific yield one year after the operation. Though top working can be done in all seasons, it is preferable to do it during the rain-free period in irrigated gardens. For rainfed situations, it may preferably be done after the receipt of pre-monsoon showers.

The top-worked trees start yielding heavily from the second year onward. About 50 percent improvement in yield is obtained in the second year (Nair, Mallika, and Swapna, 1994), and about 100 percent increase is obtained in the third year (Jose et al., 1998). Loss of crop for one year during the operation is thus compensated by bumper crops in the coming years. The main stem will continue to belong to the original plant, and the fruits borne in this area belong to the poor yielder. Better yields are, however, obtained from the fan branches of the high-yielding clone used for top working. Analysis of performance of these top-worked trees over a period of eight years indicated that these trees continued to yield heavily (Nair et al., 2002).

FLOWERING AND FRUIT SET

Cocoa does not flower uniformly throughout the year, and there are peaks of flowering during some months of the year. These periods of peak flowering are often different for the different regions, indicating strong association with climatic factors. For example, in Ghanaian cocoa, normal flowering reaches its peak during May-June, starting from March and extending through April. In addition to this normal flowering is the "crazy" flowering that may occur during any period of the year. Several factors are considered responsible for the seasonal trend.

Genetic Factors

Cocoa types differ both in flower production and seasonal pattern. The African Amelonado is generally the least floriferous and is normally out of flower or nearly so from August to December in the Ghanaian climate. During the early part of the year, Amelonado may remain without flowers generally; but stray flowering is noted in some years. The more floriferous cocoa types, in addition to producing large number of flowers, often show some flowering every month with peaks from March to May.

Environmental Factors

Moisture Stress

Cocoa flowers profusely by the onset of the wet season following a dry spell. Flowering is more profuse when the dry period is extended. In Ghana, the wet season starts normally from February following a dry period extending over December and January. On receipt of rains, a spurt of flushing is followed by the flowering period. In some instances, decreased flowering appears to result from decreased soil moisture following a period of excess moisture.

Temperature

Mean monthly temperatures below 23°C are considered to suppress flowering in this crop.

Solar Intensity

Limitation in illumination intensity of the cocoa canopy is not reported as a factor affecting flowering. In fact, in locations and seasons with continuous monsoonal rains and cloudy weather, flower production is significantly inhibited.

Internal Factors

Information available so far suggests a strong association between carbohydrate, nutritional, and hormonal reserves in the plant and seasonal patterns in flowering. Induction of flowering following fertilizer application shows that mineral nutrient status influences flowering. Flowering is reported to follow heavy flushing by the onset of the wet season in Ghanaian cocoa and the consequent rise in leaf area of plants. Increased leaf area index (LAI) would support higher photosynthetic activity and carbohydrate status. Hence, carbohydrate supply is considered as a factor responsible for inducing flowering in such situations. The fact that fruit load affects flowering intensity, there being a decrease in flowering during intense fruit development periods, is taken as another indication of carbohydrate status as a factor.

Taking the previous factors together, the seasonal flowering pattern of cocoa can be explained in most of the situations. Thus, Ghanaian cocoa is considered to enter into a period of activity from February to May with the beginning of the wet season following the dry period extending over December and January. The relief from water stress induces flushing, and thereby an increase in LAI. After about a month of flushing, when leaves harden and start photosynthetic activity, an induction of flowering occurs during this period. Competition from developing pods acts as the inhibitory factor up to November, which is the month of peak harvest in Ghana. The dry spell ensuing from December prevents flowering up to February.

Cherelle Wilt

Among the large number of flowers that are successfully fertilized, only a small percentage is carried to maturity. Until it attains a length of about 10 cm the young cocoa fruit is called a "cherelle," and more than 80 percent of the cherelles formed on a mature tree usually wilt. This phenomenon is called "cherelle wilt" and several reasons are attributed to its occurrence. Several fungal organisms and insect pests are found associated with wilting of cherelles, especially fungi. Attempts to correct it using fungicides and insecticides, although they could almost eliminate these organisms, could reduce cherelle wilt only slightly. The major cause of it is considered to be physiological, involving competition mainly for carbohydrates and, to a lesser extent, for mineral nutrients. Because of physiological reasons, cherelle wilt occurs only up to a certain stage of growth of the developing pods, there being two peaks, 50 and 70 days after fertilization. No wilting occurs after 100 days. The first wilt coincides with cell division in the endosperm, and the second with the rapid growth of an embryo.

Indications of the involvement of mineral nutrient supply as a factor affecting cherelle wilt are given by the observations of a decrease in percentage of wilt following fertilizer application. However, fertilizer supply appears to have a much larger effect on setting of fruits than on cherelle wilt.

The major factor considered responsible for incidence of nonpathological cherelle wilt is the competition for carbohydrates. It has been reported that the highest degree of wilting occurred during or just after periods of leaf flush. Another observation indicating competition as the factor is the decrease in wilting following removal of pods. Probably more important than the association with the vegetative sinks is the competition offered by older pods for the carbohydrate supply.

Indications of competition from pods as sink are evident in the rate of growth of young pods. The general trend is that the pods that set early grow fast whereas the late sets are slower in growing. The period required for wilting will be longer in early set pods than those of the late set ones. The pattern of competition discussed previously for carbohydrates would mean that if a tree has massive flowering, a large number of cherelles will set compared to when fruiting is scattered. When hand pollination is used in hybrid seed production, this fact is put into practice. The total annual yield, however, may remain the same as the trees that are heavily loaded with fruits at a time tend to show less profuse flowering and more cherelle wilt in the period that follows. In the large-scale hybrid seed production in Ghana, hand-pollinated mother trees are allowed to rest in alternate years as a safeguard against excessive carbohydrate drain and the probable consequent damage to the trees. This would also facilitate having a higher fruit load when hand pollination is resorted to.

PLANT PROTECTION

Pests

Cocoa is affected by more than 1,500 insects in different cocoa-growing countries of the world. However, only a small number is of economic importance. Among the major pests infesting cocoa, the important ones are the red borer, tea mosquito bug, mealy bug, gray weevil, cockchafer beetle, rat, striped squirrel, etc.

Insect Pests

Red borer (Zeuzera coffeae). This pest infests mainly the young cocoa plants. Larvae bore into thick shoots and into the main stem below the first jorquette along the center and cut a traverse tunnel before pupation. Small lateral galleries with openings are seen at intervals of about 25 cm on the stem. Infestation of the main stem causes drying up of the plants. The affected fan shoots show complete withering of leaves, and these subsequently break off. Plants can be protected from the red borer by spraying carbaryl 0.1 percent on the stems. Pruning and burning of the affected parts are to be done as soon as the symptoms are noticed.

Tea mosquito (Helopeltis antonii). These bugs mainly attack the pods. Circular water-soaked spots develop on the infested pods around the feeding punctures. These punctures later turn pitch black. Multiple feeding injuries cause deformation of the fruits. The pest can be controlled by spraying endosulfan 0.05 percent.

Mealybugs (Planococcus lilacinus). These bugs occur in groups on the tender shoots, flower stalks, foliage, and on the developing pods. Cherelles are often severely attacked. Maturing pods infested by the bug develop irregular sunken patches leading to the formation of scabs. This pest occurs throughout the year, but attains peak levels during July-October. It can be controlled effectively by the application of quinalphos at 0.05 percent. The insecticide may be applied only after collecting the mature pods.

Gray weevil (Myllocerus spp.). A number of species of the *Myllocerus* weevils infest cocoa in the different cocoa-growing regions of the world and cause considerable damage. The peak periods of infestation occur during July-September. Infestation is severe on the young plants. Adults occur in groups on the underside of the leaves and feed on the green matter leaving the veins intact. The flaccid young leaves are generally not affected. The entire foliage will be badly skeletonized causing retardation of growth.

Attack on young plants up to two years can be controlled by prophylactic spraying of carbaryl 0.1 percent. Spraying may be concentrated on the under surface of the leaves. Application of the insecticide may be done twice a year—once during May and again in September.

Cockchafer beetle (Leucopholis spp.). Grubs of this insect feed on the surface roots and taproots of young cocoa plants causing yellowing and drooping of the foliage. Adult plants are also sometimes seen infested. Infestation by this pest is more frequent in coconut-cocoa intercropping situations, as the same pest attacks coconut as well. Seedlings can be protected by application of carbaryl 10 percent dust in pits around the root zone. Adult plants may be drenched with chlorpyrifos 2.5 ml per l.

Citrus aphid (Toxoptera aurantii). These aphids occur mainly during June-October. They infest flower stalks, tender flaccid leaves, buds, and tender chupons. On the foliage, the aphids are confined to the lower side. Chemical control of these aphids is not essential. Severely affected plant parts are to be collected and destroyed.

Red branded thrips. Adults and nymphs of the thrips appear in colonies on the undersurface of the leaves and also on the pods. The thrips feed on the fluid exuding from the scraped tissues. Infested leaves turn pale green to pale brown and later dry up. The thrips can be controlled by the application of 0.05 percent suspension of quinalphos, phosalone, or fenthion.

Storage pests. Cocoa beans stored for more than two months are found to be damaged by many species of insect pests. The most important among these is the rice meal moth *Corcyra cephalonica.* The larvae of this moth feed on the internal contents of the beans and construct silken galleries using frass and broken-down particles of the beans. Direct application of insecticides to cocoa beans or to the containers is not recommended. The cocoa beans for long-term storage can be mixed with neem leaves at 2 percent by weight, which helps to protect the beans up to six months.

Noninsect Pests

Striped squirrels (Funambulus tristriatus). These rodents cut irregular holes on the walls of maturing pods and completely extract the contents. They feed on the mucilaginous pulp around the beans. Continuous trapping using attractants and poison baiting will be effective to check the population. Since the squirrels damage ripe fruits only, damage can be reduced by harvesting mature pods at regular intervals. Mechanical protection of the pods by covering with punched polybags (150 gauge) smeared with bitumen–kerosene mixture can also be partially effective.

Rats. The nature of damage is similar to that caused by squirrels. The holes made on the pods are surrounded by areas of endocarp exposed by scraping the fleshy portions of the pods. Rats also prefer pods in the postbronzing stage for feeding. Harvesting the pods at the right stage when the furrows start bronzing will reduce damage considerably. Baiting with rodenticides in the garden is recommended. Rainproof preparations are preferred. Fumarin bars (rainproof) tied to the base of the inner frond of coconut and setting up bamboo noose traps in coconut crowns can be quite useful to control rats.

Civet cat (Paradoxurus hermaphroditus). Unlike the rodents, civet cats gnaw holes on the pods, bite, and break the husk. Pieces of broken chunks are 2-3 cm in diameter. Infested pods show two distantly spaced (about 1 cm apart) markings caused by the canine teeth and a row of small dots representing the markings of the incisors. The civets swallow the beans, and as such no trace of beans will be visible under the tree. Instead, piles of defecated beans are seen scattered around the plantation. Civet cats can be controlled by poison baiting with carbofuran granules. Ripe bananas are split longitudinally into two halves, about 0.5 g of carbofuran is added, and the halves are closed properly. Two such bananas may be kept on the trunk at 5-6 cocoa trees per hectare.

Diseases

Cocoa is affected by many diseases. Loss due to diseases in cocoa has been estimated to be 21 percent (Hale, 1953). Diseases may debilitate or kill the tree depending upon the type of pathogen involved. The important diseases are described as follows:

Phytophthora Pod Rot *(*Phytophthora palmivora, P. capsici, P. megakarya, *and* P. citrophthora*)*

This disease is very serious during rainy seasons. Infection appears as minute, translucent, water-soaked spots on the pod surface, which turn chocolate brown, darken, and increase in size. Ultimately, the whole pod is invaded by the fungus and the pods become completely black. The beans in a ripe pod may escape partially or wholly from infection as the beans get separated from the pod husk on ripening.

Periodic removal and destruction of the infected pods will help to reduce spread of the disease. Cultural practices, such as proper pruning and regulating the overhead shade to reduce humidity and improve aeration, have been recommended for the control of the disease. Copper fungicides, such as bordeaux mixture, copper oxychloride, cuprous oxide, or copper hydroxide, are generally used for the control of the disease. Spraying of bordeaux mixture 1 percent at 15-day intervals starting from the onset of the monsoon along with periodic removal of infected pods is effective in controlling the disease in severely affected gardens. Extracts of *Allium sativum, Cinnamomum zeylanicum, Lawsonia inermis,* and *Adenocalymma allicea* have been found to be effective in inhibiting lesion development on detached cocoa pods. Antagonistic effects of *Pseudomonas fluorescens* against *Phytophthora palmivora* have also been reported.

Genetic resistance offers better prospects for the control of this disease. Selections of cocoa, such as SCA 6, SCA 12, Pound 7, Catongo, and K82, had shown some resistance to this disease. Studies conducted in Java have indicated that the cocoa types DRC 16, SCA 6, SCA 12, and ICS 6 were resistant to *Phytophthora* pod rot.

Colletotrichum Pod Rot *[*Colletotrichum gloeosporioides *(Penz.) Sacc.,* C. theobromae, C. lurcificum, *and* C. eradwickii, C. incarnatum, C. fructitheobromae, C. thobromicolum*]*

Infection starts from on the pod surface, usually from the stalk end or from the tip of the pod. The lesions develop as dark brown areas with diffused yellow halo. Infection spreads to the stalk and advances to the cushion. The internal tissues of the pod become discolored. In certain cases, infection may initiate from other parts of the pods also. Dark brown sunken lesions later coalesce to form larger ones. In severe cases of infection, the whole pod surface is infected. The pod shrinks and remains on the tree in a mummified form. Carbendazim and mancozeb are reported as promising fungicides for the control of the disease.

Botryodiplodia *Pod Rot/Charcoal Pod Rot* [Lasiodiplodia theobromae *(Pat) Griffon and Maubl.* Botryodiplodia theobromae *Pat]*

This pod rot occurs more frequently during the dry season. Symptoms initially appear as pale yellow spots on the pods, which enlarge into chocolate brown larger lesions. In general, infection originates at the stalk end or at the tip of the pod. At times, the lesion develops from other parts of the pod. In most cases, the entire pod becomes black and exhibits a sooty covering all over consisting of the spores of the fungus. Infected pods become mummified and remain attached to the plant. The disease is also found to affect young twigs causing dieback.

Since the disease mainly affects wounded pods and pods of plants under stress, better management practices will reduce the incidence of the diseases. Use of Rovral (Iprodione) 2,000 ppm at monthly intervals for six months during the dry season is the suggested control measure. Since the fungus is a wound pathogen, spraying of 1 percent bordeaux mixture along with an insecticide will also be useful.

Witches'-Broom [Crinipellis perniciosa *(Stahel) Singer]*

The disease was first reported from Suriname in 1895 and was largely responsible for the destruction of cocoa both in Suriname and Guyana. This disease has spread to Bolivia, Brazil, Colombia, Ecuador, Guyana, Peru, Suriname, Venezuela, Grenada, Tobago, and Trinidad. The characteristic symptom is the development of brooms or shoots due to the hypertrophic growth of the infected bud. Infection of axillary bud or terminal bud leads to the production of vegetative broom. Such infected buds develop into dense, curved growth with excessive lateral shoots and short internodes. Leaves remain small with swollen stalk and pulvinus. Stipules are generally larger than normal and persistent. Some shoots grow vigorously from the infected region, as a result of which a "grown-through" broom develops. Brooms develop on the fan branches as well as on the chupons. If the bud dies due to infection, the broom remains inconspicuous. Abnormality occurs also in the internal tissues. Infected flower cushions sometimes produce leafy broom with less proliferation of axillary buds or abnormal flowers known as "star broom." Pods are infected by hyphae ramifying through the stalk from the cushion or directly by the penetration of the germ tube. As a result, pods exhibit abnormal shape, such as carrot shape, or one-sided distortion, which varies with the age of the pod. Sometimes infected cherelles do not ripen normally. Larger pods become hard when dry.

The source of inoculum can be reduced by removal of the brooms. It is essential to start phytosanitation before the disease becomes severe. Thus, early removal of the brooms helps in minimizing the cost of disease control. All diseased tissues should be removed and burned, since brooms on the ground still produce basidiospores. Pruning of brooms should be done twice a year during dry periods. This disease has not been noticed in India.

*Monilia Pod Rot (*Moniliophthora roreri *Evans)*

Monilia pod rot or moniliasis is prevalent in South American countries. The fungus infects only young pods. The initial symptom appears as small water-soaked lesions. Such lesions coalesce to form dark brown necrotic spots with an irregular margin, which later spread and cover the entire pod surface. Later, the lesions become covered with whitish mycelium consisting of abundant conidia. In certain cases, the diseased pods may look healthy; but at harvest, the internal portions will be found rotten.

The disease can be controlled effectively by weekly removal and destruction of the diseased pods. Disease intensity can be reduced by improved drainage, regulation of shade, frequent and light pruning, and timely weeding of the cocoa plantation. Control of the disease through the planting of resistant/tolerant cultivars seems promising.

*Mealy Pod Rot (*Trachysphaera fructigena *Tabor & Bunting)*

This disease is is reported only from west Central Africa. Infected pods develop brown lesions similar to those of *Phytophthora* pod rot. However, this pod rot can be distinguished from *Phytophthora* pod rot based on the nature of sporulation. The white encrusted mass of spores produced on the infected area turns pink very rapidly. Vegetative parts of the plant are not affected. It has been found that fungicidal application for the control of *Phytophthora* pod rot can give good control of mealy pod rot also.

Seedling Blight [Phytophthora palmivora *(Butler) Butler*]

Symptoms of the disease develop on the leaves and stems of seedlings. On young leaves, initial symptoms appear as small water-soaked lesions on the undersurface of the lamina. These lesions are either scattered all over the leaves or seen at the distal end and margins of the leaves. Later, defoliation occurs. On mature leaves, the water-soaked lesions appear along and near the veins. These later turn to dark brown, resulting in leaf blight and defoliation. On immature stems, initial symptoms develop as water-soaked linear brown lesions, which later turn black. Infection starts from the tip of the seedlings, spreads downward, and results in defoliation and dieback. Infection at the cotyledonary region spreads both upward and downward resulting in wilting.

The disease can be controlled by improving drainage facilities in the nursery, by adjusting the shade, and by drenching or spraying the seedlings with bordeaux mixture or copper oxychloride just before the onset of the monsoon and thereafter at frequent intervals. Severely infected seedlings should be removed and destroyed. Potassium phosphonate is also found effective in reducing the severity of the disease.

Vascular Streak Dieback (Oncobasidium theobromae *Talbot and Keane*)

Vascular streak dieback is a destructive disease of cocoa in Pupua New Guinea and many Southeast Asian countries, such as Malaysia, the Philippines, southern Thailand, India, China's Hainan Island, and several provinces of Indonesia. This disease may occur on the main stem of a seedling or on a branch of an older tree. The first symptom is yellowing of the second or third leaf behind the growing tip with the development of green spots or islets scattered over a yellow background (Photo 5.7). The infected leaves fall off within a few days and, subsequently, the leaves above and below turn yellow and are shed, resulting in a distinctive situation where the leaves on the middle portion of the shoot fall. Lenticel enlargement, axillary bud proliferation, and interveinal chlorosis are the other symptoms. If the bark is peeled off from the infected region, the cambium turns rusty brown very rapidly. The xylem vessels show several brown streaks when the affected stem is split open. Hyphae of the fungus are seen in xylem vessels of the infected stem and leaves. Genetic resistance offers good prospects of controlling VSD in the long run. Cultivars of Upper Amazon and *trinitario* origin are, in general, less susceptible than Amelonado or its hybrids.

PHOTO 5.7. Vascular streak dieback disease on cocoa leaves.

Phytophthora Canker [Phytophthora palmivora *(Butler) Butler]*

This disease is reported from Sri Lanka and India. Stem canker appears in different parts of the tree, including jorquette and fan branches. The external symptom appears as grayish-brown water-soaked lesions with broad dark brown to black margin on the bark. A reddish-brown liquid oozing out from such lesions dries up and forms a rusty deposit. The tissues beneath always show a characteristic reddish-brown discoloration. Lesions in the tissues coalesce leading to extensive rotting. The infection spreads from the cortical tissues to the vascular tissues and reaches the wood. Wood infection appears as grayish-brown to black discoloration with black streaks. When canker girdles the stem, dieback occurs. Leaves wilt, turn yellow, and fall off. Pods also show wilting. Finally the whole tree dies. Spread of infection in the internal bark is faster than the spread in the surface of bark.

This disease can be managed effectively if detected in the early stages of infection. If detected early, remove the affected tissues and apply bordeaux paste, Difolatan, or any copper-based fungicide. Care should be taken to remove and destroy the infected tissues completely. Cut and remove the infected small branches. Proper measures of control of *Phytophthora* pod disease will also help in reducing the canker incidence.

Pink Disease (Corticium salmonicolor *Berk & Br*)

Fan branches and small twigs are generally infected. The first indication of the disease is the death of the branch. On the bark of the infected branch, characteristic pinkish, powdery encrustations of the fruiting bodies of the fungus can be seen. After a long time, the pinkish color turns to grayish white. Before the appearance of visible external symptoms, many fine, white, silky mycelia have already spread over the surface and into the cortex of the bark. This leads to the defoliation and death of the distal part of the branch.

Removal and destruction of the infected branches and wound sealing with bordeaux paste will check the disease. Avoidance of excess shade and provision of adequate drainage are important for better control of the disease. Because the pathogen has a wide host range, care should be taken not to use the susceptible species as shade or as a cover crop. In areas where severe incidence is recorded, the disease can be prevented by regular spraying of 1 percent bordeaux mixture, especially during the rainy season.

Thread Blight

Three types of thread blight, viz., white thread blight, horsehair blight, and *Koleroga* have been reported from different cocoa-growing countries.

White thread blight [Marasmiellus scandens *(Massee) Dennis & Reid*]. The young branches of infected plants contain white mycelial threads of the fungus, which spread longitudinally and irregularly along the surface of the stem. Under high humid conditions, the fungus grows very rapidly on the stem and enters the leaf at the nodes through the petiole. On the leaf lamina, the fungus spreads in the form of fine threads. As a result, the affected portion turns dark brown to black, ultimately leading to the death of the leaves. Such dead leaves in a branch eventually get detached from the stem, but are found suspended by the mycelial threads in a row. Extensive death of young branches and hanging leaves in mycelial threads in rows are the common field symptoms. Severity of the disease can be reduced by the removal of the dead materials, pruning of the affected parts, and proper shade regulation. In severe cases of infection, spraying of copper fungicides will be helpful.

Horsehair blight. (Marasmius equicrinis *Muell.*). Young branches of the affected plants are found to be ramified with tangles of the black hairlike growth of the fungal mycelium. The black branched mycelial network of the fungus spreads over the leaves, petioles, and twigs. The mycelia remain hanging loosely. As a result of the infection, the midribs of the leaves show a brown discoloration. From the midrib, the discoloration spreads to the veins and veinlets. Later, the affected leaves dry up. The infected twigs also show drying. Such dried-up leaves and twigs get detached from the branches and remain suspended by mycelial strands. Since horsehair blight disease is observed in neglected gardens, incidence of the disease can be reduced by proper management practices. Removal and destruction of the affected parts are quite useful.

Koleroga *thread blight.* The symptoms of *Koleroga* thread blight caused by *Pellicularia koleroga* Donk are similar to those of white thread blight except that the fungal threads are brown.

Colletotrichum *leaf blight* [Colletotrichum gloeosporioides *(Penz.)* *Sacc.*]. *Colletotrichum* infection on cocoa leaves was reported as one of the serious problems of cocoa in Columbia and Ghana. In India, *Colletotrichum* causes three types of foliar symptoms, viz., leaf blight, shot hole, and irregular leaf spot. Of these, leaf blight and shot hole are widespread and occur on plants of all age groups. Shade regulation is found to be an effective method of control of the disease in Ghana. The disease can be effectively controlled by proper pruning, and in cases of severe incidence, by spraying carbendazim or mancozeb.

Chupon blight and twig dieback [Phytophthora palmivora *(Butler) Butler]*. Infection usually initiates anywhere on the young leaf lamina, in the axils of the leaves, on the petiole, or at the tip of the twigs or chupons. The characteristic symptom is the appearance of water-soaked lesions, which later turn brown or black. In severe cases of infection, the lesions coalesce to form large lesions. Severe infection on the leaves leads to defoliation. When the lesions girdle the stem, the portion above the point of infection dies, causing twig dieback or chupon blight. Shade regulation by proper pruning is found to be the best method of reducing the disease incidence. Pruning should be done just before the rainy season. In cases of severe incidence, the disease can be controlled by spraying bordeaux mixture or any copper oxychloride preparation.

Cocoa Swollen Shoot Virus

Many strains of cocoa swollen shoot virus exist, which differ in the symptom expression, the vectors that transmit them, and the range of their other hosts. The virulent strains observed in Ghana produce various types of leaf necrosis, root and stem swellings, and dieback in Amelonado trees. Such infected trees are usually killed within two to three years. Avirulent strains rarely produce leaf symptoms and are not lethal. However, root and stem swellings are often produced.

Swollen shoot virus strain IA (new Juaben strain) produces swelling on the fan branches, chupons, and roots. This strain on young flush leaves of Amelonado causes red vein banding and vein clearing. Later, a fern leaf pattern is produced, and mature trees at this stage have a yellowish appearance. Pods on the infected tree become mottled and are smoother than normal, containing only half the normal weight of beans.

Different strains of cocoa swollen shoot virus produce different symptoms on leaves. The symptom variation on leaves produced by the different strains helps in strain identification. The Ghanaian strains are more virulent and damaging than Nigerian strains. This disease is more damaging to cocoa trees under stress.

More than a dozen species of mealybugs are reported to transmit the virus. The important mealybug species transmitting the disease are *Planococcus citri*, *Planococcoides njalensis*, *Planococcus hargreavesi*, *Ferrisia virgata*, and *Pseudococcus concavocerarii*. Spread of the disease can be arrested by removal and destruction of infected plants. However, this method does not prevent new outbreaks. The use of resistant lines is probably the ultimate solution to this malady. Apart from cocoa swollen shoot disease, other virus diseases, such as cocoa necrosis virus, cocoa mottle leaf virus, and cocoa yellow mosaic virus, have also been reported.

Ceratocystis *Wilt (*Ceratocytis fimbriata *Ellis & Halst.)*

This disease is usually associated with damage by beetle borers or pruning wounds. The disease occurs in almost all cocoa-growing countries of the world. Wilting of the whole or part of the tree followed by rapid death are the visible symptoms of the disease. In the initial stages of the disease, mature leaves change from their normal horizontal position and become pendulous. The wilted leaves then dry and remain attached to the dead branch for several weeks. The disease is always associated with borer holes on the stem made by *Xyleborus* beetles. The holes are about 1 mm in diameter, with a small amount of wood dust around it. The internal wood tissue surrounding the wound will be discolored to brownish red or purplish.

Neither chemical control of the beetle or fungus nor destruction of the infected plants have proved a useful method of control of the disease. The most practical method of preventing the disease incidence is to minimize wounding during harvesting and pruning. It has been reported that *criollo* cultivars are more susceptible than *forastero*.

Verticillium *Wilt* (Verticillium dahliae *Kleb.*)

This disease is prevalent in Uganda and Brazil. The first symptom is the drooping of the leaves without any loss of turgor, so that the leaves hang down without any flaccidity. Subsequently, the leaves dry and roll inward and later fall off. The infected small branches break off gradually. In the early stages of disease development, a marked reduction occurs in the root system and wilting of young pods. Necrosis of tap and main lateral roots occurs only after defoliation of the shoot. Discoloration of the xylem vessels of the petiole, pedicel, stem, and roots is also observed. Severe incidence of the disease, especially following stress conditions of drought or water logging can cause death of the tree within one week.

Cushion Galls

Cushion galls are important in certain countries in South and Central America. Five types of cushion galls, green point gall, flowery gall, knob gall, disc gall, and fan gall, have been identified and described. Of these, only fan gall and knob gall have been reported from India. *Cushion gall* is a collective term for a number of forms of flower cushion hypertrophy. Among the different cushion galls, green point and flowery galls are important and are reported to be caused by *Fusarium rigidiuscula*. Causal agents of the other cushion galls have not yet been identified.

Root Diseases

A number of root diseases have been reported in cocoa. Rarely do these diseases cause any substantial damage to the crop. Different pathogens are known to cause root diseases. Their primary source of inoculum comes from forest trees cleared prior to planting or infected shade trees.

The above-ground symptoms of all root diseases caused by different pathogens are sudden wilting of the leaves of the tree followed by the death of the plant. However, root diseases can be identified by the presence of fruiting bodies present on the roots or on the collar region of the affected plant. Root diseases have so far not been reported from India. The major root diseases are:

1. Brown root disease—*Phellinus noxius (Fomes noxius)*
2. White root disease—*Rigidoporus lignosus (Fomes lignosus)*
3. Black root diseases—*Rosellinia pepo* Pat.
4. Collar crack—*Armillaria mellea* Vahl.

The spread of the diseases can be checked by the removal of infected plants, including roots. Digging trenches around the infected tree is also recommended. The infected plants with their roots should be removed and burned. There is no recommendation for chemical control of the disease, except that the cut surfaces of all stumps of cocoa and shade trees should be painted with an arboricide, such as sodium arsenite or 2,4,5-T, at the time of felling to kill the stumps. If 2,4,5-T is used, a 3 percent solution of copper fungicide should be added.

HARVESTING

The pods mature in about 150 to 170 days from the day of pollination. This period varies depending on environmental conditions. A highly significant negative correlation between the number of days from pollination to harvest and the mean temperature during the period of fruit development has been established. According to Alvim et al. (1972), the days to maturity may be calculated from the equation, $N = 2,500/ T-9$, where N is the number of days from pollination to harvest and T is the mean temperature in °C.

The stage of maturity of pods is best judged by change of color of the pods. Pods that are green when immature turn yellow when mature, and the reddish pods turn yellow or orange. The change in color starts from the grooves on the pods and then spreads to the entire surface. Although pods can be harvested as color changes, the pods may remain on the tree without damage up to a maximum of about one month. The intervals between harvests can therefore be extended to one month. However, it is safer to harvest at fortnightly intervals. In areas prone to damage by mammalian pests, harvesting intervals may preferably be shorter. When black pod incidence is serious, shorter harvesting intervals are preferred for ensuring field sanitation.

As fruits are borne on the cushions, and as damage to flower cushions is to be avoided, harvesting is to be done using a knife. When trees are tall, harvesting may have to be done using knives attached to poles.

Postharvest Storage and Breaking Pods

The harvested pods can be stored for two to five days. This enhances prefermentation activity inside the pods and helps to facilitate rapid rise in temperature during fermentation, reduces acidity, and imparts stronger chocolate flavor (Arikiah et al., 1994). The harvested pods are broken by hitting against a hard surface, and beans are extracted without placenta and kept for fermentation immediately. Only mature, well-developed pods contain good beans. Pods showing symptoms of damage from black pod on the surface need not be discarded if the beans inside are unaffected. The color of the pulp will be a good indication of suitability as damaged pods show discoloration.

PRIMARY PROCESSING

Primary processing denotes production of dry cured beans for the market. This involves fermentation and drying.

Fermentation

Raw cocoa beans are covered with sugary mucilaginous pulp, and the beans with the pulp around them are called "wet beans." The kernel, which is also called the "nib," is the economically important part. Fresh nib is bitter and is not suitable for manufacture of different products. When raw, it does not have any flavor, aroma, or taste of any of the cocoa products. Chocolate flavor is developed during the two processes, viz., fermentation, which is done by the grower, and roasting, which is done by the manufacturer.

All of the standard methods of fermentation essentially involve keeping together a mass of reasonable quantity of wet beans for periods ranging from four to six days. In most of these standard methods, mixing of the mass of beans occurs, usually on alternate days. One of the consequences of fermentation is the loss of most of the pulp around the beans; however, more important is the series of biochemical reactions occurring in the beans, which are necessary for inducing the characteristics of the cocoa products.

Biochemical Changes Occurring During Fermentation

The pulp contains about 84.5 percent water, 10.0 percent glucose and fructose, 2.7 percent pentosan, 0.7 percent sucrose, 0.6 percent protein, 0.7 percent acids, and 0.8 percent inorganic acids (Hardy, 1960). The pulp is sterile initially, but the presence of sugars and high acidity (pH 3.5) provide excellent conditions for the development of microbial populations. A wide range of microorganisms infect the mass of beans through the activity of fruit flies and contamination from the fermentary. Initially, yeasts proliferate and convert sugars to alcohol. The cells of the pulp start to break down soon after the fermentation process begins either through an enzyme change or by simple mechanical pressure, and the watery contents of the pulp, which are called "sweating," drain out. This continues for 24-36 hours. The sweatings constitute 12 to 15 percent of the weight of wet beans. The activity of yeasts leads to the production of CO_2, and at this stage relatively aerobic conditions prevail and allow the development of lactic acid bacteria, which assist in the breakdown of sugars.

The activity of bacteria leads to the production of organic acids. When the sweatings have run off, the conditions become more aerobic and the acidity is reduced by the removal of citric acid. The presence of oxygen allows acetic acid bacteria to take over from the yeasts and convert alcohol to acetic acid. These reactions cause a rise in temperature in the mass of beans. A positive correlation exists between sizes of the relevant microbial populations and the amount of acids produced during fermentation (Samah et al., 1993).

The temperature increases after the first mixing to a peak of about 48 to 50°C and falls slowly till the next mixing. With the next mixing also, temperature rises again; but often to a lower peak of around 46 to 48°C, which falls again slowly toward the completion of fermentation. Variations are likely depending on the method of fermentation, location of the beans in the ferment, and environmental conditions. The rise in temperature should be taken as indicative of the necessary biochemical reactions.

The pH of the beans and pulp also varies conspicuously. The fresh cocoa bean pulp is acidic with a pH of around 3.5. The pH of cotyledons is very much higher, around 6.5. After death of the beans, components of pulp diffuse through the testa into the beans, and the acids that are synthesized from pulp move into the beans to lower the pH of the nibs still further. The pH of the nib on the third day will be around 4.8. With further progression in fermentation, pH tends gradually to increase to values around 5.0 by the end of fermentation period. Although a decrease in pH occurs in the cotyledons, the pH of the pulp increases from the initial level of 3.5 to a final value equal to the nib.

The acetic acid diffusing through the testa causes a breakdown of the polyphenol and lipid membranes of the vacuoles of the cell, and cell contents get mixed. Various enzymatic reactions take place, and polyphenols get oxidized. This reaction is partially responsible for the removal of bitter taste from the beans. The production of volatile compounds arising from the reaction of amino acids with sugars leads to induction of aroma. The exact nature of compounds responsible for this is not known, although more than 300 compounds are considered to have their influence. The most important change that occurs in the cotyledons during fermentation is the appearance of the chocolate flavor precursors. The proteins in the cotyledons undergo hydrolysis, giving rise to amino acids and conversion to insoluble forms by reaction with polyphenols. Voigt et al. (1994) found that free amino acids and oligopeptides are essential aroma precursors. The combined action of two enzymes, viz., aspartic endoprotease and carboxypeptidase on cocoa bean protein appeared to be required for the generation of cocoa-specific aroma precursors.

Factors Affecting Fermentation

Ripeness of pods. Harvesting at intervals of one to two weeks ensures quality. Only healthy, ripe pods should be harvested. Use of overripe pods may be avoided as these may contain germinated beans, which may allow the entry of molds and insects. The underripe pods do not ferment properly, and the temperature of the fermenting mass continues to remain at 35°C after an initial rise to 40°C (Knapp, 1926). Alamsyah (1991) observed a weak chocolate flavor and low pH of cured cocoa beans from unripe pods

Pod diseases. Most pod diseases can lead to complete loss of beans. Even when the beans are not destroyed, it is undesirable to use the beans for fermentation.

Type of cocoa. Criollo gets fermented in a relatively shorter period of two to three days, whereas *forastero* takes five to seven days. Hence mixing the beans of these two types should be avoided.

Quantity of cocoa. The heat generated during fermentation is retained by insulation, but this becomes more difficult to achieve with small quantities. About 50 kg beans are required for satisfactory fermentation.

Duration. The duration varies depending on the genetic structure of cocoa mass, the climate, volume, and the method adopted. The duration of fermentation ranges from 1.5 to ten days.

Turning. Turning ensures uniform fermentation. Dias and Avila (1993) recorded faster fermentation when turning was done every 24 hours. Frequent mixing (6 or 12 h intervals) produced a higher number of well-fermented beans than other treatments (Senanayake et al., 1997).

Seasonal effects. Temperature rises more slowly in wet weather in June-July. Dias and Avila (1993) recorded higher volatile acid contents in May than in June. Fermentation during the dry season was better than fermentation during the wet season.

Cocoa Bean Acidity

Cocoa products processed from some samples of cured cocoa beans are found to have detestable acid taste. This is often designated as cocoa bean acidity. Cocoa bean acidity has been reported often from Malaysian cocoa. It has also been found that the beans giving acid taste to the products generally have low pH. The low pH of cocoa beans is also strongly related to titrable acidity.

It has been established that the organic acids responsible for cocoa bean acidity are mainly acetic and lactic acids. These are produced from sugars present in the pulp during the fermentation process. Acetic acid produced during fermentation is an essential component of the fermentation process as the acids contribute to bean death, prevent colonization by putrefactive micro-organisms, and create an environment conducive to the formation of flavor and aroma precursors within the bean cotyledons. However, excessive quantities of acetic and lactic acids produce an acid taste in the cocoa products, as these are not adequately dispelled in the roasting and conching processes.

The problem of cocoa bean acidity is reported mostly from Malaysian cocoa. The best chocolates are produced from Ghanaian cocoa, and acid taste is almost never observed. Bean pH ranges from 5.3 to 5.5 for the Ghanaian cocoa in contrast to the range of 4.4 to 4.7 recorded by Malaysian cocoa.

Bean Maturation

Bean maturation is described as the process involving the removal of acid from cocoa beans by keeping the fermented beans warm, moist, and with good air supply. By maintaining this at desirable levels, it had been possible to raise the pH of beans to acceptable levels in the range from 5 to 5.5. Two methods are suggested to reduce the acidity and to improve flavor of box-fermented Malaysian cocoa:

1. *Box maturation:* The beans set for fermentation in boxes are to be mixed as usual on the third and fifth days. Five extra turns may occur on the sixth and seventh days, and the beans may be taken out on the eighth day.
2. *Drier maturation:* Beans may be kept to thickness of 25 cm and dried at 50°C. Stacking to depths lower or higher than 25 cm results in poorer quality cocoa presumably because of too fast drying of beans in the former and lack of adequate aeration in the latter.

Methods of Fermentation

The method of fermentation and its duration will depend largely on the variety of cocoa and the season. *Criollo* cocoas in general need a shorter period of fermentation as compared to the *forasteros*. Season influences the duration of fermentation mainly through its effects on temperature and humidity. At lower temperatures and high humidity, fermentation period will be usually longer.

Among the various methods adopted for fermentation in the different cocoa-producing countries, the heap, tray, and box methods are considered as the standard, widely adopted methods.

Traditional Standard Methods

Heap method. This method is widely practiced in West African countries. The heap method essentially involves heaping a mass of 50-500 kg of wet beans over a layer of banana leaves. The banana leaves are spread over a few sticks to keep them slightly raised over the ground level to facilitate the flow of sweatings. The leaves are folded and kept over the beans, and a few wooden pieces are placed on top to keep the leaves in position. The purpose of keeping the beans covered with the leaves is to conserve the heat produced during the fermenting process. The heaps are dismantled and the beans are mixed the third and fifth days. It needs about six days for the completion of fermentation, and the beans can be taken out for drying on the seventh day.

As soon as the beans are heaped, flow of sweatings starts and continues for the first two days. Color of the pulp changes on the surface of the mass of beans to a depth of about 10 cm by the third day, with the bulk of beans inside retaining the whitish color. This change of color indicates the beginning of acetic acid production, which is limited to the surface layers where aeration is adequate. With the mixing of beans on the third and fifth days, beans on the surface whose pulp has almost drained away get mixed with the other pulpy beans. Diffusion of air is thus enhanced and acid fermentation occurs deep in the bean mass also.

Tray method. This method was developed based on the early observations that when the beans are heaped for fermentation, a change of color of the beans occurs up to a depth of about 10 cm when beans are not mixed. This was taken as an indication that there will be adequate aeration of the bean mass up to this depth without mixing, and that if the beans are kept only up to this height, mixing can and probably should be avoided. Based on this, beans were filled in trays of 10 cm height, holding reasonable quantities of beans, and trial fermented. It was found that when such trays are stacked one over the other, adequate development and conservation of heat occurs and that fermentation would be over in a shorter period.

The usual size of wooden trays is $90 \times 60 \times 13$ cm. Battens or reapers are fixed at the bottom of the trays with gaps in between that prevent beans from falling through, and allow for the free flow of sweatings. Allowing for the space required for the reapers, net depth of the beans inside should be about 10 cm. The length and width of the trays could be increased to any extent theoretically, but the standard dimensions given earlier will make the size suitable for handling. Each tray can contain about 45 kg of wet beans. Thus filled, the trays are stacked one over the other. The minimum number of trays required for a stack will be about six. An empty tray is kept at the bottom to allow for drainage of the sweatings. After stacking, the beans of the topmost tray are covered with banana leaves. After 24 hours, the stack of trays is covered with gunnysacking to conserve the heat that develops. There is no need of mixing the beans. Fermentation will be normally completed in four days. On the fifth day, the beans are taken out for drying.

The minimum number of trays required to be stacked is about six, but as many as 12 trays can be used simultaneously. If it is a six-tray stack, the total quantity of wet beans required for effective fermentation will be about 270 kg. When a 12-tray stack is used, the minimum quantity will be about 540 kg.

Box method. This method is suitable for handling large quantities of beans. It is common in Malaysia where cocoa is grown in estate scale. The boxes are made of wood with a standard dimension of $1.2 \times .95 \times .75$ m. Boxes of this size can hold about one tonne of wet beans. Holes are provided at the bottom and on the sides of the box to allow flow of the sweatings and to facilitate aeration. The beans are to be mixed on alternate days, transferring the beans from one box to another at the time of mixing. This would necessitate having a minimum of three boxes. These may be arranged in a row, in which case the beans are to be transferred from boxes after lifting them. To make transfer of beans convenient, the boxes are sometimes arranged in tiers, and shutters are provided on one side of the boxes so that beans falling from the box at the top will run to the lower box on removing the shutters. The beans are mixed thus on alternate days, on the third and fifth days, and are taken out for drying on the seventh day after six days of fermentation.

Although the box method of fermentation will be convenient for handling large lots of beans, the quality of box-fermented beans is often rated as inferior to those obtained from the heap and tray methods. The factors responsible for lowering the quality are often related to inadequate aeration of the fermenting mass, which results in induction of acidity in the beans.

Other Traditional Methods

Basket method. Beans are fermented in woven baskets after lining them with a layer of banana leaves. The beans are mixed on alternate days. The fermentation period lasts about six days. Although variable quantities of beans can be used in this method, good quality beans result only when the quantity is reasonable.

Curing on drying platform. This method has been adopted in Ecuador. It consists of spreading the beans on drying trays during the day and heaping them at night. The quality of beans is reported to be reasonable in the main cocoa season, but not in the off season.

Judging the End Point of Fermentation

- *Color of the beans: Forastero* beans turn brown with a pale brown center that has a brownish ring around the outside.
- *External shell color:* The pulp, which is whitish initially, turns to pinkish white after sweatings have run off. At the end of fermentation, the shell surface will attain a reddish-brown color.
- *Smell of fermenting mass:* The fresh pulp has a faint sweet smell, which changes to a characteristic acid smell as fermentation proceeds. This odor persists until the end of fermentation. Overfermented, putrefying beans will produce an ammonia smell.
- *Development of heat:* After setting for fermentation, the temperature of the mass increases steadily to reach a peak of 47 to 49°C by the third day. Temperature falls slowly until the end of fermentation to a range of 45 to 46°C. However, in the methods of fermentation involving mixing of beans, there will be rise in temperature following mixing, which again will tend to drop steadily.
- *Plumpy nature of the beans and color of the exudate:* Well-fermented beans will appear plump and full; on squeezing, a reddish brown exudate flows out.

All beans in a fermenting lot will not be at the same stage of fermentation, and hence all the beans in a sample drawn will not show the indices detailed earlier. Hence, when 50 percent of the beans show such signs, it can be presumed that fermentation is complete.

Small-Scale Methods of Fermentation

All the standard methods of fermentation need relatively large minimum quantities of wet beans. Even in the heap method, the smallest batch size is 50 kg or the produce from about 500 pods. In areas where cocoa is grown in small holdings, a more convenient alternative would be to adopt a method of fermentation involving small quantities of beans. Development of a small-scale method is not easily accomplished, as very small quantity of beans will make it difficult to develop adequate temperature of the fermenting mass. In the standard methods of fermentation, the conditions in the bulk of the fermenting mass remain anaerobic in the early part of the fermentation period. This also is difficult to be simulated with very small quantities of beans.

Attempts were made to develop small-scale methods of fermentation using bean lots substantially smaller than those required for the standard methods. Some of these methods have been found to be successful, as judged from temperature development of the ferment, pH of the beans, and cut test (Kumaran et al., 1981).

Drying

The fermented beans will have a moisture content of about 55 percent. Such a high moisture content is unsuitable for storage of the beans as putrefaction may set in. The moisture content has to be brought down to about 6 percent for safe storage and transportation. Drying should commence immediately on cessation of fermentation. Unless the beans are skin dry within 24 hours after fermentation, molds set in and damage the beans.

Drying of fermented cocoa beans is, however, something more than just driving out the moisture, as part of the fermentation reactions continue during drying also. Biochemical oxidation of acetic acid from the beans continues during drying also. Thus, a very quick drying or excessive heating of the beans will not be suitable. A very slow drying also will not suit as the beans get moldy if they continue to remain moist for too long. The methods used to dry cocoa can be divided into two main types: sun drying and artificial drying.

Sun Drying

Sun drying is the simplest and the most popular method in most of the cocoa-producing countries. Depending upon the climatic conditions, the beans are exposed to the sun for about 12-20 days. This method generally gives good quality beans in traditional areas of cocoa production where the weather is sufficiently sunny. In West Africa, the beans can be simply laid out in the sun, spread in a thin layer on mats raised off the ground or on concrete floors. After two days, the beans are stirred and dried again. In West Indies and South America, drying is done on wooden floors with moveable roofs, referred to as *boucans* in Trinidad and *barcacas* in Brazil. Drying can also be done on moveable trays that can be pushed under a fixed roof. Attempts to improve the efficiency of sun drying have been made using solar cabinets. A rocking dryer designed by the small farmers of Ivory Coast consists of a bamboo platform with wooden edges covered with PVC sheeting that can be removed to facilitate mixing. The platform is pivoted about its center so that it can be directed toward the sun.

Artificial Drying

With the introduction of cocoa into newer areas where the climate remains unsuitable for drying in the peak season, artificial drying methods became necessary. The relatively large requirement of space also made sun drying difficult when cocoa cultivation was extended on an estate scale. Several types of artificial dryers came to be used, and some work on the best drying conditions has been done in different cocoa-growing countries. The results reveal that the major conditions recognized are temperature, rate of airflow, bean depth, and extent of bean stirring. High temperatures, high rates of airflow, lower thickness of beans, and frequent stirring logically achieve quicker and more economical drying. However, these result in high acidity. Suitable drying conditions are thus a balance between the economy in drying and bean quality. The maximum permissible temperature for drying is generally taken as about 60°C. It is better to use deeper loads of beans, the depth being convenient enough for stirring. Stirring of the beans also has been found necessary both for uniformity of drying and its efficiency. A convenient thickness could be about 12 to 15 cm when mixing is done manually.

Recovery of Cured Beans

Recovery varies from 30 to 46 percent depending upon the season and variety of cocoa. During the dry season, recovery is high. Amelonado records recovery as high as 44 percent, whereas Amazon records only 38 percent. Bean size and recovery are inversely correlated. Ripe pods give high recovery of dry beans when compared to unripe beans. To a certain extent, the grower can manipulate the rate of recovery of dry beans by harvesting fully ripe pods and postharvest storage.

Cleaning and Bagging

The dried beans are bagged in jute bags with capacity of 62.5 kg, after removing flat and broken beans.

Storage

Dry cocoa beans can be stored for long under suitable conditions. However, the period of safe storage will depend mainly on the relative humidity and temperature of the atmosphere in which the beans are stored. In the temperate climates where humidity also is low, the storage life is considered almost infinite. In the tropical regions of high humidity, it will be difficult to store the beans for a considerable length of time. As in the case of other plant products, cocoa beans also attain equilibrium moisture content in a given atmosphere either by gaining or losing moisture.

It has been found that the bean moisture content will exceed 8 percent when the relative humidity of the atmosphere reaches 85 percent. This moisture level of 8 percent in the beans is critical, as mold growth sets in when it is above this level. This means that it will be difficult to store cocoa beans without damage in atmospheres whose relative humidity exceeds 85 percent for a considerable period of the year unless special precautions are taken to prevent contact of the dry beans with air. An increase in pH of the beans during a 28-week storage was recorded by Premalatha and Mohanakumaran (1989). The damage due to molds and insects increased after 36 weeks in storage (Bopaiah, 1992). Storage of cocoa beans beyond 36 weeks requires redrying and packing to prevent deterioration.

International standards stipulate certain storage precautions:

1. the ambient humidity must not exceed 70 percent;
2. the bags must be stored at least 7 cm from the ground, normally on a duckboard to allow free air circulation;
3. there must be a passage at least 60 cm wide between the walls and the bags and between bags of different types of cocoa;
4. protection against storage pests/rodents must be ensured;
5. steps must be taken to avoid contamination by odors, off flavors, or dust; and
6. the moisture content should be checked at frequent intervals.

SECONDARY PROCESSING

Secondary processing denotes the steps involved in conversion of cured beans into different finished products, the main product being chocolate. Secondary processing of cocoa beans is done in specialized factories. Wood (1975), Wood and Lass (1985), and Mossu (1992) described the principles of chocolate manufacture in large factories. The essence of cocoa and chocolate manufacture lies in the development of flavor by roasting the beans followed by extraction of cocoa butter from the nib to produce cocoa powder, and addition of cocoa butter to the nib and sugar to produce chocolate. The major steps involved are

Cleaning and Sorting

When the beans arrive in the factory, they are cleaned to remove any foreign matter and sorted to separate the small or broken beans by passing them over a continuously vibrating screen. This is well aerated and is filled with powerful magnets. The metallic foreign matter, dust, and broken beans are removed.

Alkalization

When beans are used for manufacture of cocoa powder, the cocoa liquor is generally treated with alkali to improve color and to develop flavor. Alkalized cocoa is known commercially as "soluble cocoa." The amount of alkali used for the preparation as soluble cocoa is adjusted to bring about partial rather than complete neutralization. Saturated solutions of sodium or potassium carbonate or bicarbonates are most generally used, whereas ammonia, ammonium carbonate, magnesium oxide or carbonate or bicarbonate, or mixtures of these chemicals are favored by some manufacturers. Alkali may be introduced prior to roasting or at the nib or chocolate liquor stages. However, it is more economical to mix it with chocolate liquor.

Roasting

Roasting of cocoa beans, more correctly termed as treatment of cocoa beans in hot air, is one of the most important operations in the processing of cocoa, the degree of treatment required being adjusted to the degree of ripeness of the beans concerned and any other pretreatment which they may have undergone (Riedel, 1977). The true purpose of roasting is not only restricted to the loosening of the shells, but also to develop positive flavor as well as the removal of excess moisture and other undesirable volatile matter. Roasting enables moisture content to be reduced to 1.5-2 percent. Different methods of roasting can be employed, and they produce different end effects, some of which are more applicable to particular varieties of beans than others. The style of roasting should ensure an absolutely equal treatment of all the beans in the batch. According to Riedel (1974), the most favored temperature for proper roasting of cocoa beans for chocolate making lies between 120°C and 125°C. The optimum temperature is also to some extent dependent on the actual time allowed for roasting. The temperature and time have considerable influence on color and flavor. Chemical changes take place in the nib at a temperature of about 120°C to 135°C. Roasting can be done using direct or indirect heating, direct heating by gas, direct heating by steam pipes, or heating by hot air.

Kibbling and Winnowing

The shell is separated from the cotyledon by a process known as kibbling. The purpose of winnowing is to separate the shell and germ and to split the cocoa into its natural segments (cocoa nibs). Roasted cocoa beans can contain between 10 percent and 15 percent shell, depending on the source, and about 1 percent germ. The separation of shell and germ can be carried out separately or together, depending on the choice of commercial plant. Cocoa beans are first cracked by passing through rollers or rotating cones. An air current is then used to blow away the lighter shell. The velocity of this air stream is critical; it should be sufficient to remove the undesirable shell, but not too high to blow off the costly nib, and must be varied to suit the changing size of cocoa bean from differing sources.

Blending and Grinding

When it is necessary to have a blend of beans produced in different regions, blending is done before grinding. The composition of the beans after blending is a trade secret of each chocolate manufacturer. The cotyledons (called the "nibs" at this stage) are ground to get "mass" or "liquor." Cocoa mass contains about 5 to 58 percent fat, which is also called "cocoa butter." This butter has the characteristic of melting at body temperature. The cocoa nibs are finely ground at a relatively high temperature. Normally, cocoa is subjected to a pregrinding stage followed by fine grinding (Bauermeister, 1978). The particle size of the finished product has a pronounced effect on its suitability as an ingredient of different food products (Minifie, 1968). During grinding, heat is generated by friction, which melts the cocoa butter.

Normal grinding is by means of either cylinder rollers of three or four stages or a ball mill. The ball mill gives better overall performance in terms of fineness of grinding, and is simple to maintain (Bauermeister, 1978).

The cocoa mass can be kept as a fluid under hot conditions or molded and cooled before storage. It is the raw material for conversion into commercial cocoa. It is often made in the cocoa-producing countries and exported in this form. This mass can be used for producing cocoa butter, cocoa powder, or chocolate.

Extraction of the Butter from the Cocoa Mass

Cocoa butter is extracted from mass or liquor with the help of a hydraulic press. Screw presses have been employed on nibs but not too successfully. Another method of fat removal is solvent extraction. The powder and butter that is obtained by solvent extraction will contain solvents, which may cause undesirable flavor changes as in the case of screw pressing. A cocoa butter extractor for small-scale use was devised by Ganesan (1982), which utilizes the pressure developed by a hydraulic jack for extraction of the butter. The equipment can extract 44.8 percent of the butter by applying a pressure of 248.72 kg cm^2 at 70°C. Broadbent et al. (1997) used a Brazilian-made, small-scale, portable expeller for extracting cocoa butter.

The cocoa butter obtained by employing any one of the above methods is filtered, if necessary, neutralized and refined, deodorized, and tempered. It is then molded and cooled. At this stage it is hard in consistency, waxy, slightly shiny, pale yellow in color, and oily to touch. It melts at a temperature close to 35°C giving a clear liquid.

Making Cocoa Powder

The cake left behind at the bottom of the presses after the extraction of butter contains a further 20 percent butter. This cake is milled and sieved. Cocoa powder is of two types: high-fat powders containing 20 to 25 percent fat and low-fat powders containing 10 to 13 percent fat. High-fat powder is used in drinks, whereas low-fat powder is used in cakes, biscuits, ice creams, and other chocolate-flavored products. In Thailand, high-fat powder is used for the manufacture of cigarettes.

Production of Plain Chocolate

A large number of grossly unidentified compounds are considered to be involved in inducing the characteristic chocolate taste and aroma of cocoa products. The relative abundance of these is expected to vary, depending on each step in the process of manufacture. Precise standardization of conditions is, therefore, required to make cocoa product of standard and reproducible quality. These compounds are also considered to be responsible for the large brand-related variation in the taste of cocoa products. In simple terms, chocolate is produced by mixing sugar with nib or mass to which cocoa butter is added to enable the chocolate to be molded. The proportion of mass, sugar, and cocoa butter varies with manufacturer, and it remains a trade secret. The mixture of mass and sugar is ground at elevated temperatures to such a degree that the chocolate is very smooth. The mixture is then refined. This gives an absolutely homogenous mixture and a very fine grain size. It is carried out in cylindrical grinders, which are placed on top of the other and adjusted to operate at increasingly closer spacings, rotating at different speeds of around 200 revs/minute. The mass then becomes dry and flaky. It is kneaded again in a blender, and at this stage cocoa butter is added along with flavoring agents, if necessary. This mixture is then subjected to a process of mixing called "conchin." It is carried out in large vats—the conchs. The original conch was a shell-shaped tank, hence the name. In this conch, a roller is pushed to and fro on a granite bed for several hours or even days at temperatures ranging from 60 to 80°C. The time spent in conchs determines the texture of the chocolate. Most of the cocoa butter and lecithin needed is added at the final stage of conching. Conching removes volatile acids contained in the beans and makes the chocolate homogeneous.

Tempering

This consists of reducing the temperature to 28-30°C in automatic tempering vats.

Dressing

This includes molding, where the tempered chocolate passes into a weighing hopper that distributes it into molds; tapping, which causes the molds to be continually shaken in order to distribute the mass evenly without air bubbles; refrigeration at 7°C; and finally removing the chocolate from the molds. This is done by turning out the molds onto a felt conveyor belt, which receives the chocolate.

Packaging

The chocolates are wrapped in attractive packages. These operations are fully automated.

- *Milk chocolate:* The method of preparation remains the same as described earlier for plain chocolate. The only difference is that milk or milk powder is added at the first stage of mixing cocoa mass with sugar. The milk can be condensed with sugar; mass is then added, and the mixture is dried under vacuum. This product is called "crumb," which is ground and conched with additional cocoa butter as described earlier. A typical crumb contains 13.5 percent liquor, 53.5 percent sugar, and 32 percent milk solids.
- *White chocolate:* This is made of milk, cocoa butter, and sugar.
- *Other cocoa-based products:* A number of products are now available in the market— drinking chocolate, enrobing chocolate, chocolate flavored milk, etc.

Nutritional Value of Chocolates

Chocolate-based products have high energy value in relation to their volume. They contain a proportion of carbohydrate and protein together with B complex vitamins. Milk chocolate also contains milk protein, calcium, and other minerals. Plain chocolate contains 64.8 percent carbohydrates, 29.2 percent fat, 4.7 percent protein, sodium 11 mg/100 g, K 300 mg/100 g, Ca 38 mg/100 g, Mg 100 mg/100 g, P 140 mg/100 g, etc. A 100 g bar has an energy equivalent of 500 calories. It contains theobromine and caffeine, which are responsible for stimulatory effects. The chocolates have a restorative, energy-producing, and tonic effect on the body. Some studies indicate that plain chocolate has a cholesterol content of 1 mg/100 g and therefore plays a negligible role in cholesterol intake. However, chocolate is not advocated for diabetic patients.

By-Products

Processing of cocoa both at the primary and secondary levels results in a large quantity of waste materials. The disposal of these is one of the problems in major cocoa-growing tracts. Research on utilization of these waste materials indicates that several useful by-products can be made from cocoa waste. The important waste materials are pod husk, sweatings, germ, and shell.

Pod Husk

About 70 to 75 percent of the pod is constituted by pod husk. This is generally discarded after collection of the beans. The pod husk contains crude protein (5.69 to 9.69 percent), fatty substances (0.03 to 0.15 percent), glucose (1.16 to -3.92 percent), sucrose (0.02 to 0.16 percent), pectin (5.30 to 7.06 percent), nitrogen free extract (44.2 to 151.27 percent), crude fiber (33.19 to 39.45 percent), theobromine (0.20 to 0.21 percent), and ash (8.83 to 10.18 percent) (Nambuthiri and Shivshankar, 1987). The pod husk contains less theobromine than the cocoa shell, which makes it less dangerous as a feed stuff. Incorporation of a 20 percent pod husk in cattle feed has shown beneficial effect (Sampath et al., 1990). The nitrogen and phosphorus content of the pod husk is comparable to farmyard manure from animals. The potash content is very high (2.85 to 5.27 percent K_2O). The high fiber content of pod husk suggests its use in paper manufacture, but its low fiber length of 0.3-0.5 mm rules out this possibility. Pod husk as a source for the production of furfural (9 percent) is not comparable in yield with materials such as oat hull, corncob, and cottonseed. Hence production of furfural from pod husk is not commercially viable. The dry pod husk contains 5.3 to 7.08 percent pectin. This is high when compared to established raw materials such as orange pulp, lemon pulp, and apple pomace. The quality of endocarp pectin is superior to that of pectin from sweatings.

Mucilage

The concentration of alcohol in the sweatings is about 2 to 3 percent and the concentration of acetic acid is about 2.5 percent. The sweatings contain water (79.2 to 84.2 percent), dry substances (15.2 to 20.8 percent), citric acid (0.77 to 1.52 percent), glucose (11.60 to 15.32 percent), sucrose (0.11 to 0.92 percent), pectin (0.90 to 1.19 percent), proteins (0.56 to 0.69 percent), and salts (K, Na, Ca, Mg) (0.41 to 0.54 percent) with a pH of 3.2 to 3.5. Sweatings can be used for making jelly or jam. The pectin from sweatings show slow setting characteristics.

Shell

The availability of bean shell is of the order of 11 to 12 percent of the dry beans. It contains 2.8 percent starch, 6.0 percent pectin, 18.6 percent fiber, theobromine 1.3 percent, caffeine 0.1 percent, total nitrogen 2.8 percent, fat 3.4 percent, total ash 8.1 percent, tannins 3.3 percent, vitamin D 300 IU, etc. The yield of furfural is about 5 to 6 percent. Although it is possible to extract protein, tannin, and red color from shells, it is not economically viable. The scope for use as animal feed is limited due to high theobromine content. As fertilizer, shells act as a humus-forming base. They do not decompose readily. This can be overcome by heaping for one season. Theobromine is extracted commercially and methylated to form caffeine, which has greater demand than theobromine. As fuel, the calorific value of shell is about 7,400-8,600 BTU, which is a little higher than that of wood.

Germ

The cocoa germ has fat (3.5 percent), ash (6.5 percent), protein (24.4 percent), crude fiber (2.9 percent), and theobromine (3.0 percent). The composition of the germ varies considerably depending upon the country of origin.

RESEARCH AND DEVELOPMENT ORGANIZATIONS

It is currently estimated that 70 percent of the world's cocoa is grown by small farmers using traditional labor-intensive systems of husbandry. Improvements in productivity are therefore largely dependent on the work of research institutes in the producing countries and of the few plantation companies that have expertise and facilities. The volume of research carried out by these bodies is insufficient to provide the assurance need for the future. It is therefore essential to contribute through sponsorship of research and arranging for the results to be disseminated widely. The important research institutes on cocoa are as follows:

1. The Cocoa Research Unit, The University of West Indies, St. Augustine, Trinidad, Tobago
2. Cocoa Research Institute of Ghana (CRIG), P.O. Box 8, New Tafo Akim, Ghana
3. Centro de Pesquisas do Cacao (CEPLAC), Cacao Research Center (CEPEC), Ilheus, Bahia, Brazil
4. Fondo Nacional de Investigaciones Agropecuarias (FONAIAP)–Centro Nacional de Investigaciones Agropecaurias (CENIAP), Itabuna, CEP 45650-000,Venezuela
5. Cocoa Research Institue of Nigeria (CRIN), P.M.B. 5244, Ibadan, Nigeria
6. Papua New Guinea Cocoa and Coconut Research Institute, P.O. Box 1846, Rabaul, Papua New Guinea
7. Instituto Colombiano Agropecuario (ICA), Programa de Cacao, AA 1017, Bucaramanga, Colombia
8. Cocoa Research Station, Quoin Hill, Tawau, Department of Agriculture, Sabah, Malaysia

9. Centre de Co-operation Internationale en Recherches Agronomiques pour le Developpement (CIRAD)-IRCC, BP. 5035, 34032, Montpellier Cedex, France
10. CIRAD-IRCC, BP. 1827, Abidjan, Cote d'Ivoire
11. CIRAD-IRCC, BP. 701, 97387, Kourou, French Guiana
12. Institut de Forests, Department Cafe Cacao (IDEFOR-DCC), BP. 1827, Abidjan 01, Cote d'Ivoire
13. Central Plantation Crops Research Institute, Regional Station, Vittal, 574 243, Karnataka, India
14. Kerala Agricultural University (KAU), P.O. 680656, Thrissur, Kerala, India
15. Coffee and Cocoa Research Institute, J.P.B. Sudirman, Jember 68118, Indonesia

MANUFACTURERS OF COCOA PRODUCTS

Cocoa products are manufactured by many multinational companies, viz., Cadbury, Nestlé, Amul, Lotus, Sathe, etc.

FUTURE OUTLOOK

- Although a lot of work has been done on improvement of the crop since the 1930s and 1940s, the progress achieved in terms of yield is not very substantial. World cocoa production has remained stagnant since the 1980s. A major part of world's cocoa production comes from genotypes that are not significantly different from their wild progenitors. This is due to restricted genetic base of the crop. The need for explorative search in centers of diversity is highly essential to make any breakthrough in crop improvement.
- Cocoa cultivation in many countries is facing severe threat by major diseases, the control of which is not feasible by conventional methods. It is time to concentrate on breeding for resistance to diseases. Since the 1950s efforts have been made to identify effective resistance to major diseases and incorporate these into the varieties for commercial use. It is now generally considered that the effort has largely been ineffective, and for most of the serious diseases sufficiently strong resistance remains to be identified and incorporated. It has also been argued that the focus on disease resistance has been at the expense of the overall performance of modern varieties.

- There appears to be considerable scope for successful breeding of cocoa cultivars with satisfactory levels of resistance to one or more important diseases according to national priorities. Some achievement in crop improvement has been obtained by recurrent selection schemes with distinct subpopulations. The genotypic components of variation for all major agronomic traits are shown to be due mainly to additive gene effects, and maximum gene dispersion over subpopulations will increase the chances of detecting transgressive hybrids.
- Clonal selection after recombination would be better for short-term progress, where clones are accepted commercially. By concentrating on development and utilization of technologies, there is need to develop stable, long-term conventional breeding work.
- It is imperative that conventional breeding programs be maintained and indeed expanded for quantitative traits such as yield and horizontal resistance. The reasons for this are (1) all the desirable genes in a polygenic system cannot be assembled in a single plant in a single generation, (2) it is impractical to screen using gene markers when many genes producing small effects upon the trait are involved, and (3) quantitative traits tend to be greatly influenced by genotype/environment interactions, thus making it necessary to screen for such traits.
- In most of the breeding programs, the breeder neglects the aspect of flavor. This aspect is of paramount importance as the flavor of the finished product is determined primarily by the variety or type used. Assessment of flavor is not very easy in cocoa. However, certain simple procedures for assessment of flavor have been developed recently. Hence, flavor improvement must find an important position among the future thrusts in breeding.
- Consideration should be given to redesigning the tree architecture to improve photosynthetic efficiency and harvest index.
- Current achievements in molecular genetics appear as if they will have an impact on crop improvement of cocoa in the coming years. The progress in the development of new technologies in plant breeding has been tremendous since 1985. These techniques can make a useful contribution if the traditional breeding base is strong enough to support their integration.
- Organic farming is becoming relevant in cocoa and studies in this line may be taken up as an important reseach area in the coming years.
- Physiology of flowering and fruit set in cocoa need a detailed investigation.

- Evolution of biological control measures against serious pests and diseases is another important area for future research.
- Emphasis may be given for the development of technology for value addition at the farm level.

REFERENCES

Alamsyah, T.S. 1991. The effect of pod storage on the quality of dry cocoa beans. *Buletin Perkebunan* 22(2): 137-145.

Alvim, P.T. 1987. Cacao. In Alvim, P.T. and Kozlowski, T., Eds., *Ecophysiology of Tropical Crops*. Academic Press, New York.

Alvim, P. T., Machado, A.D., and Vello, F. 1972. Physiological responses of cacao to environmental factors. *Proceedings of the Fourth International Cocoa Research Conference*, Cocoa Research Unit, Trinidad, pp. 210-225.

Amma, S.P.K., Mallika, V.K., Manoharan, S., Namboothiri, R., and Nair, R.V. 2002. An insight into the preselection method in cocoa. National seminar on technologies for enhancing productivity in cocoa (Ext. Sum.) Central Plantation Crops Research Institute, Regional Station, Vittal, Karnataka, India, November 29-30, 2002.

Aragundi, J., Suarez, C., and Solorzano, G. 1987. Evidence of resistance of cocoa clone EET 233 to moniliasis and witches' broom. *Proceedings of the Tenth International Cocoa Research Conference*, Santo Domingo, Dominican Republic, May 17-23, 1987, Cocoa Producers' Alliance, London, UK, pp. 479-483.

Arikiah, A, Tan, Y.P., Sharma, M., and Clapperton, J.F. 1994. Experiments to determine influence of primary processing parameters and planting material on the flavour of cocoa beans in Malaysia. *Cocoa Growers' Bull.* 48: 36-46.

Bauermeister, J. 1978. Grinding of cocoa mass. *Industrie Alimentari.* 17(5): 424-428.

Bell, G.D.H. and Rogers, H.H. 1956. Cacao breeding at WACRI. *Proceedings of the Cacao Breeding Conference*. West African Cocoa Research Institute, October 1-3, 1956, West African Cocoa Research Institute, Ghana, pp. 31-49.

Blaha, G. 1996. PNG's experience on diseases of cocoa: Management for chemical treatments and for resistant plant material. *Proceedings of the Twelfth Cacao Research Conference*, Salvador-Bahia, Brazil, November 17-23, 1996, Cocoa Research Center, Bahia, Brazil.

Bopaiah, B.M. 1992. Deterioration of processed cocoa beans in storage and mycotoxin. *Ind. Cocoa Arecanut Spices J.* 16(1): 11-13.

Broadbent, J.H., Turatti, J.M., Tocchini, R.P., and Iaderos, M. 1997. Rural processing of cocoa beans in Brazil. *Trop. Sci.* 37(3): 164-168.

Cheesman, E.E. 1944. Notes on the nomenclature, classification, and possible relationship of cacao populations. *Trop. Agric. Trin.* 21(8): 144-159.

Coe, S.D. and Coe, M.D. 1996. *The True History of Chocolate*. Thames and Hudson Ltd, London, United Kingdom.

Cope, F.W. 1962. The mechanism of pollen incompatibility in *Theobroma cacao.* L. *J. Hered.* 17: 157-182.

Dias, J.C. and Avila, M.G.M. 1993. The effect of post harvest storage period of pods, turning process of the mass, and fermentation time on the acidity of cocoa beans. *Agrotropica* 5(2): 25-30.

Djekpor, E.J., Cilas, C., and Paulin, D. 1994. Spotlight on cocoa breeding for resistance to swollen shoot. *Proceedings of the INGENIC Workshop on Cocoa Breeding Strategies,* Kaula Lumpur, Malaysia, October 18-19, 1994, International Group for Genetic Improvement of Cocoa, University of Reading, UK, pp. 38-40.

ED and F Man. 2002. *Cocoa Market Report,* No 370, April.

Enriquez, G.A. 1981. Early selection for vigor of hybrid seedlings. *Proceedings of the Eighth International Cocoa Research Conference,* pp. 535-539.

Francies, R.M. 1998. Genetic analysis of certain clones, hybrids, and inbreds in cocoa. PhD thesis, Kerala Agricultural University, Thrissur, India.

Ganesan, V. 1982. Development of small scale equipment for extraction of cocoa butter and production of cocoa powder. MSc (Ag. Engg.) thesis, Kerala Agricultural University, Vellanikkara, Thrissur, India.

Glendinning, D.R. 1966. The relationship between growth and yield in cacao varieties. *Euphytica* 9(35): 1-35.

Gramacho, K.P., Luz, E.D.M.N., Lopes, U.V., and Paim, M.A. 1996. Inoculum density and inoculum suspension vehicle for evaluating resistance of cocoa seedlings to *Crinipellis pernicosa. Proceedings of the Twelfth International Cocoa Research Conference,* Salvador-Bahia, Brazil, November 17-23, 1996, International Group for Genetic Improvement of Cocoa, University of Reading, UK, pp. 91-101.

Hale, S.L. 1953. World production and consumption—1951-53. *Report of Cocoa Conferences,* London, UK, 1953, pp. 3-11.

Iwaro, A.D., Sreenivasan, T.N., and Spence, J.A. 1997. Studies on black pod resistance in Trinidad. *Proceedings of the INGENIC Workshop on the Contribution of Disease Resistance to Cocoa Variety Improvement,* Salvador-Bahia, Brazil, November 25-26, 1996, International Group for Genetic Improvement of Cocoa, University of Reading, UK, 1999, pp. 67-74.

Jose, N.R., Nair, R.V., and Unnithan, V.K.G. 1998. Growth and yield analysis of top worked cocoa (*Theobroma cacao* L.) *J. Plantation Crops* 26(1): 13-19.

Keshavachandran, R. 1979. Propagational studies in cocoa (*Theobroma cacao* L.). MSc (Hort.) thesis, Kerala Agricultural University, Vellanikkara, Thrissur, India.

Knapp, A.W. 1926. Experiments in the fermentation of cacao. *J. Soc. Chem. Ind.* 45: 140-142.

Kumaran, K., Nair, P.C.S., and Nair, R.V. 1980. Studies on the methods of curing for small quantities of cocoa beans. *Indian Cocoa Arecanut and Spices J.* 4(2): 42-44.

Kumaran, K., Prasannakumari, S., and Nair, P.C.S. 1981 Experiments on small scale fermentation of cocoa beans II. Effect of the different factors for aeration on the extent of fermentation and quality of cured cocoa beans. *Proceedings of* the International Conference on Plantation Crops, Central Plantation Crops Research Institute, Kasaragod, *PLACROSYM III, Cochin*, Kerala, India, pp. 124-137.

Laker, H.A., Sreenivasan, T.N., and Kumar, D.R. 1987. The inheritance of some cocoa clones to *Crinipellis perncosa* in Trinidad. *Proceedings of the Tenth International Cocoa Research Conference*, Santo Domingo, Dominican Republic, May 17-23, 1987, pp. 637-641.

Lamin, K., Chong, T.C., Bong, C.L., Lee, M.T., and Phua, P.K.O. 1996. Breeding for resistance to cocoa diseases in Malaysia with special reference to vascular-streak dieback. *Proceedings of the INGENIC Workshop on the Contribution of Disease Resistance to Cocoa Variety Improvement*, Salvador-Bahia, Brazil, November 25-26, 1999, International Group for Genetic Improvement of Cocoa, University of Reading, UK, pp. 189-193.

Lass, R.A. and Wood, G.A.R. 1985. Cocoa production. Present constraints and priorities for research. *World Bank Technical Paper* 39: 95.

Longworth, J.F. and Freeman, G.H. 1963. The use of trunk girth as a calibrating variate for field experiments on cocoa trees. *J. Hort. Sci.* 38: 61-67.

Mallika, V.K., Amma, S.P.K., Abraham, K., Nair, R.V., and Binimol, K.S. 2000. Evolution of cocoa varieties resistant to vascular streak die back through hybridization. *Abstract of the International Conference on Plantation Crops. PLACROSYM XIV*, December 12-15, 2000, Hyderabad, India, Rethenaim, P. (ed.), Central Plantation Crops Research Institute, Kasaragod, Kemala, India, pp. 6-7.

Mallika, V.K., Amma, S.P.K., and Nair, R.V. 2002a. Crop improvement in cocoa (*Theobroma cacao* L.). *National Seminar on Technologies for Enhancing Productivity in Cocoa (Ext. Sum.)*, Central Plantation Crops Research Institute, Regional Station, Vittal, Karnataka, India, November 29-30, 2002, Bhat, R. Balasimha, D., and Jayasekhar, S. (eds.), Central Plantations Crops Research Institute, Kasaragod, Kerala, India, pp. 19-27.

Mallika, V.K., Amma, S.P.K., and Nair, R.V. 2002b. An overview of inbreeding in cocoa. *National Seminar on Technologies for Enhancing Productivity in Cocoa (Ext. Sum.)*, Central Plantation Crops Research Institute, Regional Station, Vittal, Karnataka, India, November 29-30, 2002, Bhat, R. Balasimha, D., and Jayasekhar, S. (eds.), Central Plantations Crops Research Institute, Kasaragod, Kerala, India, pp. 50-51.

Mallika, V.K., Amma, S.P.K., Nair, R.V., and Namboothiri, R. 2002. Cross compatibility relationship within selected clones of cocoa. *National Seminar on Technologies for Enhancing Productivity in Cocoa (Ext. Sum.)*, Central Plantation Crops Research Institute, Regional Station, Vittal, Karnataka, India November 29-30, 2002, Bhat, R., Balasimha, D., and Jayasekhar, S. (eds.), Central Plantation Crops Research Institute, Kasaragod, Kerala, India, pp. 44-46.

McKelvie, A.D. 1957. The polythene sheet method of rooting cacao cuttings. *Trop. Agric. Trin.* 34: 260-265.

Minifie, B.W. 1968. Special cocoas—Their manufacture and uses. *Confectionary Production* 34(12): 778-784.

Mossu, G. 1992. *The Tropical Agriculturist—Cocoa.* Macmillan Press Ltd., London, UK.

Nair, R.V., Mallika, V.K., and Amma, S.P.K. 2002. Top worked trees: An appraisal of field performance. *National Seminar on Technologies for Enhancing Productivity in Cocoa (Extended Summary)* Central Plantation Crops Research Institute, Regional Station, Vittal, Karnataka, India, November 29-30, 2002, Central Plantation Crops Research Institute, Kasaragod, Kerala, India, pp. 74-75.

Nair, R.V., Mallika, V.K., and Chandrika, C. 1995. Clonal propagation of cocoa. *Indian Cocoa, Arecanut, and Spices J.* 19(4): 120-122.

Nair, R.V., Mallika, V.K., and Swapna, M. 1994. A procedure for top working in cocoa. *International Symposium Plantation Crops (PLACROSYM XI)* (Abstr.) National Research Center for Spices, Calicut, Central Plantation Crops Research Institute, Kasaragod, Kerala, India, p. 54.

Nair, R.V., Mallika, V.K., and Swapna, M. 1996. Response of cocoa to shade. *J. Plantation Crops* 24(Suppl): 99-101.

Nair, R.V., Virakthmath, B.C., and Mallika, V.K. 1994. Management of cocoa. In Chadha, K.L., Ed., *Advances in Horticulture Plantation Crops and Spices,* Part 1, Volume 9. Malhothra Publishing House, New Delhi, pp. 563-570.

Nambuthiri, E.S. and Shivashankar, S. 1987. Cocoa waste and its utilization. *Indian Cocoa, Arecanut, and Spices J.* 8: 78-80.

Ngatchou, N.J. and Lotode, R. 1971. Variability in precocity of the first hybrids obtained in the Cameroon Republic and search for a correlation between girth at a given age and precocity of the hybrids. *Trop. Abstracts* 26(4): 232.

N'Goran, J.A.K., Kebe, I., Tahi, M., Paulin, D., Clement, D., and Eskes, A.B. 1996. Pathology and breeding research on resistance to black pod in Cote d'Ivoire. *Proceedings of the INGENIC Workshop on the Contribution of Disease Resistance to Cocoa Variety Improvement,* Salvador-Bahia, Brazil, November 25-26, 1996, International Group for Genetic Improvement of Cocoa, University of Reading, UK, pp. 135-139.

Nyasse, S., Bidzanga, N.L., Blaha, G., Berry, D., Cilas, C., Despreaux, D., and Eskes, A.B. 1996. Update on the work on resistance of cocoa to *Phytophthora megakarya* in Cameroon. *Proceedings of the INGENIC Workshop on the Contribution of Disease Resistance to Cocoa Variety Improvement.* Salvador-Bahia, Brazil, November 25-26, 1999, International Group for Genetic Improvement of Cocoa, University of Reading, UK, pp. 41-49.

Paulin, D., Mossu, G., Lachenaud, P., and Cilas, C. 1993. Cocoa breeding in Cote d' Ivoire. Performance analysis of sixty-two hybrids in four localities. *Café, Cacao, Thé* 37(1): 3-20.

Phillips-Mora, W. 1996. Studies on resistance to black pod disease *(Phytophthora palmivora)* at CATIE. *Proceedings of the INGENIC Workshop on the Contribution of Disease Resistance to Cocoa Variety Improvement,* Salvador-Bahia, Brazil, November 25-26, 1999, International Group for Genetic Improvement of Cocoa, University of Reading, UK, pp. 41-49.

Pittier, H. 1933. Degeneration of cacao through natural hybridization. *J. Hered.* 36: 385-390.

Posnette, A.F. 1951. Progeny trials with cacao in the Gold Coast. *Emp. J. Agr.* 19(76): 242-251.

Pound, F.J. 1931. The genetic constitution of the cacao crop. *Ann. Rep. Cacao Res. Trinidad* 1: 10-24.

Premalatha, T. and Mohanakumaran, N.1989. Effect of storage before fermentation and after curing on quality attributes of cocoa. *South Ind. Hort.* 37(5): 277-281.

Purseglove, J.W. 1968. *Theobroma cacao* L. In Purseglove, J.W., Ed., *Tropical Crops—Dicotyedons* 2. John Wiley & Sons Inc., London, pp. 571-599.

Richards, A.J. 1986. *Plant Breeding Systems*. George Allen & Unwin, London, pp. 189-232.

Riedel, H.R. 1974. Effects of roasting on cocoa beans. *Confectionery Production* 40(5): 193-194.

Riedel, H.R. 1977. Roasting of cocoa beans. *Confectionery Production* 43(6): 240-241.

Rudgard, S.A. and Lass, R.A. 1985. Some economic information on cocoa production in South American countries affected by witches'-broom disease. *Report of the Second Workshop of International Witches' Broom Project*, Wageningen.

Sackey, S.T. 2000. Novel technologies for disease indexing and screening for CSSVD resistance. *Proceedings of the International Workshop on New Technologies and Cocoa Breeding*, October 16-17, 2000, Kota Kinabalu, Sabah, Malaysia, International Group for Genetic Improvement of Cocoa, University of Reading, UK, pp. 149-156.

Samah, O.A., Ibrahim, N., Alimon, H., and Karim, M.I.A. 1993. Comparative studies on fermentation products of cocoa beans. *World J. Microbiol. Biotech.* 9(3): 381-382.

Sampath, S.R, Kumar, M.N.A., and Sundareshan, K. 1990. Evaluation of cocoa bean by-products for ruminant feeding. *Ind. J. Animal Nutr.* 7(1): 45-48.

Senanayake, M., Jansz, E.R., and Buckle, K.A. 1997. Effect of different mixing intervals on the fermentation of cocoa beans. *J. Sci. Food Agri.* 74(1): 42-48.

Sridevi, R. 1999. Estimation of genetic parameters from specific crosses of cocoa (*Theobroma cacao* L.). MSc (Hort.) thesis, Kerala Agricultural University, Thrissur, India.

Tan, G.Y. and Tan, W.K. 1990. Additive inheritance to pod rot caused by *Phytophthora palmivora* in cocoa. *Theor. Appl. Genet.* 80: 258-264.

Verghese, R. 1999. Standardization of selection criteria for cocoa hybrids. MSc (Ag.) thesis, Kerala Agricultural University, Thrissur, India.

Voigt, J., Biehl, B., Heinricha, H., and Voigt, G. 1994. Formation of cocoa specific aroma precursors from reserve globulins of cocoa seeds. *Symposium Tropische Nutzpflanzen*, Hamburg, Germany, September 22-24, 1993, Angewandte Botanik Berichte No. 5, 1994, pp. 138-147.

Wahid, P.A., Kamalam, N.V., Ashokan, P.K., and Nair. R.V. 1989. Root activity pattern of cocoa (*Theobroma cacao* L.). *J. Nuclear Agric. Biol.* 18: 153-156.

Wheeler, B.E.J. and Mepstead, R. 1988. Pathogenic variability amongst isolates of *Crinipellis pernicosa* from cocoa (*Theobroma cacao*). *Plant Path.* 37: 475-488.

Williams, J.A. and Akinwale, S.A. 1994. Breeding strategies adopted in Nigerian cocoa collections. *Proceedings of the INGENIC Workshop on Cocoa Breeding*

Strategies, Kuala Lumpur, Malaysia, October 18-19, 1994, International Group for Genetic Improvement of Cocoa, University of Reading, UK, pp. 33-37.

Wood, G.A.R. 1975. *Cocoa,* Third Edition. Longman Group Ltd., London.

Wood, G.A.R. and Lass, R.A. 1985. *Cocoa,* Fourth Edition. Longman Publications, New York.

Young, A.M. 1994. *The Chocolate Tree: A Natural History of Cacao.* Smithsonian Institution Press, Washington, DC.

Chapter 6

The Coconut Palm

M. A. Foale
G. R. Ashburner

ORIGIN AND EVOLUTIONARY HISTORY

Unique Among Palms

The coconut stands out among palms with its high degree of continuity and consistency of flower and fruit production, month after month, year after year. This capability suggests evolution in an environment free from severe seasonal or episodic constraints on growth. Such an environment would most likely have resembled those few isolated niches where the coconut presently thrives unaided by any human intervention or management. On the island of North Keeling for example, southwest of Java in the Indian Ocean, the coconut dominates a dense woodland growing on low sand-cays underlain by fresh water (Leach et al., in press). On the east coast of Cape York Peninsula in Australia, where coconut has been introduced since European settlement in the nineteenth century, pockets of "coconut woodland" have also formed.

Seeds cannot naturally move landward beyond this zone because their great size defies carrying by birds and most mammals, so the only avenue

We would like to acknowledge the kind support of David Friend of Bundaberg, and Diana Wood of the Botanic Garden, Cooktown, both in Australia, who made available unpublished material. We thank also Hugh Harries of CICY in Mexico, Roland Bourdeix of CIRAD in Cote d'Ivoire, Pons Batugal of COGENT, Renick Peries of the Victorian government service, Australia, and Francois of Burotrop for generously providing information and discussing ideas. Brian Leach of Perth, Professor K. V. Peter of the Kerala Agricultural University, and Gabrielle Persley of Glasgow provided very helpful comments on the completed manuscript. Thanks go to CSIRO Sustainable Ecosystems for its logistic support while the work was being done. The artwork in Figure 6.2 was done by Lisette Ackhurst (CSIRO, Toowoomba) and the images in Photos 6.2 and 6.4 through 6.8 were taken by Simon Foale, both of whom are also thanked.

for dissemination is via the ocean to other coasts with fringing sand cays. Humans have made extensive use of the coconut near the coast but also at great distances inland. It succeeds well only where rainfall is plentiful (more than 2,000 mm per year) and well distributed, or where irrigation is carried out.

Evolution on Drifting Coastlines

The general course of evolution of this prolific modern palm, emerging from the primal ancestral palms of Mesozoic Gondwana, remains obscure, but has been the subject of curious speculation and limited molecular exploration by scientists (Lebrun et al., 1999). There is a remarkable lack of fossil record for the coconut, which can be attributed, in part at least, to the instability and "migration" of the coastal fringe environment. Even within the past 120,000 years, there have been no fewer than six cycles of glaciation, when the sea level has fallen between a few tens of meters and more than 100 m in one case around 20,000 years ago (Veeh and Veevers, 1970; Chappell, 1983). The corresponding movements of the coastal zone by many kilometers, in some cases, would leave the coconut zone "high and dry" during the phase of falling sea level and progressively inundated during the rising phase. Potential fossil material would be destroyed by oxidation through exposure to aerobic conditions in the first phase and disintegrated by wave action in the second. It is likely that the only "recent" fossil to be found for coconut are those where humans have planted it adjacent to a swampy environment, where it would not have been established by natural dispersal (e.g., Spriggs, 1984). In this case the fossil remains were found to be a little over 5,000 years old.

On a geological timescale, the ancestors of the coconut possibly began to emerge from the palm branch of the tree of plant evolution around 100 million years ago (–100 Myr). One theory (Harries, 1990) has the ancestral coconut inhabiting the "north" coast of Gondwana as that great landmass of the southern hemisphere began to break up around –80 Myr. Huge crustal plates carrying exposed or partly submerged land surfaces, which now comprise Australia, India, and Arabia, and smaller fragments such as New Zealand and Madagascar, began to drift northward.

The expanse of ocean, between these wandering land masses and associated islands, is referred to as the Tethys Sea (Harries, 1990). This sea is likely to have been warm and stormy, delivering high rainfall and periodic cyclonic wind gales onto the neighboring coastlines. Recent experience of extremes of weather suggests that a warmer ocean spawns more intense cyclones. It is suggested that natural selection occured in the palm population for characters resistant to extreme wind, as described later.

Development of Wind-Resistant Traits

The conditions believed to have prevailed in the Tethys Sea would favor the emergence, on the shoreline, of a tropically adapted palm with a flexible, wind-resistant trunk that could flex rather than break during episodes of violent wind. The trunk of the modern palm has a "tubular" structure relating to the density of its cortex and interspersed vascular fibers, forming a thick outer "wall" surrounding a softer core that undergoes compression without fracture when the trunk flexes in the wind.

The coconut palm exhibits a further adaptation to survive strong wind, which is a capacity progressively to shed older fronds (Photo 6.1). This brings about a reduction of wind pressure, thereby reducing the risk of damage to the heart of the palm crown. The coconut also has the ability to con-

PHOTO 6.1. Surviving palms on Rennell Island of the Solomon Islands following a severe cyclone that stripped away many older fronds reducing the pressure on the crown, but also breaking off the crown of some palms.

tinue to grow even after the trunk has fallen flat on the ground, with new
trunk growth resuming a vertical attitude (Marty et al., 1986).

The Sea-Going Coconut Fruit

Fossil remains of the coconut ancestor include a very small nut found in
New Zealand, less than 50 mm dia and unlikely to contain sufficient liquid
endosperm to survive a long seagoing episode (Ashburner, 1994). Dispersal, by means of a hardy floating seed with a thick husk that allowed it to
float high in the water, was a later adaptation that allowed the coconut to
move on the ocean and occupy coastal niches throughout the favorable climatic zone. The coconut seed of such wild places is still capable, in our
time, of being picked up in large number by a tidal surge or in lesser number
by falling from seaward-leaning palms, and surviving three or four months
at sea. Such a period is long enough to travel distances up to some thousands of kilometers depending on wind and current. This open "pooling" of
genetic diversity would have counteracted the natural tendency toward separate paths or pockets of evolution fostered by isolation. Natural diversity in
coconut appears to be limited indeed, in contrast to most species of "land"
plants in which genuine physical isolation of subpopulations has occurred.
This has given rise, in the case of *Macadamia* populations for example, to
easily distinguished species, subspecies, and varieties, determined by both
morphological traits and molecular methods (Aradhya et al., 1998).

The question of how far the coconut dispersed naturally by means of its
floating seed has given rise to intense speculation, much of it failing to ap-
preciate the long time frame in which dispersal has taken place. The major
tropical landmasses and islands of the Indian and Pacific oceans and the
South China Sea have been in place for some millions of years. There would
appear to have been ample time for the ocean-dispersed coconut palm to
spread far and wide, both aided and disrupted, during frequently recurring
episodes of glaciation, by rising and falling sea level. Narrowed ocean barriers between neighboring islands and landmasses would at least favor colonization and further mixing of populations at a more regional level.

During periods of falling sea level, the coconut would have become
widespread, not only on coastlines as such but also across the "temporary"
landscape between the coast at the original sea level and the coast at the
eventual low extreme. It has the capacity to compete quite strongly with
other vegetation in the short term, but would eventually have been over-
shaded by tall forest species. When the great northern hemisphere ice sheets
receded, the sea level rose once more and coconut palms in the low-lying areas would have succumbed to inundation. A phase of rising sea level would

most likely have been conducive to mobilization of seeds onto the ocean. In common with the fringing coral reefs, the fringing vegetation of the strand, which included the coconut, would have followed the fluctuating shorelines of the tropics back and forth.

True Palm Characters

Although the coconut appears to have "branched off" from other palms early in its evolutionary history (Ashburner, 1994), it retains the essential physical traits of a true palm. In common with all mature palm trunks, radial (diameter) expansion of the coconut trunk is fixed once the attached fronds have developed, confining trunk growth to the axial (length- or height-wise) dimension. A "unit" of growth in the mature coconut palm comprises a section of trunk, supporting an attached frond and inflorescence. In a well-watered, warm environment the coconut sustains a most remarkably uniform and stable growth pattern through time, giving rise to a trunk increment, a new frond, and a new bunch about every 25 days. A frond generally persists from two to three years and detaches cleanly from the trunk (except in especially dry atmospheric environments) along with the remnant of the fruit bunch.

Association of the Coconut with Human Activity

The bounty of the coconut provided a most convenient and welcome food source as human settlers eventually arrived at southern and southeast Asian coastlines. It is entirely obscure as to when human settlers (proto-Melanesian people) first appeared in this region, although migration as far east as Papua New Guinea and Australia can be located between 60,000 and 100,000 years ago. These migrants appear to have preferred the refuge of secure and isolated mountainous hideaways, far from the convenience but also the danger of the coast. One might even speculate that *Homo erectus* or "Java Man" could have made use of the coconut 1.2 million years ago.

It appears to have been only within the past 5,000 to 10,000 years that coast-dwelling, seagoing peoples known as Austronesians have formed a close interdependence with the coconut palm. In contrast to the earlier, hunter-gatherer inhabitants, there emerged farming people who began to domesticate plants such as taro and banana as well as coconut. Not only did the coconut provide a source of food and drink for daily living but it became a staple commodity in travel at sea including long-range exploration, as it does to this day in many island communities. Perhaps the greatest age of maritime discovery in human history was the age of Polynesian exploration and colonization reaching out from Southeast Asia through Melanesia to Hawaii, the Marquesas, and Easter Island. The domesticated form of coconut even appears to have reached the Pacific coast of Central America, perhaps by canoe, accompanied or not by surviving people, who may have perished on the journey or following arrival or else been absorbed by the local inhabitants "without trace." This era of exploration is believed to have begun around 4,000 to 5,000 years B.P. and extended well into the second millenium of the modern era, culminating with the arrival of Polynesian peoples in New Zealand less that 1,000 years ago. Some Austronesians traveled westward as evidenced by strong linguistic ties between Sumatra/Borneo and Madagascar. These voyagers may have arrived 3,000 years B.P. taking with them food crops such as coconut and banana (Simmonds, 1976).

The evidence of the role of the coconut in providing food and water from long Polynesian voyages is impressive. Almost without exception, wherever Polynesian colonization took place there is to be found, in modern times, authentic domesticated forms of the coconut. These forms, characterized by large size of fruit and relatively thin husk, are highly efficient and convenient water containers with a "shelf life" at maturity of many weeks. Such fruit taken on the voyage would eventually germinate and sprout, but would remain as a vital source of food, and also drink, until the haustorium eventually absorbed it all. Whenever a voyaging Polynesian party colonized a new place, some of the remaining coconut supplies from the journey would have been available for planting among the native wild coconut palms of the place. Most modern Polynesian communities in the tropics have inherited various forms of domesticated coconut, well-preserved through ongoing selection. Recent random fragment length polymorphism (RFLP) molecular analysis has shown a close affinity of coconut identity from Malaysia to Panama (Lebrun et al., 1999).

The coconut has been associated with human settlement in Sri Lanka for at least 2,300 years, and in southern India for even longer, being mentioned in post-Vedic literature that is 3,000 years old (Thampan, 1982). That a distinct population of coconut has developed in the south Asian coastal region is attested by clear differences in DNA detected by RFLP analysis (Lebrun et al., 1999). There appears, however, to be little evidence in India of selection for large nuts with thin husks and high water content, although the variety Kappadam has thin husk. Kappadam is the exception, perhaps because a good supply of husk has been vital as a source of household fiber and fuel in southern India since ancient times. The coconut had probably spread naturally to some of the islands of the Indian Ocean, and through trade or migration to Seychelles, Madagascar, and the coast of East Africa. DNA analysis has also revealed that southeast Asian germplasm entered the Indian ocean via Madagascar, giving rise to intermediate forms which also extended to the East African coast (Lebrun et al., 1999).

It was only with the arrival of a Portuguese expedition led by Vasco da Gama into the Indian Ocean in 1497/1498 that the first coconut seeds were taken from India for planting in the tropical lands of the Atlantic ocean, beginning at the Cape Verde Islands. From there nuts spread to the West African mainland as well as being taken in the mid-sixteenth century to the Carribean islands, and thence to all coasts of Central America and tropical South America (Harries, 1978). Thus, from its obscure and untraceable early origins on the coasts of migrating continents and Southeast Asian islands, the distribution of coconut finally encircled the globe about 400 years ago. The present distribution can best be described by summarizing production data, which are presented in Table 6.1.

TABLE 6.1. The 86 coconut-producing countries of the world, grouped into eight categories by scale of output.

Category	Indian Ocean, Southeast Asia, Pacific	West Africa, Caribbean, Americas
< 1 kt	Angola, Cocos Islands (Australia), Mauritius, Nauru, Niue, Oman, Seychelles, Singapore, Somalia, Tokelau, Tutuila (U.S.), Tuvalu, Wallis and Futuna	Barbados, Benin, Cameroon, Cape Verde, Central African Republic, Gabon, Guadaloupe, Martinique, Puerto Rico, Senegal, St. Kitts, Democratic Republic of the Congo
1-5 kt	Caroline Island, Cook Island, Maldives, New Caledonia	Belize, Costa Rica, Cuba, Dominica, Equatorial Guinea, Grenada, Guinea, Guinea-Bissau, Honduras, Liberia, Nigeria, Panama, Peru, Sao Tome and Principe, Sierra Leone, St. Lucia, St. Vincent, Suriname, Togo
5-10 kt	China, Comoros, Guam, Kenya, Palau, Tonga	Ecuador, Guyana, Haiti, Trinidad and Tobago
10-50 kt	Bangladesh, Fiji, French Polynesia, Kiribati, Madagascar, Mozambique, Myanmar, Solomon Islands, Vanuatu, W. Samoa,	Colombia, Ghana, Ivory Coast, Jamaica, Nicaragua, Venezuela
50-100 kt	Papua New Guinea, Tanzania	
100-500 kt	Malaysia, Sri Lanka, Thailand, Vietnam	Brazil, Mexico
500 kt-1 Mt	India	
> 1 Mt	Indonesia, Philippines	

Note: Oil equivalent—1 kt = 1,000 tonnes; 1 Mt = 1,000,000 tonnes.
Source: Burotrop, 1992.

BOTANY

Classification

The coconut stands alone in the *Cocos* genus of the Arecaceae or palm family of the Cocoeae tribe, and has no close surviving relatives (see the section on evolution). In brief, the coconut palm is considered to have evolved on the strand of the ever-changing coastline of islands and landmasses fringing the Tethys Sea as they drifted north from Gondwana (Harries, 1990). The "melting pot" of the ocean provided frequent (in geological time) opportunities for mixing of the diversifying progeny of many small populations of palms, occupying many contrasting strand environments, which evolved finally into the modern coconut, *Cocos nucifera* L.

Forms of Coconut Palm

Two distinct forms of coconut palm are recognized, the tall and the dwarf. The essential difference is in the rate of trunk elongation which is at least twice as rapid in the tall. The trunk diameter of the tall is also generally 1.5 to 2.0 times greater than that of the dwarf, giving a cross-section area two to four times greater. There is one important subgroup of the dwarfs (Niu Leka or Fiji Dwarf), however, which has a similar trunk diameter to the tall but a trunk extension rate even less than the other dwarf form. The frond length of the tall is around 6 m, compared with 4 m for the dwarf, resulting in a much larger crown on the tall palm. Another major difference is that the tall is predominantly cross-pollinated and therefore heterozygous, while the dwarf is predominantly mostly self-pollinated and therefore largely homozygous.

Morphology

The Trunk

The coconut trunk comprises an outer dense zone or tube surrounding a central "rod" of much lower density, although both zones become more dense with age with the outer zone reaching a maximum of around 1.1 t/cubic meter. Except for the zone immediately below the crown, which comprises a low-density trunk formed in the last year or two, the trunk is very tough and relatively flexible. The upper portion of a trunk 15 m long is capable of bending almost parallel to the ground, which allows critical relief of wind-induced mechanical shear. The crown adopts a position presenting a streamlined shape to the wind, minimizing pressure and potential damage.

The trunk of the tall form typically has a bole (basal region of large diameter tapering from about 1 m height to the standard diameter at 2 m height) that develops between the age of three years and the initiation of the first inflorescence. It can be understood in terms of the amount of assimilate produced by the canopy that is available to support the development of a "sink" or plant organ. During that period, prior to any demand by the reproductive system, the early growth of the tall trunk is very well supported by photosynthetic assimilate supply. The trunk diameter tapers down gradually as the palm diverts assimilate to meet the demand from its developing inflorescences and later the fruit-bearing bunches. In the dwarf, on the other hand, flowering is initiated much earlier and there is practically no development of a bole.

The trunk of both the tall and the dwarf forms elongates quite rapidly during early fruit production. However, as yield increases to a high level, competition for assimilate gradually reduces the rate of trunk elongation in the tall from its early peak of greater than 1 m/yr to 50 mm/yr at age 60 years (Photo 6.2), whereas the diameter diminishes only by 30 percent over that

PHOTO 6.2. Frond bases packed tightly in the crown of an old palm. Sixteen fronds are "packed" onto about 100 mm of trunk.

period (Foale, unpublished data). A period of severe stress due to water or nutrient deficit may cause a restriction of the diameter of the trunk, which often reverts to its former diameter when the stress is relieved.

The properties of the wood in a mature trunk are such that valuable timber can be milled out for both structural and ornamental purposes. Special mechanical sawing materials (Stellite or tungsten teeth) and techniques (injection of cold water into the active cutting zone) are adopted to prevent heat damage to the saw blade and to help clear fibers that are released by the blade. The outer zone of the trunk is milled separately from the inner, lower-density sector.

The Root System

In common with most monocotyledonous plants the coconut possesses an adventitious root system. There are many primary roots attached to the base of the trunk and "radiating" thence to occupy, in an ideal light-textured soil, a hemispherical root zone. Where the soil is shallow or contains a compacted B horizon, and where the water table is close to the surface, downward extension will be limited. The root, of diameter 6 to 10 mm, is capable of extending 5 to 7 meters outwards from the palm base (Thampan, 1982). The vertical extension down the soil profile is usually 1 to 1.5 m, but greater in sandy soil. Significant overlap exists between the root systems of neighboring palms in a plantation, which suggests that fertilizer applied anywhere across the interrows can be accessed for uptake. Pomier and Bonneau (1984) found that root depth was much reduced by a clay content of 30 percent compared to a profile dominated by sand.

The primary roots sometimes have first order (major) branches of 4 to 5 mm in diameter. Second (2 mm) and third (0.5 to 1 mm) order branches fill the role of feeder roots (Avilan et al., 1984). There are no root hairs on coconut roots and no reports of nutrient-scavenging fungal micorrhizae, although these are common on many forest species that are also devoid of root hairs. Closer study may yet reveal that coconut roots do have an association with supportive microorganisms, as it is hard to explain their resilient performance underwater and nutrient deficits on the basis of what has been discovered thus far.

Extending a few mm from the upper surface of primary roots can be found small, whitish, pointed organs known as pneumatophores which evidently evolved to maintain oxygen supply to the root tip during diurnal submersion in the water table.

The Frond

In common with most palms, the spear-shaped frond emerges vertically from the single terminal-growing point. The leaflets lack chlorophyl while tightly packed together, but are transformed rapidly, becoming green as they unfold. The angle of phyllotaxis (arrangement around the axis of the stem) of the coconut is close to 140 degrees, either clockwise or counter-clockwise. The fifth frond above or below any chosen frond subtends an angle of 20 to 30 degrees with that frond, as shown in Figure 6.1. That is, each frond is only slightly aside from being placed in vertical alignment with fronds that are five positions above or below. This relationship enables very rapid counting of the number of fronds in a palm crown, which is convenient in research work.

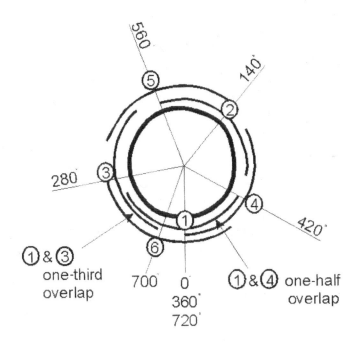

FIGURE 6.1. Representation of the way in which the base of each frond is "wrapped" one-third of the way around the trunk. The relative positions of fronds are shown progressing down the trunk from number 1 to number 6, which is off-set only slightly to the left of number 1, thereby overlapping almost completely. (*Note:* Center circle represents the palm trunk. ① is the youngest frond. Cumulative angular separation of the frond sequence is shown.)

Collectively the fronds form the crown of the palm. Some important aspects of the dimensions of the frond base and the height interval on the trunk between successive fronds influence the shape and behavior of the crown at different stages in the life of the palm. These effects are crucial to forming a realistic expectation for productivity as palms grow older.

The Crown

The drawing in Figure 6.1 and the image of frond bases attached to a trunk in Photo 6.3 show the sideways overlap between neighboring fronds of the thick wad of tissue, referred to henceforth in this discussion as the

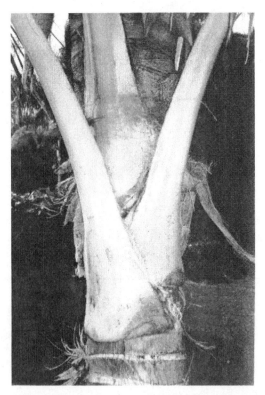

PHOTO 6.3. The lowest frond shows a clear overlap with frond 3 to its right and major overlap with frond 6 in the center. The spacing betgween scars here is about 4 cm, which is sufficient to allow the frond to retain a "trunk-hugging" position. Further reduction of the spacing will increase the outward pressure of the upper fronds on neighbors below.

"base pad" (Foale et al., 1994). This section of the base of a frond adds mechanical strength to resist the tendency of the long axis of the frond to rotate around the pivotal zone of firm attachment to the trunk. The attached surface is actually wrapped one-third of the way around the circumference of the trunk, providing a powerful "grip" to resist any pivoting action or "lowering" of the frond from its almost vertical position, hugging the trunk.

Photo 6.4 shows the thickness of the base pad on the frond of a young palm to be about 7 cm. On the lower trunk of the palm, up to 5 m in height, the average vertical spacing (interval between leaf scars) of fronds has a maximum value around 7 cm (Photo 6.5). The spacing diminishes to around 4 cm at 25 years, causing the upsweeping attitude seen in all the fronds of a young palm, as shown in Figure 6.1, to change. Gradually, over 10 to 20

PHOTO 6.4. Longitudinal slice through the center of the frond base, showing the zone of attachment (vertical) and the thickness of the base pad, which in this case is about 7 cm (the scale is marked in cm).

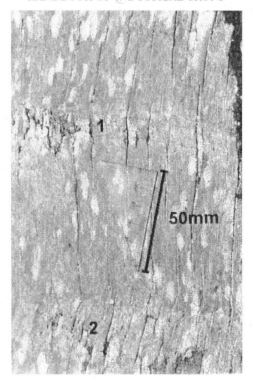

PHOTO 6.5. Scar spacing 2 m above ground level in an old palm—these scars are about 8 cm apart.

years or so, the behavior of the frond changes. The frond, pivotting on its attachment to the trunk, droops or descends more rapidly from its early vertical position. In time, half of the fronds hang at angles below the horizontal and the other half are above the horizontal plane, giving the crown a distinctly spherical shape—the classic iconic shape so loved by tourists. This change can be explained in terms of mechanical pressure exerted between the base pads of neighboring fronds.

The rearrangement of the "angle of repose" of the frond is simply a response to internal pressure on its base pad as adjacent younger fronds expand. Angular movement or drooping of the frond axis of the older (outer) frond about its point of attachment relieves this pressure. When the average interval on the trunk between fronds is 3 to 5 cm, there is firm pressure due to contact between the base pads of any frond and its neighbors two positions up and two position down from it. The frond has one-third overlap with each of these neighbors (Figure 6.1 and Photo 6.2). The pressure increases as trunk extension diminishes so that eventually the base pads of fronds 1 and 4, which have a one-half overlap, are also in firm contact. It appears that the resulting sustained outward pressure weakens the attachment of older fronds so that these frond progressively give way as shown by their drooping (Photo 6.6). The base of the inflorescence would also contribute to this overcrowding at the surface of the trunk, generating even more outward pressure. In extremely old coconut palms, the overcrowding is so severe that the frond base is almost wrenched from the trunk, remaining loosely attached by elongated, fibrous-looking, xylem bundles, leaving a sunken "scar" on the trunk (Photo 6.7).

PHOTO 6.6. Vertical slice through the crown shown in Photo 6.5. The progressive forcing of the frond bases away from the trunk is clearly demonstrated.

PHOTO 6.7. Crowded scars on the upper trunk of a 60-year-old palm. The two large scars bottom left and right center are separated by a gap equal to the 50 mm scale. The interval between these scars represent ten fronds, giving an average scar spacing of 5 mm, which is similar to that in Photo 6.2.

Other economic palms, such as oil palm and date, differ from coconut in that the trunk is of greater diameter and the width of basal frond attachment is much less than that of coconut, so that interfrond pressure does not come about until the palms are very old indeed.

There are two important consequences of this evolution of the shape of the coconut crown. Young palms up to 25 years of age have a crown that evolves during that period from an inverted cone shape to hemispherical. This range of shapes achieves maximum light interception, and the incident energy is fairly well distributed over all fronds, ensuring high photosynthetic efficiency (Foale, 1993a). As the crown takes on a spherical shape, however, with a decreasing proportion of the fronds angled upward, light interception falls until, when the fronds are "half up and half down," only 50 percent of solar radiation is intercepted at a standard plantation density. Progressive reduction in length of the frond with age serves to reduce interception even further. The obvious consequence of the declining rate of capture of solar energy is that potential biomass production falls in proportion, following a downward spiral of reducing photosynthetic infrastructure, and also increasing the intensity of competition for resources between developing frond, inflorescence, and trunk.

The second consequence of the loosening of older frond in the coconut crown is that these fronds can be shed more readily in a destructive wind. The coconut minimizes risk of destruction by its ability to reduce the size of the crown, which complements its ability to adopt a streamlined position, already mentioned in the section on origin. Young palms are more prone to wind damage because the crown is more robust and also because there is insufficient length of trunk to bend downwind; but still many survive (Photo 6.1).

A further consequence of falling light interception is that in a natural coconut woodland more light reaches the young palms below struggling to capture sufficient energy to become productive.

The Inflorescence

Every frond of a healthy mature palm has an associated inflorescence in its axil, which emerges almost 12 months after the frond expands. First the encasing spathe (a somewhat leathery sheath) appears and extends to its full length, then its lower side splits open and the multibranched "flower bunch" is released. Several thousand small male flowers (4-6 mm long) are borne on the 20 to 40 branches and begin to shed pollen progressively from the distal ends for about three weeks. The fairly rare "spicata" form of coconut has only one or two short branches bearing just a few dozen male flowers. The female flowers, which expand to about 25 mm diameter during the period of opening of male flowers, are borne toward the proximal end of the mid and lower branches, usually singly or in pairs. The male phase starts with pollen shedding on the day of inflorescence opening and continues for 18-21 days. However, the duration of the female phase is generally only five to eight days, and the individual stigma will remain receptive for one to three days.

The female flowers of the tall form become receptive to pollination after pollen shedding is complete, although this is dependent on genotype and environmental conditions. However, on the dwarf, pollen is still abundant within the inflorescence during female receptivity. Thus the tall is generally outbreeding and heterozygous (diverse characters between palms), whereas the dwarf is generally inbreeding and homozygous (uniform characters). There is some inbreeding in the tall, where a favorable environment (which is seasonal for plentiful rainfall in many coconut regions) speeds up the rate of emergence of fronds and inflorescences. The result is that pollen from the next youngest inflorescence is released while some female flowers of the older one are still receptive. Some outcrossing occurs in the dwarf because pollen from other palms can be brought by wind or insects to the female flowers to compete with the local pollen.

The Fruit

The number of female flowers, which varies with season and various stresses experienced by the palm, is usually well in excess of the number of fruits that develop. The interval between pollination and maturity is around 12 months—less in very warm environments, and as much as 15 months at higher latitude where there is a cool season. At higher altitude within the tropics, the delay in rate of maturity will be in proportion to the drop in mean temperature below the coastal mean of 28°C. After six months the fruit has reached full size and its vacuole is completely full of water (liquid endosperm). Thenceforth the volume of the vacuole diminishes as the 10 to 15 mm layer of kernel (solid endosperm) is formed on the inner surface of the shell.

Maturity of the fruit is usually indicated by the entry of air into the vacuole, so that the water "splashes" audibly when disturbed. In an environment with a large atmospheric vapor pressure deficit, early loss of nut water takes place, and the splash in the vacuole can be generated before maturity. In that case one should avoid interpreting the splash as an indicator of maturity.

Both low seasonal temperature and soil water deficit appear to influence the shape of the shell within the fruit. A common response is reduction of the diameter of the nut, giving it an elongated appearance in contrast to the dominant spherical shape of most varieties (Photo 6.8 and Foale, unpublished data; and Photo 6.9, Bourdeix, personal communication).

The size of the fruit differs greatly with genotype, and the weight of mature fruit ranges between 1 kg and 3 kg, although there are examples of fruit outside this range. The size of fruit varies with the number borne on the palm at one time, and the contrast is especially notable between bunches carrying few and many fruit respectively. The coconut fruit is the very largest of seeds in the plant kingdom after the coco de mer (*Loidecea seychellarum* L.). The seed provides an almost unique long-term source of energy for the emerging seedling, which enables it to cope with both water and nutrient deficit during the first year of its life (Foale, 1968b). This capacity was an important contributor to the survival and competitiveness of the coconut in newly colonized habitats, and has earned it "weed status" in some environmental reserves and semiurban beach environments where it has proliferated and suppressed other strand vegetation (Foale, unpublished).

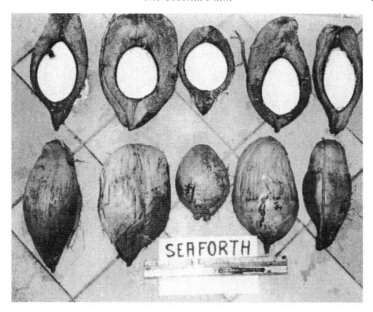

PHOTO 6.8. An example of elongated nut shapes taken from a coconut population at latitude 20 degrees on the east coast of northern Australia. During a period of about three winter months the mean temperature is below 21 degrees, which may be responsible for this "stressed" nut shape.

The Seed and Seedling

The seed comprises the shell and its contents, with the husk filling both protective and dispersal roles. However, the entire fruit is usually left intact, except for trimming some husk from the region where the shoot is expected to emerge, when selected for seedling production. Following maturity the single embryo present in the nut begins to develop by first emerging through the germ pore. In "domesticated" Pacific coconut types, germination is rapid, often taking place before the fruit falls from the palm, whereas in other populations there may be a delay of several weeks, or the need for added moisture to stimulate the embryo, before germination takes place. The presence of the husk prevents the observation of true germination, as there is about two months delay before a sprout develops sufficiently to emerge through the husk.

Germination can be controlled by dehusking and storing the nut, unbagged and free from moisture. A nut in this state will not sprout until the

PHOTO 6.9. The shape of nuts of the same genotype grown in different environments. (*Source:* Bourdeix, personal communication.)

atmosphere is vapor saturated by placement under moist mulch or in a sealed bag with enclosed moisture (Foale, 1993b). The embryo expands through the germ pore, developing within ten days into a rounded mass of soft white tissue. The "internal" end of the embryo expands to form the haustorium, an enzyme-secreting organ that breaks down and absorbs the kernel and progressively fills the vacuole. Absorption of the energy-rich tissue of the kernel supports the continuing expansion of seedling tissue and its differentiation, which commences after 14 days or so, eventually forming a shoot on the upper side and a root initial below.

Growth of the coconut seedling could continue for months in the dark as absorption of the kernel proceeds, but at three months or so from true germination the first small leaf appears and photosynthesis begins. Over the next several months, as leaf area expands, a gradual transition from total dependence on the kernel to complete independence occurs around 12 months of age, when uptake of the kernel is complete (Foale, 1968b). During the early months, the endosperm-assisted growth rate of the seedling is very high compared with small-seeded plants. If microorganisms invade the endosperm within the first six months from germination, the growth rate of the seedling falls sharply, usually leading to rejection of the seedling.

Cytogenetics

Despite early research of a descriptive nature on the cytology of the coconut (e.g., Nambiar and Swaminathan, 1960; Ninan et al., 1960; and Abraham et al., 1961), interest in this field has waned. However, Louis and Rethinakumar (1988) presented a useful report titled "Genetic Load in Coconut Palm," identifying mechanisms, arising from the heterozygous nature of coconut, that resulted in the elimination of undesirable recessive genes. The advent of technology for the unravelling of the genetic code and identification of specific genes and linkages has now somewhat sidelined the discipline of cytogenetics.

GENETIC IMPROVEMENT

Source of Diversity

The diversity of coconut populations and the association of readily-identifiable variants with particular dispersed ethnic groups, especially across Southeast Asia and the Pacific, reveal that coconut improvement has been an objective for a long period of human history. Large-fruited populations, for example, are widespread wherever there are Polynesian peoples, who have detectable linguistic and cultural links with island peoples close to the Southeast Asian landmass. The large fruit fulfills the human objective of a convenient source of food and water for sea voyages. A variant of the large fruit has especially long fibers, valuable in the manufacture of rigging for seagoing vessels. Since 1950 there has been an industrial initiative for genetic improvement of the coconut, to raise the level of productivity of oil and the profitability of plantations.

All genetic improvement is based on the diversity present in the species. However, in order to be useful to plant breeders, this diversity needs to be characterized reliably.

Characterization of the Genome of Coconut

Coconut populations have been distinguished since antiquity on the basis of appearance and utility. Tall coconut palms show variation in fruit color, size, and shape, and humans have shown a preference for larger fruit, especially for water and food supply at sea, but also for convenience in processing the kernel for home use. Dwarf palms were preferred for their ease of harvest to provide drinking water to the household, but their kernel was less palatable. Tall palms, on the other hand, were more difficult to manage because of outbreeding and heterozygosity. As a consequence, tall populations have provided greater challenges in their characterization, to establish whether real and heritable differences exist (Liyanage and Sakai, 1960).

Fruit Component Analysis

This method was first used widely as a characterization tool by Pieris (1935), who confined his measurements to the dehusked nut, and later by Whitehead (1966) and Harries (1978). It was subsequently adopted by many other researchers with considerable success. Fruit component analysis allowed identification of similarities and differences between populations with a fair degree of confidence, based on the low sensitivity of fruit composition to most changes in the environment. Harries (1978) made use of differences in fruit components in developing a general theory of the evolution and dissemination of the coconut around the world.

The basis of the method is to weigh the whole newly mature fruit and then its dissected components. The fruit is selected when only a trace of the fresh color remains in the husk. The moisture content of the husk would be expected to be similar in fruit from different populations, at around 20 percent, and water content of both shell and kernel would be fairly stable. Broad categories were established, named *domestic* and *wild* types at the extremes, with a low and high proportion of husk respectively. Many populations were of intermediate proportions, and these were described as introgressed. It is not possible to determine true genetic relatedness or difference between diverse populations that are identified relying solely on fruit component analysis. Such distinction required a molecular approach. This has only recently become available.

Molecular Markers

An array of molecular tools is now at the disposal of the coconut geneticist, with different price labels and degrees of efficacy to characterize the coconut genome. The basic entity in this process is known as a molecular marker, defined as "an inherited chemical trait that can be used to distinguish between individuals, groups of individuals or positions on a chromosome" (Ashburner, 1999). Until the early 1990s, attention was given to "gene products," such as isozymes, to detect genetic difference, but a direct assay on the DNA itself has proven to be more effective. A review of the application of molecular markers to coconut improvement was published by Ashburner (1999), but knowledge in this field is changing very rapidly.

A comprehensive study of a very wide range of germplasm covering the global variation of coconut germplasm, was undertaken by CIRAD (Centre de coopération internationale en recherche agronomique pour le développement) of France during the 1990s and reported by Lebrun et al. (1999). Their report reveals the groupings of coconut genotypes based on RFLP analysis. The study showed, for example, that all germplasm in populations extending from Southeast Asia eastward right across the Pacific, including the Pacific coast of Central America, could be grouped together as quite distinct from that of southern Asia and the Caribbean. An intermediate group was apparent that took in Madagascar and the east coast of Africa. Dwarfs were found to be genetically similar to the talls of their geographic region.

Molecular markers are used in many resource management applications, including diversity analysis based on at least 20 individuals per population. This will provide data on genetic distance both within and between populations, genetic fingerprinting—such as establishing true crosses from interpopulation hybrids, and outcrossing analysis—which can establish with certainty the mating system of a coconut population (Ashburner, 1999). Many interesting hybrid combinations have been tested in coconut, selected in the early days using a mix of both intuition and geographical separation. Molecular characterization of coconut populations will make possible choice of future combinations with a higher probability of significant improvement, although a great deal more research would be needed to validate this strategy.

The First Breeding Work

The coconut fruit, which had been a primary energy food in the diet of tropical peoples for millennia, served as the primary source of vegetable oil for the European and U.S. markets during the late nineteenth and early twentieth centuries, experiencing high demand and little competition. It was only when other sources became competitive and the price began to decline, following World War II, that these global economic forces aroused concerted interest in genetic improvement of coconut as an oil-producing crop.

Improvement of coconut populations has been occurring unwittingly for thousands of years. However, it was during the twentieth century that a serious attempt was made to focus research into the crop in a plantation context. This took place at different times in different regions, but the earliest attempt was made with the establishment in India of four research institutes in 1916, of which the coconut research institutes at Kasaragod and Nileswar became the most important (Nair and Nampoothiri, 1993; Iyer and Damodaran, 1994).

Hybrid Vigor

The discovery of hybrid vigor in coconut by Patel (1937) between West Coast Tall × Chawghat Green Dwarf was a significant landmark in the history of coconut improvement. Since then, a number of hybrids both D × T and T × D were evolved and released from India. The important hybrids from India are Lakshaganga (Laccadive Ordinary × Gangabondam), Anandaganga (Andaman Ordinary × Gangabondam), Keraganga (West Coast Tall × Gangabondam), Kerasree (West Coast Tall × Malayan Yellow Dwarf), Kerasawbhagya (West Coast Tall × Straight Settlement Apricot), Kerasankara (West Coast Tall × Chawghat Orange Dwarf), Chandrasankara (Chawghat Orange Dwarf × West Coast Tall), Chandralaksha (Laccadive Ordinary × Chawghat Orange Dwarf), VHC-1 (East Coast Tall × Malayan Green Dwarf), and VHC-2 (East Coast Tall × Malayan Yellow Dwarf). Drought tolerance studies conducted at CPCRI, Kasaragod revealed the possibility of identifying desirable traits of drought tolerant cultivars under field conditions (Rajagopal et al., 1988). The promising drought-tolerant varieties/hybrids were West Coast Tall × West Coast Tall, FMS (Federal Malay State), Java Giant, Fiji, Andaman Giant and Laccadive Ordinary × Chawghat Orange Dwarf.

In Fiji, a hybrid was generated by crossing the indigenous dwarf known as Niu Leka with the Red Malayan Dwarf in the 1920s (Marechal, 1928). Worldwide depression and war interrupted follow-up research on these pioneer efforts. The Coconut Research Institute in what was then Ceylon (established in 1929 as the Coconut Research Scheme, becoming an Institute in 1950) contributed to the early work (e.g., Pieris, 1935). A broad foundation of understanding was developed in those early decades of the opportunities and the constraints presented to those seeking to improve the yield of coconut.

Overcoming Constraints on Coconut Breeding

The constraints to breeding for yield improvement of coconut are formidable compared to most other crops. One such constraint is rapid height increase, such that a proven high-yielding palm is many meters high, which makes for time-consuming access to perform pollination. Another constraint is the large size of an individual, such that the breeder requires 8 ha to conduct a genetic trial (Bourdeix et al., 1993). Yet another constraint is the long delay in developing a successful technique for harvesting a manageable amount of pollen from the inflorescence, finally achieved in the 1960s (Whitehead, 1963). In addition, there are the biological constraints of few seeds per bunch and per palm; the extended period of serial receptivity of the female flowers within a bunch, requiring multiple visits for pollination; and the long turn-around time of six to ten years between generations of talls. Genetic constraints include a high degree of heterozygosity within any population of tall palms and different levels of combining ability between genotypes. There is also the constraint of lack of a clonal propagation method for outstanding selected or bred genotypes, which are otherwise constrained by the low number of seeds generated (Santos, 1999).

Progress in Selection

Locally, tall populations of coconut have been distinguished mostly by fruit characters, and regionally or globally by place of origin. The latter approach seemed somewhat arbitrary until collections of different populations were assembled for research purposes. It was sometimes found that the introduced germplasm often was severely attacked by (one or more) insect pests that were known locally as a mild nuisance or by a disease that had not previously been known in the area. Examples of this are the attack on Malaysian Tall in Solomon Islands by the leaf beetle *Brontispa* sp.; the outbreak on the leaves of Polynesian Tall in Solomon Islands of the fungus *Drechslera* sp.; and the attack on hybrids that had a West African Tall parent, planted in Indonesia and the Philippines, by a *Phytophthora* strain that differed from strains that were tolerated in West Africa. In India there is a devastating disease complex known locally as "root (wilt)" and elsewhere as Kerala coconut decline, of which the exact etiology still remains unknown after decades of effort to understand it.

A most unusual experience in Vanuatu was the appearance in many different introduced populations of the foliar decay virus. Although at first it was feared that the new pathogen had arrived with the introductions, it was eventually found that the virus was actually present all along in Vanuatu.

The indigenous population possessed tolerance, which enabled it to remain free of symptoms, and the pathogen had been transmitted to the nontolerant introductions with devastating results.

These examples point to the fact that many coconut populations have adapted to a local environment with its particular pests, diseases, and even climate (such as extreme cold episodes in Hainan Island; Zushum, 1986), but the adaptation is not expressed in any obvious morphological trait. This is an aspect of population identity to which molecular markers are beginning to make an important contribution. Breeding for improvement in yield or other product characters, while retaining "invisible" traits such as pest resistance or disease tolerance, will be feasible only with the aid of molecular tools.

Researchers were slow to develop confidence about the possibility of breeding for yield improvement, partly because of the difficulty of demonstrating its heritability (Liyanage and Sakai, 1960), and partly because of a negative correlation between the number of fruit borne and the copra content. However, beginning in the 1950s, yield became a major objective for breeding efforts in India, Sri Lanka, Philippines, Ivory Coast, and several Pacific countries, whereas in Jamaica the main objective was to overcome the ravages of the devastating lethal yellowing disease. In India and Sri Lanka, on the one hand, emphasis was at first on mass selection within a population (even though the first report of dwarf × tall hybrid vigor was made in India in 1937), whereas others concentrated on interpopulation hybrids, especially in the 1960s.

Progress Using Hybrids

Much interest was aroused in Fiji and Solomon Islands in the 1960s by the great diversity of the F2 generation from hybrids produced in the 1920s in Fiji (Marechal, 1928) between Malayan Red Dwarf and Niu Leka (Fijian Dwarf).

In parallel with work going on in India and Sri Lanka, the use of hybrids in the Pacific and in Ivory Coast has yielded rapid progress for some combinations, such as the Malayan Yellow Dwarf × West African Tall and the Malayan Red Dwarf × Rennell Tall, which became commercial cultivars in the 1970s. In each case the yield of the first generation of hybrids was raised about 30 percent above the best performing tall population (Foale, 1993b). Outstanding yield increase has also been demonstrated for tall × dwarf hybrids in India (Nair and Nampoothiri, 1993) and Sri Lanka (Peries, 1993) in recent decades. Because of the risk of a particular parent lacking tolerance to a potential natural enemy in locations far from the source of the parent genotypes, local industries have latterly moved to include at least one local parent in any tall × dwarf hybrids tested.

Apart from the general combining ability between tall and dwarf populations, subsequent testing of many combinations in Ivory Coast has revealed that some are significantly more productive than others (Baudouin, 1999). This author reports, for example, that hybrids from 15 Rennell Tall (RLT) palms combined with each of three "testers"—West African Tall (WAT), Malayan Red Dwarf (MRD), and Cameroon Red Dwarf (CRD)—produced very different yields of copra ranging from 15 to 27 kg/yr. The mean value (kg) for the 15 palms in each group was (by tester): MRD, 19; WAT, 21; and CRD, 24.

Other combinations of "genetically distant" tall parents have proved to be very interesting. Geographic separation, supported by isozyme and latterly RFLP markers (Lebrun et al., 1999), has enabled three major groups to be defined within the global coconut population. Crosses between subpopulations from Group I (Southeast Asia/Pacific) and Group II (southern Asia/West Africa/Caribbean) gave a higher yield than any combinations within these two groups (Baudouin, 1999). A most interesting finding of this work, however, was that an intermediate group (III) could be identified from Madagascar and East Africa, from which the Mozambique Tall hybrid with WAT equaled the best combinations between the geographically more separated Groups I and II. It is expected that molecular markers will play an important role in identifying outstanding combinations for hybridization in the future.

Producing Hybrid Seed Commercially

The preferred method for production of a large quantity of hybrid seed for plantation development, for example, or as a marketer of improved seednuts, involves interplanting of the two parents. A dwarf × tall cross is most readily done with the dwarf as the female parent, requiring emasculation or removal of all the male flowers before pollen is shed. The dwarf palms, planted in two or three rows per row of the pollen parent must be checked daily. Any inflorescences that have begun to emerge from the opening spathe are dealt with by complete removal of all male flowers using purning shears. The male flowers must be removed entirely from the field so that the only possible source of pollen is the tall parents nearby. Multiple male parents can be included in such seed gardens to provide the possibility of producing different hybrids as required, but only one hybrid is available by open pollination at any given time. Controlled pollination of bagged inflorescences, to produce small lots of seed, is also possible but obviously more costly and less successful.

Where crossing is to be done on a smaller scale with a few selected parents, the female parent is bagged after emasculation, and the chosen pollen blown into the bag for several days in a row. This expensive procedure is only feasible for research purposes. Commercially viable hybrid seed gardens have been set up in many countries; for example, 2 million hybrid seeds are produced each year in India and in Sri Lanka the capacity to produce hybrids is rising rapidly. Nursery managers make use of the fact that both the red and yellow colors of widely used female parents are due to a recessive gene, which, if expressed in a progeny from the seed garden, is evidence of self-pollination of the mother palm. Any seedling expressing the red or yellow color on the petiole is therefore rejected.

In summary the difficulties with hybrid seed production in coconut are that (1) they are expensive; (2) the number produced seems always to fall short of demand; and (3) hybrid palms do not breed true to type, risking problems if their fruits are used in turn for seed, which some farmers might be inclined to do.

In Vitro Propagation

Development of this potential tool has challenged many researchers for at least two decades, yet only one commercial application has been achieved—the propagation of embryos from makapuno nuts. A makapuno nut is filled with a jelly-like substance that is rich in coconut flavor but incapable of stimulating and supporting the germination of the embryo. The raw makapuno product is sufficiently valuable that there is commercial support for culture of embryos to ensure production of makapuno nuts in the progeny. In vitro propagation has been considered promising and presented as such to funding bodies (Santos, 1999) but other viable applications elude researchers (Harries, 1999).

It appears that objectives other than yield of oil are necessary in order to excite any truly commercial interest in in vitro propagation. In the meantime, limited propagation of outstanding elite palms to form the basis of a seed garden might possibly be justified although not yet technically feasible on a suitable scale. The long-term possibility that genetic transformation can be performed on coconut callus, which is then differentiated into propagules, such as has been achieved in many other agricultural crops, still entices researchers to persevere in the quest to streamline this technique. No major change in the present uncertain prospect of in vitro-propagated coconut palms contributing to yield improvement, or quality enhancement, is anticipated in the near future. Embryo culture has found a place in limited exchange of germplasm, although even here the technique is not yet suitably advanced for ease of use by technicians of intermediate skill.

GROWING CONDITIONS: AGRONOMY

Adaptation to Diverse Soils

In the course of its evolution on the "tropical edge" of the freed landmasses drifting north from Gondwana, the coconut would have experienced highly diverse strand environments. Its basic need for freely available soil-water supplied either by frequent rainfall or via fresh groundwater could have been met in a wide range of soils: highly alkaline sands derived from the erosion of coral; silica sands; deltaic silts and loams; black and red clays formed from volcanic ash; and highly acidic larva sands. Adaptation to such diverse soils appears to reside in most modern tall populations, reinforcing the idea that a great deal of natural mixing of genotypes between subpopulations from these diverse environments has taken place over geological time. The result of this is that the coconut has thrived in the tropics far from its shoreline home, wherever its need for a regular supply of available soil water has been met.

The extremes of soil environment are the coral atoll with sand and gravel at a pH of 8.3, where some critical nutrients are rendered difficult to take up, and the organic peat soils of Sumatra, for example, where the pH falls to 4.5 or even lower. This fall is because drainage brings about the oxidation of accumulated sulfur, transforming it to sulfuric acid, which releases aluminium and in some cases manganese in toxic forms from clay minerals.

Water

The coconut accesses the life-giving freshwater "lens" or reservoir found under atoll soil or the water table of larger landmasses that commonly flows to the sea under coastal sand dunes. In other situations the soil must be capable of retaining sufficient water for the coconut to survive the longest seasonal intervals between significant rainfall events.

Deep, well-structured clay and clay loam soils, which hold up to 250 mm of plant-available water could, from a fully saturated starting condition, sustain the growth of the coconut palm for two dry months. Such soil is found on the uplifted coralline "benches" of many South Pacific islands and on river plains and deltas worldwide. Coconut is also planted widely on banks between irrigated fields, where the water level is controlled by farmers irrigating other crops, thereby providing an accessible water table.

Although the coconut appears to thrive in the vicinity of the swamp, which often occupies the swale adjacent to a coastal dune, the palm actually is very sensitive to waterlogging. Where the water level is static, and low in oxygen, coconut roots become inactive. On the other hand, the sort of diurnal, tide-induced vertical oscillation of the water level in the freshwater lens of an atoll or under a coastal dune is ideal. The root has pneumatophores (a sort of air snorkel), attached to short vertical branchlets which supply oxygen to the main root while its physiologically active root-tip region is temporarily submerged. It has been found necessary to provide good drainage for coconut on heavy-textured lowland plains to avoid water-logging during sustained rainfall.

As mentioned earlier, the coconut does best in the absence of severe water deficit, but there does appear to be some adaptation to sustained periods of deficit. Whereas in most island environments the coconut was rarely moved far from the coast prior to the plantation era of the late nineteenth century, it has evidently been grown in the subcoastal zone and further inland in India since antiquity. This provided an opportunity for adaptation to more severe episodes of water deficit than would be experienced elsewhere. This development is supported by evidence that the modern West Coast Tall, and related populations in East Africa, West Africa, and the Caribbean, are more drought-resistant than most populations from Southeast Asia and the Pacific. An exception may exist in the Pacific on some of the dryer atolls such as those of northern Kiribati, where the level of salinity in the freshwater lens rises sharply toward the end of long periods of low rainfall. The local tall population generally shows less sign of stress (collapsing lower fronds, premature nut fall, failure of flower emergence) than material introduced from better-watered environments.

Nutrients

The coconut requires all the nutrients common to the plant kingdom, including a relatively high need for chlorine for which most other species have a low requirement (von Uexkull, 1972). Legends have long existed in coconut cultures about the need of the coconut to be near the sea, or at least, when grown at a distance from the coast, to have seawater or salt sprinkled on the soil occasionally. The folklore was finally shown to have substance when the substantial need for chlorine was proved. This nutrient is deficient most commonly beyond the range of cyclic salt, delivered near coastlines in rainfall, and where the soil is readily leached during intense seasonal rainfall.

Since 1950 there has been extensive investigation of the suspected failure of yield to reach the potential of the local environment, and limiting nutrients have been identified using the classic type of fertilizer experiment. Numerous case studies of the association of a particular nutrient deficiency with certain soil characters or landscape history have enabled some findings to be extrapolated to similar locations. For example, on atolls it is found that when nitrogen is not limiting, the trace elements iron and manganese become important, rendered poorly available by the high pH of the soil. Another case concerns sulfur, which has been found to be limiting in many areas where coconut was planted on former grassland. Frequent natural or human-induced burning of the grass has led to a cumulative loss of sulfur to the atmosphere and, in time, exhaustion of the reserve in the soil (Southern, 1969).

Nitrogen deficiency is most common in dryer environments, especially on sandy soil, which leaches readily during the wet season. Where rainfall is high and the dry season is short or absent, nitrogen is rarely in short supply as the rapid recycling of nutrients from the considerable body of organic residue being returned from the palms and the undergrowth meets the needs of the entire plant community. Exported nitrogen is replaced by the accession of nitrate in rainfall and from any nitrogen-fixing herbs or shrubs. In this well-watered environment, potassium commonly becomes limiting because of the large amount that is exported in the fruit, as well as likely loss through leaching. When rainwater is high in cyclic salt (the mineral salts present in seawater, usually present near the coast) sodium displaces much of the potassium cation from the clay, allowing it readily to be leached beyond the root zone, to the detriment of the coconut palm.

A low level of available phosphorus is common in many soils but the problem is especially acute where the soil is high in iron oxide (ferralitic or lateritic soils) which binds phosphorus in an insoluble form. The application of a remedial dose of phosphorus fertilizer is therefore best done in a band so that the binding capacity of the soil adjacent to the band becomes saturated. Magnesium has been found to be limiting in some environments, especially after other limiting factors have been dealt with. On some clay soils situated on uplifted coral benches, low magnesium was evidently due to loss induced both by high calcium and sodium.

Tissue Analysis

A comprehensive review of the mineral nutrition of the coconut palm was published by Manciot et al. (1979a,b, 1980). These authors affirm the value of tissue analysis as a tool for the identification of nutrient deficiency or imbalance. The coconut lends itself particularly well to nutrient investigation based on tissue analysis because of the regular production of foliage and fruit throughout the year. Leaf analysis is used most commonly as material of a similar stage of maturity, such as from the fourteenth frond, can be used as a standard source for collection of samples. So-called "critical levels" of nutrient have been proposed, providing a very useful guide for further detailed investigation of palms that are evidently falling short of potential yield. The critical levels in coconut leaf proposed by Manciot et al. (1979a,b, 1980) and de Taffin (1993) are summarized in Table 6.2.

The prescription of fertilizer based on either nutrient diagnosis or fertilizer experiments is an exercise in economics depending on the relativity of input costs to the increased market return from increased yield. Coconut is

TABLE 6.2. Critical values for concentration of mineral elements in the leaf tissue (fourteenth-youngest frond) of adult tall coconut.

Major elements	% Dry matter	Comments
N	1.8-2.0	In tall × dwarf hybrid—2.2
P	0.12	
K	0.8-1.0	In tall × dwarf hybrid—1.4
Mg	0.20-0.24	Strong inverse sensitivity to extremes of K
Ca	0.30-0.40	Some inverse sensitivity to extremes of K
Na	Not essential	Substitutes for K in case of deficiency
Cl	0.5-0.6	
S	0.15-0.20	

Trace elements	Parts per million	Comments
B	10	
Mn	>30	Difficult to fix value as very interactive with Fe in strongly alkaline soil; potentially toxic in extreme acid soils
Fe	50	Deficient only in strongly alkaline soils
Cu	5-7	Deficiency very rare, not very certain
Mo	0.15	Common value—no response observed yet
Zn	20	Common value—no response observed yet
Al	>38	Nonessential element but always present; potentially toxic at values well in excess of this common level

Source: Based on Manciot et al., 1979a,b, 1980; de Taffin, 1993.

generally slow to respond to fertilizer, although correction of acute potassium and chlorine deficiency can result in a yield increase within six months due to increased kernel per nut (Foale, 1965; Manciot et al., 1979b). Clearly early detection of limiting nutrients is vital to achieving the most economical yield potential of the local environment.

Management of Coconut-Based Mixed Crop Regimes

The coconut plantation is very commonly a place of dual cropping, with pasture, cacao, diverse fruit trees, maize, pineapple, coffee, root crops, banana, and many others being grown, sometime in more diverse mixtures, to capture the solar energy that escapes the coconut canopy. For reasons outlined in the section on botany, the coconut canopy follows an evolutionary trend in its capacity to intercept light. In a healthy crop of plant density 180 palms/ha light interception ranges from low values preflowering to around 90 percent at ten years, continuing up to 25 years (D. Friend, unpublished data), and then gradually declining to around 45 percent at 50 years of age. The brief preproduction phase is often quite productive for short-term crops, especially if the land has been newly cleared, providing a plentiful supply of nutrients released from the great bulk of residue generated by the destruction of the previous vegetative cover.

In some cases coconut density may be lowered, either uniformly or by "hedgerow" planting, in order to allow continuous intercropping, but the potential yield of the coconut stand will be proportionately reduced. Intercrops under old and so-called "declining" coconut plantations frequently become more productive with any increase in the amount of solar energy which bypasses the coconut canopy.

Seed and Seedling Management

The almost unique large seed of coconut provides special challenges and opportunities in providing the best possible seedlings for field planting. Strong interest has long existed in "seedling vigor," since recognition in India and Sri Lanka of its correlation with the yield potential of the adult palm (Liyanage and Abeywardena, 1958). One challenge in recognizing relative seedling vigor is to provide a clearly defined starting point for a batch of newly germinated seeds. Some populations, particularly in Southeast Asia and the Pacific, have seeds that germinate very quickly after maturity, some tending to sprout well before the fruit falls naturally. On the other hand, most other populations have a brief "dormancy" after maturity before being ready to germinate when the dry husk is moistened. If a batch of the latter comprises fruit that has been harvested at a comparable stage of maturity, a firm starting point is achieved. On the other hand the "fast germinating" seeds may contain a mix of maturity when harvested, with some embryos already emerged from the germ pore, but remaining hidden for many weeks within the husk. Such seeds also are subject to the risk of misorientation in the seed bed when the still-hidden shoot is placed pointing downward. This can lead to protracted delay or even fatal entrapment as the shoot changes its direction of growth upward once more.

Germination Rate

Time taken from true germination (expansion of the embryo through the germ pore) to sprouting (emergence of the sprout from the husk) depends both on the vigor of the seedling and on the thickness of the husk. As these characters are quite variable in a tall population the time lapse could be six to ten weeks (Foale, unpublished data). A reasonable comparison of vigor, at least within small batches of seedlings, can be achieved, however, by taking all the nuts that sprout within one week, for example, and keeping them together for a within-group comparison in the nursery. Checks done at time intervals thereafter will allow the seedlings to be sorted into fast, medium, and slow growers on the basis of leaf length and area, robustness of the collar, and general appearance.

Another potential influence on the growth rate of the seedling is the amount of kernel present in the nut. It has been shown that most seedlings continue to draw energy from the kernel for around 12 months and, paradoxically, smaller nuts within a population support a slightly higher early growth rate (Foale, 1968b). Evidently the haustorium makes more rapid and effective contact with the kernel when the vacuole is smaller. As both genetic and environmental factors influence the amount of kernel in a seed nut, it would appear that the response of seedling growth rate to nut size would introduce a small bias into early seedling selection.

Polybag Seedlings

Traditionally in many coconut cultures the ability of the seedling to recover from "bare root planting," aided as it is by the energy supply from the kernel, has resulted in that method being still in use. However, a great gain is achieved by raising seedlings in polybags (sized 40 cm tall and radius 25 cm for eight-month planting out—larger for older ones). At transplanting to the field there is no root damage and the growth continues unchecked (Foale, 1968a). Widespread use is made of polybag seedlings in many areas, in spite of the greater logistic challenge of carrying into the field the seedling and polybag full of soil weighing 20 to 25 kg. Comparative health and vigor can usually be judged well by eight months of age, but polybags provide the opportunity to hold seedlings in the nursery for up to 12 months. A 50 percent increase in the size of polybag would be advisable in this case, thereby saving a few months of more expensive field maintenance. It is important, however, to spread older seedlings further apart in the nursery to avoid slowing the growth rate due to mutual shading.

Seedling Selection

It has long been considered good practice to select only the 50 to 60 percent most vigorous seedlings from an open-pollinated tall population, but one would expect to discard very few seedlings from a batch of hybrid seed. Where a high proportion of seedlings is expected to be discarded, this should be done as early as possible, so that selected seedlings can be transferred from an open nursery bed to polyethylene bags without a severe setback. Based on correlation studies, the seedling (one year old) selection criteria developed in India include collar girth (10-12 cm), number of leaves (six to eight), and early splitting leaves. The nuts that do not germinate within six months of sowing should be removed from the nursery (Iyer and Damodaran, 1994).

Rigorous protection of the dense population of seedlings in a nursery is required against insect pests and leaf spot diseases, using appropriate chemicals. Choice of plant density for the field planting depends on the genotype, with talls at 160 to 180/ha and hybrids at 180 to 220/ha for sole coconut production. Different densities and palm arrangements other than isometric (triangular) are chosen when intercropping is planned.

Early Field Management

In a seasonally dry environment it is common to plant the seedling with the nut a little below the ground surface. On atolls, where the water table may be 1 m or more below the surface, sometimes the seedling is placed deep so that its roots reach moist soil quickly. Once palms have been established in the field, competition from weeds needs to be kept to a minimum to allow the palm to develop and begin fruit production according to its potential. A common procedure is to maintain a weed-free circle, by hand weeding or herbicide application, enlarging the circle as the "spread" of the young palm increases. Under ideal conditions in the South Pacific some tall populations (e.g., Rennell Tall) flower within four years, and their hybrid with Red Malayan Dwarf flowers almost one year earlier, especially when polybag seedlings are used. Populations in India and most of Africa must endure annual water deficit in the dry season. Talls tend to flower after about seven years, but their tall × dwarf hybrids flower two years earlier.

Productive Palms

The management of the crop once production begins is highly dependent on what is to be processed and marketed. In areas of high human population density and small coconut holdings, every part of the mature fruit is value added for the market, being transformed to a variety of foods (milk, cream, milk powder, desiccated coconut, oil), drink (fresh or bottled coconut water), fuel and charcoal, fiber (for a rich range of possible uses), and "coco peat" from coir dust, not to mention a great range of utensils and curios based on the fruit and shell. At the other extreme many tens of thousands of hectares of coconut plantations still exist from which the sole product is oil via the "copra pathway." Increasingly, the oil is separated for export in the country of production, and the residual cake is either exported or fed to livestock locally.

In some situations control measures are used against pests and diseases, especially where a fatal outcome is likely if nothing is done. For example, phosphoric acid treatment can arrest the development of *Phytophthora* bud rot, and oxytetracycline is used to protect particularly valuable palms against lethal yellowing. Bourgoing (1991) published recommendations for controlling insect pests, including hygiene practices such as removing old coconut logs in which pests may breed.

ADAPTATION TO BIOTIC FACTORS

Degree of Exposure

The coconut was exposed to diverse insect and microorganisms throughout its evolution, especially those populations established next to diverse rainforests on the coastline of large islands or continental landmasses. Establishment on such coasts might not always have succeeded, such as in the case of northern Australia. An indigenous rat species (the white-tailed rat, *Uromyces caudimaculatus*) is capable of chewing through husk and shell to feed on the kernel. In recent millennia (perhaps a period of 60,000 years) colonizing hunter-gatherer human populations would have contributed to failure by the coconut as they assiduously collected for food any coconut seeds and seedlings lying on the beach. The tribal language of the Cook-town and Lockhart River regions in the far northeast coast of Australia contains separate words for a plain nut (on the beach) and a nut with an haustorium and emerged seedling (Tucker, 1983; Diana Wood, personal communication).

The examples presented in this section add to those mentioned earlier in this chapter of adaptation in the form of tolerance or resistance to attack or invasion by insects and microorganisms. A range of types of organisms is presented here to convey the apparent broad attractiveness of coconut tissues and organs as food or habitat.

Diversity of Coconut Pests

There are many scenarios for the association of the coconut with other biota, some swiftly fatal such as infestation with the palm weevil (*Rhyncho-phorus* sp.) which lives and multiplies within the upper, softer part of the trunk and growing point, consuming soft tissues until the palm dies. Likewise, the nematode *Rhadinaphelencus cocophilus* (red ring) enters the phloem of the coconut trunk, causing a slower death than the weevil as the result of clogging the tissue. The Polynesian rat *(Rattus exulans),* a native of Indochina, is widespread on many coconut islands, feeding on immature nuts that fall once damaged. In the western Pacific, a highly specialized "coconut crab" *(Birgus latro)* may have evolved parallel with the coconut. It posesses a very powerful claw capable of tearing off the husk and smashing the shell in order to feed on the kernel. The termite also attacks the trunk of coconut consuming tissue inward from beneath the outer layer and eventually proving to be fatal to the tree. Moving from "fauna" to microflora, a diverse group of organisms is responsible for fatal diseases, ranging from a trypanosome (heart rot; Dollet, 1999), through fungal *Phytophthora* (bud rot; Dollet, 1999), phytoplasma (lethal yellowing and many related variants; Harrison et al., 1999), and virus (foliar decay; Randles et al., 1999), to a viroid (cadang-cadang; Rodriguez, 1999).

Insect Pests

In the case of insect pests, evidence exists of adaptation of populations taking the form of resistance to some pests. The different intensity of attack of the leaf beetle *Brontispa* sp. was mentioned earlier in this chapter. This insect is endemic to Papua New Guinea, Solomon Islands, and Vanuatu, and is spreading in northern Australia. Any exotic genotypes brought to the first three of those countries requires concerted protection, at least during the first two to three years in the field. Local populations, on the other hand, while needing protection in the intense environment of the nursery, are rarely attacked in the field. The Papua New Guinea rhinoceros beetle *Scapanes australis* also shows a preference for exotic germplasm, but is still inclined to inflict serious damage on indigenous populations, especially when no choice is available. No record exists of resistance to many other insect pests, such as *Oryctes rhinoceros,* which has an almost global distribution, and various locusts and stick insects. Both species of rhinoceros beetle often disfigure the coconut canopy without fatal results, but their damage commonly "opens the door" to a *Rhyncophorus* sp., which infests with quite deadly results.

There are very many other insect pests of coconut, including scale, mite, white fly, locust and stick insect, leaf miner, and hemipterous nut-fall bugs, for which no observed adaptation has been reported.

Disease Organisms

Adaptation to disease, where it is manifest, usually takes the form of tolerance, but the situation is often very hard to evaluate. The pathologist first concentrates effort on investigating plants that are not doing well, often identified by very specific symptoms such as discoloration of the foliage or distortion of the inflorescence (e.g., Rodriguez, 1999). By the time such symptoms are visible the plant will be heavily infected with the organism, which can be identified positively using appropriate miscroscopy and molecular probing, and often the investigation rests there. Other palms in a population might also be infected, but either have the capacity to hold the proliferation of the organism in check, or the infection is recent, and therefore they do not show classical symptoms. In the case of actual tolerance, there might be some loss of vigor. If that is noted it usually would be attributed to some other environmental effect such as the limiting supply of one nutrient or another.

Molecular tools for the detection of "subclinical" pathogens would be useful to dispel any uncertainty in cases of nonlethal or even nonexistent reaction to an invading organism. Tolerant individuals are identified, whereas previously all survivors of a disease outbreak would have been lumped together as "escapes."

Whither Adaptation?

It is early yet in our understanding of adaptation and tolerance to disease in coconut. As molecular tools become more affordable and widely applied, the great uncertainties that bedevil management of outbreaks of disease can be dealt with. At present, the survival of some members of a population otherwise decimated by a phytoplasma-induced disorder or by bud rot, for example, gives rise simply to speculation. Do the survivors possess tolerance? or resistance? are they lucky? or are they particularly robust due to an excellent local environment imparting some sort of vigor-related resistance? The case of the phytoplasma which causes lethal yellowing of coconut in the Caribbean, Florida, and Central America illustrates some of these issues very well.

There is known and widely exploited tolerance (or maybe resistance) to the phytoplasma in Red Malayan Dwarf and Panama Tall, which are widely used in Jamaica and elsewhere. This is a quantitative trait of high heritability (Ashburner and Been, 1997). Some other genotypes also show tolerance, though usually less than either of the two discussed earlier (Zizumbo et al., 1999). However, there are recent reports that the hybrid has succumbed to the disease in some environments, and it is now known that there are many other phytoplasma-induced lethal diseases worldwide. The organism has been shown, by molecular probing, to have a significantly different form in each of those other locations around the globe. Phytoplasmas have also been reported in palms free of severe symptoms in Southeast Asia. This is the region of origin of the Red Malayan Dwarf, which poses the question, "did the Red Malayan Dwarf acquire tolerance of phytoplasmas through natural selection during its evolutionary history?"

It appears then that adaptation to this major scourge of coconut, in lands where it has been grown for a limited time—if 3,000 years in India can be considered limited—has actually developed in the truly ancient heartland of the coconut species. The adaptation of the organism itself is part of the problem however, as, in common with so many other pathogens, it too has undergone change through mutation and selection. The convergence of adapting host and adapting pathogen may be clarified in the foreseeable future, however, as molecular tools are brought to bear. Such tools would be used to identify markers associated with tolerance on the coconut side and pathogenicity on the side of the pathogen (Cardena et al., 1999). However, in order to achieve this, an understanding of the genetics of resistance/tolerance is required.

An Unbeatable Pathogen?

Another very serious pathogen, the coconut cadang-cadang viroid (CCCVd) of northern Philippines (and a related form in Guam), is less well studied than the phytoplasmas at this stage. This is due to its limited distribution and slow rate of spread. It proved far more recalcitrant in the stage of detection and proof of its pathogenicity because of the molecular minuteness of the causal organism (it is the smallest infectious particle known to biological science), the absence of any identified transmission agent, and the existence of nonpathogenic near-analogues in coconut populations worldwide (Hodgson and Randles, 1999; Rodriguez, 1999). There is still no evidence of tolerance to CCCVd, so attention is focussed on early detection of infected palms using molecular probes, and elimination of these palms in the hope that viroid-free zones might be created in the affected regions.

Adapting the Coconut to the Market

Human interest in the coconut is expressed through activity in "the market," even in cases where the palm is preserved mainly as a graceful icon that beautifies the environment. The amenity added to that environment by the coconut palm enhances its value in the tourist market or the real-estate market! The coconut is so highly valued in some instances that a "safe" form is sought for busy tourist centers, which does not produce large and "dangerous" fruit that require expensive pruning. As the fruitless coconut is diametrically opposite to the fruitful adaptation sought by all the coconut farmers of the world, its development is not likely to be included soon in the program of coconut industry research institutes.

Yield Potential

High yield of fruit and oil in particular is still the primary driver of effort to bring about genetic change in the coconut. Related to increased yield expressed in this way is also yield of edible products and coconut water, shell, and fiber, so increased yield covers most of the market interest in bringing about genetic change. Associated characters that have been sought within the broad thrust for increased yield are

- earliness of onset of fruit production (precocity);
- large nuts to facilitate processing by hand;
- uniformity of nut dimensions to facilitate mechanization of processing;
- higher oil content (often associated with enhanced flavor);
- ease of harvesting (e.g., dwarfs for production of drinking nuts);
- special aroma or taste characters (e.g., edible husk and perfumed water);
- makapuno (jellied endosperm for edible treats), and so on.

The flavor and quantity of toddy or "coconut nectar" tapped from the inflorescence are also potential breeding objectives not presently identified.

Quality Traits

Within particular markets are often specific quality traits that add value to the product. For example, variations exist in quality and length of fiber for use in the vast array of products derived from coir. There is also interest in possible variants in the mix of fatty acids in coconut oil. Marketers of unsaturated oils have had some temporary success in displacing coconut oil from traditional markets by branding saturated coconut oil, among many saturated vegetable and animal fats as especially a danger to heart health. The label "artery-clogging saturated oils" was coined by the saturated oil lobby after some flawed research in which a deficiency of essential fatty acids in a saturated-oil diet drove up serum cholesterol. Recent research underlines three great dietary strengths of coconut oil, as it provides a readily-digestible source of energy, critical cell wall lipids, and the precursors to monolaurin and monocaprin which have outstanding antimicrobial and antiviral properties (Enig, 1999, 2000).

About half the fatty-acid component of coconut oil is lauric acid, which has high market value for manufacturing detergents and other industrial chemicals. Surprisingly, the huge increase over recent decades in supply of lauric oil from palm kernels and from genetically modified canola seems to have stabilized this high demand, working to the advantage of coconut oil.

Fatty-Acid Mix

Some variability exists in the relative mix of the six main fatty-acid components of coconut oil but the range is fairly narrow for each one. If a particularly good market for one component were to be found it might best be met by increased overall production rather than seeking increased yield of that component by genetic modification (F. Rognon, personal communication).

Coconut, a Food

Immense possibilities exist for coconut in domestic and global markets as food. Canned or tetra-packed coconut water of different grades, including tender and intermediate stages up to mature water, and different degrees of dilution and sugar adding, enjoys increasing popularity, along with water served directly from the fresh nut. In China coconut water is labelled "official drink of the Chinese State Banquet" and is widely available in the coastal cities. Demand for coconut milk and cream also will continue to rise as its flavor and nutritional merits become more widely known outside the centers of production. Canned coconut nectar is an attractive sweet drink increasingly consumed in Asia.

A need exists for concerted argument against the marketplace enemies of coconut oil products, who seek to capture market share by the dubious tactic of casting doubt, or even making strong and questionable negative assertions about its nutritional qualities. Much recent research has shown coconut oil to possess some outstanding qualities which contribute to better health by suppressing the activity of some viruses (Enig, 2000). Such "fringe benefits" must be proclaimed alongside high quality standards in all processed products to enhance the overall reputation of coconut as food. From a breeding point of view, however, all effort is directed to the general aim of raising yield, and any change in quality traits at this stage is purely fortuitous.

COCONUT RESEARCH AND DEVELOPMENT ORGANIZATIONS (R&D)

Wherever the coconut is grown there is R&D, because coconut farmers, like all farmers throughout the world, constantly seek ways to improve yield, protect their palms against hazards, and where possible improve the quality of the products. Nevertheless, institutions in many countries are funded in order to apply scientific tools and inquiring minds to the problems and possibilities presented in coconut production. Further, many regional alliances exist, especially where many small countries are clustered, as in the South Pacific and the Caribbean. In addition, some institutions exist with a global view of coconut improvement, such as the International Plant Genetic Resources Institute (IPGRI), a UN agency, and its specialized offshoot the Malasia-based Coconut Genetic Resources Institute (Cogent). The French research agency CIRAD also takes a very broad view of the needs for coconut research worldwide.

Funding bodies have been formed in some "donor countries" from which monetary aid is directed to support coconut research. For example, based in Europe, with a small secretariat in France, the Bureau for the Development of Research on Tropical Perennial Oil Crops (BUROTROP) is an international nonprofit association that was formally established in 1989. Its current mandate is to assist, strengthen, and further develop research on coconut and oil palm. It helps in the transfer of research results to the production sector, to benefit the small-scale farmer in the form of improved self-sufficiency and of capacity to produce a cash crop for local or regional consumption.

The BUROTROP Board of Administrators consists of seven representatives from regional organizations and producer countries in Africa, Asia, Latin America, and the Pacific, and seven others from European donor countries and international agencies. It is operating, until September 2003, with the financial support from the Directorate General (DG) of Research of the European Commission.

Global Coordination

Cogent, referred to earlier, has a very specific mandate to facilitate the description and exchange of coconut germplasm, and its use in generating populations with improved performance around the world. Cogent has linkages with a high proportion of all the coconut-producing countries of the world and promotes standardization of morphological descriptors for coconut as well as providing training in germplasm collection, breeding strategy, and technology. It has obtained funding to support regional collections of coconut germplasm in Latin America (Brazil), Africa (Ivory Coast), India (Kerala), Southeast Asia (Indonesia), and South Pacific (Papua New Guinea).

A list of coconut country institutes and centers linked to the work of Cogent is included in the Appendix to this chapter. This was kindly supplied by Dr. Pons Batugal, director of Cogent, which is attached to the International Plant Genetic Resources Institute (IPGRI).

National Research Centers

Several outstanding national centers of coconut research exist around the world, the oldest being in India (since 1916), now under the banner of the Central Plantation Crops Research Institute based at Kasaragod in Kerala. Research objectives over the years have concentrated on improved soil management, increased yield through genetic improvement with a focus on drought tolerance, and seeking a solution to the disease syndrome (root [wilt]) now known as Kerala coconut decline, which has been a scourge for many decades.

India and Sri Lanka

In Sri Lanka, the Coconut Research Institute was established in 1929. It performed outstanding work on factors controlling yield in coconut, and grappled with the challenges of breeding to improve a highly heterozygous population. In both India and Sri Lanka, much effort was and remains directed toward multiple cropping associated with coconut, having achieved significant progress over the decades. Many problems beset coconut production in many other countries early in the twentieth century, for example, various coconut declines in the Caribbean, and insect attacks on almost every part of the palm in diverse populations around the world, but attempts to deal with them were short lived. Higher yield was sought through replacement of tall coconut trees with dwarfs on plantations in what was then the Federated Malay States, and the generation of a dwarf hybrid in Fiji.

Latter Half of the Twentieth Century

From 1950 onward interest in coconut improvement became much more widespread as competition from other oilseed crops began to threaten the profitability of plantations. The Philippine Coconut Authority was formed to oversee research and has an outstanding track record of improving coconut technology and collaboration with other players, especially in development of high-yielding hybrids.

France, through an agency now known as CIRAD (formerly Institute for Research for Oils and Oil Seeds [IRHO]), set up a major facility in Ivory Coast in the early 1950s—extending to Vanuatu, French Polynesia, Cambodia, and Brazil, and the British did likewise in Jamaica, as did Australia in Papua New Guinea, all in the 1950s. CIRAD continues to play a major role in coconut research with high-tech laboratory work done in Montpellier, and staff seconded long-term to programs in Ivory Coast and Vanuatu and short-term to many other countries.

Unilever, a major private investor, initiated research in 1951 in Solomon Islands that was subsequently supported by the colonial government but eventually closed down in the 1970s after releasing a very promising hybrid that was used to replant the island's entire plantation.

GTZ of Germany has been active in supporting coconut research, especially in Tanzania. Other institutes that are playing very important roles are: Wye College of UK, pioneers of tissue culture; University of Adelaide, which provides outstanding expertise in viroid and virus research; Max-Planck-Institut of Köln, Germany, contributing molecular expertise; and the Center for Scientific Investigation at Yucatan (CICY), Mexico, which is a leader of research into phytoplasma disorders.

PROSPECTS FOR THE FUTURE OF COCONUT

The Background

There are many reasons to be optimistic for coconut, which has for many millennia been truly "the tree of life" even though a sector of the global human economy in the twentieth century has dealt with it harshly. In a sense, the tree of the people was hijacked by world business to meet a desperate trade shortage of vegetable oil that began in the 1850s and extended to the 1950s. Other oilseeds were established closer to the foreign markets toward the mid-twentieth century, and demand for coconut began to decline. Quite ruthlessly competing oil producers took early, and often flawed, research into the health effects of different vegetable oils, and interpreted them to the detriment of coconut. Without a matching promotional effort by the coconut industry to provide balance in the information reaching consumers, some of its best markets disappeared, particularly for food uses. Happily, the immensely superior properties of lauric oil for soap and detergent making have provided some stabilizing influence on the global trade price in recent years, although even here lauric oil is now being produced in transgenic canola.

A Health Food

Great progress has been made on the health front in establishing many properties of coconut oil that are highly desirable (Enig, 1999, 2000), but generating adequate resources to gain broad community recognition of the true value of coconut in the diet may require a prolonged campaign. In a comprehensive review titled "Coconut: In support of good health in the twenty-first century," Mary Enig (1999) reported some outstanding findings with very important long-term implications. As an example:

> Monolaurin [formed in the body from lauric acid in coconut oil] is the antiviral, antibacterial, and antiprotozoal monoglyceride used by the human or animal to destroy lipid-coated viruses such as HIV, herpes, cytomegalovirus, influenza, various bacteria including *Listeria monocytogenes* and *Helicobacter pylori,* and protozoa such as *Giardia lamblia.*

Many other research results, cited by Enig (1999, 2000), directly contradict the questionable interpretation of other findings from projects funded by rival vegetable oil industries. These give hope that coconut will be restored to

its rightful place as a valuable food for all. The challenge now lies in education of the consumer more than in the need for further research.

Protection of the Production Base

Once the future of the market for coconut is more secure, attention can be
directed to issues on the production and processing sides, which should also
give rise to optimism. As outlined in this chapter, coconut management is
well understood and much progress has been made in adaptation of the crop
to different environments, especially those where periodic water deficit is
important. The use of molecular markers now offers the possibility of deal-
ing with the many disease entities of coconut with confidence and increasing precision. The phytoplasma group of organisms has been a scourge
whose geographical span was not realized until the new molecular technologies became available. Testing worldwide is still incomplete, so the distribution "map" for phytoplasmas in coconut is also incomplete (Harries, personal communication).

With the aid of molecular techniques, phytoplasma-infected palms can
be identified before symptoms develop (which allows early clearing out of
infected palms), and molecular markers linked to resistance or tolerance
can be used to identify desirable individuals and populations. Much remains to be done to perfect techniques and reduce costs, and the capacity of
an entity such as the phytoplasma to evolve into more virulent forms cau-
tions science to sustain its vigilance. Other diseases need to be added to the
list for molecular testing, but the outlook is more hopeful on this front than
it has ever been. *Phytophthora* is another widespread disease (bud rot)
where there are diverse strains that might better be dealt with using the same
molecular technology, but little is being done about it at present.

Several other serious diseases are more localized than the phytoplasmas
and bud rot, where molecular methods also have great potential. The list includes coconut cadang-cadang viroid in Philippines and its variant in
Guam, the virus foliar decay of Vanuatu, and the trypanasome heart rot in
the Caribbean (Dollet, 1999).

Improvements in Processing Technology

Much cause for optimism exists on the processing side as technology for
extraction of coconut oil improves over the "copra pathway." On a large
scale, copra has yielded an oil requiring costly refining, bleaching, and de-
odorizing to be marketable. There is an especially good opportunity in extracting oil from shredded kernel dried only to 12 percent. The avoidance of

high temperature and pressure provides a far more attractive, aromatic oil for food and cosmetic uses (Etherington and Mahendrarajah, 1998; Etherington et al., 1999). In general, manufacturers are seeking to adopt accepted food standards in the production of coconut milk, cream, and desiccated products. The Asian Pacific Coconut Community (APCC, 1997) has published codes and standards for aqueous coconut products, which will assist in standardizing the range of products and raising consumer confidence. Concerted promotion of such standards by industry leaders is required to achieve widespread adoption.

The many other products from coconut, ranging from high-value coconut "nectar" and alcoholic and nonalcoholic drinks—all derived from toddy—through fiber, shell, and wood derivatives, have sound market prospects when high quality is maintained. Science will continue to be called upon to generate the knowledge required to grow and protect the coconut in the wide range of environments where it presently prospers. If that support is forthcoming, we can look to the ingenuity of farmers and processors to sustain the outstanding status that coconut deserves in the human and environmental economies.

It is well to remember that the coconut provides, in the field, a long-term crop that lends stability to the biophysical environments where it is a major component of the vegetation community (Persley, 1992).

APPENDIX: CONTACT INFORMATION FOR RESEARCH CENTERS AND INSTITUTES

The following list contains contact information for research centers and institutes that are linked to Cogent, the coconut network of International Plant Genetic Resources Institute.

Bangladesh

Senior Scientific Officer
Horticultural Research Centre, Pomology Division
Bangladesh Agricultural Research Institute (BARI)
GPO Box 2235, Joydepur Gazipur—1701
Bangladesh
Tel: 880-2-9800441 / 9332340
Fax: 880-2-841678
E-mail: baridg@bttb.nct.bd

Benin

Director, Institut National
Des Reserches Agricoles Du Benin (INRAB)
Station de Recherche sur le Cocotier
BP Cotonou, Benin
Tel: 229-240101
Fax: 229-250266
Fax: (225) 20 226985 / 21 248872

Brazil

Head of Coconut Germplasm Bank, EMBRAPA/CPATC
Av. Beira-Mar, 3250, CEP 49025-040, Aracaju-SE Brazil
Tel: 55-79-2171300
Fax: 55-79-2319145
E-mail: tupi@cpatc.embrapa.br

China

Director, Wenchang Coconut Research Institute
Chinese Academy of Tropical Agriculture Science
Wenchang Erli
Wenchang City, 571321 Hainan Province, China
Tel: 86-898-3222416 / 3224820
Fax: 86-898-3230949 / 3223918
E-mail: cricatas@public.hk.hi.cn

Cook Island

Ministry of Agriculture
Government of the Cook Islands
PO Box 96 Rarotonga, Cook Islands
Tel: (682) 28711
Fax: (682) 21881
E-mail: cimoa@oyster.net.ck

Costa Rica

Director, Coconut Research
Ministerio de Agricultura y Ganaderia Siquirres
Frente, Servicentro Siquirres—Limon, Costa Rica

Tel: (506) 718-60-92
Fax: (506) 718-7191 / 768-8410

Cote d'Ivoire

Director, Station de Recherche Marc Delorme
Centre National De Recherche Agronomique (CNRA)
Port Bouet, 07 PO Box 13, Abidjan 07, Côte d' Ivoire
Tel: (225) 21 248872 / 248067

Cuba

Deputy Director, Ministerio de La Agricultura
Instituto de Investigaciones de Citricos y Otros Frutales (IICF)
Ave. 7ma. #3005 entre, 30 y 32, Miramar, Playa
Havana 10600, Cuba
Tel: 537-293585 / 225526 / 246794
Fax: 537-246794 / 537-335217
(National Citrus Corporation)
E-mail: iicit@ceniai.inf.cu

Fiji

Principal Agronomist
Ministry of Agriculture, Fisheries and Forests
Private Mail Bag, Raiwaqa, Suva, Fiji
Tel: 679-477044 ext. 263
Fax: 679-400262
E-mail: krsinfo@is.com.fj

Ghana

Director, Oil Palm Research Institute
PO Box 74, Kade, Ghana
Tel: (031) 804-710229 (Director) / 710226 / 710228
Fax: (031) 233-46357
E-mail: csur@ghana.com

Guyana

Program leader (roots and tubers)
National Agriculture Research Institute (NARI)
Mon Repos, East Coast Demerara, Guyana
Tel: (592) 202841 / 42/43
Fax: (592) 204481
E-mail: nari@guyana.net.gy

Haiti

Coordinator, Ministry of Agriculture
Centre de Recherche et de Documentation Agricoles (CRDA)
Damien, Republique d'Haiti
Tel: +509 (22) 4503
Fax: +509 (45) 4034

India

Director, Central Plantation Crops Research Institute
Indian Council of Agriculture Research
Kasaragod 671 124, Kerala, India
Fax: 91-499-430322
Tel: 91-499-430333
E-mail: cpcri@x400.nicgw.nic.in

Indonesia

Director, Department of Forestry and Estate Crops
Agency for Forestry and Estate Crops Research and Development
Research Institute for Coconut and Palmae
P O Box 1004, Manado 95001, Indonesia
Tel: (62) 431-812430
Fax: (62) 431-812587
E-mail: balitka@mdo.mega.net.id

Jamaica

Director of Research, Coconut Industry Board
18 Waterloo Road, PO Box 204
Kingston 10, Jamaica

Tel: 1-876-9261770
Fax: 1-876-9681360
E-mail: conchar@cwjamaica.com / cocindbrd@cwjamaica.com

Kenya

Regional Research Centre (RRC)
Mtwapa Box 16 Mtwapa
Kenya
Tel: (254-11)485842 / 39
Fax: (254-11)486207

Kiribati

Chief Agriculture Officer, Division of Agriculture
Ministry of Natural Resources and Development
PO Box 267
Bikenibeu, Tarawa, Kiribati
Tel: 686-28-139 / 108
Fax: 686-28-139 / 21-120

Malaysia

Station Manager
Stesen MARDI Kemaman, Batu 11
Jalan Air Putih PO Box 44, 24007
Kemaman Terengganu, Malaysia
Tel: +60 09-8646361 / 148
Fax: +60 09-8646361
E-mail: abo@mardi.my

Marshall Island

Chief of Agriculture
Ministry of Resources and Development
Agriculture Division
PO Box 1727, Majuro, Marshall Islands
Tel: (692) 625-3206 / 0740
Fax: (692) 625-3005
E-mail: agridiv@ntamar.com

Mexico

Researcher
Centro de Investigacion Cientifica de Yucatan, A.C. (CICY)
KM7 Antigua Carretera A Progresso, Ex-Hacienda
Xcumpich, Apartado Postal 87, 97310 Cordemex
Merida, Yucatan, Mexico
Tel: 52-99-813923 / 813966
Fax: 52-99-813900 / 813941
E-mail: cos@cicy.mx

Mozambique

Research Assistant – Entomology
National Agriculture Research Institute (INIA)
Entomologist Crop Protection Section
Box 3658, AV. Das FPLM, INIA-MAPUTO, Mozambique
Tel: +258 (1) 460097
Fax: +258 (1) 460074
E-mail: sancho@zebra.uem.mz

Myanmar

Acting Director General
Department of Agriculture
Planning Ministry of Agriculture
Thiri Mingalar Lane, Off Kaba Aye Pagoda Road
Yankin PO, Yangon, Myanmar
Tel: 095-1-665750
Fax: 095-1-663984 / 651184
E-mail: dap.moai@mpt.maid.net.mm

Nigeria

Chief Research Officer
Nigerian Institute for Oil Palm Research (NIFOR)
PMB 1030 Benin City, Nigeria
Tel: 52-440130
Fax: 52-248549

Pakistan

Director Horticulture
Pakistan Agricultural Research Council
Plat No. 20, G-5/1, Post Box 1031
45500 Islamabad, Pakistan
Tel: 92-51-920-7402
Fax: 92-51-920-2968 / 920-240908
E-mail: hashmi@reshem.sdnpk.undp.org / mtahir@isb.comsats.net.pk

Philippines

Deputy Administrator
Agricultural Research and Development Branch
Philippine Coconut Authority, Don Mariano
Marcos Avenue, Diliman, Quezon City, Philippines
Tel: (632) 920-0415 / (632) 426-1398
Fax: (632) 920-0415
E-mail: cbcarpio@mozcom.com

PNG

Acting Director, PNG Cocoa and Coconut Research Institute
P O Box 1846
Rabaul, East New Britain Province
Rabaul, East New Britain, PNG
Tel: (675) 983-9108 / 983-9131 / 983-9185
Fax: (675) 983-9115
E-mail: ccri@datec.com.pg

Samoa

Acting Assistant Director of Research
Ministry of Agriculture, Forests, Fisheries and Meteorology
Research Division
PO Box 1587, Apia, Samoa
Tel: (685) 23416 / 20605
Fax: (685) 23426 / 20607 / 23996
E-mail: apeters@lcsamoa.net

Seychelles

Director General
Crop Development and Promotion Division
Ministry of Agriculture and Marine Resources
Grand Anse, PO Box 166, Victoria Mahe, Seychelles
Tel: 248-378252 / 378312
Fax: 248-225425
E-mail: antmoust@seychelles.net

Solomon Islands

Director of Research for Permanent Secretary
Dodo Creek Research Station
Ministry of Agriculture and Fisheries
PO Box G 13, Honiara, Solomon Islands
Tel: 677-31111 / 31191 / 31037
Fax: 677-31039 / 21955 / 31037
E-mail: ibsram@welkam.solomon.com.sb

Sri Lanka

Director, Coconut Research Institute
Bandirippuwa Estate, Lunuwila, Sri Lanka
Tel No: 94-31-57391 / 253795 / 55300
Fax No: 94-31-57391
E-mail: rescri@sri.lanka.net / dircri@sri.lanka.net

Tanzania

Director, Mikocheni Agricultural Research Institute (MARI)
Ministry of Agriculture and Co-operatives
P.O. Box 6226
Dar es Salaam, Tanzania
Tel: ++255 51-700552 or 74606
Fax: ++255 51-75549 or 116504
Mobile: ++255-812-784031
E-mail: arim@africaonline.co.tz

Thailand

Director, Horticulture Research Institute
Department of Agriculture
Chatuchak, Bangkok 10900 Thailand
Tel: +66 (2) 579-0583 / 579-0508 / 561-4666
Fax: (662) 561-4667
E-mail: hort@doa.go.th

Tonga

Head of Research
Ministry of Agriculture and Forestry
Vainani Research Division, Nuku' alofa
Kingdom of Tonga
Tel: 676-23038
Fax: 676-32132 / 24271 / 23093

Trinidad and Tobago

Ministry of Agriculture
Cental Experimentation Centend V 117
Arima PO 52 La Florissante Garden
Dabadie, Trinidad and Tabago
Tel: 1-809-642-8552 / 642-0718
Fax: 1-809-622-4246

Tuvalu

Acting Director of Agriculture
Ministry of Natural Resources and Environment
Department of Agriculture, Private Mail Bag
Vaiaku, Funafuti Atoll, Tuvalu
Tel: (688) 20-825 or 186
Fax: (688) 20-826

Vanuatu

Head of Coconut Division
Vanuatu Agricultural Research Centre
PO Box 231, Espiritu Santo, Vanuatu

Tel: (678) 36320 / 36130
Fax: 678-36355
E-mail: labouiss@pop.vanuatu.com.vu / CARFV@vanuatu.com.vu

Vietnam

Coconut Scientist, Oil Plant Institute of Vietnam (OPI)
171-175 Ham Nghi St.
District 1, Ho Chi Minh City, Vietnam
Tel: 84-8-8297336 / 8243526
Fax: 848-8243528
E-mail: opi.vn@hcm.vnn.vn

REFERENCES

Abraham A., Mathew P.M., and Ninan C.A. 1961. Cytology of *Cocos nucifera* L and *Areca catechu* L. *Cytologia* 26: 327-332.

APCC (Asia Pacific Coconut Community). 1997. APCC codes and standard for aqueous coconut products. Asian Pacific Coconut Community, Jakarta.

Aradhya M.K., Yee L.K., Zee F.T., and Manshardt, R.M. 1998. Genetic variability in *Macadamia*. *Genetic Resources and Crop Evolution* 45: 19-32.

Ashburner, G.R. 1994. Characterization, collection, and conservation of *Cocos nucifera* L. in the South Pacific. PhD thesis, School of Agriculture and Forestry, University of Melbourne.

Ashburner, G.R. 1999. The application of molecular markers to coconut genetic improvement. In Oropeza, C., Verdeil, J.L., Ashburner, G.R., Cardega, R., and Santamarma, J.M., Eds., *Current Advances in Coconut Biotechnology*. Kluwer Academic Publishers, Boston.

Ashburner, G.R. and Been, B.O. 1997. Characterization of resistance to lethal yellowing in *Cocos nucifera* and implications for genetic improvement of this species in the Caribbean region. In Eden-Green, S.J. and Offori, F., Eds., *Proceedings of an International Workshop on Lethal Yellowing-Like Diseases of Coconut*. NRI Chatham, Kent, UK, pp. 173-183.

Baudouin, L. 1999. Genetic improvement of coconut palm. In Oropeza, C., Verdeil, J.L., Ashburner, G.R., Cardega, R., and Santamarma, J.M., Eds., *Current Advances in Coconut Biotechnology*. Kluwer Academic Publishers, Boston.

Bourdeix, R., Sangare, A., Le Saint, J.P., Meunier, J., Gascon, J.P., Rognon, F., and Nuce de Lamothe, M. 1993. Coconut breeding at IRHO and its application in seed production. In Nair, M.K., Kahn, H.H., Gopalakrishnan, P., and Bhaskara Rao, E.V.Y., Eds., *Advances in Coconut Research and Development*. International Symposium on Coconut Research and Development, Oxford and IBH Publishing Co PL, New Delhi.

Bourgoing, R. 1991. Coconut—a Pictorial Technical Guide for Smallholders. Edited D.R.A. Benigno. IRHO/CIRAD Paris, p. 301.

Burotrop. 1992. Coconut and Oilpalms: World production and consumption. Map. BUROTROP, Montpellier, France.

Cardena, R., Ashburner, G.R., and Oropeza, C. 1999. Prospects for marker assisted breeding of lethal yellowing-resistant coconuts. In Oropeza, C., Verdeil, J.L., Ashburner, G.R., Cardega, R., and Santamarma, J.M., Eds., *Current Advances in Coconut Biotechnology*. Kluwer Academic Publishers, Boston.

Chappell, J. 1983. A revised sea level for the past 300,000 years from Papua New Guinea. *Search* 14: 99-101.

de Taffin, G. 1993. *Le Cocotier* (The coconut). Maisonneuve et Larose, Paris, p. 166. (French language only)

Dollet, M. 1999. Conventional and molecular approaches for detection and diagnosis of plant diseases: Application to coconut. In Oropeza, C., Verdeil, J.L., Ashburner, G.R., Cardega, R., and Santamarma, J.M., Eds., *Current Advances in Coconut Biotechnology*. Kluwer Academic Publishers, Boston.

Enig, M. 1999. Coconut: In support of good health in the 21st century. In *Promoting Coconut Production Products in a Competitive Global Market: Proceedings of the Thirty-Sixth COCOTECH Meeting of the Asia Pacific Coconut Community*, Pohnpei. APCC Jakarta, p. 366.

Enig, M. 2000. *Know Your Fats: The Complete Primer for Understanding the Nutrition of Fats, Oils, and Cholesterol*. Bethseda Press, Silver Spring, MD.

Etherington, D.M. and Mahendrarajah, S. 1998. Economic benefits of Direct Micro Expelling coconut oil in the South Pacific. In Topper, C.P., Caligari, P.D.S., Kullaya, A.K., Shomari, S.H., Kasuga, L.J., Masache Khao, P.A.L., and Mpumami, A.A., Eds., *Trees of Life: The Key to Development: Proceedings of the International Cashew and Coconut Conference*, Dar-es-Salaam. BioHybrids International Ltd, Reading, United Kingdom, pp. 457-468.

Etherington, D.M, Zegelin, S., and White, I. 1999. Oil extraction from grated coconut: Real-time moisture content measurement and its impact on oil production efficiency. *Tropical Science* 38: 10-19.

Foale, M.A. 1965. Nutrition du jeune cocotier dans les Iles Russell (Archipel des Salamon). *Oleagineux* 20: 1-4.

Foale, M.A. 1968a. The growth of coconut seedlings. *Oleagineux* 23: 651-654.

Foale, M.A. 1968b. The growth of the young coconut palm (*Cocos nucifera* L.). 1. The role of the seed and of photosynthesis in seedling growth up to 17 months of age. *Australian Journal of Agricultural Research* 19: 781-789.

Foale, M.A. 1993a. Physiological basis for yield in coconut. In Nair, M.K., Khan, H.H., Gopalasundaram, P., and Bhaskara Rao, E.V.V., Eds., *Advances in Coconut Research and Development*. Oxford and IBH Publishing Co. Pty Ltd., New Delhi, pp. 181-189.

Foale, M.A. 1993b. The effect of exposing the germpore on germination of coconut. In Nair, M.K., Khan, H.H., Gopalasundaram, P., and Bhaskara Rao, E.V.V., Eds., *Advances in Coconut Research and Development*. Oxford and IBH Publishing Co. Pty Ltd., New Delhi, pp. 247-252.

Foale, M.A., Ashburner, G.R., and Friend, D. 1994. Canopy development and light interception of coconut. In Foale, M.A. and Lynch, P.W., Eds., *Coconut Improvement in the South Pacific.* Australian Centre for International Agricultural Research, Canberra, pp. 71-73.

Harries, H.C. 1978. The evolution, dissemination, and classification of *Cocos nucifera* L. *The Botanical Review* 44: 265-320.

Harries, H.C. 1990. Malesian origin for a domestic *Cocos nucifera*. In Baas, P., Payme, J., and Purwaningsih Saridan, A., Eds., *The Plant Diversity of Malaysia*. Kluwer, Dordrecht, Netherlands, pp. 351-357.

Harries, H.C. 1999. Does clonal coconut material have a potential use in any agricultural system? In Oropeza, C., Verdeil, J.L., Ashburner, G.R., Cardega, R., and Santamarma, J.M., Eds., *Current Advances in Coconut Biotechnology*. Kluwer Academic Publishers, Boston, pp. 431-436.

Harrison, N., Cordova, I., Richardson, P., and DiBonito, R. 1999. Detection and diagnosis of lethal yellowing. In Oropeza, C., Verdeil, J.L., Ashburner, G.R., Cardega, R., and Santamarma, J.M., Eds., *Current Advances in Coconut Biotechnology*. Kluwer Academic Publishers, Boston, pp. 184-196.

Hodgson, R.A.J. and Randles, J.W. 1999. Detection of cadang-cadang viroid-like sequences. In Oropeza, C., Verdeil, J.L., Ashburner, G.R., Cardega, R., and Santamarma, J.M., Eds., *Current Advances in Coconut Biotechnology*. Kluwer Academic Publishers, Boston, pp. 227-246.

Iyer, R.D. and Damodaran, S. 1994. Improvements of coconut. In Chadha, K.L. and Rathinam, P., Eds., *Advances in Horticulture,* Volume IX: *Plantation and Spice Crops,* Part 1. Malhotra Publishing House, New Delhi, pp. 217-242.

Leach, B.J., Foale, M.A., and Ashburner R.A. (in press). The wild-type coconut population of the Cocos (Keeling) Islands, Indian Ocean. Genetic resources and Crop Evolution.

Lebrun, P., Grivet, L., and Baudouin, L. 1999. Use of RFLP markers to study the diversity of the coconut palm. In Oropeza, C., Verdeil, J.L., Ashburner, G.R., Cardega, R., and Santamarma, J.M., Eds., *Current Advances in Coconut Biotechnology*. Kluwer Academic Publishers, Boston, pp. 73-88.

Liyanage, D.V. and Abeywardena, V. 1958. Correlation between seednut, seedling, and adult palm characters in coconut. Bulletin No. 16, Coconut Research Institute, Ceylon.

Liyanage, D.V. and Sakai, K.I. 1960. Heritabilities of certain yield characters of the coconut palm. *Journal of Genetics* 57: 245-252.

Louis, I.H. and Rethinakumar, L. 1988. Genetic load in coconut palm. In Silas, E.G., Aravindakshan, M., and Jose, A.J., Eds., *Coconut Breeding and Management: Proceedings of the National Symposium on Coconut Breeding and Management*. Director of Extension, Kerala Agricultural University, Trichur, India.

Manciot, R., Ollagnier, M., and Ochs, R. 1979a. Nutrition minerale et fertilisation du cocotier dans le monde (Mineral nutrition and fertilization of the coconut around the world) *Oleagineux* 3: 499-515.

Manciot, R., Ollagnier, M., and Ochs, R. 1979b. Nutrition minerale et fertilisation du cocotier dans le monde II: Etude des differents elements. (Mineral nutrition and fertilisation of the coconut around the world II: Study of different elements.) *Oleagineux* 3: 563-580.

Manciot, R., Ollagnier, M., and Ochs, R. 1980. Nutrition minerale et fertilisation du cocotier dans le monde II: Etude des differents elements—suite. (Mineral nutrition and fertilisation of the coconut around the world II: Study of different elements—continued). *Oleagineux* 35: 13-22.

Marechal, H. 1928. Observations and preliminary experiments on the coconut palm with a view to developing improved seednuts for Fiji. *Fiji Agricultural Journal* 1(2): 16-45.

Marty, G., Le Guen, V., and Fournial, T. 1986. Cyclone effects on plantations in Vanuatu. *Oleagineux* 41: 64-69.

Nair, M.K. and Nampoothiri, K.U.K. 1993. Breeding for high yield in coconut. In Nair, M.K., Khan, H.H., Gopalasundaram, P., and Bhaskara Rao, E.V.V., Eds., *Advances in Coconut Research and Development.* Oxford and IBH Publishing Co. Pty Ltd., New Delhi, pp. 61-69.

Nambiar, M.C. and Swaminathan, M.S. 1960. Chromosome morphology, microsporogenesis, and pollen fertility in some varieties of coconut. *Indian Journal of Genetics* 20: 200-211.

Ninan, C.A., Pillai, R.V., and Joseph, J. 1960. Cytogenetic studies on the genus *Cocos*—I. Chromosome number in *C. australis* and *C. nucifera* L. vars *spicata* and *androgena. Indian Coconut Journal* 13: 129-134.

Patel, J.S. 1937. Coconut breeding. *Proceedings of Association of Biologists* 5: 1-16.

Peries, R.R.A. 1993. Seedling selection in coconut based on juvenile and adult palm correlations. In Nair, M.K., Khan, H.H., Gopalasundaram, P., and Bhaskara Rao, E.V.V., Eds., *Advances in Coconut Research and Development.* Oxford and IBH Publishing Co. Pty Ltd., New Delhi, pp. 181-189.

Persley, G.J. 1992. *Replanting the Tree of Life: Toward an International Agenda for Coconut Palm Research.* CAB International, Wallingford, UK.

Pieris, W.V.D. 1935. Studies on the coconut palm II: On the relation between the weight of husked-nuts and the weight of copra. *Tropical Agriculturist* 85: 208-220.

Pomier, M. and Bonneau, X. 1984. The development of the coconut root systems depending on environmental conditions in Cote d'Ivoire. *Oleagineux* 42(11): 409-421.

Rajagopal, V., Voleti, S.R., Kasturi Bai, K.V., and Sivashankar, S. 1988. Physiological and bio-chemical criteria for breeding for drought tolerance in coconut. In Silas, E.G., Aravindakshan, M., and Jose, A.I., Eds., *Proceedings of the National Symposium on Coconut Breeding and Management,* Director of Extension, Kerala Agricultural University, Thrissur, India, November 23-26, pp. 136-143.

Randles, J.W., Wefels, E., Hanold, D., Miller, D.C., Morin, J.P., and Rohde, W. 1999. Detection and diagnosis of coconut foliar decay disease. In Oropeza, C., Verdeil, J.L., Ashburner, G.R., Cardega, R., and Santamarma, J.M., Eds., *Current Advances in Coconut Biotechnology.* Kluwer Academic Publishers, Boston, pp. 247-258.

Rodriguez, M.J.B. 1999. Detection and diagnosis of coconut cadang-cadang. In Oropeza, C., Verdeil, J.L., Ashburner, G.R., Cardega, R., and Santamarma, J.M.,

Eds., *Current Advances in Coconut Biotechnology*. Kluwer Academic Publishers, Boston, pp. 221-226.

Santos, G. 1999. Potential use of clonal propagation in coconut improvement programs. In Oropeza, C., Verdeil, J.L., Ashburner, G.R., Cardega, R., and Santamarma, J.M., Eds., *Current Advances in Coconut Biotechnology*. Kluwer Academic Publishers, Boston, pp. 419-430.

Simmonds, N.W., Ed. 1976. *Evolution of Crop Plants*. Longman, London.

Southern, P.J. 1969. Sulphur deficiency in coconuts. *Oleagineux* 24(4): 211-220.

Spriggs, R. 1984. Early coconut remains from the South Pacific. *Polynesian Society Journal* 93: 71-77.

Thampan, P.K. 1982. *Handbook on Coconut Palm,* Second printing. Mohan Primlani, Oxford and IBH Publishing Co, New Delhi, p. 311.

Tucker, R. 1983. *The Palms of Subequatorial Queensland*. Palm and Cycad Society of Australia, Milton, Queensland.

Veeh, H.H. and Veevers, J.J. 1970. Sea level at –175 m off the Great Barrier Reef 13,000 to 17,000 years ago. *Nature* (London) 226: 536-537.

Von Uexkull, H.R. 1972. Response of coconuts to (potassium) chloride in the Philippines. *Oleagineux* 27(1): 13-19.

Whitehead, R.A. 1963. The processing of coconut pollen. *Euphytica* 19: 267-275.

Whitehead, R.A. 1966. Sample survey and collection of coconut germ plasm in the Pacific Islands, 30 May–5 September 1964. Ministry of Overseas Development (Overseas Research Publication No 16), London, p. 78.

Zizumbo, D., Fernandez, N., Torres, N., and Oropeza, C. 1999. Lethal yellowing resistance in coconut germplasm for Mexico. In Oropeza, C., Verdeil, J.L., Ashburner, G.R., Cardega, R., and Santamarma, J.M., Eds., *Current Advances in Coconut Biotechnology*. Kluwer Academic Publishers, Boston, pp. 131-144.

Zushum, M. 1986. An investigation on meteorological indices for coconut cultivation in China. *Oleagineux* 41: 119-128.

Chapter 7

Coffee

J. M. Njoroge
C. O. Agwanda
P. N. Kingori
A. M. Karanja
M. P. H. Gathaara

HISTORICAL BACKGROUND

Coffee is the most important nonalcoholic beverage in the world trade. It commands a turnover of about U.S. $10 billion annually, thereby making it the second most traded commodity after petroleum. Its production forms the backbone of more than 50 developing nations (de Graaf, 1986; Kushalappa, 1989), with its contribution to the total foreign currency earnings reaching as high as 80 percent in some countries, such as Uganda. Historically, the use of coffee has evolved from the original chewing of leaves and beans of the plant to relieve pain, hunger, and fatigue to the present more sophisticated uses, such as espresso and decaffeinated coffees. The consumption of coffee apparently started between the fifteenth and sixteenth centuries in Arabia. The habit has since spread to various parts of the world, first reaching Europe through Turkey in the seventeenth century (Charrier and Eskes, 1997). Consumption by the producer countries has, however, remained negligible except for Brazil, which consumes about 10 percent of its own production. From the center of origin in Ethiopia, Arabica coffee was introduced in Yemen between the fifteenth and sixteenth century. Thereafter, the species spread to the Malabar Coast of India and Ceylon (present-day Sri Lanka) during the last decade of the seventeenth century. Coffee further spread to Java (1706), Martinique, and South America via Amsterdam and Paris. Unlike Arabica coffee (*Coffea arabica* L.), robusta coffee *(C. canephora)* has not gone through a history of extensive diaspora. First found in a state of semicultivation in the Democratic Republic of the Congo (DRC), the species reached Southeast Asia in the early twentieth century where it replaced Arabica coffee in the lower altitude zones of Java and Sumatra due to *Hemileia vastatrix* menace. The species has a rather dispersed center of genetic diversity within the west and central subtropical regions of Africa. Its center of genetic diversity extends from Guinea and Liberia to Uganda and Sudan. Maximum genetic diversity is, however, apparently found in the Democratic Republic of the Congo. Most of the coffee spread around the world resulted from just a few trees. Consequently, the majority of the present-day cultivated varieties are based on a very narrow genetic base (van der Vossen, 1985). The resulting genetic uniformity is a major problem to plant breeders since it predisposes most of the varieties grown worldwide to the risk of epidemics whenever new diseases arise. At the same time, it offers little opportunity for selection to stress factors such as diseases and insect pests. At present, the commercial production of coffee relies on two species, *C. arabica* and *C. canephora* (van der Vossen, 1985). In terms of volumes, Arabica coffee commands more than 75 percent of the coffee traded in the world market (van der Vossen, 1985). Its predominance in the world trade is attributed to the superior liquor quality inherent in the species. The exclusive production of the species world over has nevertheless been limited due to a number of factors, notably, susceptibility to diseases, drought susceptibility, and low tolerance to high temperatures (van der Vossen, 1985; Agwanda, 1997). Its main competitor, *C. canephora,* is relatively tolerant to some of these stress factors, especially those prevalent in the low-altitude coffee-growing zones, for example, *H. vastatrix* and high temperatures.

GENETIC RESOURCES

Coffee belongs to the family Rubiaceae, genus *Coffea.* The genus is further organized into three sections, Mascarocoffea, Eucoffea, and Paracoffea (more correctly referred to as Coffea) of which the first two are natives to Africa, whereas Paracoffea is endemic to East Asia. Most of the known coffee species are found within the sections Coffea and Mascarocoffea. Based on factors such as tree height, leaf thickness, fruit color, and geographical distribution, the section Coffea has been subdivided into five subsections, namely, Nanocoffea, Pachycoffea, Erythrocoffea, Melanocoffea, and Mozambicoffea. The cultivated species *C. canephora* and *C. arabica* belong to the subsection Erythrocoffea, whereas *C. liberica* belongs to Pachycoffea subsection.

Africa is considered to be the home of most coffee species, with the centers of genetic diversity situated in the region covering southwest Ethiopia and southeast Sudan for Arabica coffee, and the humid forest of Central and West Africa, including Uganda and Madagascar, for *C. canephora* and *C. liberica.* With the exception of *C. arabica,* all the other naturally occurring species are diploids with the basic chromosome number of $2n = 2x = 22$. On the other hand, Arabica coffee is an allotetraploid with a chromosome constitution of $2n = 4x = 44$. The ancestry of Arabica coffee has, for some time been controversial, but with a general consensus that *C. eugenioides* was one of the parents. Recent molecular studies have now confirmed *C. canephora* to be the other parent of Arabica coffee (Lashermes et al., 1999).

Germplasm Conservation

The main collections presently under conservation resulted from collection efforts made during the 1960s and include the Food and Agriculture Organization (FAO) collection mission of 1964 and the ORSTOM (Institute Francais de la recherche scientifique pour les developpement en cooperation) collection missions of 1966/1978. Good representations of these germplasms are conserved in Ethiopia, Costa Rica, Cameroon, Ivory Coast, Colombia, and Madagascar. In addition, a number of working collections are also conserved in various national institutes entrusted with coffee research. Traditionally, long-term conservation of the genus *Coffea* is realized through live field gene banks. This is because of the recalcitrant nature of the seeds of the species within the genus *Coffea.* Other complementary germplasm conservation techniques, such as pollen storage (Walyaro and van der Vossen, 1977), have only been limited to short-term application in breeding activities.

Field collections, although the most widely used at present, have a number of limitations. The collections are prone to natural hazards such as drought and disease incidences. Wild fires are also a major threat, whereas civil unrests are increasingly becoming a major consideration in the African continent. Increased pressure on agricultural land due to increased population in the third-world nations has also led to the encroachment of forest land for human settlement. It is for these reasons that the recent developments in the field of cryopreservation of seeds and in vitro cryopreservation of zygotic embryos (Dussert et al., 1997, 1998, 2001) will be a major contribution toward securing the world's coffee collections. The methods are still not widespread, however, with the first applications being in Costa Rica.

In general, the majority of the stored germplasm has not been exploited for breeding or selection purposes. Exploitation has been limited to breeding for resistance against diseases in which only a few genotypes are involved. One reason for this scenario is the fact that most of the germplasm is so far not comprehensively described. The other reason is that most of the genes of interest are located in wild species for which barriers preventing crossing of such genotypes with the commercial species exist.

BREEDING AND SELECTION

Breeding Objectives

Until recently, most breeding and selection efforts were directed toward enhancing yield, quality, and adaptability to agroecological conditions (van der Vossen, 1985). A number of high-yielding, good quality varieties were developed in the early breeding efforts, which mainly involved individual plant selection. These include varieties such as 'Caturra', 'Mundo Novo', and 'Catuai' in Brazil, 'Typica' in Colombia and Central American countries, SL selections in Kenya, and N 39 in Tanzania among others. These cultivars, usually referred to as traditional varieties, have since formed the backbone of most coffee industries throughout the world.

The advent and subsequent spread of diseases, such as coffee leaf rust, *H. vastatrix* (Berkley, 1869, as cited in Kushalappa, 1969), and coffee berry disease (CBD), *Colletotrichum kahawae* (McDonald, 1926), have however, made the continued use of the traditional varieties quite expensive as this entails intensive use of fungicides to protect the crop. In an attempt to ensure that coffee production remains economical in the face of such diseases, a number of breeding programs were initiated during the early 1960s to incorporate disease resistance into the traditional varieties. Comprehensive reviews of such programs have been given by van der Graaff (1981), van der Vossen and Walyaro (1981), van der Vossen (1985), Agwanda and Owuor (1989), Carvalho and Monaco (1969), Echeverrie and Fernandez (1989), Bouharmont (1995), and Charrier and Eskes (1997). Although varied, the present-day coffee-breeding programs share one goal: to develop varieties which, under defined socioeconomic and agroecological conditions, would maximize the net earnings of the farmer. To realize this goal, higher yield, superior quality, disease resistance, and adaptability of varieties to agroecological conditions have been the main objectives of coffee breeding, and are discussed briefly in the following sections.

Yield

From a commercial perspective, yield is measured as the amount of clean coffee, that is, coffee beans dried to a moisture content of 10 to 12 percent (van der Vossen, 1985), per unit area of land. Two approaches have been used by breeders to improve yield potential. Yield was improved by selecting for vigorous trees with improved yield potential of individual trees. This approach is, however, not amenable to intensive land use, and has since been replaced by the development of varieties with compact growth habits. Compact growth allows for increased plant density, which in turn leads to dramatic increases in yield per unit of land. High density is known to increase yield even in the robusta-type coffee varieties (Mitchell, 1976; Browning and Fisher, 1976; Walyaro, 1983; Njoroge and Mwakha, 1983; Njoroge and Kimemia, 1993). To realize the full potential of compact varieties, it is, however, desirable that such varieties be resistant to the most prevalent diseases. Conditions of crowding would otherwise create microclimatic conditions favorable to disease development, thereby negating the benefits associated with compact varieties.

Quality

Quality is the main driving force behind coffee consumption. Indeed, as pointed out by Charrier (1982), the amount and sustainability of coffee consumption depend on the desirability of the coffee brew and the pleasure derived from its consumption. From a breeding point of view, two levels of quality are important. The first consideration is given to the quality of beans prior to and after roasting. The bean quality, also referred to as bean grade, is important from a technological point of view since it acts as an indicator of the suitability of a given coffee for roasting purposes. Raw quality assessment results can also be used to explain liquor quality defects arising from suboptimal agronomic and processing practices. The second quality criterion involves the organoleptic assessment of the quality of coffee brew (liquor or cup quality). Several quality traits are assessed at this stage to determine the desirability of the coffee brew for human consumption.

Bean quality. Bean quality is judged based on shape, size, weight, and appearance. Classification of coffee beans into various grades is largely done mechanically (see van der Vossen, 1985, for details). A study by Walyaro (1983) indicated that most of the size, shape, and weight parameters of bean quality are highly heritable and therefore easily improved through simple selection procedures such as mass selection and back-cross-breeding methods. To enhance the rate of genetic gain, attention should be given to the general agronomic practices and the nature and timing of environmental stress factors, such as moisture stress. Optimum environmental conditions during expansion and bean-filling stages have been shown to enhance differential expression for bean grades, thereby making it possible to select for superior genotypes with increased precision (Agwanda, 1997).

Liquor quality. In addition to being the driving force behind coffee consumption, the cup quality of coffee determines the price offered at the world market. An important consideration when breeding for improved quality is the consistency in performance both between locations and across years, as this is of fundamental interest to the producer nations in capturing and retaining given market sectors of the world trade.

Unlike bean grades, liquor quality is determined subjectively based on organoleptic procedures and by panels of experienced coffee liquors. Three traits are usually used in determining the quality of liquor: acidity, body, and flavor (Devonshire, 1956; Moreno et al., 1995). The traits are heavily influenced by nongenetic factors such as edaphoclimatic conditions, cultural practices, postharvest processing and storage, roasting, and preparation of the liquor. Nevertheless, reasonable progress from selection could still be realized by subjecting the test materials to recommended agronomic conditions and following standard sampling and liquor preparation procedures (Afnor, 1991). Efficiency of selection for overall liquor quality could also be improved by identifying the traits that could be assessed organoleptically with greater accuracy, and which are strongly correlated to the overall quality of the brew (Agwanda, 1999). To enhance the differential expression between test genotypes for the various traits, it is also necessary to manipulate the moisture regimes in the selection fields in relation to the various bean growth phases (Agwanda, 1997).

Other quality criteria. The coffee market is a continuously evolving arena with new methods of coffee preparation and uses appearing in the market on short intervals. Similarly, consumer taste and preference are equally evolving, though at a slower rate. In order to keep pace with the changing circumstances, additional quality traits may have to be introduced in the coffee-breeding programs. For instance, percentage of soluble solids, caffeine content, and chlorogenic acid could be used more routinely as selection criteria. Although considered minor at present, such biochemical traits are expected to continue attracting the interest of breeders, especially in Arabica coffee. This is particularly so because a number of coffee-breeding programs are presently exploiting the Hibrido de Timor germplasm or its derivatives. Timor derivatives are known to have higher levels of these traits. Consequently, their use as donors in Arabica coffee-breeding programs is expected to increase the three traits to levels above those characteristic of Arabica coffee and hence affect the organoleptic integrity of the newly developed varieties. Hibrido de Timor in particular is known to impart poor taste to its progenies (Walyaro, 1983). To realize faster progress, studies aimed at understanding the inheritance mechanism of the traits and developed methods for relating their levels to the perceived organoleptic quality will be necessary.

Disease Resistance

Disease resistance is the main preoccupation of most breeding programs today. Two main diseases, namely, coffee berry disease caused by *C. kahawae* and coffee leaf rust caused by *H. vastatrix*, are targeted. The two diseases are confined to Arabica coffee and are therefore more important to the Arabica coffee-producing countries. Elaborate screening techniques have been developed for the two diseases. In the case of CBD for example, selection for resistance is based on inoculation of eight-week-old seedlings (van der Vossen et al., 1976) or on a combination of seedling resistance and field expression (van der Graaff, 1981). Screening at the seedling stage has a number of advantages. First, it allows for a large number of lines and genotypes within lines to be assessed at the same time. The ability to assess a large number of plants per cross is of significance to coffee breeders since it allows for better visualization of segregation patterns. Since coffee has a long juvenile phase, the resulting long generation cycle is a major handicap to breeding and selection in the species. The use of a seedling inoculation technique greatly reduces the time required for the assessment of a given line thereby minimizing the impact of a long generation cycle on the rate of turnover of new varieties. Breeding for rust resistance has similarly benefited from advanced selection techniques, including detached leaf, leaf disc inoculation (Eskes, 1983), and inoculation of intact leaves on young seedlings. The impact of the techniques on progress in breeding for resistance to the two diseases has been considerable, with resistant varieties being released to the farmers in countries such as Colombia (Castillo and Moreno, 1988) and Kenya (Anonymous, 1985) after a relatively short period of time. The other coffee diseases caused by pathogens, such as *Fusarium stilboides* and *Gibberella xylarioides,* are considered minor and have not attracted any breeding attention. Problems related to nematodes, particularly *Meloidogyne exigua,* is now attracting some breeding attention especially in Central America. With the increasing neglect of coffee farms, such as in the Arabica coffee-producing nations, and the shifting weather patterns presently being experienced, the diseases hitherto considered minor may gain importance as is already the case with tracheomycosis in robusta coffee. Given the time taken between initiation of a breeding program and the realization of results, it may be prudent for the breeders to develop screening techniques for the minor diseases, identify genes of resistance against them, and, where possible, transfer the identified genes on genetic backgrounds that could be used for practical breeding.

Stability of Performance

Stability of performance in crop plants refers to the ability of a variety or plant to adjust its phenotypic state in response to transient environmental fluctuations in such a way as to give high and stable economic returns (Allard and Bradshaw, 1964). It can be the result of phenotypic plasticity of individual genotypes (individual plant buffering), differences in phenotypic plasticity between genotypes (population buffering), or the result of different genes being switched on and off (Allard and Bradshaw, 1964; Yamada, 1991; Gallais, 1992). Any of these mechanisms could be targeted in breeding for improved stability in Arabica coffee. Recent results (Agwanda, 1997) have demonstrated that population buffering in Arabica coffee could be realized by selecting for individuals that show complementary performance under different environmental conditions but are otherwise similar in the important agronomic traits. Since coffee is perennial, year-to-year fluctuations in climatic conditions could be considered as part of the transient environmental fluctuations. The combination of years and locations could thus enable the breeders to accommodate the phenomenon of biennial bearing, which is characteristic of Arabica coffee.

Combined selection. To realize the goal of maximization of economic returns to the farmers, it is necessary to simultaneously improve the various economic characters that directly impact on net returns to the farmers. To this end, it is necessary to define, within practical limits, what would constitute an ideal variety. In principal, such a variety should contribute toward reducing the cost of production and maximize net returns to the farmer while preserving both human and environmental health. Key factors to consider when attempting to construct such varieties include responsiveness to farm inputs, such as fertilizer use, high productivity per unit of land, resistance to prevailing diseases, and the production of good bean and liquor qualities. The construction of varieties combining all these traits require multicriteria selection techniques, such as index selection or marker-assisted selection (MAS) (Melchinger, 1990). Index selection methods have been shown to be useful in concurrent selection for yield and quality (Agwanda, 1997) and in predicting long-term performance (Walyaro and van der Vossen, 1979). More recent developments in molecular marker methods for coffee diseases (Agwanda et al., 1997; Lashermes et al., 1993, 1996, 1997) could also greatly improve the efficiency of selecting for such ideotypes through MAS.

Insect Resistance

Little attention has been given to breeding for insect resistance both in Arabica and robusta coffee. The main explanation to this could be attributed to the fact that the insect menace tends to be highly weather dependent and more easily predictable, and could therefore be easily contained using insecticides. This scenario is changing, however. For example, the problem of berry borer is now a more pressing issue in Colombia and the surrounding coffee-producing countries than any other pest. Similarly, nematodes have become an economically important constraint to coffee production in Central America. The two problems, although earlier considered minor, may now warrant full-fledged breeding programs. In light of these considerations and with increasing demand for organically grown coffees, insect resistance may soon be an important breeding objective in coffee breeding.

BREEDING PROGRAMS

Breeding for Disease Resistance

Systematic breeding of coffee dates back to 1925 in India and Brazil. The Indian program concentrated on rust resistance in addition to yield and quality, whereas in Brazil most of the breeding activities concerned genetic or cytogenetic evaluations. In Africa, breeding activities commenced around 1934 in Tanzania and around 1944 in Kenya. At their inception, the programs concerned the improvement of quality, yield, and adaptation. With increased leaf rust and CBD menace, breeding priorities shifted to the development of varieties resistant to the disease. The programs characteristically involved three phases. The first stage involved the introduction of new varieties with the aim of identifying lines with resistance to the disease. In extreme cases, total replacement of the variety with other economically viable commodities were considered, as was the case when *H. vastatrix* first hit Sri Lanka in 1869. In the absence of resistant varieties, the use of fungicides to combat the disease become the second line of action. This can be exemplified by the elaborate spray programs developed in Kenya to control both CBD and leaf rust. Breeding for disease resistance as the final line of action against new diseases is the development of new varieties through systematic breeding. For both CBD and leaf rust, elaborate breeding programs are now in place in a number of countries.

Coffee Leaf Rust

At present, coffee breeding is predominated by two diseases, namely coffee berry disease and coffee leaf rust. Coffee leaf rust, which has now attained international status, remains the most important coffee disease at the global level. The disease first manifested its full potential around 1869 when it arrived in Sri Lanka, leading to a near total collapse of the coffee industry in the country. The disease has since spread to all coffee-growing countries, reaching Brazil in 1970 and establishing itself as the most important coffee disease in Latin American. The disease was nevertheless less disastrous in these countries compared with its impact on Sri Lanka, thanks to the efficient spray programs already developed in countries such as Kenya at the time the disease reached the South American continent.

Through international collaboration, coordinated by the Centro de Investigação das Ferrugens do Cafeeiro (IFC), Oeiras, Portugal, coffee rust has undergone elaborate studies with a number of genes for resistance and genes for compatibility, being described in detail in host and the pathogen respectively. Such genes have been exploited extensively in Latin America and India, giving rise to resistant varieties, such as the Colombia variety (Castillo and Moreno, 1988). The breeding methods used involved initial crossing of the desired progenitor with the commercial cultivar followed by a series of selfing generations to fix the resistance genes.

Coffee Berry Disease (CBD)

Unlike leaf rust, CBD is still restricted to the continent of Africa where it causes considerable loss in Arabica coffee. Reported first in Kenya (McDonald, 1926), the disease is now present in virtually all Arabica-coffee-growing countries in the continent. Control of the disease can be achieved through chemical methods, the use of resistant varieties, and, to some extent, through cultural methods. Chemical control is very expensive, however. In Kenya, for example, the cost of controlling CBD is estimated at U.S. $500 per hectare of susceptible variety. With a total of 120,000 ha under coffee at national levels, the cost of CBD is therefore astronomical and presents a major drain in the much-needed foreign currency for the producer countries. Furthermore, losses in excess of 40 percent can still occur during bad weather despite the use of chemicals. Consequently, breeding for CBD resistance is now considered the only way to sustainable and economic production of Arabica coffee on a long-term basis.

The main coffee-breeding programs dedicated to CBD resistance are found in Kenya and, to some extent, Ethiopia and Tanzania. Systematic breeding for CBD resistance in Kenya started in 1972 with the objective of introgressing CBD resistance genes into the traditional commercial cultivars, mainly SL 28 and SL 34. The program exploits the resistance genes identified in the accessions Rume, Sudan, Hibrido de Timor and its derivatives, and one commercial cultivar, K7. To combine all the resistance genes in a single cultivar, the various sources of resistance are crossed to the commercial cultivars (SL 28 or SL 34). The single crosses are then crossed together in various combinations giving rise to multiple crosses. A series of backcrosses are then conducted with the commercial cultivars as the recurrent parent. The objective of the backcross program is to restore the genetic background of the traditional varieties, which are renowned for the production of very fine mild coffees. After every backcross, selfing of the backcross progenies is carried out to realize homozygosity at the resistance loci. Screening for CBD resistance is carried out to identify the progenies to form the next generation. So far, up to four generation of backcrosses have been realized.

As a way of exploiting the early generation selections for commercial use, the Kenyan program embarked on the production of hybrid varieties. Superior breeding lines are used as pollen donors, whereas selected 'Catimor' progenies, originally from Colombia, are used as the maternal parents. The use of 'Catimor' as a mother population in the hybrid program has a number of advantages. First, the 'Catimor' lines bring with them the rust resistance of Hibrido de Timor but in a genetic background which is commercially more acceptable than that of the original Timor hybrid. 'Catimor' also possesses one of the CBD resistance genes also originating from Hibrido de Timor. The final advantage is the short stature, a character which is more or less fixed in the variety. The variety resulting from the hybrid program, Cultivar Ruiru 11, combines all these advantages with the superior quality attributes of the elite breeding lines as well as the CBD resistance genes originating from Rume, Sudan, and 'K7'.

Bacterial Blight of Coffee (BBC)

Bacterial blight of coffee (BBC) caused by *Pseudomonas syringae* pv. *garcae* is the only bacterial disease of economic significance in coffee. The spread of the disease is highly restricted, being present mainly in Brazil and Kenya. The disease is considered inconsequential in Brazil, whereas its coverage amounts to about 3 percent of the total coffee hectarage in Kenya. The disease is gaining importance in Kenya, however, since it is endemic in areas with great potential for coffee expansion. In addition, since increased hectarage is being covered by CBD-resistant cultivars, BBC epidemiology may be expected to change as the use of fungicides, some of which are known to have bactericidal effects, diminishes. In addition, shifts in weather patterns and a general increase in the neglect of farms due to poor prices are becoming frequent phenomena and will undoubtedly increase the importance of the disease. Work carried out in Brazil indicates that variation for resistance to the disease exists in Arabica coffee introductions from Ethiopia, with resistance levels ranging from complete susceptibility to immunity. These introductions could therefore form the basis for breeding against the disease.

Other Minor Diseases

Minor diseases in coffee can be described as those diseases that, under optimal agronomic conditions, can easily be put under check without having to resort to expensive control procedures. A number of coffee diseases belong to this category, and include diseases such as *Fusarium* wilt diseases and tracheomycosis. Situations in which minor diseases evolve into major problems, are not uncommon, however, and may be due to a number of reasons, including deterioration in general agronomic practices, change in farming systems, and shifts in weather patterns. A recent example is the case of tracheomycosis in Uganda and the DRC, where the disease has since become a major threat to economic and sustainable production of coffee in the two countries. In anticipation of such problems, breeders could at least initiate activities aimed at identifying sources of resistance, and, if possible, transfer the genes to genetic backgrounds where they could easily be incorporated into breeding programs whenever the need arises.

FIELD MANAGEMENT

Environmental Requirements

Both Arabica and robusta coffee are tree crops, which produce yields two to three years after planting with a long economic life beyond 30 years depending on local conditions and husbandry. Coffee is sensitive to excessive heat or cold or rapid change of temperature. Each type requires different growing conditions, with *C. arabica* preferring temperatures of 15-24°C, whereas *C. canephora* prefers warmer conditions of 24-30°C with less contrasting dry and rainy seasons. Both require an average rainfall of 1,800 mm per annum for healthy growth and satisfactory productivity. In the equatorial and tropical zones, Arabica does well at altitudes of 1,200-

1,800 m above sea level, whereas it does well below 600 m above sea level in the subtropical zones. *Coffea canephora,* on the other hand, does well in the warmer zones of the plains. The ideal soils are light, deep, well drained, loamy, slightly acidic, and rich in humus and exchangeable bases, especially potassium.

Establishment

Coffee seedlings are usually raised from seed or as rooted cuttings in the nursery or through tissue culture and grafts. After germination or establishment of the cuttings/tissue culture material, they are transplanted (potted) into black polyethylene bags measuring 23 × 17 cm to 30 × 17 cm, previously filled with a mixture of topsoil, manure, phosphatic fertilizer, and a suitable insecticide. They are ready for field transplanting in 12-18 months after potting. Use of biological, cultural, and physical control of insects should be encouraged at both the nursery and the field stages.

In the field, seedlings are planted in well-prepared land free of tree stumps to avoid soilborne diseases, such as root disease caused by *Armillaria mellea,* which affect coffee trees and removal of difficult weeds such as couchgrass for good management of the young seedlings. Suitable soil conservation measures should be taken, such as bench terraces whose terrace sides should be well protected with grass, or suitable trees, such as the multipurpose trees, e.g., *Leucaena, Sesbania, Calliandra,* etc. Planting holes are dug three months before planting to minimize soilborne diseases, e.g., those caused by *Fusarium* spp. They are dug along the contour as an erosion control measure. The holes are then refilled one month before planting with topsoil mixed with well-rotted organic manure, phosphate fertilizer (type depending on the acidity of the soil), and an appropriate insecticide. Agricultural lime can also be added depending on the soil reaction (pH). The seedlings are planted after the onset of the rains. Application of mulch along the planted coffee row or around the seedlings helps to preserve moisture, suppress growth of weeds, assist in soil conservation, and improve the soil structure. The type of mulch material is important as continued use of one type may cause soil nutrient imbalance. For example, use of napier grass mulch can increase soil potassium if used for a long period, thereby disturbing the Ca+Mg/K ratio. Soil amendments to rectify this anomaly then become necessary. Table 7.1 gives nutrient analysis of some of the possible mulching material and organic manures.

TABLE 7.1. Nutrient content of organic manures and mulches.

Source	Percent						ppm				
	N	P_2O_5	K_2O	CaO	MgO	SO_3	B	Cu	Fe	Mn	Zn
Boma manure	1.32	1.08	1.66	0.92	0.35	0.45	25	68	27,500	916	99
Cattle manure	2.50	1.12	6.70	1.43	1.00	0.37	31	45	11,200	1,040	95
Coffee husks	0.48	0.07	0.40	0.31	0.08	0.15	4	96	100	32	52
Coffee pulp	3.73	0.40	6.51	0.99	0.30	0.85	18	35	880	226	18
Goat manure	2.66	3.89	4.87	1.36	1.17	0.37	45	30	1,940	256	88
Maize stover	2.11	0.35	1.95	1.08	0.32	0.15	26	8	442	57	8
Napier grass	1.51	0.62	4.23	0.27	0.25	0.32	18	25	1,300	157	132
Pig manure	2.34	5.27	0.96	4.23	1.55	0.52	25	211	7,100	648	440
Pineapple tops	0.86	0.21	1.69	0.29	0.15	0.20	14	13	100	167	180
Poultry manure	3.54	3.24	1.55	5.06	0.98	0.57	30	190	6,030	363	225
Rice husks	0.45	0.39	0.47	0.20	0.12	0.12	11	310	11,500	326	142
Rice straw (immature)	1.04	0.39	1.49	0.55	0.52	1.05	19	14	4,950	965	59
Rice straw (mature)	2.59	0.69	0.60	1.13	0.45	0.22	16	39	2,750	902	51
Sawdust	0.36	0.11	0.16	0.24	0.03	0.10	15	70	5,950	314	153
Sisal waste	1.15	0.28	1.52	5.81	0.98	0.22	36	242	1,875	279	378
Sugarcane filter mud	1.34	2.38	0.73	3.27	0.36	—	45	60	4,695	969	168
Vlei grass	1.39	0.30	1.06	0.66	0.25	0.32	19	72	12,950	846	85

Source: Based on data from Chemistry section, Coffee Research Foundation, Kenya.

Spacing

The common coffee spacings for the traditional tall cultivars are 2.74 × 2.74 m (1,329 trees per ha), 2.74 × 1.37 m (2,658 trees per ha), and 1.37 × 1.37 m (5,320 trees per ha). For the compact cultivars, such as Ruiru 11 hybrid in Kenya, the spacing adopted is 2 × 2 m (2,500 trees per ha) and/or 2 × 1.5 m (3,333 trees per ha) (Njoroge, 1991), or 1 × 1 m for Catimor and Colombia cultivars in Colombia. To facilitate efficient use of machinery, large coffee estates can practice different spacings while maintaining the above tree densities. High densities assist in soil conservation, especially through raindrop impact, control of runoff through litterfall, and weed suppression. Due to the good soil surface cover in high-density plantings, most of the solar energy is utilized by the coffee trees, while the well distributed mass of roots utilize effectively the soil nutrients. During the first two years of coffee establishment, before the canopy closes up, nurse crops or intercrops can be introduced in the coffee interrow spaces. These help provide soil conservation, weed suppression, and an economic return to the farmer when an economic annual crop is used apart from food security. Early maturing annual crops, such as legumes, cereals, and vegetables can be used, including nonclimbing beans, peas, tomatoes, kales, carrots, irish potatoes, soy beans, millet, sorghum, etc., depending on the ecological zone where the food crop can do well. The legumes and nonlegumes should be alternated seasonally or on alternate coffee interrows. Cover crops, such as *Desmodium* spp. and sweet potatoes, may strangle the young seedlings though they protect the soil.

Nutrition

Most of the soils on which coffee is grown have low plant nutrients, especially nitrogen. On average, one tonne of coffee beans can remove 46 kg N, 8 kg P_2O_5, and 58 kg K_2O; parchment 2.3 kg N, 0.3 kg P_2O_5, and 1.9 kg K_2O; and pulp 15.3 kg N, 3.7 kg P_2O_5, and 27.4 kg K_2O. This, together with losses through parchment, pulp, erosion, and leaching, leaves the soils seriously exhausted. It is therefore necessary to return some of the losses through recycled prunings when used as mulch. Hence fertilizers are needed for both vegetative growth of tree and production of high-quality coffee beans. To apply the correct type and rate of fertilizer, and thus avoid toxicity and nutrient imbalances in the soil environment, fertilizer recommendations should best be based on soil and coffee leaf analysis results where the necessary facilities are available.

Nitrogen is the most limiting element, and Arabica coffee has responded positively to nitrogen application rates of 50-100 kg N per ha per year in Kenya (Njoroge, 1985). Responses of 300-400kg N per ha has also been recorded, especially under very high yields and in soils with the right soil reaction for coffee. The proportion of large-sized beans has been shown to decrease with increased nitrogen rates of application unless balanced with phosphate fertilizer (Njoroge, 1985). Better response to nitrogen has been observed with split application than with single dose application (Njoroge, 1985). This would also reduce leaching of nitrogen to underground water, thereby polluting the surface water and groundwater. Studies in Kenya have also shown no positive yield increases to phosphorous application alone despite observed low soil phosphorous (Keter, 1974). In most cases, the soils are well supplied with potassium, a major nutrient in coffee production, especially during berry expansion period. The types of fertilizers used include straight, compound, and foliar fertilizers. Efforts should be made to minimize excess nutrients, especially nitrogen in the soil to avoid it being washed or leached to surface or underground water.

Organic manures, mainly cattle manure, have been used for a long time in coffee. Such manures increase the soil organic matter, thus improving the water-holding capacity and physical characteristics of the soil, and release plant nutrients upon decomposition (Oruko, 1977). Their use has been reported to lead to increased coffee yields and improved quality, especially on very poor soils (Mitchell, 1970).

As the organic manures are formed from different sources, they have varying nutrient composition, and their continuous use may lead to nutrient imbalances, which may affect coffee bean quality (Northmore, 1965). However, organic manures can be used to substitute, to some extent, inorganic fertilizers and thus reduce production costs. Organic manures can be prepared easily at homestead level as boma manure or composting of farm trash. Increased use of organic manures would improve soil fertility as a whole especially of the degraded and overused soils in the coffee-growing zones. Studies in Kenya have shown the possibility of substituting the inorganic fertilizer requirements in Kenyan coffee with two 13 kg tins of well-rotten cattle manure per annum. However, this would depend largely on the source of the manure and calls for nutrient analysis of the various manures (chicken, pig, goat, etc.). Decomposed coffee pulp, sludge from methane gas plants, etc., are increasingly being utilized. Results from Kenya indicate the possibility of using green manures as sources of coffee nutrients by use of plants such as lucerne *(Medicago sativa),* and multipurpose trees, such as *Leucaena* spp., *Sesbania* spp., and *Calliandra* spp. (Kimemia, Chweya, and Nyabundi, 1999).

Green manure from permanent cover crops, such as *Desmodium* spp., appear to mineralize very slowly and may not be very useful in the short run but can be used to control soil erosion on sloping coffee farms. Planting multipurpose trees for this purpose on the bench terraces and wastelands may also improve sustainability of the soil environment.

Weed Control

The weed species in coffee can be classified into annuals, perennials, and sedges. Weeds have been shown to reduce coffee yields by more than 50 percent, as well as the coffee quality (Njoroge and Kimemia, 1989) as compared to free-situation weed. Due to this reduction of yield by weeds, various weed control methods are used. The most common methods used are digging using forked hoes, slashing, mulching, and the use of herbicides. A wide range of herbicides are used in coffee (Njoroge, 1994), which include contact, systemic, and soil-acting herbicides. Some of those commonly used are depicted in Table 7.2. Low rates and volumes of recommended herbicides can be made effectively to control annual weeds in coffee at the one to four leaf stage using low volume nozzles (Njoroge and Kimemia, 1992). Continued use of this technology can reduce the amount of herbicides released to the environment, thereby reducing environmental pollution. The use of one type of herbicide has led to the development of herbicide tolerance by some weeds, such as the tolerance of black jack (*Bidens pilosa* L.)

TABLE 7.2. Some commonly used herbicides in coffee.

Soil acting	Contact	Systemic
Atrazine (50 and 80% wp)	Actril DS (70% EC) (mixture of loxyil and 2,4-D)	Ametryne (80% wp)
Candex (65% wp) (mixture of asulam and atrazine)	Amitrole (25 or 50% Ml)	2,4-D amine
Diuron (48% EC, 80% wp)	Diquat	Asulam (40% SL)
Flumeturon (80% wp)	Paraquat	Dalapon (74 and 85% wp)
Linuron (50% wp)	Glufosinata-ammonium (20 and 14 SL)	Fluazifop butyl (25% EC)
Oxyfluorfen (24% EC)		Glyphosate (various)
Simazine (50 and 80% wp)		Haloxyfop ethoxyethyl
		MCPA (various)
		Tordon 101 (picloran plus 2,4-D)

to paraquat observed in Kenya (Njoroge, 1986). Integration of the different herbicides and other methods of control is therefore recommended. Slashing creates a carpet of weeds, which may help reduce soil erosion during the rains, whereas forking may encourage rainwater acceptability and reduce runoff. A long-term approach to weed management through integrated weed management (IWM) would be the best option for efficient weed management and reduction in environment degradation through soil erosion, nutrient leaching, and environment pollution.

Mulching

Mulching is the covering of the soil with a layer of dry vegetation materials. The benefits are soil moisture preservation for better nutrient absorption by the plant, prevention of soil erosion, thereby avoiding the delivery of nutrients, pesticides, etc., to surface and underground water, improvement of soil structure, supply of mineral nutrients on decomposition, minimizing use of inorganic nutrients, regulation of soil surface temperature, suppression of weeds leading to reduced herbicide use, and reduction of thrips incidence, thereby reducing insecticides use. Due to these benefits, mulch helps to increase coffee yield and quality. The main mulching materials are napier grass, maize, banana stover, coffee prunings, and any other dry vegetation material. The mulch material is applied in alternate coffee interrows and then alternated in the following years. There is no doubt that mulch enhances environment sustainability. Use of live mulch such as *Desmodium* spp. may be very useful, especially on sloppy grounds, with the main shortcomings being low biomass and production and possible moisture competition with coffee trees (Njoroge and Mwakha, 1983; Snoeck et al., 1994). *Desmodium* mineralizes very slowly.

Pruning

Coffee pruning involves the removal of unwanted branches and old stems. The main reasons for pruning coffee trees are to maintain a suitable crop/leaf ratio for good cropping level and maintenance of a high proportion of large beans to open the tree centers to light, to facilitate disease and pest control, as well as harvesting. This leads to efficient utilization of the sun's energy by the coffee plant and chemicals used to control the pests, thus avoiding excess pesticides in the environment.

The pruning system depends on the type of tree training adopted. Training is the modification of the natural habit of the coffee trees to suit the particular conditions under which they are grown. There are basically two training systems: single system and multiple stem system, which is either capped or uncapped. The change of cycle to raise new stems is carried out after six or seven years, especially for the multiple system. This improves coffee quality and may reduce disease problems, as the trees are healthier leading to less use of pesticides. Most smallholders allow their coffee to grow freely without capping, whereas most largeholders cap their coffee for ease of mechanization. The compact cultivars, such as Ruiru 11 hybrid, are currently recommended on uncapped coffee and require "stumping" after every five to seven years to replace the old stems with new ones (Njoroge et al., 1992). Little pruning is carried out in some countries in southern Africa where foliar diseases pose no major problem. Growers prefer replanting rather than raising new stems at the time of cycle change. Prunings are best left in situ to act as mulch, which would return nutrients to the soil. However, where fuel energy is increasingly becoming scarce, these prunings are used as firewood.

Irrigation

The importance of soil moisture lies in the fact that it is the matrix through which the soil nutrients, including nitrogen, are transported to the plant roots for absorption. Second, adequate soil moisture levels encourage high root proliferation and consequently allow a bigger volume of the soil to be exploited by the plant, and third, a correct soil moisture is required for trees to maintain a proper level of turgidity for growth and various synthetic processes. Therefore, conservation or addition of soil moisture is a precedent to the realization of high coffee yields and quality. Excessive and untimely irrigation can, however, affect yields adversely. Coffee trees require about four to eight weeks of dry season to build up internal water stress prior to breaking of flower dormancy. The frequency of irrigation is determined by the rate of evapotranspiration from the coffee field and varies according to weather conditions. Different methods of irrigation, such as overhead irrigation, need further evaluation on their efficiency and effects on pest management.

Intercropping and Shading in Coffee

Coffee is grown mainly as a monocrop in most countries, the primary reason being that quality of coffee might be affected adversely if farmers ignore coffee in favor of intercrops. These could be due to competition for nutrients, water, and light between coffee and the intercrops. However, coffee farmers, particularly small-scale farmers, have been intercropping their coffee with various food, fruit, and tuber crops, especially at the establishment and change of cycle periods and even during the production phases. Large-scale estates have also been observed to move in this direction.

Since coffee occupies a substantial amount of the high-potential land, available land for food crop planting is becoming limited, and hence more intercropping is expected to occur in most countries as in Kenya. In Ethiopia, the indigenous home of coffee, the crop is grown mostly in a multistory cropping system with trees in the upper story followed by coffee along with food crops, such as maize, sorghum, and legumes (e.g., beans, peas, and lentils). The ground floor is covered by root crops, such as yam and taro, vegetables, such as cabbages and peppers, and spices, such as ginger and cardamom (Awoke, 1997). In Kenya, preliminary results have indicated that it is possible and economical to intercrop young Arabica coffee with some food crops during the first two years after establishment (Njoroge et al., 1993). It is also possible to intercrop coffee with dry beans during the change of cycle phase (Mwakha, 1980). More studies are encouraged on this line to maximize on available land, improve food availability, and achieve higher incomes. This would also help to sustain the coffee and farmers in periods of low coffee prices. Intercropping also assists in protecting the soil from the vagaries of soil erosion before the coffee canopy closes up. There is also better utilization of solar energy and effective weed management.

Several tree species have been grown in coffee mainly as shade trees or as windbreaks, such as *Cordia* spp., *Grevillea robusta*, *Albizia* spp., *Leucaena leucocephala*, and *Cypress* spp. (Njoroge and Kimemia, 1993). In Ethiopia, indigenous trees such as *Albizia gummifera*, *Allophylus abyssinica*, *Celtis africana*, *Cordia africana*, *Ekebergia capensis*, *Ficus sur*, *F. sycomorus*, *F. vasta*, *Millettia ferruginea*, *Macaranga kilimandscharica*, and *Croton machrostachys* are left as shade trees (Awoke, 1997). Use of shade trees has been shown to help even out erratic yields caused by periodic overbearing and also reduces crinkling of coffee leaves, commonly known as "hot and cold" disease and hail damage. Shade has also been shown to reduce infection of bacterial blight of coffee due to reduced hail injury on the coffee trees, thereby reducing pesticide usage. Apart from such benefits, shade trees help to recycle soil nutrients from deep in the soil to the coffee rooting topsoil through litter fall; leguminous trees fix atmospheric nitrogen, assist in controlling soil erosion and weed control. In multistory intercrop systems, most nutrients are held in the vegetative mass, which is returned to the soil with litter fall. This system also may utilize the soil more efficiently.

DISEASES

The economic production of coffee is influenced greatly by several major diseases whose relative importance vary depending on the locality and variety involved. The majority of these diseases attack *Coffea arabica* L., which represents 75 percent of the total world coffee, both in terms of production and trade (Wrigley, 1988). All parts of the coffee tree, including foliage, berries, stems, and roots are affected by one or the other of the various diseases. They will be grouped and discussed in general according to the part of the plant attacked.

Foliage Diseases

Diseases affecting foliage reduce the photosynthetic capacity of the coffee plant and eventually its productive potential, but only those which damage large areas of leaf tissue or cause leaf shedding, such as coffee leaf rust *(Hemileia vastatrix)* and South American leaf spot *(Mycena citricola)* and, to some extent, brown eye spot *(Cercospora coffeicola)*, have major effects on the plant. Photosynthesis provides carbohydrates for both the developing berries and vegetative growth. Since developing berries provide the strongest physiological sink for carbohydrates, any reduction in photosynthesis on heavily bearing trees will result in carbohydrate starvation of shoots and roots (Cannell, 1970). Because the current season's new growth carries the following season's crop, the main effect of foliage diseases is to reduce the next season's crop. Where major leaf diseases continue unchecked over a number of seasons, progressive decline in yield and vigor occurs. Severe leaf diseases on trees carrying a crop may result in photosynthesis being unable to meet the demands of the developing crop. Carbohydrates are in such cases withdrawn from the remaining leaves and young vegetative tissue, resulting in leaf loss, overbearing stress, and dieback of young shoots and roots (Cannell, 1970). Often a large proportion of the crop on such trees fail to mature properly; the berries appear dull rather than glossy and are particularly prone to berry diseases (Waller, 1987). Yellow ripening is another characteristic symptom, and a large proportion of light and empty beans are produced with the accompanying loss of quality.

Among the foliage diseases, coffee leaf rust is by far the most important, with distribution covering almost all Arabica coffee-growing regions. Disease is favored by warm, humid conditions (such as those found in equatorial regions below 1,500 m above sea level), which support a short latent infection period leading to faster development of epidemics (Waller, 1972; Kingori and Masaba, 1994). For this reason such regions are mostly unsuitable for growing of Arabica coffee. Indeed, much of the Arabica coffee introduced into Sri Lanka and Indonesia was decimated by leaf rust epidemics during the last two decades of the nineteenth century. They were replaced by robusta introductions from Central Africa where they became successful, especially at lower altitudes (de Graaf, 1986). Control of coffee rust is based largely on the use of fungicidal sprays because earlier attempts to utilize resistance were frustrated by the occurrence of many different races of the pathogen; until recently, resistance to all these was not available (Rodrigues et al., 1975). Copper-based fungicides have been found to be universally effective and the cheapest in terms of cost. Several systemic fungicides, such as triadimephon and other triazoles, have been used with mixed success partly due to cost and variable field performance (Figueiredo et al., 1981). Currently, the results of programs using the Grade A (complete) resistance derived from 'Catimor' are being used in Columbia and Kenya where commercial cultivars have been released. This resistance has so far proved durable.

Minor leaf diseases include *Ascochyta tarda* Stewart—a pathogen of high altitude areas, where it can cause death of young leaves and dieback of shoot tips. Predisposing conditions such as wounding and physiologic damage due to "hot and cold" diseases are usually required for infection (Firman, 1965). A similar disease occurs in high-altitude coffee zones of central and northern Latin America, and is attributed to *Phoma costaricensis* Echandi; symptoms and conditions for infections are very similar to those of *Ascochyta tarda* (Echandi, 1958).

Colletotrichum spp. also cause leaf lesions on coffee, again usually following damage by some other agents. Hocking (1966) has shown that *Colletotrichum* can be a primary pathogen on coffee leaves, and there are records of severe defoliation of *Coffea canephora* and *C. excelsa* being associated with infection by *Colletotrichum* spp. Saccas and Charpentier (1969) and Muthappa (1970) reported a stalk rot of leaves caused by *Colletotrichum* in India. The *Colletotrichum* state of *Glomerella cingulata* has also been implicated in the etiology of "weak spot" (Shaw, 1977) and *mancha mantecosa* (oil spot) in South America (Vargas and Gonzales, 1972).

Berry Diseases

The coffee berry is the harvestable portion of the plant. Diseases affecting berries can therefore cause direct loss of yield even though they may not reduce the vegetative vigor of the plant or its subsequent productive potential. Several fungal pathogen can attack coffee berries. Coffee berry disease caused by *Colletotrichum kahawae* Waller & Bridge is a particularly destructive disease that affects developing berries, causing them to rot and shed from the plant before the beans are formed inside. Coffee berry disease occurs only in Africa, but a less virulent form of the same fungus occurs worldwide; this form attacks only ripening berries causing "brown blight," and the mature beans inside are not destroyed. The disease causes pulp to stick to the bean, making wet processing difficult, and can reduce quality (Waller, 1987).

Extensive research work was done on the CBD pathogen and disease in the late 1960s and 1970s, and this was reviewed by Firman and Waller (1977). The source of the pathogen's spores are acervuli on the maturing bark of young twigs and on diseased berries, and their production, dispersal, germination, and infection depend on water (Waller, 1972). Subsequent development of the disease depends on the rainfall distribution and cropping pattern of the coffee trees. In Kenya, where overlapping crops result from two rainy seasons, diseased berries of the first flowering are still present when berries from the second flowering are just developing, and are therefore most susceptible. These observations resulted in the recommendation, still current in Kenya, to spray monthly in the rainy season from February to July with the aim of providing a constant protective layer of fungicide on the berry surfaces (Anonymous, 2001). Copper-based fungicides are the most popularly used, as they also control coffee leaf rust and reduce the intensity of bacterial blight of coffee. Other protective fungicides include Chlorothalonil, Dithianon and Anilazine (Anonymous, 2001).

The use of systemic fungicides, which were at the beginning very successful, was discontinued after a few years of intensive use of benomyl and Carbendazim due to development of resistance in the field. The fungicide-resistant strains are very stable and easily detectable, even without selection pressure of the fungicide (Kingori and Masaba, 1991), and have been found to interfere with control of the disease by contact fungicides (Masaba et al., 1990). Chemical control of CBD and other coffee diseases account for up to 30 percent of the cost of production (Nyoro and Sprey, 1986). A majority of the small-scale farms apply those fungicides less frequently than recommended for reasons of economy. These occasional sprays induce higher levels of CBD than would occur in their total absence (Griffiths, 1972). To provide a sustainable long-term control of the disease, a breeding program to combine resistance to CBD and leaf rust, with the high yield and high quality of Kenya coffee (van der Vossen and Walyaro, 1981), was undertaken resulting in the production of Ruiru 11 hybrid, which has been released commercially.

Another common but seldom important disease is berry blotch caused by *Cercopsora coffeicola*. The disease is characterized by dark-brown blotchy lesions, which are often confused with lesions caused by the CBD fungus (the latter are typical sunken anthracnose lesions) (van der Vossen and Cook, 1975). The normal resistance of healthy green berries is reduced when the tree is under physiological stress or when the berries are wounded and other fungi infect the immature fruit. Some yeastlike fungi (*Nematospora* spp.) infect berries attacked by *Antestiopsis* spp. and can induce an internal rot of the bean.

Other fungi commonly isolated from damaged berries include *Fusarium stilboides, Phoma* spp., and *Colletotrichum gloeosporioides. Botrytis cinerea* may also infect coffee berries to produce "Warty disease" in wet cool conditions. Both *Cercospora coffeicola* and *Fusarium stilboides* are major secondary pathogen attacking berries weakened by overbearing stress, and there have been several instances when the true reasons for infection have been overlooked.

Dieback Diseases

Dieback of shoots may be a secondary effect of root, trunk, or foliage diseases that cause general debilitation of the tree, but some are caused by pathogens that infect shoots directly. Physiological dieback is often the result of overbearing due to the tree carrying more crop than the photosynthesis capacity of the tree can support. Pathological dieback diseases occur most commonly at high altitudes and are usually the result of progressive infection by the minor leaf pathogens (e.g., *Ascochyta* and *Phoma* spp.), which can infect immature coffee stems. Dieback disease may not be sufficiently severe to have much effect on the crop already on the tree (which is carried on growth produced in the previous season) but it can reduce the cropping potential of the tree by restricting growth of young stems so that the following season's crop may be much reduced. Bacterial blight of coffee, the most severe pathological dieback in Kenya, is caused by the bacterium *Pseudomonas syringae* pv. *garcae.* The bacterium is an infectant on coffee shoot, but under wet cool conditions it will gain entry through young tissues to produce water-soaked, dark, necrotic lesions on the leaves, twigs, and berries. When terminal buds are attacked, infection spreads backward from the shoot and tips of twigs causing dieback, which is distinctly different from overbearing dieback as leaves remain attached to the dead twigs. In extreme cases, a whole sucker or even a large part of the main stem may be infected and killed.

The disease is the most prevalent on exposed slopes as opposed to sheltered valleys. Shade has been observed to reduce the incidence of BBC (Thorald, 1945) probably by reducing wind and storm damage, which assist the entry of bacteria. Control of the disease in Kenya has been by intensive, straight foliar applications of copper-based bactericides to reduce the bacterial population on the coffee trees. Kairu et al. (1991) pointed out that the disease is stimulated when organic fungicides for control of CBD, which occurs together with BBC, are applied. Continued use of coppers in the control of BBC is also threatened by the recent observations of presence of copper-resistant strains of *Pseudomonas syringae* on farms where copper has been used routinely (Kairu et al., 1988). The sustainability of the current control program for the disease may not be possible considering the cost of bactericides and farm machinery vis-à-vis coffee prices. For this reason work aimed at identifying sources of genetic resistance to the pathogen in the coffee germplasm has been started in Kenya.

Trunk and Branch Diseases

These diseases cause general debilitation of coffee trees by disrupting the translocation of substances between roots and shoots. Initially, these diseases cause leaf wilting, shedding, or chlorosis, but they may also infect berries and leaves directly as they spread along branches. On trees showing these early symptoms, the crop may fail to mature properly producing many light and empty berries. Cessation of growth and shoot dieback then occurs and diseased trees or large parts of them are killed. These diseases usually occur sporadically on individual trees or groups of trees and do not cause rapidly spreading epidemics. Some are greatly influenced by soil and climatic conditions, causing severe damage only if trees are under climatic stress. Others are primarily wound pathogens and become severe only on damaged trees. The most prominent are *Fusarium* bark disease caused by *F. stilboides* Wollen and web blight caused by *Corticium salmonicolor* or *C. koleroga*. The former is characterized by stem cankers usually at the bases of suckers (Storey's disease) (Storey, 1932), on mature main stems, especially at pruning wounds and primary branch bases (scaly bark), and around the bases of mature stems at or just above soil level. The cankers enlarge insidiously under the bark to girdle the stem, eventually killing the affected tree. Other symptoms include berry and leaf lesions (Siddiqi and Corbett, 1963). The disease is considered a major factor limiting coffee production in Malawi, Tanzania, and Zimbabwe (Siddiqi and Corbett, 1983; Clowes and Logan, 1985). In 1970 (Baker, 1970), the disease in Kenya was largely restricted to the southeastern district of Taita Taveta, which is well isolated from other coffee growing areas. The distribution has since widened progressively to cover key coffee-growing districts (Kingori, in press). Control of the disease is achieved mainly by restricting its spread to healthy trees. Pruning tools easily disseminate the spores during field operations and should be sterilized using a suitable disinfectant after using them on diseased trees. Wounds caused by boring insects and during weeding operations form common entry points for the pathogen, whereas mulch and unchecked weeds near the base of the stem create a good environment for infection and sporulation by the fungus.

Earlier work on screening coffee varieties for resistance (Siddiqi and Corbett, 1965) failed to identify varieties with consistent resistance. This could be due to the low success of inoculation tests with pure cultures of the pathogen reported by Siddiqi and Corbett (1963).

Wilt Diseases

The most serious wilt disease in coffee is *Fusarium* wilt disease (tracheomycosis) caused by *Fusarium xylarioides* Heim & Saccas. It occurs in several African countries, and has been particularly troublesome on *Coffea canephora* in West and Central Africa and Uganda (Flood and Brayford, 1997). A similar disease occurs in South America caused by *Ceratocystis fimbriata* Ellis & Halsted (Waller, 1987). This disease also gains entry through wounds at the base of the trunk and causes sunken necrotic cankers that girdle the trunk and kill the tree.

Root Diseases

As with other perennial tropical crops, basidiomycete root pathogens may be troublesome, especially on newly cleared forest land. *Armillaria mellea* (Valhl ex Vries) Karsten is the most widespread root pathogen of coffee in such conditions. Moribund stumps or roots provide a food source from which the fungus can spread to infected coffee. Symptoms usually appear as a rapid debilitation wilting and death of affected trees. Creamy white mycelial strands can be seen beneath the bark and cluster of characteristic mushroomlike sporophores occur on the bases of recently killed trees.

Black root rot caused by *Rosellinia* spp. also occur on coffee, particularly in Latin America and the Antilles. Symptoms of general debilitation followed by collapse are broadly similar to *Armillaria* except that the mycelia occur as broad-spreading fans rather than rhizomorphs and the fruiting bodies are minute spherical ascocarps (Waller, 1987). *Fusarium solani* causes a sporadic but lethal root disease of Arabica coffee in East Africa. No visible external mycelia are present with this disease, but a distinct purple brown discoloration of the wood at the crown of the tree is characteristic of infection (Baker, 1972).

A range of nematode species attack coffee roots. The root-knot nematodes *Meloidogyne exigua* and *Pratylenchus* spp. have been particularly troublesome in some Latin American countries and Tanzania. General debilitation with leaf chlorosis, stunting, and dieback are the major aboveground symptoms.

Nursery Diseases

Damping-off of young coffee seedlings is a common problem in coffee nurseries. This is usually caused by *Rhizoctonia solani, Fusarium solani,* and *F. stilboides* either singly or in combination. Nematodes can also attack plants in the nursery and may eventually cause debilitation of mature coffee trees in the field.

PESTS

Coffee has a large number of potential pests, which have been fully documented by Le Pelley (1968). Most of them have a restricted distribution, and a few are of economic importance. They can, however, be serious locally or in certain seasons. Since most of them are indigenous to the areas where they occur, they are usually accompanied by effective natural enemies. Localized outbreaks can nevertheless result in significant yield losses and quality decline when they happen. The major pests include boring beetles, scale insects, mealybugs, sucking bugs, leaf miners, and defoliators. A brief discussion of some of the economically important ones follows.

Coffee Berry Borer (Hypothenemus hampei *Ferri)* (*Coleoptera: Scolytidae*)

The berry borer, a small black beetle, is the only serious field pest on coffee. Its distribution covers East and Central Africa, Brazil, Java, Peru, Equador, Colombia, Central America, and Mexico (Wrigley, 1988). The first incidence in Kenya was reported in 1928 (Wilkinson, 1929) with infestation levels of less than 10 percent. More recent infestation levels of up to 80 percent have been recorded. The symptom of attack by berry borer is a single small hole, the entry point by the female beetle, at the apex of otherwise sound-looking green or ripe berries. The beetle tunnels within the bean and lays eggs, which hatch into larvae. These larvae continue to feed on the beans further damaging them. Damaged beans have a distinct blue-green staining (Anonymous, 1989). If attacked, young berries usually detach and fall to the ground. Infested mature beans are of no commercial value. The distribution of *H. hampei* is affected by the altitude, tending to be more severe at a lower zone below 1,370 m. According to Wrigley (1988) the pest is rare at 1,525 m and absent at 1,680 m above sea level.

Successful control of the pest depends on field cultural practices that encourage populations of natural enemies. In Kenya, these include two parasitic wasps, *Prorops nasuta* Waterston and *Heterospilus coffeicola* Schmiedeknecht. The parasitism levels by both wasps is low (Mugo, 1994). Heavy shade from shade trees on inadequately pruned coffee bushes creates conditions unsuitable for natural enemies and should be removed. To prevent build up of pest populations, picking of berries should be carried out at least fortnightly during fruit peaks and at least monthly at other times. No ripe or dried berries should be left on the ground or on the trees. All infested berries should be destroyed by burning, deep burying, or rapid drying on trays. The old crop should, if possible, be stripped completely just before the main flowering (Anoymous, 1989). Insecticide sprays should only be regarded as a supplement to these cultural measures. Recommended insecticides include Dursban 48 percent emulsion concentrate and Nuvellawcl. C. (Anonymous, 1989).

Antestia Bug (**Antestiopsis** *spp.*) (*Hemiptera: Pentatomidae*)

Antestiopsis spp. is an important pest of Arabica coffee in Africa with an average of one to two bugs per tree causing appreciable loss (Anonymous, 1989). Several species including *Antestiopsis orbitalis bechuana, A. ghesquierei, A. intricata,* and *A. facetoides* (Greathead, 1965) occur in Kenya. The antestia bug has a flattened body with striking black, orange, and white markings. The female lays eggs on the underside of a coffee leaf. After hatching, the nymphs undergo five stages before maturing (Le Pelley, 1932). Damage by the adult antestia bug is through its feeding on green berries, buds, and green twigs. The piercing of the young berries causes them to be shed while the older ones remain attached. However, the berries develop sunken discolored patches on their surfaces due to the introduction of yeastlike fungi, *Nematospora coryli* and *N. gossypii,* vectored by the bug. On pulping such berries yield zebra-pattened beans of low appeal to customers (Le Pelley, 1968). Antestia bugs also feed on shoot tips, causing a number of small shoots to develop at the end of the primary branches (fan branching) (Anonymous 1989). Such branches are difficult to prune and do not produce crop. When flowers are attacked, they fail to develop, and in most cases turn black and die. According to Le Pelley (1968) two antestia bug egg parasitoids, *Telenomus seychellensis* and *Hadrontus antestiae,* are common with a parasitism percentage of 80 to 90 percent. Several nymph and adult parasitoids have also been reported (Wanjala, 1980). Coffee bushes should be kept open by regular pruning so as to make them unsuitable habitats for the pest and a suitable one for parasitism (Mugo, 1994). Antestia bugs can be controlled through pesticides (see Anonymous, 1989, for a recommended list). Spraying should be done when the average population (adults plus nymphs) exceed two to a tree in the drier areas or one to a tree in the wetter areas.

Scales and Mealybugs

The nymphs and adults of these insects are of great economic importance on coffee. They are dependent on either roots, branches, leaves, or flowers on which they live a mainly settled life and suck the plant sap seriously affecting yield and general health of the coffee tree.

Mealybugs are soft bodied insects covered with a white waxy coat and capable of movement. Scale insects on the other hand become attached to the plant and develop a tough scaly covering once they begin to feed. Their numbers increase rapidly during dry weather and tend to decline during prolonged wet spells. The most important in Kenya are the root mealybug (*Planococcus citri* Risso), Kenya mealybug (*Planococcus kenyae* Le Pelley), green scales *(Coccus viridis),* white waxy scale (*Ceroplastes brevicaudas* Hall), and fried egg scales (*Aspidiotus* spp).

The root mealybug is the most common mealybug in all coffee-growing areas worldwide (Mugo, 1995). Apart from coffee it also attacks the roots of citrus (Le Pelley, 1968). The severity of attacks tends to be higher on coffee growing on poor soils deficient in calcium and with pH below 5.0 (Baum, 1968). Affected plants yield poorly, develop yellow leaves, and eventually die. In areas prone to infestation, soils should be mixed with Furadan or Temik at recommended rates during planting and replanting (Anonymous, 1989).

The Kenya mealybug originated from Uganda and northern Tanzania where a complex of natural enemies kept it under control (Mugo, 1994). Its numbers in Kenya were brought down by a parasitic wasp, *Anagyrus kivuensis,* imported from Uganda (Le Pelley, 1968). On occasion this control has been disrupted by attendant ants, but trunk banding of infested trees using recommended insecticides offers a good solution (Anonymous, 1989).

Green coffee scales were first described from Sri Lanka but now occur in all countries where coffee is grown. Its host range also includes tea, cassava, citrus, guava, and mango (Wrigley, 1988). Heavy attacks, especially in dry areas, lead to arrested tree growth and reduced flowering. Green scales are kept under control by their many insect parasites and predators provided ants are prevented from attending the scales (Mugo, 1994).

Coffee Leaf Miner (Leucoptera spp.) (Lepidoptera: Lyonetiidae)

These small moths, the larvae of which mine the leaves of the coffee plant, are widely distributed in South America and Africa (Wrigley, 1988). The symptoms of leaf miner attack are irregular brown blotches on the upper side of leaves. The removal of the top skin of the blotch reveals fresh mines and several white catapillars up to 12 mm long (Anonymous, 1989). Severe attacks cause major destruction of leaf tissue and serious leaf fall, reducing the amount of photosynthates badly needed by developing berries. Where this combines with coffee leaf rust, twig dieback may result. According to Bess (1964), the rise of leaf miner to major pest status was due to an upset in the ecosystem caused by the application of mulch and increased use of nutritional and fungicidal sprays since 1954. Mulching reduces predation by ants on leaf miner larvae, which drop to the ground to pupate in the crevices in the soil (Bardner, 1978). Bess (1964) established that increased nutritional and fungicide spray improve leaf retention, creating a humid microclimate in the average coffee bush favoring breeding of leaf miner, which prefer old thick leaves to young thin ones. Where attacks are serious, chemical control is possible using an insecticide with translaminar systemic action to get into the mines and a narrow spectrum of activity so that the pest's natural enemies will not be destroyed.

Stem and Branch Feeders

Stem and branch feeders are common in the tropics, and form some of the most serious pests of coffee, several of which are capable of causing the death of coffee trees. The major ones among them are the white borer (*Anthores leuconotus* Pascoe), yellow-headed borer (*Dirphya nigricornis* Olivier), and West African coffee borer (*Bixadus sierricola* White). The white borer is widely distributed in Africa, being present in Angola, Cameroon, and East Africa (Coste, 1968). It is particularly severe on Arabica coffee especially at lower altitudes (below 1,500 m). Most damage is done by the larvae, which bore into the trunk and roots exuding woody shavings from its burrows. Young trees up to two years old are frequently killed. Older trees wilt and the foliage becomes chlorotic. If they survive, yields are much reduced and the trees are prone to infection by *Fusarium stilboides*

(Anonymous, 1989). Control is by banding or spraying the stem about 18 inches from the soil level using one of the recommended insecticides (Anonymous,1989).

Yellow-headed borer has been reported on coffee in Senegal, Malawi, Kenya, and Tanzania (Le Pelley, 1968). In Kenya, the pest is particularly serious in Taita Hills (Clowes, 1949) where early attempts to grow the crop were abandoned due to the ravages of the borer. The larvae bore through the wood weakening its structure and disrupting the activities of the vascular system. Affected trees therefore break easily, have low resistance to water stress and yield poorly. Wounds caused by the borer are common entry points for bark pathogens, such as *Fusarium stilboides*.

EFFECTS OF CONSUMER NEEDS ON ADAPTATION

Coffee is grown mainly in the southern hemisphere and consumed in the developed countries in the northern hemisphere. The developed countries account for about 80 percent of the total world incomes and preferences. As such, the consumer incomes and preferences in the developed countries not only determine the level of coffee consumption but also the types of coffee grown as well as the production and processing practices adopted by coffee producers.

Coffee (be it mild, Brazilian, or canephora) as a beverage has been traditionally consumed mainly as ground coffee, which represents 80 percent of the sales in the United States and Europe. Ground coffee is usually offered as roasted coffee in various blends, such as instant, flavored, or decaffeinated. All these coffees have to undergo various processing, the most important being roasting. Current roasting techniques require exact knowledge of green coffee characteristics, not only to identify the species, varieties, and types from various production areas, but also their particular characteristics, which depend on grading, storage, and nature of impurities and defects. These concerns of the roaster coupled with the ultimate consumer needs dictate the types and origins of coffee uses.

Consumer dynamics, in terms of choice and presentation as well as health concerns, have added a new dimension in terms of coffee production and processing. The need for speciality or gourmet coffees has risen in America and Europe. The speciality concept is derived from the belief that there are certain coffees which exhibit unique cup quality and hence satisfaction as they are grown in special areas and processed in a special manner. Such coffee includes Inter Alia Supremos from Colombia, AA grade from Kenya, and Blue Mountain from Jamaica. These coffees are grown mainly in high altitude areas with rich soils and high rainfall, which are ideal for bean development. Furthermore, these coffees are processed mainly through the wet method and dried in natural conditions. Due to the high premium price attracted by such coffees, there is a concerted effort to develop varieties, which can do well under these highland conditions.

Since the early 1990s, consumer health concerns mainly in regard to pesticide residues, have necessitated the introduction of "organic" or environmentally friendly coffee. Coffee qualifying for the organic "green" label is usually grown in special conditions with minimum use of pesticides, be they fertilizers, fungicides, or herbicides. Such coffee is also processed separately under natural conditions. This development has seen the adaptation of the coffee crop to minimal use of inorganic inputs. Consumer preferences in favor of shade coffee have also been on the increase since 1975. This has necessitated the growing of coffee under all manner of shade trees, a necessity that calls for coffee varieties that do well under shade.

RESEARCH AND DEVELOPMENT ORGANIZATIONS

Most major coffee-producing countries have institutions or organizations that carry out research and coordinate training in various aspects of the coffee industry. These institutions fall under different categories. They are run either independently or as departments in government ministries. Academic institutions such as universities are also involved in research in coffee, usually in collaboration with other research organizations. Examples of coffee research organizations include the Coffee Research Foundation in Kenya, which is funded mainly by Kenya's coffee farmers; the Federacion Nacional de Cafeteros de Colombia, and the Fundacao Instituto Brasiliero de geografica e Estadistica of Brazil, which are run by their respective governments. Other coffee research programs are within a wider research mandate within their respective government research centers.

In addition to the research institutions, several regional and international bodies and organizations are involved in research and development of coffee. Such bodies include the European Union (EU), International Coffee Organization (ICO), World Bank, and African Coffee Research Network (ACRN). These organizations are involved in facilitating or funding collaborative research. They also organize forums such as workshops, seminars, or conferences where researchers can exchange ideas and set research priorities.

FUTURE PROSPECTS

The main problem facing coffee production worldwide is the continuous decline of net earnings by farmers, whereas the consumers continue to pay increasingly more. This is the biggest threat to the production of good quality coffee and is the result of two main factors. The main cause is the poor prices offered by the buyers at the world market. The poor prices are in turn the result of the large quantities of poor quality coffees presently flooding the world market from certain countries, especially Vietnam. As would be expected, buyers have capitalized on the situation, preferring to buy more of the poor quality coffees cheaply and upgrading the quality through mechanical means or through blending. The second contributor to the poor net earnings by the producers is the increasing cost of agrochemicals, particularly pesticides, thereby making disease and insect control very expensive.

Under the circumstances of poor prices and high cost of production, the only viable option to sustainable production of coffee is to produce "more with less." This means that superior varieties combining resistance to major diseases and pests with high yields and good quality should be developed and made available to producers. In most cases, high yields and good quality are already inherent in the old varieties, whereas disease resistance is already being exploited in a number of countries. The concern of coffee breeders should therefore focus more on improving the stability of the resistance base and develops ways of responding rapidly to new diseases or new races of old pathogens. In this regard, two areas of research come to light. First, the germplasm resident in the various world collections needs to be characterized comprehensively with the objective of identifying genes of economic significance, particularly those controlling disease resistance, quality, and insect resistance. Such genes should be consequently accumulated in genetic backgrounds, which are more attractive for breeding work through prebreeding. The second area, which needs some consideration, is the development of transformation procedures to be put in place as an option for rapid transfer of genes of interest. At present transgenic varieties are not popular with the consumers. However, by limiting coffee transformation to genes within the same family or genus, consumer acceptance should be realized with less effort.

Research efforts in coffee to date have resulted in increased production, productivity, and quality of coffee. However, many old and new challenges still need to be overcome to make coffee cultivation sustainable in the future. Biotechnology has strong prospects for applications in coffee breeding. Other important considerations for sustainable products include: breeding for specialty coffees, e.g., decaffeinated coffee; organic production of coffee;

biocontrol of pests and diseases and integrated pest management (IPM) considerations; research into genetic modifications for pest resistance and enhanced quality or specific tastes; and further reduction in production costs and subsequent increase in production. Also important are environmental considerations and impact of coffee cultivation, e.g., the use of chemicals and land-use systems.

REFERENCES

Afnor, 1991. Contrôle de la qualité des produits alimentaires analyse sensorielle, Fourth Edition. Association Francaise de normalisation (AFNOR-DGCCRF), Paris.

Agwanda, C.O. 1997. Amélioration du Caféier *Coffea arabica* L. au Kenya; optimisation des stratégies des sélection en présence d'interaction génotype X milieu. PhD thesis, University of Montpellier, France.

Agwanda, C.O. 1999. Flavour: An ideal selection criterion for the genetic improvement of liquor quality in Arabica coffee. *Proceedings of the Nineteenth International Scientific Colloquium on Coffee,* Helsinki.

Agwanda, C.O., Lashermes, P., Trouslot, P., Combes, M.-C., and Charrier, A. 1997. Identification of RAPD markers for resistance to coffee berry disease, *Colletotrichum kahawae* in Arabica coffee. *Euphytica* 97: 241-248.

Agwanda, C.O. and Owuor, J.B.O. 1989. Toward a more sustainable and economic production of Arabica coffee (*Coffea arabica* L.) in Kenya through the exploitation of improved cultivars: A review. *Kenya Coffee* 54(638): 735-743.

Allard, R.W. and Bradshaw, A.D. 1964. Implications of genotype-environment interactions in applied plant breeding. *Crop Sci.* 4: 503-508.

Anonymous. 1985. Notes of the month: New CBD/Leaf rust resistant variety launched. *Kenya Coffee* 50: 378-382.

Anonymous. 1989. *An Atlas of Coffee Pests and Diseases.* Coffee Research Foundation, Ruiru, Kenya.

Anonymous. 2001. Control of coffee berry disease and leaf rust in 2001. Coffee Research Foundation, Technical Circular No. 804.

Awoke, T.C. 1997. The culture of coffee in Ethiopia. *Agroforestry Today* 9(1): 19-21.

Baker, C. J. 1970. Coffee bark disease in Kenya. *Kenya Coffee* July: 226-228.

Baker, C.J. 1972. *Fusarium solani* associated with a wilt of *Coffea arabica* in Kenya. *E. Afr. Agric. For. J.* 337: 137-140.

Baum, H. 1968. The coffee root mealybug complex. *Kenya Coffee* 33: 175-178.

Berkley, M.J. 1869. Notes. Gardeners (cited in Kushalappa, 1989).

Bess, H.A. 1964. Populations of the leaf miner *Leucoptera meyricki* and its parasites in sprayed and unsprayed coffee in Kenya. *Bulletin of Entomological Research* 55: 59-82.

Bouharmont, P. 1995. La sélection du caféier Arabica au Cameroon (1964-1991). Document de travail, No. 1-95 (40 p), CIRAD, France.

Browning, G. and Fisher, N.M. 1976. High density coffee: Yield results for the first cycle from systematic plant pacing designs. *Kenya Coffee* 41(483): 209-217.

Cannell, M.G.R. 1970. The contribution of carbohydrates from vegetative laterals to the growth of fruits on the bearing branches of *Coffea arabica* L. *Turrialba* 20: 15-19.

Carvalho, A. and Monaco, L.C. 1969. The breeding of Arabica coffee. In Ferwerda, F.P. and Wit, F., Eds., *Outlines of Perennial Crop Breeding in the Tropics.* Miscellaneous Paper 4. Landbouw, Wageningen, pp. 198-216.

Castillo, Z.J. and Moreno, R.G. 1988. *La variedad Colombia* (The Colombia Variety). Cenicafé publication.

Charrier, A. 1982. Quelques réflexions sur les possibilités d'amélioration génétique de la qualité des cafés. In *Tenth International Colloquiumon Coffee,* Salvador, pp. 369-374. Association Scientifique International de Cafes, Paris. de couverture dans les cafeieres au Burundi. Cafe Cacao The xxxviii: 41-48

Charrier, A and Eskes, A.B. 1997. Les Caféiers. In Charrier, A., Jacquot, M., Harmon, S., and Nicolas, D., Eds., *L'Amelioration des plantes tropicales.* CIRAD and ORSTOM, pp. 171-196.

Clowes, M. St. J. and Logan, W.J.C. 1985. *Advances in Coffee Management and Technology in Zimbabwe 1980-1985.* Cannon Press, Harare, Zimbabwe.

Clowes T.J. 1959. The yellow-headed borer. *Kenya Coffee,* October.

Coste, R. 1968. *Les Caféiers.* Maisonneuve and Larose, Paris.

de Graaf, J. 1986. The economies of coffee. In *Economics of Crops in Developing Countries,* No 1. Pudoc, Wageningen, Netherlands.

Devonshire, C.R. 1956. Explanation of the coffee report form. *Coffee Board of Kenya Monthly Bull.* 21: 186-187.

Echandi, E. 1959. La quema de las cafetas causada por *Phoma costarricensis* n.sp. *Rev. Porol. Trop.* 5: 81-108.

Echeverri, J.H. and Fernandez, C.E. 1989. The PROMECAFE program for Central America. In Kushalappa, A.C. and Eskes, A.B., Eds., *Coffee Rust: Epidemiology, Resistance, and Management.* CRC Press, Inc., Boca Raton, FL.

Eskes, A.B. 1983. Incomplete resistance to coffee leaf rust *(Hemileia vastatrix).* PhD thesis, Wageningen Agricultural University, Netherlands.

Figueiredo, P., Mariotto, P.R., Bonini, R., de Olieveira, N.L. Filho, and Olieveira, D.A. 1981. Efeito do pyracarbolid e oxicarboxin aplicados em misturas e in intercalados con fungicida cuprica no controle de ferrugem do cafeeiro (*Hemileia vastatrix* Berk and Br.). *Biologico* 47: 237-244.

Firman, I.D. 1965. Some investigations on a disease of *Coffea arabica* caused by *Ascochyta tarda. Trans. Brit. Mycol. Soc.* 48: 161-166.

Firman, I.D. and Waller, J.M. 1977. Coffee berry disease and other *Colletotrichum* diseases of coffee. Phytopath. Pap No. 20 C.M.I.

Flood, J. and Brayford, D. 1997. Re-emergence of *Fusarium* wilt of coffee in Africa. Proceedings of the Seventeenth ASIC conference, July 20-25, 1997, Nairobi, Kenya.

Gallais, A. 1992. Bases génétiques et stratégie de sélection de L'adaptation générale. *Le Sélectionneur Français* 42: 59-78.

Greathead, D.J. 1965. Work in progress for the Coffee Research Foundation. The Commonwealth Institute of Biological Control, Kawada, Uganda. *Kenya Coffee,* January.

Griffiths, E. 1972. Negative effects of fungicides in coffee. *Trop. Sci.* 14: 79-89.

Hocking, D. 1966. Brown blight (*Colletotrichum coffeanum* Noack) of Arabica coffee in East Africa. *Ann. Appl. Biol.* 58: 409-421.

Kairu, G.M. 1991. Plant pathology. The Coffee Research Foundation Annual Report 1991/92.

Kairu, G.M., Nyangena, C.M.S., and Crosse, J.E. 1988. The effect of copper sprays on bacterial blight and coffee berry disease in Kenya. *Plant Pathology* 34: 207-213.

Keter, J.K.A. 1974. A study of soil phosphorous forms in Kikuyu red loam and their surface activity. MSc thesis, University of Nairobi.

Kimemia J.K., Chweya, J.A., and Nyabundi, J.O. 2000. The effect of green manure application to coffee plants growth, yield and quality in Kenya. In *Proceedings of the Eighteenth International Colloquium on Coffee Science,* Helsinki, Finland, pp. 426-429.

Kingori, P.N. and Masaba, D.M. 1991. Distribution and persistence of benonyl (Benlate) resistance in populations of *Colletotrichum coffeanum* in Kenya. *Kenya Coffee* 56: 1071-1074.

Kingori, P.N. and Masaba, D.M. 1994. Current status of coffee leaf rust in Kenya. *Kenya Coffee* 59: 1977-1987.

Kushalappa, A.C. 1989. Importance. In Kushalappa, A.C. and Eskes, A.B., Eds., *Coffee Rust: Epidemiology, Resistance, and Management.* CRC Press, Inc., Boca Raton, FL, pp. 2-11.

Lashermes, P., Agwanda, C.O., Anthony, F., Combes, M.-C., Trouslot, P., and Charrier, A. 1997. Molecular marker-assisted selection: A powerful approach for coffee improvement. In *The Seventeenth International Colloquium on Coffee,* Nairobi. ASIC, Paris.

Lashermes, P., Combes, M.-C., Robert, J., Trouslot, P., D'Hont, A., Anthony, F., and Charrier, A. 1999. Molecular characterisation and origin of the *Coffea arabica* L. genome. *Mol. Gen. Genet.* 261: 259-266.

Lashermes, P., Cros, J., Marmey, P. and Charrier, A. 1993. Use of random amplified DNA markers to analyze genetic variability and relationships of *Coffea* species. *Genet. Res. Crop Evol.* 40: 91-99.

Lashermes, P., Trouslot, P., Anthony, F., Combes, M.-C., and Charrier, A. 1996. Genetic diversity for RAPD markers between cultivated and wild accessions of *Coffea arabica. Euphytica* 87: 59-64.

Le Pelley, R.H. 1932. On the contuly Antestia Lineaticollis. Star in coffee in Kenya colony. *Bul. Ento. Res.* 23: 217-223.

Le Pelley, R.H. 1968. *Pests of Coffee.* London, Longmans, Green & Co.

Masaba, D.M., Kingori, P.N., and Muthangya, P.M. 1990. Control of coffee berry disease *(C. coffeanum)* using tank mixture of coppers and organic fungicides. *Kenya Coffee* 55: 949-954.

McDonald, J. 1926. A preliminary account of a disease of green coffee berries in Kenya colony. *Trans. Brit. Mycol. Soc.* 11: 145-154.

Melchinger, A.E. 1990. Use of molecular markers in breeding for oligogenic disease resistance. *Plant Breeding* 104: 1-19.

Mitchell, H.W. 1970. The effect of manure and fertilizers on Arabica coffee in Kenya. l: Effect on yield and quality. Manure and fertilizer use seminar. National Agricultural Laboratories. Coffee Research Foundation, pp. 13-58.

Mitchell, H.W. 1976. Research on close spacing systems for intensive coffee production in Kenya. *Ann. Report 1974/75*. Coffee Research Foundation, Kenya, pp. 13-57.

Moreno, G., Moreno, E., and Cadena, G. 1995. Bean characteristics and cup quality of the Colombian variety *(Coffea arabica)* as judged by international tasting panels. In *The Sixteenth International Colloquium on Coffee*, ASIC, Paris, pp. 574-583.

Mugo, H.M. 1994. Coffee insect pests attacking flowers and berries in Kenya. A review. *Kenya Coffee* 59(691): 1777-1783.

Mugo, H.M. 1995. A review of insect pests attacking coffee roots, stem, branches, and leaves in Kenya. *Kenya Coffee* 60(708): 2095-2102.

Muthappa, B.N. 1970. Studies on the role of *Colletotrichum coffeanum* in causing stalk rot of leaves and berries of Arabica coffee. *Indian Coffee* 34: 263-264.

Mwakha, E. 1980. Intercropping dry beans in high density Arabica coffee l. Preliminary observations on bean growth and yield. *Kenya Coffee* 45: 187-192.

Njoroge, J.M. 1985. Effect of nitrogen rates and frequency of application on coffee yield and quality in Kenya. *Acta Horticulturae* 158: 283-291.

Njoroge, J.M. 1986. New weeds in Kenya coffee—A short communication. *Kenya Coffee* 51: 353-355.

Njoroge, J.M. 1991. Management of Ruiru 11—A review. *Kenya Coffee* 56: 1027-1035.

Njoroge, J.M. 1994. Weeds and weed control in coffee. *Exptl. Agric.* 30: 421-429.

Njoroge, J.M. and Kimemia, J.K. 1989. A comparison of different weed control methods in Kenya coffee. *Kenya Coffee* 55: 863-870.

Njoroge, J.M. and Kimemia, J.K. 1992. Use of foliage applied herbicides at low rates and volumes on annual weeds control in coffee. *Kenya Coffee* 57: 1273-1278.

Njoroge, J.M. and Mwakha, E. 1983. Observations on the effects of weeding and cover crops on coffee yield and quality. *Kenya Coffee* 48: 219-224.

Njoroge, J.M., Waithaka, K., and Chweya, J.A. 1992. The influence of tree training and plant density on growth, yield components, and yield of Arabica coffee cv. Ruiru 11. *J. Hort. Sci.* 67: 695-702.

Njoroge, J.M., Waithaka, K., and Chweya, J.A. 1993. Effects of intercropping young plants of compact Arabica coffee hybrid cultivar Ruiru 11 with potatoes, tomatoes, beans, and maize on coffee yields. *Exptl. Agric.* 29: 373-377.

Northmore, J.M. 1965. Some factors affecting the quality of Kenya coffee. *Turrialba* 15: 184-193.

Nyoro, J.K. and Sprey, L.H. 1986. Introducing Ruiru 11 to estates and small holders. *Kenya Coffee* 51: 7-28.

Oruko, B.A. 1977. Yield responses of Arabica coffee to fertilizers in Kenya: A review. *Kenya Coffee* 42: 227-239.

Rodrigues, C.J., Jr., Bettencourt. A.J., and Rijo, I. 1975. Races of the pathogen and resistance to coffee rust. *Ann. Rev. Phytopath.* 13: 49-70.

Saccas, A.M. and Charpentier, J. 1969. L'antracnose des cafeiers robusta et excelsa due a Colletotrichum coffeanum Noack en republic centrafricaine. *Café—Cacao—Thé* 13: 221-230.

Shaw, D.E. 1977. Report on continuing eradication of coffee rust in two areas in Papua New Guinea. *Papua New Guinea Agric. J.* 28: 27-32.

Siddiqi, M.A. and Corbett, D.C.M. 1963. Coffee bark disease in Nyasaland: Pathogenicity, description, and identity of the causal organism. *Trans. Brit. Mycol. Soc.* 46(1): 91-101.

Siddiqi, M.A. and Corbett, D.C.M. 1965. *Fusarium* bark disease of coffee: Report on research carried out in Malawi. *E. Afr. Agric. For. J.* 31(1): 11-15.

Snoeck, D., Bitoga, J.P., and Barantwaririje. 1994. Advantages et inconvenients des divers modes.

Storey, H.H. 1932. A bark disease of coffee in East Africa. *Ann. Appl. Biol.* 19: 172-184.

Thorald, C.A. 1945. Elgon dieback disease of coffee. *E. Afr. Agric. J.* 10(4): 198-206.

van der Graaff, N.A. 1981. Selection of Arabica coffee types resistant to coffee berry disease in Ethiopia. PhD thesis, Wageningen Agricultural University, Netherlands.

van der Vossen, H.A.M. 1985. Coffee selection and breeding. In Clifford, M.N. and Wilson, K.C., Eds., *Coffee Botany, Biochemistry, and Production of Beans and Beverage.* Croom Helm, London.

van der Vossen, H.A.M. and Cook, R.T.A. 1975. Incidence and control of berry blotch caused by *Cercospora coffeicola* on Arabica coffee in Kenya. *Kenya Coffee* 40: 58-61.

van der Vossen, H.A.M., Cook, R.T., and Murakaru, G.N.W. 1976. Breeding for resistance to coffee berry disease caused by *Colletotrichum coffeanum* Noack (Sensu Hindorf) in *Coffea arabica* L. I. Methods of preselection for resistance. *Euphytica* 25: 733-745.

van der Vossen, H.A.M. and Walyaro, D.J.A. 1980. Breeding for resistance to coffee berry disease in *Coffea arabica* L. II. Inheritance of the resistance. *Euphytica* 29: 777-791.

van der Vossen, H.A.M. and Walyaro, D.J.A. 1981. The coffee breeding program in Kenya: A review of progress made and plan of action for the coming years. *Kenya Coffee* 46(541): 113-130.

Vargas, E. and Gonzales, L.C. 1972. La mancha mantecosa del café causada pas Colletotrichum spp. *Turrialba* 22: 129-135.

Waller, J.M. 1972. Water borne dispersal in coffee berry disease and its relation to control. *Ann. Appl. Biol.* 69: 1-18.

Waller, J.M. 1987. Coffee diseases: Current status and recent developments. *Rev. Trop. Plant Path.* 4: 1-33.

Walyaro, D.J.A. 1983. Considerations in breeding for improved yield and quality in arabica coffee (*Coffea arabica* L.). PhD thesis, Wageningen Agricultural University, Netherlands.

Walyaro, D.J. and van der Vossen, 1977. Pollen longevity and artificial cross-pollination in *Coffea arabica* L. *Euphytica* 26: 225-231.

Walyaro, D.J.A and van der Vossen, H.A.M. 1979. Early determination of yield potential in Arabica coffee by applying index selection. *Euphytica* 28: 465-472.

Wanjala, F.M. 1980. Effect of infesting green coffee berries in different population levels of *Antestiopsis lineaticollis* Stal (Hemiptera; Pentatomidae, in Kenya. *Turialba* 30(1): 109-110.

Wilkinson, H. 1929. Report Entomologist. Report Dept. of Agric. of Kenya, Nairobi, pp. 172-186.

Wrigley, G. 1988. *Coffee Tropical Agricultural Series*. Longmans Scientific and Technical Series, London.

Yamada, Y. 1991. Interaction: Its analysis and interpretations. *Proceedings of the 1991 International Society for Oil Palm Breeding, Malaysia, International Workshop on Genotype-Environment Interaction Studies in Perennial Tree Crops*, K.L. Hilton, Malaysia.

Chapter 8

Oil Palm

M. Wahid Basri
Siti Nor Akmar Abdullah
Maizura Ithnin
Norman Kamaruddin

INTRODUCTION

The oil palm (*Elaeis guineensis* Jacquin) is economically important for its oil and has become one of the major oil crops in the world. The most productive oil-bearing crop, the oil palm yields about three times the oil yield of coconut, seven times that of rapeseed, and ten times that of soybean. To date, in terms of world production, palm oil maintains second place after soybean as a source of vegetable oil and has overtaken other crops such as rapeseed, sunflower, groundnut, and cottonseed. Under good agricultural management, a hectare of oil palm yields 5 to 7 tonnes of oil per year (Henson, 1991).

The world production of major oils and fats in 1999 was 1.09 billion tonnes. Soybean oil accounted for 24.7 million tonnes (22.6 percent) followed by palm oil, 20.4 million tonnes (18.0 percent); rapeseed oil, 12.9 million tonnes (11.8 percent); sunflower seed oil, 9.3 million tonnes (8.5 percent); tallow, 8.1 million tonnes (7.4 percent); lard, 6.6. million tonnes (6.0 percent); and butter, 5.8 million tonnes (5.3 percent). World exports of major oils and fats amounted to 34.0 million tonnes in 1999. Palm oil was the major oil traded with 13.6 million tonnes (40.2 percent) followed by soybean with 7.5 million tonnes (22.3 percent), sunflower seed with 2.9 million tonnes (8.7 percent), and tallow with 2.2 million tonnes (6.7 percent) (Figure 8.1).

Malaysia, the world's largest producer and exporter of palm oil and its products, produced 10.5 million tonnes (51.5 percent) followed by Indonesia, which contributed 6.2 million tonnes (30.5 percent) of world palm oil production (Figure 8.1). Other countries, such as Nigeria, Colombia, Thai-

335

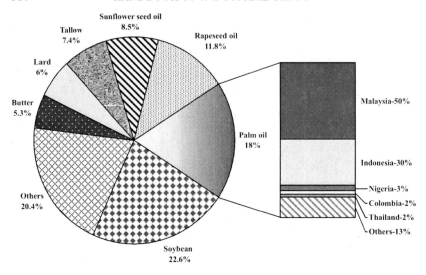

FIGURE 8.1. World production of major oil and fats in 1999 and major producers of palm oil. (*Source:* Based on data from Oil World, <www.oilworld.biz>.)

land, Ivory Coast, Papua New Guinea, Equador, and Costa Rica, produced about 3.5 million tonnes of palm oil (Table 8.1). In terms of palm oil export, Malaysia contributed 8.8 million tonnes (65.1 percent), Indonesia 3.1 million tonnes (24.2 percent) and others (10.7 percent).

The number of countries importing palm oil has increased from 67 in 1970 to more than 120 in 1999. At least 24 major palm products are available for export from Malaysia. The major importers of Malaysian palm oil are listed in Table 8.2. The main destinations of palm oil exports from Indonesia are Netherlands (44 percent), Germany (12 percent), Italy (9 percent), Spain (5 percent), and others (Kenya, United States, Greece, and United Kingdom) (8 percent) (Wakker, 2000).

Palm oil and palm kernel oil can be used for both edible (90 percent) and nonedible (10 percent) purposes. Food uses include cooking, frying, salad dressing, making margarines, shortenings, vanaspati, ice cream, confectioneries, and other emulsion-based products. For nonedible purposes, palm oil and palm kernel oil can be used directly for making soaps or indirectly as oleochemicals (fatty alcohols, fatty amines, fatty amides, fatty nitrogen, and glycerols). Carotenoids, vitamin E (tocopherols and tocotrienols), and sterols are minor components of crude palm oil that can be pretreated and encapsulated for pharmaceutical applications (Choo and Yusof, 1996). Oil palm biomass can be processed to produce wood-based

TABLE 8.1. Major world producers of palm oil: 1994 to 2003 (in thousand tonnes).

Country	1994	1995	1996	1997	1998	1999	2000	2001	2002	2003
Malaysia	7,403	7,221	8,386	9,069	8,319	10,554	10,842	11,804	11,909	13,354
Indonesia	3,421	4,008	4,540	5,380	5,361	6,250	7,050	8,030	9,200	9,750
Nigeria	645	640	670	680	690	720	740	770	775	785
Colombia	323	353	410	441	424	501	524	548	528	543
Cote D'Ivorie	310	300	280	259	269	264	278	220	240	251
Thailand	297	316	375	390	475	560	525	620	600	630
Papua New Guinea	223	225	272	275	210	264	336	329	376	325
Ecuador	162	178	188	203	200	263	222	201	217	247
Costa Rica	84	90	109	119	105	122	138	138	140	144
Honduras	80	76	76	77	92	90	97	108	110	112
Brazil	54	71	80	80	89	92	108	110	118	132
Venezuela	21	34	45	54	44	60	73	80	80	79
Guatemala	16	22	36	50	47	53	65	70	81	91
Others	1,265	1,676	815	869	844	832	879	919	922	940
Total	14,304	15,210	16,282	17,946	17,169	20,625	21,877	23,947	25,236	27,383

Source: Based on data from Malaysian Paklm Oil Board,<www.mpob.gov.my>.

products (Mohamad, 2000). The major ones are pulp and paper, particleboard of various kinds, and medium-density fiberboard. Recently, interest has been shown in oil palm fiber-plastic board as thermoplastic sheets and thermosetting boards due to their suitability for manufacturing various car components.

This chapter focuses on research and development that will ensure the sustainability of oil palm as a plantation crop. Emphasis is given to breeding and selection, and on the prospect of biotechnological intervention toward production of premium products such as oleate and stearate oils for the oleochemical industry. The need for managing the oil palm in adjusting to the changing environment is also highlighted.

ORIGIN, HISTORY, AND DISTRIBUTION

The center of origin of the oil palm is the tropical rain forest region of West and Central Africa (Zeven, 1964). The main oil palm belt of Africa runs through the southern latitudes of Sierra Leone, Liberia, Ivory Coast,

TABLE 8.2. Export of palm oil to major destinations (in thousand tonnes).

Countries	1990	1997
Pakistan	702	1,132
China	737	1,065
European Union	553	729
Australia	58	95
Bangladesh	25	141
Egypt	346	333
India	494	963
Indonesia	—	100
Japan	274	358
Jordan	45	230
Korea	215	182
Myanmar	—	158
Singapore	742	308
South Africa	—	183
Saudi Arabia	86	134
Taiwan	—	72
Turkey	191	237
United Arab Emirates	—	106
United States	143	113
Mexico	—	25
Others	1,109	813
Total	5,720	7,477

Source: Oil World Annual, 2000.

Ghana, Togo, Nigeria, and Cameroon and into the equatorial region of the Congo and Angola between 10°N and 10°S (Zeven, 1967). Tribal migration or intergroup exchanges spread oil palm across Africa (Smith et al., 1992). Oil palm was taken to Congo and East Africa before the arrival of Europeans. It was introduced to Sudan about 5,000 years ago (Clark, 1976) (Figure 8.2). The sporadic occurrence of oil palm on the East African coast (Uganda, Kenya, Tanzania, Rwanda, and Burundi) was probably due to Arab slave traders. The Africans probably brought oil palm to Madagascar in the tenth century (Pursglove, 1972).

Slave traders brought oil palm to the New World where it was cultivated in Bahia, Brazil in the fifteenth century. *Elaeis guineensis* was grown in European conservatories until 1973. It was in Calcutta in 1836 and must have reached Mauritius at an earlier date. In 1848, the Dutch imported oil palm

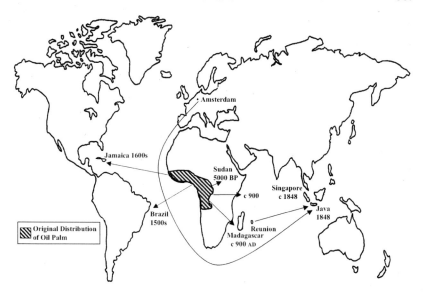

FIGURE 8.2. Historical spread of oil palm.

seeds from West Africa. Four oil palm seedlings were planted at Buitenzorg Botanical Garden (now Bogor), Indonesia, two of which came from Amsterdam Botanic Garden and two from Bourbon Island (now Reunion) (Pursglove, 1972). This introduction laid the foundation for the oil palm industry in southern Asia, especially in Malaysia and Indonesia. The Singapore Botanic Garden obtained oil palm seeds from Java around 1870 (Smith et al., 1992). Later seeds were distributed to Malayasia where they were grown as ornamental trees. The seeds were also sent to Sumatra, which later also received seeds from Buitenzorg. Palms from these sources were planted at Deli, Sumatra, and evolved into the Deli dura populations. These are the populations currently being utilized as part of genetic foundation for oil palm improvement in Malaysia and Indonesia.

TAXONOMY AND BOTANY OF OIL PALM

The genus *Elaeis* belongs to the palm family, Palmae, an important member of the monocotyledonous group in the order Spadiciflorae. It is included in the Cocoineae tribe together with the genus *Cocos* (Uhl and

Dransfield, 1987; Latiff, 2000). *Elaeis* is derived from the Greek word *elaion* which means oil; *guineensis* points to the oil palm origin: the Guinea coast (Hartley, 1988). Within the genus *Elaeis,* two species are distinguished: the economically important oil palm *Elaeis guineensis* and the oil palm of American origin, *Elaeis oleifera.* These species hybridize readily, suggesting a close relationship in spite of their origin in two different continents (Hardon and Tan, 1969).

The oil palm has a crown of 35 to 60 pinnate fronds arranged on a vascular stem. It has a single bud in the base of the crown where fronds and inflorescences originate. The palm may reach a height of between 15 to 30 meters and can live up to 300 years. The palm is monoecious with male or female inflorescences occurring separately. This enforces cross-pollination. Wind and insects assist pollen dispersal (Pursglove, 1972; Hartley 1988). The main pollinating insect is the *Elaeidobius* species (Syed, 1980).

On an average, the fronds are produced at the rate of two per month in a regular sequence. The length of the frond is typically about 7 meters, and each frond consists of a petiole, which is 150 cm long, and a rachis, which bears 250 to 350 leaflets. Each leaflet may be up to 130 cm long. The leaflets are arranged on two lateral planes. An inflorescence primordium forms in each frond axil. Male and female flowers are produced at different times in separate inflorescences. Beirnaert (1953) showed that the oil palm flower primordium has both male and female organs. In a potentially female flower primordium, the two accompanying male flowers are suppressed and remain rudimentary. In a potentially male flower primordium, the female organ is suppressed. Rarely, both the gynoecium (female) and androecium (male) may develop to give a hermaphrodite flower. The process of sexual differentiation occurs at 28 months before anthesis (Wood 1984; Hartley 1988).

The male inflorescence has an approximately 40 cm long stalk, with 100 to 300 fingerlike spikelets. Each spikelet hs about 600 to 1,500 yellow flowers. Pollen production ranges from 10 to 30 grams per inflorescence. The basic structure of the female inflorescence is similar to that of the male but the spikelets are shorter. Each spikelet bears 5 to 30 flowers that are receptive for two to three days. Fertilized female flowers produce fruits that grow and ripen over about six months. Oil palm fruits are sessile drupes borne on a large compact bunch. Each fruit consists of a hard kernel inside a shell (endocarp) that is surrounded by a fleshy mesocarp. The mesocarp contains about 49 percent palm oil and the kernel about 50 percent palm kernel oil. A mature bunch contains a few hundred to a few thousand fruits and its weight ranges from 5 to 50 kg depending on the age of the palm and environmental and genetic factors.

The external appearance of the oil palm fruit varies considerably. The most common type, known as *nigrescens,* is deep violet/blackish in color before ripening. An uncommon type is green and is called *virescens.* The virescens is dominant over nigrescens. Another type is *albescens,* which is characterized by the small amount of carotene in the mesocarp and has a low yield. Albescens fruit can be of the nigrescens or virescens type and is referred to as albo-nigrescens and albo-virescens respectively (Pursglove, 1972; Hartley, 1988).

In internal structure, the most important difference is thickness of the shell. A single gene controls the shell thickness of the oil palm (Beirnaert and Vanderweyen, 1941). The recessive homozygote, the pisifera (sh⁻, sh⁻), is shell-less, whereas the dominant homozygote, the dura (sh⁺, sh⁺), has a thick shell. When the dura crosses with the pisifera, a heterozygote, the tenera (sh⁺, sh⁻), which has a thin shell surrounded by a ring of fibers in the mesocarp, is produced. The most common cultivated fruit form is the high-yielding tenera.

GROWING CONDITIONS

Oil palm is grown mainly in tropical countries, such as those found in Southeast Asia, Central and Western Africa, and South America (Hartley 1988). Isolated groves can be found from as far north as 16°N in Senegal to 13°S in Malawi, and 20°S in Madagascar (Piggott, 1990). However, it is grown commercially in more than 20 countries with most areas within 10° north and 10° south of the equator (Goh, 2000). In this respect, Malaysia and Indonesia are the world's main producers and exporters of palm oil. In general, oil palm can be grown in tropical soils with good irrigation.

Climate

The suitable climate for growing oil palm is high rainfall throughout the year, i.e., between 2,000-2,500 mm per year; at least 150 mm of rain each month and no distinct drought season or months with less than 100 mm rain (Hartley, 1988; Piggott, 1990; Goh, 2000). A water deficit greater than 300-400 mm per year will significantly reduce fresh fruit bunch (FFB) yield (Mohd Haniff, 2000).

Optimum temperatures range between 22-33°C, with the lowest temperature supporting oil palm close to 20°C (Hartley, 1988; Goh, 2000). However, the growth rate of young seedlings is inhibited at temperatures of 15°C or lower, but the maximum temperature of 32°C did not affect palm growth

(Goh, 2000). The daily requirement of sunlight is between five to seven daylight hours and at least 2,000 hrs of sunshine a year (Hartley, 1988; Piggott, 1990). In Malaysia, the high rate of annual growth is the result of high levels of annual light interception (Mohd Haniff, 2000). Relative humidity should be between 75 to 100 percent (Piggot, 1990).

Soil

The soil type may not be very critical, provided it does not have an unusually high proportion of clay, which inflicts waterlogging during rains and cracking when dry (Piggott, 1990). The suitable soil texture is sandy loam of more than 75 cm depth. Lateritic, sandy, or peat soils are problematic soils that need proper manuring and maintenance for optimal palm growth (Hartley, 1988). Ideally, oil palm should be grown in flat areas. The elevation and slope of an area proposed for oil palm cultivation are important factors that determine its suitability. In general, oil palms are not recommended for planting in areas with an elevation of more than 200 meters (Paramananthan, 2000). In undulating areas, cost for maintenance, harvesting, and transportation would be higher. The planting density varies according to soil types. For inland soils (sedentary), planting is done in a triangular form, with a distance of 8.8 meters, giving 148 palms per hectare. However, with coastal alluvial soils, the recommended density is 136 palms per hectare, and for peat soils 160 palms per hectare (Hartley, 1988).

Nutrients

For palms planted in inland soils, between one quarter to one half of the nutrient requirements of nitrogen (N), phosphorous (P), and potassium (K) have to be supplied as fertilizers, but in fertile alluvial soils, the nutrients can be obtained directly from the soil (Tarmizi, 2000). Malaysian soils need between 0.5-1.1 kg/palm per year of N, 1.1 kg/palm per year of P_2O_5, and between 0.5-2.0 kg/palm per year of K_2O to replenish the soil nutrient supply after considering the expected losses of applied nutrients (Tarmizi, 2000). The nutrients to be applied to the oil palm are essentially for balancing soil immobilization and leaching losses against the recycling of nutrients from pruned fronds and palm oil mill effluents. Responses of high K fertilizer application on inland soils showed that the ratio of oil to bunch is decreased. N fertilizer increased the bunch number, bunch weight, and total oil produced, whereas P fertilizer increased bunch weight but not the other bunch parameters.

GERMPLASM COLLECTION

In the past, a number of expeditions were mounted to collect oil palm genetic materials from its center of origin. Some oil palm workers in the Belgian Congo started oil palm collection at a few sites after World War II (Vanderweyen 1952; Pichel 1956). In the early 1960s, Nigerian oil palm breeders collected 72 open-pollinated progenies from the eastern part of Nigeria. These materials were planted and evaluated at the National Institute for Oil Palm Research (NIFOR) main station. Some outstanding palms were selected and introduced into their current breeding program (Okwuagwu, 1985).

In Cameroon, Blaak (1967) sampled oil palm materials from the Bamenda highlands. Some of the palms were planted at Lobe, Cameroon, while the rest were distributed elsewhere. Between 1974 and 1975, the Institute de Researches pour les Huiles Oleagineux (IRHO) prospected *Elaeis guineensis* materials from the western region of Cameroon. The French oil palm workers systematically evaluated natural stands of oil palms in Ivory Coast, and selected palms were progeny tested and utilized as new foundation materials in their breeding scheme (Meunier, 1969; Meunier and Baudouin, 1985). It was reported that IRHO also selected four palms in Malaysia and another 38 palms in Benin and used them as their original tenera stock.

Recent Oil Palm Germplasm Collection

Initially, the oil palm breeding program utilized four Deli dura palms as parents. As such, the genetic base of oil palm breeding in the Far East was extremely narrow. A number of systematic expeditions to collect oil palm genetic materials were mounted by researchers in Malaysia. The objective of the exploration was not only to broaden the genetic base of the current oil palm breeding materials but also to ensure conservation of oil palm genetic resources for posterity. The first attempt was made in Nigeria (Rajanaidu, 1985), and it was followed by collections from other countries in Africa (Figure 8.3), namely Cameroon (1983), Democratic Republic of the Congo (DRC) (1984), Tanzania (1986), Madagascar (1986), Angola (1991), Senegal (1993), Gambia (1993), Sierra Leone (1994), Guinea (1994) (Rajanaidu, 1994), and Ghana (1996) (N. Rajanaidu, personal communication).

During the prospection in each country, the sites and palms were chosen at random. One bunch was harvested from each of the sampled palms and the fruits from each bunch were kept separate. The mean coefficient of variation (cv) of the traits scored in each country was computed (Rajanaidu,

FIGURE 8.3. Map of Africa indicating countries from which oil palm materials were collected.

1985). Following are the details of oil palm germplasm collections from several countries mounted by the Malaysia Palm Oil Board (MPOB).

Nigeria

Genetic materials were collected from 45 sites distributed throughout the oil palm growing areas in Nigeria in 1973 (Rajanaidu, 1985; Rajanaidu and Rao, 1987; Rajanaidu, 1994). Five to ten palms were sampled from each site depending on the rainfall pattern, soil type, and density of oil palm groves. The total number of palms sampled during the prospection was 919, which consisted of 595 duras and 324 teneras. The pisifera was virtually absent in all populations (Rajanaidu, 1985).

Cameroon

The prospection was carried out with the cooperation of Pamol, a subsidiary of Unilever, in 1984. Samples were collected from 32 sites. One to 15 palms were sampled at each site. A total of 95 (58 duras and 37 teneras) palms were taken with an average of three palms per site (Rajanaidu, 1985). About 19,000 seeds were brought to Malaysia (Rajanaidu, 1985).

Democratic Republic of the Congo

The collection of oil palm genetic materials in DRC was made at 56 sites with the cooperation of Plantation Lever Zaire (PLZ) between April and July 1984 (Rajanaidu, 1985). At most sites, five to ten palms were obtained. The number of bunches collected was 369, consisting of 283 duras and 85 teneras. A total of 73,800 seeds were dispatched to Malaysia after the intermediate quarantine (Rajanaidu and Jalani, 1994c).

Tanzania

The collection of oil palm material in Tanzania was carried out in 1986 with the cooperation of Ministry of Agriculture, Tanzania, and financially supported by the International Board for Plant Genetic Resources (IBPGR). Half of the samples collected were deposited with the Ministry of Agriculture, Kigoma, Tanzania. The materials were sampled from 13 sites near Kigoma, along Lake Tanganyika. At each site, one to seven palms were chosen randomly, and a total of 60 samples (42 duras and 18 teneras) were collected (Rajanaidu, 1986a).

Madagascar

Since the distribution of oil palm in Madagascar was very sparse, only 17 palms were collected from four sites. One to six palms were sampled from each site. This expedition was carried out in 1986 with the collaboration of the Ministry of Agriculture in Madagascar and sponsored by the International Board for Plant Genetic Resources (IBPGR) (Rajanaidu, 1986a). The palms observed were very poor in growth as compared to those found in Nigeria, Cameroon, DRC, and Tanzania (Rajanaidu and Jalani, 1994c).

Angola

Oil palm genetic materials in Angola were taken from eight sites in 1991. Only the coastal areas were covered. At each site, two to 14 samples were collected. A total of 54 bunches (42 duras and 12 teneras) were collected (Rajanaidu et al., 1991).

Senegal

Collection of genetic materials in Senegal was carried out in the months of July and August 1993 with the cooperation of Ministry of Agriculture, Senegal. Palms were sampled from 13 sites. Five to ten palms were selected from each site and a total of 104 accessions (all duras) were collected (Rajanaidu and Jalani, 1994b).

Gambia

The oil palm germplasm collection in Gambia was obtained with the collaboration of Gambia's Ministry of Agriculture and Forestry. A total of 45 palms were sampled from six sites. At each site, five to ten palms were selected. Only dura palms were encountered in this country (Rajanaidu and Jalani, 1994a).

Sierra Leone

Oil palm germplasm collection in Sierra Leone was carried out in April and May 1994 with the cooperation of Sierra Leone's Ministry of Agriculture. A total of 56 samples (52 duras, three teneras, and one pisifera) were obtained from 14 sites in the country. At each site, two to six palms were sampled. In terms of fruit color, 54 were nigrescens, one virescens, and one albescens. The seeds were divided equally between the ministry and MPOB (Rajanaidu and Jalani, 1994c).

Guinea

In May 1994, oil palm genetic materials were collected from 14 sites in Guinea. This prospection was mounted jointly by MPOB and the ministry of Agriculture, Guinea. At each site, three to five palms were sampled. A total of 61 samples (58 duras, 3 teneras) were collected. All the fruits are nigrescens (Rajanaidu and Jalani, 1994c).

A general comparison was made between the populations collected in the countries mentioned earlier. Data on bunch weight, single fruit weight, and mesocarp to fruit for duras and teneras are summarized in Table 8.3. The mean bunch and fruit weight recorded for the Cameroonian populations were higher than that observed for the Nigerian populations. Materials from Tanzania, which were sampled from a fringe population, showed bunch and fruit qualities comparable to samples taken from Nigeria, Cameroon, DRC, and Angola. A trend toward increase for these characters from

TABLE 8.3. A summary of bunch characters of oil palm genetic materials collected from Africa.

Country	Dura			Tenera		
	Bunch weight (kg)	Single fruit weight (g)	Mesocarp to fruit (%)	Bunch weight (kg)	Single fruit weight (g)	Mesocarp to fruit (%)
Nigeria	11.8	7.9	47.3	10.9	8.5	70.9
Cameroon	16.8	10.3	39.7	17.3	8.6	62.4
DRC	17.6	14.2	43.9	17.4	12.6	64.1
Tanzania	18.4	16.9	46.7	13.7	15.5	70.8
Angola	21.4	14.2	48.9	16.0	11.7	70.9
Senegal	5.9	2.6	35.1	—	—	—
Gambia	5.7	2.3	33.4	—	—	—
Sierra Leone	21.4	14.2	48.9	—	—	—
Guinea	11.4	6.4	35.0	—	—	—

Source: Rajanaidu et al., 2000.

347

Nigeria to Tanzania was observed. For dura palms collected from Angola, the percent mesocarp to fruit and bunch weight was higher compared to Nigeria, Cameroon, and DRC. The mean fruit weight was similar to DRC. As for the teneras, the percent of mesocarp to fruit recorded for Angola was similar to Nigerians and Tanzanians and higher than that obtained from Cameroon and DRC. The dura bunch weight and mesocarp to fruit of the Gambian populations were the lowest compared to other materials. The values of these characters increased from Senegal to Angola, whereas the mean fruit weight increased from Senegal to Tanzania. For Gambia, the mean fruit bunch and mesocarp to fruit were similar to that observed for Senegal (Rajanaidu et al., 2000).

Elaeis oleifera

Elaeis oleifera genetic materials were collected from six countries, namely Colombia, Panama, Costa Rica, Honduras, Brazil, and Suriname in 1981-1982 (Rajanaidu, 1986b). This American species is attractive to the oil palm breeders because it possesses a number of desirable traits such as slow height increment, high iodine value, and resistance to diseases such as *Fusarium* wilt.

GERMPLASM CHARACTERIZATION

Morphological Variation

The oil palm germplasms collected were planted and maintained in the form of open-pollinated families at Kluang MPOB Research Station, Johore, Malaysia. The palms were planted in several experimental designs, such as cubic lattice, randomized complete block design (RCBD), and completely randomized design (CRD), to study their phenotypic characters. Data on yield, oil and kernel content, fatty acid composition of oil, physiological parameters, and flower characters were recorded and analyzed. The palm oil was also screened for iodine value and carotene content.

The performance of Nigerian germplasm collection was examined and a total of five populations (12, 13, 14, 16, and 19) were identified. These populations were from the east-central part of Nigeria. The traits of interest found in these populations are as follows (Rajanaidu, 1994):

1. High yield and dwarfness: Some palms (teneras) gave a yield of more than 10 tonnes per hectare per year. In addition, they were short with

annual height increment of 15 to 25 cm compared to 45 to 75 cm of the commercial planting materials. Palms that possessed these traits were sampled from populations 12, 13, and 14 (Table 8.4).

2. High iodine value: Positive correlation exists between iodine value (IV) and the level of unsaturation of palm oil. Palms having high iodine value will have highly unsaturated oil and subsequently produce more liquid oil. A number of Nigerian palms had IV more than 60, which is higher than that observed in the DxP commercial (I.V. values of 52-53). The prospects of marketing such palm oil in temperate countries are bright.

3. High kernel content: A few Nigerian families had mean kernel-to-bunch ratios of more than 15, which is higher than that of current breeding populations (four to eight) (Rajanaidu and Jalani, 1994d). It was shown that by maximizing the level of kernel per bunch in oil palm, it is possible to get better economic returns.

TABLE 8.4. Performance of dwarf Nigerian duras and teneras.

Trial Number	Family	Palm Number	Oil/palm/ year (kg)	Oil/ hectare/ year (tonnes)	Height increment (cm/year)
Duras					
0.149	14.07	2703	38.4	5.6	21.0
0.149	14.07	2705	40.2	5.9	17.5
0.149	12.01	14483	43.9	6.5	19.1
0.149	12.01	14676	40.1	5.9	24.0
Teneras					
0.149	28.17	12724	83.8	12.1	23.1
0.149	19.11	12279	75.9	11.2	21.5
0.149	13.05	12094	76.2	11.2	24.0
0.150	16.21	4352	70.3	10.4	24.9
0.150	19.13	3759	71.5	10.5	22.5
0.151	14.03	128	59.0	8.7	25.7
0.149	12.01	2577	62.4	9.2	19.0
0.149	12.01	1704	62.7	9.2	17.5
0.149	12.01	11525	64.3	9.5	14.0
Current planting material				5.0	45-75

Source: Rajanaidu et al., 2000.

Yield in the Cameroon population ranges from 12 to 116 kg per palm per year, bunch number ranges from 2.5 to 32, and average bunch weight ranges from 2.2 to 11.6 kg. High-yielding palms were from populations 22 and 28 (Rajanaidu et al., 2000).

The yield recorded for populations collected from DRC was slightly better than that of Cameroon but lower than the Nigerian populations. Analysis of yield parameters showed that the populations had yielded 11.5 to 183.4 kg per palm per year, bunch number 2.3 to 28.8, and average bunch weight 2.8 to 15.4 kg. Populations 30 and 31 had higher yields and bunch numbers (Rajanaidu et al., 2000).

Isozyme Variation

Eighteen populations of *Elaeis guineensis* germplasm from nine countries in Africa: Angola, Cameroon, Guinea, Madagascar, Nigeria, Senegal, Sierra Leone, Tanzania, and DRC and one Deli dura population were studied using seven enzyme systems. On an average, 75 percent of the loci were polymorphic. The mean expected heterozygosity was 0.177. The genetic differentiation between populations was high ($F_{ST} = 0.384$) indicating that only 62 percent of the isozyme variation was among progenies within populations. The dendrogram constructed with obtained data showed that the Deli dura population was closely related to Sierra Leone populations (Hayati et al., 2000).

Molecular Variation

The extent of genetic variability between and within oil palm populations was estimated using restriction fragment length polymorphism (RFLP) analysis. Oil palm germplasm materials collected from Nigeria, Cameroon, DRC, Tanzania, Madagascar, Angola, Senegal, Sierra Leone, Gambia, and Guinea were screened (Maizura, 1999). An advanced breeding population, Deli dura, was included as a reference. DNA extracted from each sample was digested with five restriction enzymes and hybridized with four oil palm cDNA probes. A total of 111 bands were observed and 58 loci were identified.

Based on percentage of polymorphism, a considerable difference was found between population 12 (high-yielding dwarfs) and population 39 (low-yielding talls) of the Nigerian origin. Correlation analysis showed that a high-degree relationship exists between level of polymorphism and the mean oil yield (r = 0.77). This pointed to the possibility of using this parameter to select genetically divergent populations. Among the collections, the

Nigerian populations exhibited the highest value for all genetic variability parameters with the most number of rare alleles, suggesting that Nigeria may likely be the center of diversity for oil palm. Compared to Deli dura, six oil palm germplasm collections exhibited higher level of polymorphism. These collections are from Nigeria, Cameroon, DRC, Tanzania, Angola, and Senegal. These collections should be given priority for characterization and utilization. Oil palm populations from Tanzania showed the closest genetic relationship to Deli dura, indicating that the breeding population could have originated from there.

A total of 687 accessions belonging to 11 African countries were screened with 8 primer combinations of amplified fragment length polymorphisms (AFLPs). Nigerian materials recorded the highest number of polymorphic bands. The estimated mean genetic diversity showed lower genetic variability in Deli dura; the populations from Cameroon had the highest variability. Overall results supported Nigeria, more specifically the east-central part as the origin of oil palm. A considerable amount of mixing of natural population was noticed in the study. This was probably due to human disturbance (Kularatne et al., 2000).

Characterization of Elaeis oleifera *Germplasm Collection*

The same parameters were measured to characterize *Elaeis oleifera* germplasm materials. Significant differences existed between the six countries (Panama, Costa Rica, Colombia, Honduras, Brazil, and Suriname) for yield ratio of oil to bunch, height, total economic product, and iodine value. In general, populations from Suriname showed the lowest value for all traits except ratio of oil to bunch (Mohd. Din and Rajanaidu, 2000).

Hardon (1969) pointed out that *Elaeis oleifera* has more unsaturated oil than *E. guineensis*. Fatty acid composition of the two species has been studied (Rajanaidu et al., 1984). It was found that the oil characteristic of *E. oleifera* was quite close to olive oil. The collections from Colombia, Panama, and Costa Rica had iodine values of more than 90. The C18:1 level ranged from 52 percent to 66 percent and the C18:2 level varied between 15 percent to 23 percent. The iodine value for the populations from Brazil ranged from 76 to 81 and the level of C18:1 was lower than in accessions from Colombia, Panama, Costa Rica, and Honduras. The fatty acid composition of the Suriname genetic materials was rather unique. It has the highest level of C18:1 and the lowest of C18:2 when compared to materials from other countries (Rajanaidu et al., 1994).

Utilization of Oil Palm Germplasm

Oil palm germplasm is being utilized for oil palm improvement in several ways (Rajanaidu et al, 2000):

1. *Direct selection of individuals:* Individuals which possessed interesting traits and acceptable yield were selected and introgressed into another advanced material. In the Nigerian populations, about 3 percent of the teneras had oil yields comparable to that of the current planting materials. One third of these had annual height increments significantly less than the commercials. These elite palms were cloned by tissue culture techniques.
2. *Progeny testing:* A few of the outstanding Nigerian teneras were progeny tested with a range of Deli duras available in the Malaysian industry and MPOB. The progenies and parents were selfed and would be used for seed production following the reciprocal recurrent selection procedure (Jacquemard et al., 1981).
3. *Broadening the genetic base of Deli duras and teneras:* Crossing the Deli duras with Nigerian duras could broaden the overall genetic variability of current Deli duras. Such crosses would provide the basis for further selection and breeding. The selected Nigerian teneras could also be introgressed into the current *tenera* breeding populations.
4. *Foundation breeding program:* The outstanding dura and pisifera palms from the Nigerian collection are being used to initiate an entirely new breeding program aimed at producing superior alternatives to the current Deli duras and modern pisiferas.
5. *Development of elite planting materials:* Using the selected Nigerian palms, MPOB developed several commercial oil palm planting materials. These materials possessed high yield, dwarfness, high iodine value, and high kernel content.

GENETIC IMPROVEMENT THROUGH BREEDING AND SELECTION

The main objectives of oil palm improvement are:

- To increase oil yield per hectare,
- To produce better oil quality,
- To reduce annual height increment, and
- To develop palms resistant to pests and diseases.

In addition, oil palm improvement also includes development of planting materials with other traits, such as high and early yield, high kernel yield, high carotene and vitamin E levels, and exploitation of genotype × environment interaction.

Breeding for Dwarf Palms (PS1)

The main emphasis of oil palm improvement is toward higher oil yield. Significant progress in this regard has been reported (Table 8.5). From 1962 to 1988, the oil yield increased from an average of 5.0 to 9.6 tonnes per hectare per year, representing a boost of 93.2 percent and a yearly increase of 3.6 percent or 0.2 tonnes per hectare per year (Lee and Toh, 1991; Lee 1996; Mukesh and Tan, 1996). However, the increase was based on the Deli dura that originated from the four Bogor palms planted in 1848. They have narrow genetic base and the additive variation left on the Deli dura after generations of selection is low (Thomas et al., 1969). Crossing with other dura populations, such as those from the Nigerian populations, would increase genetic variability.

Current oil palm planting materials grow at a rate of 40 to 75 cm/year. After 20 years, the palm becomes too tall and harvesting is difficult. Efforts to breed for short palms were initiated. Sources of dwarf genes are Dumpy E206 palm (selected in 1920-1921 at Elmina Estate, Kuala Selangor, Malaysia), Dumpy-AVROS pisifera (created by HRU and Golden Hope), *Elaeis oleifera,* and, recently, palms from the Nigerian germplasm collection. The availability of these sources led to the development of PS1 (high-

TABLE 8.5. Yield performance of oil palm planting materials.

Materials	Year planted	Number of progenies	Fresh fruit bunch (tonnes/ hectare/ year)	Oil to bunch (%)	Projected oil yield (tonnes/ hectare/ year)
DD × CI	1962	32	22.0	22.2	4.9
DD × UAC	1962	15	24.6	20.6	5.1
DD × AVROS	1964	22	31.0	23.5	7.3
DD × AVROS	1968	16	31.1	22.1	6.9
DD × AVROS	1970	29	31.6	24.2	7.6
DD × AVROS	1979	5	34.5	25.8	8.9
DD × Yangambi	1988	66	34.9	25.9	9.6

Sources: Lee and Toh, 1991; Lee, 1996; Mukesh and Tan, 1996.

yield dwarf) planting material. The dwarf mutant E206 selected by Jagoe (1952) had low yield compared to other materials, making it unattractive for exploitation. Planting materials produced from Deli dura × Dumpy-AVROS are 20 percent shorter with yields comparable to the current planting materials (Kushairi et al., 1999).

Elaeis oleifera was crossed to *Elaeis guineensis* to produce short-stem hybrids. However, due to poor fruit set (thick shell and thin mesocarp) the oil yield of these hybrids was disappointing. Selected teneras from the Nigerian germplasm collection were progeny tested with the industry Deli duras. At the same time, these parents were selfed and based on the progeny test results, the respective selfs could be used for commercial seed production. Progeny test results of these crosses showed an increase of oil-to-bunch yield and mean yield from 25 percent to 32.9 percent and 180 kg to 245 kg, respectively. In general, the progenies are 15 percent to 20 percent shorter than the current planting materials (Rajanaidu et al., 2000).

The Nigerian duras were introgressed into Deli dura populations to broaden the genetic base of the latter. These intercrosses are shorter and have oil yields comparable to or higher than the commercial teneras. Further exploitation of the Nigerian materials involved population 12. This population is unique: high yielding, short, and compact. Various families of population 12 were progeny tested as D × T to create a new foundation stock. Some of the crosses are high yielding and shorter compared to the control (commercial D × P). T × T crosses within population 12 were also made to develop dwarf pisiferas (Rajanaidu et al., 2000).

Breeding for High Unsaturation Oil (PS2)

An increase in unsaturation of oil is desirable due to consumer demands for monounsaturated and polyunsaturated dietary oils and fats. A higher unsaturation level in palm oil enables penetration into the liquid oil market. Currently, palm oil is semisolid at room temperature (28°C) and can be fractionated into 70 percent liquid olein and 30 percent solid stearin. The palm oil quality can be improved by reducing the stearin and increasing the olein levels.

Three families of the Nigerian duras were chosen for high yield and high iodine value (unsaturation). The palms were selfed, and early observations showed that Iodine Value (IV) is transmitted from parents to the progenies. These palms were progeny tested with industry pisiferas. The progenies had low yield and oil-to-bunch yield but high IV (57-59) compared to the control (53). These results were based on crosses in which one parent had a higher iodine value (more than 60), which can be expected if both parents were selected for the trait. Along this line, the high IV duras are being progeny tested with the high IV Nigerian tenera. Selected palm of the same collection was selfed to generate pisifera for the next generation of high IV planting materials known as PS2 (Kushairi et al., 1999).

The iodine value for the *Elaeis oleifera* germplasm collections is about 80, much higher than that of the African oil palm. The interspecific hybrids revealed iodine values of 65. However, the yield of the hybrids was too low for commercial exploitation (Rajanaidu et al., 2000).

Breeding for High Lauric Oil (PS3)

Coconut and oil palm are the major traditional sources of lauric oils. An economic analysis showed that breeding for higher levels of kernel content realizes the highest return from oil palm. An increase of only 5 percent to 10 percent of kernel to bunch would bring about a net profit of U.S. $300 from every 100 tonnes of fresh fruit bunch (Rajanaidu et al., 1996). Cultivation of high kernel planting materials (PS3) could be a gainful venture.

The current planting materials have 5 to 10 percent kernel to bunch. Populations taken from the northern part of Nigeria have higher levels of kernel to bunch (more than 10 percent). Selected Nigerian duras are being progeny tested with AVROS pisifera. Nigerian teneras and pisiferas and Serdang pisiferas are other sources of high kernel genes that could be exploited for production of PS3.

Breeding for High Carotene, Vitamin E, and Sterol

The carotene content for the Nigerian palms ranged from 273 to 3,512 ppm, whereas the current Deli duras had 500 ppm. A special breeding program was developed to exploit the Nigerian population with high carotene content. *Elaeis oleifera* has low oil yield (less than 0.5 tonnes per hectare per year), but it has much higher carotene (4,300 to 4,600 ppm), vitamin E (700 to 1,500 ppm), and sterol (3,500 to 4,000 ppm) contents compared to *Elaeis guineensis* (500 to 700 ppm, 600 to 1,600 ppm, and 1,100 to 1,250 ppm respectively) (Jalani et al., 1997).

Breeding for Disease Resistance

Breeding for tolerance to vascular wilt caused by *Fusarium oxysporum* is an important program in West Africa. This disease does not occur in the Far East. It causes death of young and adult palms. Tolerant palms and screening techniques for identifying tolerance in a population are available

(Meunier et al., 1979). Basal stem rot caused by *Ganoderma boninense* is a serious problem for older palms planted in the Far East. A screening technique is being developed (Ariffin et al., 1995) to breed for disease resistance. The American oil palm *(Elaeis oleifera)* was utilized for developing palms tolerant to "sudden wilt" and "little bud rot," which affected oil palm plantation in Central and South America. A few hybrids exhibited tolerance to this disease (Meunier 1989).

SCOPE OF BIOTECHNOLOGICAL INTERVENTIONS

The fatty acid composition of vegetable oils determines their physical and chemical properties and their applications. The wide range of applications of vegetable oils, 90 percent edible and 10 percent nonedible, reflect diversity in their fatty acid composition (Stobart et al., 1993). Vegetable oils are renewable resources that can serve as feedstock to produce environment-friendly industrial products, such as lubricants, paints, detergents, and body care products. Demand for these oils in the nonfood sector is likely to increase to augment depleting mineral oils (Murphy, 1994; Topfer et al., 1995).

The oils of six major crops, which include soybean *(Glycine max),* oil palm *(Elaeis guineensis),* rapeseed *(Brassica* spp.), sunflower *(Helianthus annuus),* cottonseed *(Gossypium* spp.), and groundnut *(Arachis hypogaea),* contain limited numbers of fatty acids. These include unsaturated oleic (18:1) and linoleic (18:2) acids and generally lower levels of α-linolenic (18:3) and saturated stearic (18:0) and palmitic (16:0) acids (Stumpf, 1987). Hundreds of different fatty acids have been identified and characterized in the plant kingdom (Hilditch and Williams, 1964), but most are not available for economic uses. Genetic engineering provides the means to tap these vast resources by producing fatty acids of economic importance in storage lipids of oil crops (Murphy, 1994; Ohlrogge, 1994). Achievements through breeding alone or in combination with mutagenesis, such as in the development of rapeseed oil with low content of erucic acid (22:1) (Stefansson, 1983) and sunflower oil with high level of oleic acid (Garces and Mancha, 1990), indicate that plants can tolerate wide variations in fatty acid composition of storage lipids.

Extensive research efforts in altering fatty acid composition of storage oils in seed oil crops by genetic engineering during the 1990s have produced several significant achievements. These include modifications involving chain length (introduction of medium-chain fatty acids; Voelker et al., 1992), level of unsaturation, increased in saturated fatty acids (Knutzon et al., 1992) and positioning on the glycerol backbone of fatty acids (increase in erucic acid at the sn-2 position; Brough et al., 1996). In several cases, the analysis of fatty acids and other relevant studies carried out on the transgenic plants have provided new insights into the regulation of fatty acids in plants.

Oil Production in Oil Palm Fruits

Two different types of storage oils are obtained from the oil palm fruits: palm oil from the mesocarp and kernel oil from the kernel. The two oils are different in composition, physical properties, and usability (Hartley, 1988). The kernel oil is rich in medium-chain saturated fatty acid: 51 percent lauric acid (12:0) and 18 percent myristic acid (14:0), which serve as important feedstock for the oleochemical industry (Pantzaris and Yusof, 1990; de Man and de Man, 1994). Palm oil, which contains about 50 percent saturated fatty acids, 40 percent monounsaturated fatty acids, and 10 percent polyunsaturated fatty acids, is a semisolid fat at room temperature. Its fatty acid composition consists of 44 percent palmitic acid (16:0), 5 percent stearic acid (18:0), 39 percent oleic acid (18:1), and 10 percent linoleic acid (18:2). The main applications of palm oil are in the edible food industry field, mainly as solid fat for margarine, shortening, and cooking oil. Nonedible or technical applications, however, are substantial and enlarging day by day. These applications include soaps, oleochemical production, and automobile energy sources (Ngo, 1991).

Young fruits during the first ten weeks of development contain very little lipids, i.e., around 5 to 8 percent per fresh weight. Storage oil synthesis in the oil palm mesocarp can be detected as early as 12 weeks after anthesis. High rate of oil accumulation begins at 16 weeks and stops when the fruits ripen at about 20 weeks after anthesis (Oo et al., 1986). Oil is stored in oil bodies found in the cytoplasm of mesocarp cells of ripe fruits. Studies on mesocarp cell morphology by Aziz et al. (1990) showed that small-size oil bodies can already be observed at around 13 weeks after anthesis when oil synthesis begins. Oil accumulation in the kernel starts at around 12 weeks after anthesis and stops at 14 weeks. During this period, the kernel gradually solidifies (Hartley, 1988).

The fatty acid in the mesocarp of young fruits consists mainly of polyunsaturated linolenic acid (18:3) and linoleic (18:2), which are components of membrane lipids. When rapid oil accumulation begins, the level of linolenic drops to insignificant value while the level of linoleic also drops but is stably maintained at 10 percent in ripe fruits. The highest increase observed is in the level of oleic acid from 22 percent at 13 weeks to 39 percent in ripe fruits (Aziz et al., 1986).

Strategies for Modifying Palm Oil Composition

Over the years, various strategies to modify the oil composition of palm oil have been employed with the aim of producing more liquid oil. This is considered of great importance because it provides the means to diversify the use of palm oil. The objective is achievable by increasing the level of unsaturation or iodine value, thus changing the physical properties of palm oil (Rajanaidu et al., 1993).

Physical and chemical processes involving two levels of fractionation and transesterification showed promising results in changing oil composition but were not viable for commercialization. In addition, such approaches could produce triglycerides different from the ones obtained from natural processes (Wong et al., 1991).

The traditional breeding approach has been very successful in producing high-yielding planting materials after several years of research on selecting components relating to yield (Rajanaidu, 1987). Even though the main emphasis of current research is still on improving yield, combining yield with other traits of importance, such as adaptability to various soils and climatic conditions, resistance to diseases, slow vertical growth to assist harvesting, and high levels of fatty acid unsaturation, has been given attention (Yong and Chan 1996).

The South American species *Elaeis oleifera* produces mesocarp oil with an IV between 77 and 88, which is much higher than that of *E. guineensis,* which has an IV of about 53 (Sambanthamurthi et al., 1996). The interspecific hybridization of *E. oleifera* and *E. guineensis* was carried out by various groups to introgress the high oil unsaturation and slow vertical growth of *E. oleifera* into the high-yielding commercially grown *E. guineensis.* The F1 hybrids are backcrossed to the *E. guineensis* parent to improve the yield and vegetative characters. The genomic in situ hybridization (GISH) technique, which can differentiate the genomes of the two species, was developed for the oil palm (Madon et al., 1999). This molecular technique is a valuable aid to breeders for determining the genomic composition of progenies so that those having a high proportion of the *E. guineensis* parents can be selected. Another breeding strategy employed was to introgress the high IV characteristics of *E. guineensis* genetic material from Nigeria into commercial breeding material. Restriction fragment length polymorphism analysis detected high levels of genetic variability in the natural population within *E. guineensis* species, which can be further exploited to improve existing planting materials (Maizura et al., 1996). However, for oil palm, which has a long generation time, it is believed that the achievement of goals through these breeding programs will take a long time and require a lot of space and manpower (Rajanaidu et al., 1993).

Production of novel high-value products by genetic engineering provides the opportunity to diversify use and increase the economic value of palm oil. Production of specialty oils for industrial applications would be a very attractive proposition because oil palm is the most productive oil crop. Achievements in manipulating fatty acid biosynthetic pathways using recombinant DNA technology leading to the production of transgenic oil crops especially rapeseed with modified oil compositions have made this technology very attractive for oil palm. Reported success in raising the levels of lauric acid (Voelker et al., 1992) and stearic acid (Knutzon et al., 1992) in rapeseed oil proved that both fatty acid chain length and the level of fatty acid unsaturation can be modified. Field studies indicated that characteristics of the transgenic plants were similar to normal plants and that oil production was not affected by substantial modification in fatty acid composition (Kridl et al., 1993). The main advantage for a dicotyledon such as rapeseed is that it has a simple and efficient *Agrobacterium*-mediated transformation method. The successful application of the microprojectile bombardment method in transforming monocotyledons such as rice (Christou et al., 1986) and wheat (Vasil et al., 1992) provides hope for application of the technique for oil palm.

The oil palm genetic engineering program consists of various projects, each contributing toward the production of transgenic oil palm, which can produce oil containing high levels of monounsaturated oleic acid. It was envisaged that such an oil will be industrially useful for producing chemical derivatives that can serve as alternatives to petrochemical feedstocks. The strategy is to manipulate the fatty acid biosynthesis pathway in the mesocarp to reduce production of palmitate and channel it toward increasing oleate (Cheah et al., 1995).

The pathway for palmitic and oleic acid synthesis in oil palm is similar to other plants (Oo, 1988). An understanding of the biochemical pathway is critical in order to choose the most suitable gene(s) that should be manipulated in order to modify the pathway to produce the desired oil composition. In plants, de novo fatty acid synthesis occurs in the chloroplasts and the nongreen plastids in nonphotosynthetic tissues as shown in Figure 8.4. The first committed step is catalyzed by acetyl CoA carboxylase, which produces malonyl CoA from acetyl CoA and bicarbonate. A group of dissociable enzymes that make up the fatty acid synthase (FAS) complex catalyze the sequential addition of 2-carbon units derived from malonyl CoA to a growing acyl chain esterified to acyl carrier protein (ACP). The primary

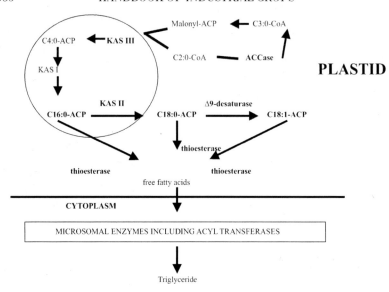

FIGURE 8.4. Fatty acid biosynthesis in plants.

products of this type II synthase are 16:0-ACP and 18:0-ACP. Elongation of palmitoyl-ACP (16:0) to stearoyl-ACP (18:0) is catalysed by β-ketoacyl-ACP synthase II (KAS II), one of the condensing enzymes from the FAS complex. Stearoyl-ACP desaturase inserts the first double bond into stearoyl-ACP to produce oleoyl-ACP. The acyl chains are then released into the cytoplasm by soluble acyl-ACP thioesterases. With the exception of plastidial desaturation, further elongation and desaturation of fatty acids occur in the cytoplasm.

Biochemical studies have indicated that the levels of palmitic and oleic acid in the oil palm mesocarp are controlled by the enzymes acyl-ACP thioesterase and KAS II. Thioesterase activity, both in the crude extract (Sambanthamurthi and Oo, 1991) and in a distinct fraction after purification by anion exchange chromatography, showed a strong preference for palmitoyl-ACP as a substrate (Abrizah, 1995). The studies showed that the activity of type synthase II increased with ripening and it correlates positively with the level of unsaturation in the crude oil extract (Cheah et al., 1995; Umi and Sambanthamurthi, 1996). It was observed that the oil palm has an active oleoyl CoA desaturase that readily converts oleate (18:1) to linoleate (18:2). It may be necessary to downregulate the expression of the gene encoding this enzyme for increasing oleic acid (Sambanthamurthi et al.,

1999). About 70-fold purification of glycerol 3-phosphate acyl-transferase was achieved from oil palm tissue cultures and mesocarp acetone powder (Arif and Harwood, 2000). This was an arduous task since glycerol 3-phosphate acyltransferase is a membrane-bound enzyme. Biochemical studies of acyl transferases are important because they determine the species of fatty acid incorporated into the glycerol backbone during triacylglycerol assembly and hence the fatty acid composition of vegetable oils.

Full-length or near full-length cDNA clones for genes encoding key enzymes in oil synthesis including stearoyl-ACP desaturases (Siti Nor Akmar et al. 1999), palmitoyl-ACP thioesterase (Abrizah et al., 1999) and acyl-carrier proteins (Rasid et al., 1999) have been isolated and characterized. The cDNAs encode precursor proteins containing N-terminal leader peptide for targeting to the plastid that is the site of de novo fatty acid synthesis. Down-regulation either by co-suppression or antisensing of oil palm palmitoyl-ACP thioesterase and stearoyl-ACP desaturase are strategies that will be taken for producing high oleic oil and high stearic oil, respectively. The identity and specificity of the encoded product of the isolated thioesterase gene is being confirmed biochemically using enzymes obtained by over-expressing the genes in bacterial systems. Consistent with observation in other plant species, more than one gene coding for acyl carrier protein was found in oil palm. Two differentially regulated stearoyl-ACP desaturase genes were identified. One is constitutively expressed, suggesting a house-keeping role in membrane lipid biosynthesis. Expression of the second is induced in mesocarp and kernel tissues in phase with oil synthesis indicating its direct involvement in oil synthesis. Polyclonal antibodies were raised against the oil palm stearoyl-ACP desaturase and used to study posttranscriptional regulation using Western blot analysis. Together, these provide useful background information for mesocarp oil modification by genetic engineering.

A major breakthrough in oil palm transformation using the micro-projectile bombardment method was reported. The production of transgenic oil palms containing herbicide-resistant genes was reported (Parveez, 1998). Regulatory sequences controlling tissue and temporal-specific gene expression are essential for the genetic engineering effort. Comparison of proteins synthesized in vitro from mRNA of mesocarp at different ages of fruit detected differential gene expression during development of this tissue (Siti Nor Akmar et al., 1994). Differential screening and subtractive hybridization techniques have been employed for the isolation of cDNA clones specifically expressed in the mesocarp during the period of oil synthesis (Siti Nor Akmar, 1999). These clones will be used for obtaining the desired promoters to be used in modifying mesocarp oil composition by genetic engineering (Siti Nor Akmar et al., 1995).

OIL PALM TISSUE CULTURE

Roles and Relevance of Tissue Culture

Exponential growth in the world's population will elevate demand for oils and fats (Zamzuri et al., 1999). Oil palm, being the most efficient oil-producing crop, can play a very important role in meeting the increasing world demand. Production of annual oil crops, such as soybean, sunflower, and rapeseed, are subjected to planting decision in major growing countries whereas oil palm is a perennial crop whose production is more reliable. Even though the conventional breeding approach has been very successful in producing high-yielding planting materials and also for producing varieties having other useful attributes, such as dwarfness and high oil unsaturation, this technique has limitations. The selection cycle for perennial crops is very long (seven to ten years), and if several selection cycles are required to produce desired phenotypes, breeding objectives can be achieved only in the long term (Sambanthamurthi et al., 2000).

Successful vegetative propagation of oil palm by tissue culture technique was reported in the mid-1970s (Jones, 1974; Rabechault and Martin, 1976). Currently, about 20 oil palm laboratories are in operation throughout the world with seedling-production capacities ranging from 10,000-200,000 plantlets per year (Zamzuri et al., 1999). Oil palm tissue culture has significant advantages over conventional breeding (Sogeke, 1998). Tissue culture allows rapid multiplication of uniform planting materials of desired characteristics. It provides the opportunity for improving planting materials using existing individuals that have all or most of the desired qualities, such as good oil yield and composition, slow vertical growth, and disease resistance as a source of explants. It also opens new avenues for oil palm biotechnology because tissue culture is the means for regeneration of tissues transformed with genes for traits of interest.

Research on tissue culture of oil palm has been carried out extensively for several reasons. First, it is used as a means for producing good tenera palms for commercial planting. High demand exists for high-quality planting materials for oil palm plantations not only in Malaysia but also in other countries. Based on current demand for oil palm seeds, it is estimated that a ready market exists for more than 100 million tissue culture plantlets annually (Zamzuri et al., 1999). Second, tissue culture is performed to multiply good parents (both dura and pisifera) for seed production. Third, tissue culture is also used as a means to expedite the exploitation of genetic potential of progenies from interspecific *E. oleifera* × *E. guineensis* crosses. Interspecific hybridization with South American species *E. oleifera* was carried out by various breeding groups to introgress the high oil unsaturation and slow vertical growth characters into high-yielding, commercially grown *E. guineensis.* The F1 hybrid was shown to have an intermediate IV but was not viable commercially due to its low yield and undesirable excessive vegetative vigour. Even after using various strategies of backcrossing with *E. guineensis,* progenies with an improved IV suitable for commercialization have not been obtained (Yong and Chan 1996). Finally, in Costa Rica, for example, tissue culture is also used to salvage diseased palms (Guzman, 2000).

Tissue Culture Process

It is absolutely essential that the ortets selected from the field have superior qualities. Tissue culture laboratories are linked to effective oil palm breeding and improvement programs that can supply the large reservoir of explants required. Ideally, an ortet should possess several desirable heritable traits, such as high oil yield, low height increment, and high oil quality. Ortets selected are supported by at least four years of field data showing good performance on yield, vegetative measurements, and physiological traits, such as bunch index and oil characteristics (Rohani et al., 2000). Oil yield is determined by the oil extraction rate (OER) or oil-to-bunch ratio (O:B) and weight of fresh fruit bunch (FFB). Since O:B has been demonstrated to be highly heritable and transmitted from ortets to ramets, it is given emphasis in ortet selection. Leaves, inflorescenses, and roots have been used as explants for oil palm tissue culture. Young leaf spears are preferred in most laboratories. Leaf explants can be surface sterilized easily, and give higher clonability rates (Rajanaidu et al., 1997).

The process of oil palm tissue culture can be divided into several different stages. Callus is initiated from the explant, followed by embryogenesis, shoot and root regeneration, hardening of ramets for nursery, and field evaluation. The regeneration process through oil palm tissue culture takes two to four years depending on genotype. Growth conditions for the different stages are typically at $28°C \pm 2°C$ with equal light and dark photoperiods, except during callogenesis, where the culture is maintained in complete darkness. The explants are placed in media containing either 2, 4-D (2,4-dichlorophenoxy acetic acid) or NAA (naphthalene acetic acid) for 12-20 weeks for callus formation. The calli are maintained in media containing lower concentrations of 2,4-D or NAA for up to 12 months for multiplication and embryogenesis (Rohani et al., 2000). The rates of callusing and embryogenesis in oil palm have been demonstrated to be genotype dependent (Ginting and Fatmawati, 1995). For example, tissues of variety La Me origin produced as high as to 60 percent callus, whereas tissues from Yangambi produced only 5 to 20 percent. The average rate of embryo-genesis ranges between 3 to 6 percent for leaf-derived calli (Wooi, 1995; Rajanaidu et al., 1997). A few oil palm clones produce embryoids after one month, whereas others may take as long as 24 months (Ginting and Fatma-wati, 1995). It was shown that the prospensity of callusing and the rate of embryogenesis are not influenced by ortet age (Wooi 1995). Small clumps of polyembryoids are transferred onto a basal nutrient medium and kept for at least three months for shoot induction. Shoots obtained are separated from the polyembryoids and placed onto solid shoot development media containing low concentrations of NAA in culture tubes (takes two to three shoots) or flasks (about 15 shoots). When the shoot height reaches 5 cm, they are transferred into liquid root-initiation media. Shoot development and rooting that involve many culture transfers are labor intensive. Based on the experience of in vitro culture in ornamentals and banana, Zamzuri (1998) introduced the double-layer rooting technique for oil palm. In this technique, the solid shoot development medium is overlayed with liquid root initiation media. This improves the productivity of the worker 18-fold and reduced the cost of rooting by about 94 percent. The plantlets are trans-planted into small polybags containing 1:1 ratio of soil and sand and kept for three to four months under shade with relative humidity of more than 70 percent for acclimitization before transferring to field nursery. In the nurs-ery, the plantlets are treated like seed derived plants.

Clonal Abnormality

The occurence of fruit and floral abnormalities in clonal palms were re-ported in the late 1980s (Hartley, 1988; Paranjothy, 1989). The abnormality referred to as mantling affects flower development involving feminization of the male parts of the flowers; the vegetative parts appear normal. Consid-erable variation exists in the severity of abnormality and such variation may influence oil yield because of the effect on fruit set. It was suggested (Obasola et al., 1978) that the abnormalities may be due to physiological factors whereby an increase in the concentration of hormones in the palms could cause biochemical disorders leading to an expression of abnormality. Studies in other plants showed correlation of increased 2,4-D levels with in-creased ploidy levels (Sogeke, 1998) and increased DNA methylation, hence affecting genome expression (Rohani et al., 2000). Sogeke (1998) used NAA which is a milder auxin, in oil palm tissue culture media instead of 2,4-D. This technique, which was able to reduce callusing time and hasten production of embryoids successfully produced normal plants in the field.

Field Performance

Field performance data from different sources indicate improvement of oil yield between 20 to 30 percent of clonal material over seedling planting materials (Soh, 1986; Hardon et al. 1987). The difficulty in selction for resistance to *Ganoderma, Fusarium,* and *Blastobacter* lies in achieving true-to-type reproduction of plants selected as ortets, especially with the incidence of mantled abnormality. The percentage of abnormality in the field has generally been reduced to less than 5 percent and maintained at this tolerable level (Maheran et al., 1995; Wong et al., 1996; Khaw and Ng, 1997). Maheran et al. (1995) suggested the need for prudent selection of good ramets at both the in vitro and nursery stages. Their observations indicated that clones appearing normal in both stages gave low levels of abnormality in the field (about 2.2 percent). Economic analyses reported by Maheran and Chan (1993) indicated that the higher productivity of clonal materials will not only compensate the high initial investment but will also give greater returns than conventional planting materials. It was estimated that the initial investment will be covered in the sixth year. After that period, returns from the clonal materials will be much higher than normal D × P planting materials. Zamzuri et al. (1999) studied the commercial feasibility of producing clonal palm planting material using four different production systems working under single or double shifts to produce between 90,000 to 700,000 plantlets for more than 30 years. For all the models, it was concluded that the venture would be attractive if the plantlets are sold at RM 20. Since an oil palm tree with productive life of more than 20 years that will produce about three tonnes of FFB in total was estimated to be worth RM 1,200, it would be worthwhile to put this initial investment on oil palm cloning.

PESTS AND DISEASES

All the insect pests of oil palm in Malaysia are of local origin and they have adapted to the crop ever since its introduction close to 100 years ago. In general, oil palm pests can be classified into insects and vertebrates. The insect pests consist of the leaf defoliators, bagworms, and nettle caterpillars; the crown attacker, rhinoceros beetle; and the bunch moth. Other insect pests which attack the nursery stages are cockchafers and grasshoppers.

The vertebrate pests include rodents, wild boar, porcupines, and elephants (Chung, 2000).

Insect pests are generally kept in balance by biotic and abiotic factors. The former include natural enemies (parasitoids, predators, and diseases caused by bacteria, fungi, and viruses). The latter are mainly environmental factors, such as rainfall, temperature, and humidity. Nevertheless, the balance may sometimes be disrupted by injudicious use of insecticides, which often triggers outbreaks. It is now imperative that chemical use is minimized to avoid the problems of insect resistance, persistence in environment, resurgence of secondary pests, and effects on natural enemies, beneficial organisms, and nontargets. However, when the need arises, especially in controlling outbreaks, it is important to ensure that pest numbers are kept below a certain economic threshold and not to eliminate the pest completely. This is in line with the concept of integrated pest management (IPM) where natural control is integrated with chemicals (Wood, 1976). However, despite this approach, outbreaks of some pests, such as bagworms and nettle caterpillars, are quite common, causing yield loss of up to 40 percent (Basri, 1993). It is important to find factors governing outbreaks because with this information, a pest can be more effectively and efficiently managed in an environmentally friendly manner.

Basic research in biological control approaches (i.e., parasitoids, predators, *Bacillus thuringiensis,* fungal and viral pathogens) has been intensified, because of their role in substantially reducing the use of chemical insecticides (Basri et al., 1995). In controlling leaf-eating pests, the propagation of natural enemies by planting beneficial plants is widely recommended. *Cassia cobanensis* was found the most suitable in sustaining the longevity of parasitoids in the laboratory (Basri et al. 1999).

Attacks by rhinoceros beetles have become common at present because of the extensive replanting programs currently undertaken in Malaysia. Other agronomic practices such as application of empty fruit bunches (EFB) as mulch, underplanting of oil palm, and zero burning have also aggravated the problem (Basri and Norman, 2000). Earlier research efforts were directed to the use of chemical insecticides (Toh and Brown, 1978; Chung et al., 1991) and pheromones (Chung, 1997; Norman et al., 1999b) for integrated control of this pest. Information on the ecology of the rhinoceros beetles, especially in the zero-burning environment (Samsudin et al., 1993; Norman et al., 1999a), will further enhance IPM approaches for *Oryctes rhinoceros.* Greater emphasis is placed in the development and use of microbial pathogens *(Metarhizium anisopliae, Bacillus thuringiensis)* for control of *Oryctes rhinoceros* (Ho, 1996; Ramle et al., 1999) For general reviews on insect pests of oil palm, see Wood (1968, 1976), Turner and

Gillbanks (1974), Varghese (1981), Hartley (1979), Chung et al. (1995), and Basri and Norman (2000).

The oil palm is also susceptible to disease from seed germination to field planting. The most common organisms causing oil palm disease are fungi. In general, there are five major diseases, namely, vascular wilt, basal stem rot, bud and spear rot, red ring disease, and sudden wilt. In Malaysia, the most important disease is basal stem rot (BSR) caused by *Ganoderma* spp. The other diseases mentioned earlier are not present in Malaysia but are very serious in other countries in Africa and South America.

The symptoms of BSR in young palms are retarded growth, one-sided yellowing of the lower fronds, and pale green foliage (Ariffin, 2000). In older palms, the canopies of infected palms are pale green with multiple unopened spears. This is a typical symptom resulting from decreased water uptake due to the rotting stem. As the disease progresses, fruiting bodies will start to appear; initially as small white buttons, later developing into bracket-shaped sporophores (Ariffin, 2000). When the fruiting bodies appear, the palm is likely already in an advanced stage of decay and close to its death.

The symptoms of BSR disease usually manifest after the palms are between 10 and 12 years old (Gurmit, 1991). In areas where oil palm is planted after coconut, the disease symptoms occur much earlier, within one to two years of planting. Serious incidence of BSR occurs more in coastal clay soils compared to inland soils (Khairuddin, 1990; Gurmit, 1991; Benjamin and Chee, 1995).

Control measures for BSR include injecting systemic fungicides into diseased palm trees (Chung, 1990; Khairudin, 1990; Ariffin and Idris, 1997). However, chemical control has not been demonstrated to be viable (Ariffin, 2000). The best method of controlling is by clean clearing, completely removing the infected palms, and destroying the bole and root masses that harbor the pathogen. This is particularly important as even the detached infected root segments can still transmit the disease and cause outbreaks (Gurmit, 1991; Ariffin, 2000).

The latest breakthrough on the BSR is the identification of four species of *Ganoderma*: *G. boninense, G. zonatum, G. miniatocinctum,* and *G. tornatum.* Although the first three species are pathogenic, *G. boninense* is the most aggressive. *Gandoderma tornatum* is nonpathogenic (Idris et al., 2000).

OIL PALM PLANTATIONS AND THE ENVIRONMENT

Conservation of the environment is of utmost importance for ensuring its sustainability. A large proportion of the agricultural land in Malaysia (51.9

percent) is planted with oil palm, which fulfills varied needs (environmental, biological, and industrial) (Figure 8.5). Intensive research and development efforts have been undertaken since 1975 and the findings have been implemented.

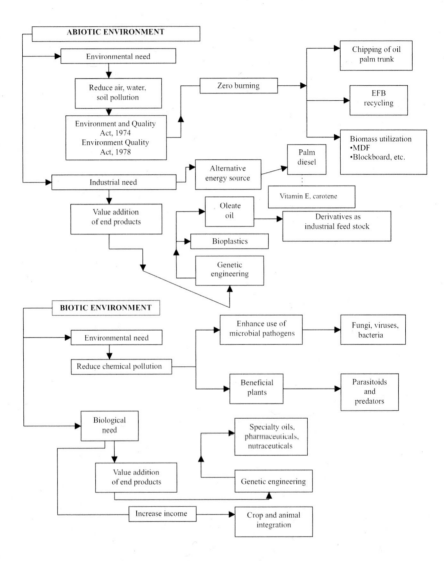

FIGURE 8.5. Adapting the oil palm to the environment.

Under the Environment and Quality Act of 1974 and the Environment Quality (Clean Air Regulations) Act of 1978 open burning of felled palms during replanting is prohibited in Malaysia, and zero burning has been practiced by virtually all plantations in the country. Zero burning involves shredding of oil palm trunks with an excavator (5-10 cm thick) and stacking the shredded trunks in the interrows (Mohd Hashim et al., 1993). Decomposition of the shredded biomass takes place within two years (Norman and Basri, 1997), after which it does not remain a suitable breeding substrate for the rhinoceros beetles. In addition, this technique has merits in terms of recycling large quantities of plant nutrients through decomposition and improving soil physical properties. Mohd Hashim et al. (1993) reported that the quantities of available nutrients from oil palm trunks were 219.6 kg/ha of N, 21.2 kg/ha of P, 314 kg/ha of K, and 52.6 kg/ha of Mg. Chan et al. (1981) reported the presence of 146 kg/ha of Ca in the felled palms. Theoretically, these reserves could provide to the palms N, K, and Mg for six to seven years and P for about two years. Thus, by employing zero-burning practices, the requirement for inorganic fertilizer can be reduced by 20 to 30 percent for first four years. However, much research is needed to optimize fertilizer inputs during replanting with zero burning (Zin Zawawi, 2000).

Traditionally, empty fruit bunches were incinerated to produce bunch ash, which is a good source of potash fertilizer. Under the two acts mentioned earlier, incineration is prohibited because it causes air pollution. To circumvent the problem, EFB are applied in plantations as mulch within palm circles and interrows as partial sources of nutrients. Chan et al. (1981) estimated that the quantities of available nutrients from EFB were 5.4 kg/ha of N, 0.4 kg/ha of P, 35.2 kg/ha of K, 2.7 kg/ha of Mg, and 2.3 kg/ha of Ca. This has become a standard agronomic practice within plantations.

Oil palm has an economic life of about 25 years, and during replanting vast amount of palm biomass is available. In Malaysia for instance, because of massive replanting programs (yielding 4 million tonnes of trunks and 2.9 million tonnes of EFB per year for the period 2001-2003), opportunities other than agronomic applications have been considered. Intensive research has been undertaken since 1980, and technologies are currently available in converting oil palm biomass into value-added products such as pulp and paper, particleboard, medium-density fiberboard, and thermoplastic (Mohamad, 2000). However, further study is required to determine the balance between the quantum of biomass removed for value-added products and the quantum that needs to remain in the plantation to maintain organic matter content and soil fertility.

Although the usage of pesticides in oil palm plantations has been minimal, research has been intensified for the use of microbial pathogens and natural enemies for control of major insect pests, such as the rhinoceros beetle and bagworm. Basri and Norman (2000) reported that of three strains of *Beauveria bassiana* isolated from the bagworm *Metisa plana,* one strain was highly effective against the bagworm yet safe against the pollinator, *Elaeidobius kamerunicus* (Ramle et al., 1995). Nuclear polyhedrosis virus (NPV) was also isolated from the bagworm. Since the infection was only tertiary, there is little scope of using the virus for the biological control of this pest (Siti Ramlah et al., 1996). An effective strain of *Metarhizium anisopliae* for the control of rhinoceros beetle has been field evaluated, and current efforts are focused on its mass production for application in plantations.

Based on the studies of Basri et al. (1999), a number of major plantation companies in Malaysia are planting beneficial plants, in particular *Cassia cobanensis,* within their plantations. These plants provide a good source of nectar to the parasitoids of the bagworm and other pests, thereby extending the lifespan of the natural enemies in the oil palm ecosystem. It is anticipated that with the use of microbial pathogens and parasitoids, the use of chemicals will be reduced at least for insect pest control.

There is also some degree of awareness among plantations on the need to grow palms that meet basic standards for fulfilling environmental demands. A number of plantations have been accorded ISO 14001 certification. This ensures production in a clean envionment. This has important implications on the trade of palm oil in the future in the sense that advanced countries would prefer importing palm oil from companies that take environmental concerns into account.

The value of oil palm needs to be enhanced by as many folds as possible. This can be made possible through genetically engineering the palm so that it could produce high-value products, such as oleate oil, specialty oil, bioplastics, pharmaceuticals, and nutraceuticals. The derivatives of oleate oil can be used as industrial feedstock for the oleochemical industry. As mentioned earlier, major breakthroughs have been made toward producing oleate oil, and research in this area is currently being intensified. Intensive joint research is also being undertaken between the Malaysian Palm Oil Board and the Massachusetts Institute of Technology toward the production of bioplastics. The ultimate aim would be in the production of palms that will produce different premier products for certain targeted markets.

Instances can arise whereby the price of palm oil can decline below the cost of production. These instances involved low demand by importing countries and oversupply of the world's oil and fats brought about by good growing conditions in other oil-producing countries. Under that sort of situation, it is worth converting part of the stock for the production of palm diesel as fuel for vehicles. The technology is available and it has been proven on the road by vehicles made to run on the palm diesel for a few hundred thousand kilometers. Even if the price is favorable in relation to the cost of production, the project is still economically viable because the process also produces products of high value, in particular carotene and glycerol.

The oil palm needs to be adapted to the presence of other crops during its growth, either during its immaturity period or throughout its life span depending on the types of crop chosen. This essentially involves crop and animal integration. In essence, the project is aimed at increasing total income to the grower without affecting the overall production of palm oil. The types of economical plants being studied by MPOB include yam, timber trees, banana, pineapple, and fruit trees; whereas the animals under study include cattle, sheep, and village chicken. These studies are still at the initial stage.

FUTURE OUTLOOK

The inevitable increase of the world's population in the future will translate into increasing demand for oils and fats. According to the Malaysian Palm Oil Board, to meet the increase in demand, oil and fat production is projected to grow by 24.6 percent by 2020; palm oil is expected to contribute 4.4 percent annually to account for the 24.6 percent of the total by the year 2020. As such, palm oil has an important role to play in the scenario of world's oils and fats.

Other factors can also lead to better demand for palm oil in the future. First, changes in planting policies of other types of oil crops such as a reduction in area, brought about by low prices, would increase the demand for palm oil. Second, depleting mineral oils will also be in favor of palm oil derivatives, which can be substituted for industrial feed stocks.

As such, research and development (R&D) on oil palm constitute an integral component of the palm oil industry. The R&D has to ensure that the industry remains well sustainable and competitive. Further, the R&D has to be aligned to a direction that delivers products and processes which serve ever-changing consumers' needs.

The availability of novel genes from MPOB germplasm collections has made it possible for the development of new oil palm planting materials (PS1: high yield, dwarf; and PS2: high yield, high iodine value). Such availability of genetic variability could also be exploited for the production of planting materials with high carotene, vitamin E, and sterols. By 2020, major changes are likely to occur in the types of planting materials used by commercial plantations.

The potential of genetic engineering in the production of novel crop plants should be fully exploited by the oil palm industry in order to remain

competitive. The production of the first transgenic oil palm containing herbicide resistance genes indicated that the key milestone for future utilization of this technology has been achieved. Currently, research is geared toward producing high oleic oil palm, and field testing of these genetically
engineered materials is expected. When this is realized, the production of
other novel and value-added products will be easily achieved because, in essence, the same tools and techniques will be applied. Other targets of interest for genetic manipulation include industrial oils, such palmitoleic and
ricinoleic acids, thermoplastics, and nutraceuticals. Indeed the oil palm has
two storage tissues, the kernel and the mesocarp, that can be the targets for
accumulating the genetically engineered products. The substrates and intermediates in these tissues may be channeled to alter the levels of existing
products or to produce novel value-added products without deleterious effects on the plants. In comparison to other oil crops, the application of transgenic technology to oil palm offers certain advantages. First, it is conceiv-
able that the genetically engineered products can be produced at high levels
because oil palm is very productive. Second, being perennial the oil palm
will not face the same problem of instability of transgene inheritance as
with the annual oil crops because the same crops remain in the field for 25
to 30 years.

Even though the early reports on clonal abnormality slightly dampened
the interest in oil palm tissue culture, there is renewed interest in this technique for mass propagation of planting materials. Increasing evidence indicates that clonal materials from elite explants are more uniform and superior in peformance compared to D × P seedlings. Heritable traits, such as
fruit and vegetative characters, were shown to be heritable from ortets to
clones. Good household practices by prudent selection at various stages,
minimizing the time in culture and reducing hormone levels in culture media, are able to substantially reduce the rate of abnormality. In addition,
studies are being carried out on abnormality to understand underlying
mechanism causing somaclonal variation and to develop markers for early
detection. The present system, which requires repeated subculturing of
embryogenic tissues, is labor intensive, expensive, and time consuming.
Media and culture protocols that are applicable to a wide range of genotypes are being developed so that large scale production will be cost effective, opening the possibility of automation. Some laboratories have initiated
liquid cultures, which offer several advantages, including the ability of embryogenic cells to proliferate faster, requiring less subculture, and greater
savings in labor and space requirements.

Commitment on other critical and long-term areas of research will continue, particularly those which have an important bearing on the environment. These areas include the utilization of beneficial plants for parasitoids

in the field, precision agriculture, the use of geographic information systems to enhance the management of oil palm plantations, the management of oil palm residues, biological control of *Ganoderma,* and enhancement of mechanization.

REFERENCES

Abrizah, O. 1995. Partial purification and characterization acyl-ACP thioesterase from oil palm mesocarp *(E. guineensis).* MSc thesis, Universiti Kebangsaan Malaysia.

Abrizah, O., Lazarus, C.M., and Stobart, A.K. 1999. Isolation of a cDNA clone encoding an acyl-acyl carrier protein thioesterase from the mesocarp of oil palm *(Elaeis guineensis). Journal of Oil Palm Research* Special edition: 81-87.

Arif, M. Manaf and Harwood, John L. 2000. Purification and characterization of acyl-CoA: glycerol 3-phosphate acyltransferase from oil palm *(Elaeis guineensis)* tissue. *Planta* 210: 318-328.

Ariffin, D. 2000. Major diseases of oil palm. In *Advances in Oil Palm Research,* Volume 1. Malaysian Palm Oil Board, pp. 596-622.

Ariffin, D. and Idris, A.S. 1997. Chemical control of *Ganoderma* using pressure injection. *Proceedings of the PORIM-Industry Forum.* Bangi, Malaysia, pp. 104-106.

Ariffin, D., Idris, A.S., and Marzuki, A. 1995. Development of a technique to screen oil palm seedlings for resistance to *Ganoderma. Proceedings of the 1995 PORIM National Oil Palm Conference Technology in Plantation—The Way Forward,* pp. 132-141.

Aziz, A., Rosnah, M.S. Mohamadiah, B., and Wan Zailan, W.O. 1986. Relationship of the formation of storage oil with fatty acid composition in developing oil palm fruit. *Proc. Mal. Biochem. Soc. Conf.* 12: 147-151.

Basri, M.W. 1993. Life history, ecology, and economic impact of the bagworm, *Metisa plana* Walker (Lepidoptera: Psychidae) on the oil palm *Elaeis guineensis* Jacquin (Palmae) in Malaysia. PhD thesis, University of Guelph, Ontario, Canada.

Basri, M.W. and Norman, K. 2000. Insect pests, pollinators, and barn owls. In Yusof, B., Jalani, B.S., and Chan, K.W., Eds., *Advances in Oil Palm Research.* Malaysian Palm Oil Board, Kuala Lumpur, pp. 466-541.

Basri, M.W., Norman, K., and Hamdan, A.B. 1995. Natural enemies of the bagworm *Metisa plana* (Lepidoptera: Psychidae) and their impact on host population regulation. *Crop Protection* 14(8): 637-645.

Basri, M.W, Simon, S., Ravigadevi, S., and Othman, A. 1999. Beneficial plants for the natural enemies of the bagworm in oil palm plantations. In Ariffin, D., Chan, K.W., and Sharifah, S.R.S.A., Eds., *Proceedings of the 1999 PORIM International Palm Oil Congress—Emerging Technologies and Opportunities in the Next Millenium.* Palm Oil Research Institute of Malaysia, Kuala Lumpur, pp. 441-455.

Beirnaert, A. 1953. Introduction à la biologie florale du palmier à huile *Elaeis guineensis* Jacq. *Publis I.N.E.A.C. Serie Sci.* No. 5.

Beirnaert, A. and Vanderweyen, R. 1941. Contribution à letude genetique et biometrique des varieties *d'Elaeis guineensis.* Jacquin. *Publs. Inst. Natn. Etude Agron Congo Belge. Ser. Sci.* No. 27.

Benjamin, M. and Chee, K.H. 1995. Basal stem rot of oil palm—A serious problem on inland soils. *MAPPS Newsletter* 19: 3.

Blaak, G. 1967. Oil palm prospection tour in the Bamenda highlands of West Cameroon. Internal Report. Unilever, London.

Brough, C.L., Coventry, J.M., Christie, W.W., Kroon, J.T.M., Brown, A.P., Barsby, T.L., and Slabas, A.R. 1996. Toward the genetic engineering of triacylglycerols of defined fatty acid composition: Major changes in erucic acid content at the *sn*-2 position affected by the introduction of a 1-acyl-*sn*-glycerol-3-phosphate acyltransferase from *Limnanthes douglasii* into oilseed rape. *Molecular Breeding* 2: 133-142.

Chan, K.W., Watson, I., and Lim, K.C. 1981. Use of oil palm waste material for increased production. *Planter* 57: 14-37.

Cheah, S.C., Sambanthamurthi, R., Siti Nor Akmar, A., Abrizah, O., Mohamad Arif, A.M., Umi Salamah, R., and Ahmad Parveez, G.K. 1995. Toward genetic engineering of oil palm (*Elaeis guineensis* Jacq.). In Kader, J.C. and Mazliak, P., Eds., *Plant Lipid Metabolism.* Kluwer Academic Publishers, Dordrecht, Netherlands, pp. 570-572.

Choo, Y.M. and Yusof, B. 1996. *Elaeis oleifera* palm for the pharmaceutical industry. Palm Oil Research Institute of Malaysia, Technology Transfer No. 42. Kuala Lumpur, Malaysia.

Christou, P., Corley, R.H.V., Lee, C.H., Law, I.H., and Wong, I.W. 1986. Abnormal flower development in oil palm clones. *Planter* 62: 141-240.

Chung, G.F. 1990. Preliminary results on trunk injection of fungicides against *Ganoderma* basal stem rot in oil palm. In *Proceedings of the Ganoderma Workshop.* Palm Oil Research Institute of Malaysia, Bangi, Kuala Lumpur, Malaysia, pp. 81-97.

Chung, G.F. 1997. The bioefficacy of the aggregation pheromone in mass trapping of *Oryctes rhinoceros* (L) in Malaysia. *Planter* 73: 119-127.

Chung, G.F. 2000. Vertebrate pests of oil palm. In Yusof, B., Jalani, B.S., and Chan, K.W., Eds., *Advances in Oil Palm Research.* Malaysian Palm Oil Board, Kuala Lumpur, Malaysia, pp. 542-595.

Chung, G.F., Basri, M.W., and Ariffin, D. 1995. Recent development in plant protection of the Malaysian oil palm industry—1990 to 1995. In Jalani, S. and Ariffin, D., Eds., *1995 PORIM National Oil Palm Conference—Technologies in Plantations—The Way Forward,* Kuala Lumpur, Malaysia, pp. 105-124.

Chung, G.F., Sim, S.C., and Tan, M.W. 1991. Chemical control of rhinoceros beetles in the nursery and immature oil palms. *Proceedings of the PORIM International Palm Oil Development Conference Module II—Agriculture,* Kuala Lumpur, Malaysia, pp. 380-395.

Clark, J.D. 1976. Prehistoric population and pressures favoring plant domestication. In Harlan, J.R., de Wet, J., and Stemler, A.B.L., Eds., *Plant Population, Genetics, Breeding, and Genetic Resources*. Sinauer Associates, Sunderland, MA, pp. 98-115.

de Man, L. and de Man, J.M. 1994. Functionality of palm oil, palm oil products, and palm kernel oil in margarine and shortening. *PORIM Occasional Paper* 32: 1-16.

Garces, R. and Mancha, M. 1990. In Quinn, P.J. and Harwood, J.L., Eds., *Lipid, Biochemistry, Structure, and Utilization*. Portland Press, Wye, London, pp. 387-389.

Ginting, G. and Fatmawati, R.S. 1995. Propagation methodology of oil palm at Marihat. In Rao, V., Henson, I.E., and Rajanaidu, N., Eds., *Proceedings of the 1993 ISOPB International Symposium on Recent Developments in Oil Palm Tissue Culture and Biotechnology*. Palm Oil Research Institute of Malaysia, Bangi, pp. 114-121.

Goh, K.J. 2000. Climatic requirements of oil palm for high yields. *Proceedings of the Seminar on Managing Oil Palm for High Yields: Agronomic Principles*. Malaysian Society of Soil Science, Kuala Lumpur.

Gurmit, S. 1991. *Ganoderma*—The scourge of oil palms in coastal areas. *Planter* 67: 421-444.

Guzman, N. 2000. The present status of clonal propagation of oil palm (*Elaeis guineensis* Jacq.) in Costa Rica by culture of immature inflorescences. *In Vitro Cellular & Developmental Biology* 31: 79A.

Hardon, J.J., Corley, R.H.V., and Lee, C.H. 1987. Breeding and selecting the oil palm. In *Improving Vegetatively Propagated Crops*. Academic Press, London, pp. 64-81.

Hardon, J.J. and Tan, B.K. 1969. Interspecific hybrids in the genus *Elaeis* 1. Crossability, cytogenetics, and fertility of F_1 hybrids *E. guineensis* × *E. oleifera*. *Euphytica* 18: 312-379.

Hartley, C.W.S. 1979. *The Oil Palm*, Second Edition. Longman, London.

Hartley, C.W.S. 1988. *The Oil Palm* (*Elaeis guineensis* Jacq.), Third Edition. Longman Group Limited, New York.

Hayati, A., Wickneswari, R., Rajanaidu, N., and Maizura, I. 2000. Genetic diversity of oil palm (*Elaeis guineensis* Jacq.) germplasm collections using isozyme markers. *Proceedings of the International Symposium on Oil Palm Genetic Resources and Utilization*, June 8-10, 2000, Kuala Lumpur, Malaysia.

Henson, I.E. 1991. Estimating potential productivity of oil palm. In Rao, V., Henson, I.E., and Rajanaidu, N., Eds., *Proceedings of the 1990 ISOPB International Workshop on Yield Potential in the Oil Palm*. Palm Oil Research Institute of Malaysia, Bangi, pp. 98-108.

Hilditch, T.P. and Williams, P.N. 1964. *The Chemical Constitutions of Natural Fats*. Wiley, New York, pp. 332-343.

Idris, A.S., Ariffin, D., Swinburne, T.R., and Watt, T.A. 2000. The identity of *Ganoderma* species responsible for BSR disease of oil palm in Malaysia—Pathogenicity test. No. 77b. Malaysian Palm Oil Board.

Jacquemard, J.C., Meunier, J., and Bonnot, F. 1981. Etude genetique de la production du croisement chex le palmier à huile *Elaeis guineensis. Oleagineux* 36(7): 543-552.

Jagoe, R.B. 1952. Deli oil palms and early introductions of *Elaeis guineensis* to Malaya. *Malay. Agric. J.* 35: 4-11.

Jalani, B.S., Cheah, S.C., Rajanaidu, N., and Darus, A. 1997. Improvement of palm oil through breeding and biotechnology. Paper presented at the 88th American Oil Chemists Society (AOCS) Annual Meeting and Expo, May 11-14, 1997, Seattle, Washington.

Jones, L.H. 1974. Propagation of clonal oil palm by tissue culture. *Oil Palm News* 17: 1-8.

Khairuddin, H. 1990. Results of four trials on *Ganoderma* basal stem rot of oil palm in Golden Hope estates. In *Proceedings of the Ganoderma Workshop*. Palm Oil Research Institute of Malaysia, Bangi, pp. 67-80.

Khaw, C.H. and Ng, S.K. 1997. Performance of commercial scale clonal oil palm (*Elaeis guineensis* Jacq.) planting in Malaysia. Paper presented at the International Society of Horticulture Science Symposium, Brisbane, Australia, p. 8.

Knutzon, D.S., Thompson, G.A., Radke, S.E., and Johnson, W.B. 1992. Modification of *Brassica* seed oil by antisense expression of a stearoyl-acyl carrier protein desaturase gene. *Proc. Natl. Acad. Sci. USA* 89: 2624-2628.

Kridl, J.C., Davies, H.M., Lassner, M.W., and Metz, J.M. 1993. New sources of fats, waxes, and oils: The application of biotechnology to the modification of temperate oilseeds. *AgBiotech News and Information* 5: 121N-126N.

Kularatne, R.S., Shah, F.H., and Rajanaidu, N. 2000. Investigation of genetic diversity in African natural oil palm populations and Deli *dura* using AFLP markers. *Proceedings of the International Symposium on Oil Palm Genetic Resources and Utilization,* June 8-10, 2000, Kuala Lumpur, Malaysia.

Kushairi, A., Rajanaidu, N., Jalani, B.S., Mohd Rafii, Y., and Mohd Din, A. 1999. PORIM oil palm planting materials. *PORIM Bull.* 38: 1-13.

Latiff, A. 2000. The botany of the genus *Elaeis.* In: Yusof, B., Jalani, B.S., and Chan, K.W., Eds., *Advances in Oil Palm Research.* Malaysian Palm Oil Board, Kuala Lumpur, pp. 1346-1412.

Lee, C.H. 1996. Yield potential of Golden Hope D×P oil palm planting materials. Paper presented at the Seminar on Sourcing Oil Palm Planting Materials for Local and Overseas Joint Venture, July 22-23.

Lee, C.H. and Toh, P.Y. 1991. Yield performance of Golden Hope D×P planting materials. *The Planter* 47: 317-324.

Madon, M., Clyde, M.M., and Cheah, S.C. 1999. Application of genomic in situ hybrization (GISH) on *Elaeis* hybrids. *Journal of Oil Palm Research* (Special Issue): 74-80.

Maheran, A.B. and Chan, C.Y. 1993. Produktiviti klon lebih baik memberi pulangan yang lebih tinggi walaupun kos awalnya lebih mahal. *Kemajuan Penyelidikan FELDA* 21: 24-27.

Maheran, A.B., Teng, A.W., Othman, A.Z., and Weng, C.C. 1995. Vegetative propagation of oil palm *(Elaeis guineensis)* from laboratory to field—FELDA experience. In Jalani, B.S., Ariffin, D., Rajanaidu, N., Tayeb, M.D., Paranjothy, K., Basri, M.W., Henson, I.E., and Chang, K.C., Eds., *Proceedings of the 1993 PORIM International Palm Oil Congress—Update and Vision.* Palm Oil Research Institute of Malaysia, Bangi, Kuala Lumpur, Malaysia, pp. 99-113.

Maizura, I. 1999. Genetic variability of oil palm (*Elaeis guineensis* Jacq.) germplasm collections using RFLP markers. PhD thesis, Universiti Kebangsaan Malaysia.

Maizura, I., Cheah, S.C., and Rajanaidu, N. 1996. Genetic variation of oil palm (*Elaeis guineensis*) germplasm based on restriction fragment length polymorphism. *Proceedings of the Second National Congress on Genetics, Genetics Society of Malaysia,* Kuala Lumpur, Malaysia, pp. 338-344.

Meunier, J. 1969. Etude des populations naturelles d'*Elaeis guineensis* en Cote d'Ivoire. *Oleagineux* 24(2): 195.

Meunier, J. 1989. Advances in oil palm breeding: Progress and prospects. *International Conference on Palms and Palms Product.* National Institute for Oilpalm Research (NIFOR), Nigeria, p. 15.

Meunier, J. and Baudouin, L. 1985. Evaluation and utilization of Yocoboue population of *Elaeis guineensis. Proceedings of the International Workshop on Oil Palm Germplasm Utilization,* pp. 144-152.

Meunier, J., Renard, J.L., and Quillec, G. 1979. Heredite de le resistance à la furiose chez la palmier *Elaeis guineensis* Jacq. *Oleagineux* 34: 555-559.

Mohamad, H. 2000. Utilization of oil palm biomass for various wood-based and other products. In: Yusof, B., Jalani, B.S., and Chan K.W., Eds., *Advances in Oil Palm Research.* Malaysian Palm Oil Board Kuala Lumpur, pp. 1346-1412.

Mohd Din, A. and Rajanaidu, N. 2000. Evaluation of *Elaeis oleifera,* interspecific hybrids, and backcrosses. *Proceedings of the International Symposium on Oil Palm Genetic Resources and Utilization,* June 8-10, 2000, Kuala Lumpur, Malaysia.

Mohd Haniff, H. 2000. Yield and yield components and their physiology. In Yusof, B., Jalani, B.S., and Chan, K.W., Eds., *Advances in Oil Palm Research,* Volume 1. Malaysian Palm Oil Board. Kuala Lumpur, Malaysia,

Mohd Hashim, T., Teoh, C.H., Kamaruzaman, A., and Mohd Ali, A. 1993. Zero burning—An environmentally friendly replanting technique. In Jalani, B.S., Ariffin, D., Rajanaidu, N., Mohd Tayeb, D., Paranjothy, K., Mohd Basri, W., Henson, I.E., and Chang, K.C., Eds., *Proceedings of the 1993 PORIM International Palm Oil Congress,* Palm Oil Research Institue of Malaysia, Kuala Lumpur, pp. 185-194.

Mohd Nasir, A. and Ramli, A. 2000. Marketing of palm oil. In Yusof, B., Jalani, B.S., and Chan, K.W., Eds., *Advances in Oil Palm Research.* Malaysian Palm Oil Board, Kuala Lumpur, pp. 1462-1493.

Mukesh, S. and Tan, Y.P. 1996. An overview of oil palm breeding program and performance of D×P planting materials at United Plantation Berhad. Paper presented at the Seminar on Sourcing Oil Palm Planting Materials for Local and Overseas Joint-Venture, July 22-23.

Murphy, D.J. 1994. Future perspectives for oil crops. In Murphy, D.J., Ed., *Designer Oil Crops: Breeding, Processing, and Biotechnology.* VCH, Weinheim, pp. 297-310.

Ngo, V. 1991. Palm oil marketing issues and developments in the 1990s. *Proceedings of the PORIM International Palm Oil Conference.* Module IV Promotion and Marketing, pp. 102-106.

Norman, H.K. and Basri, M.W. 1997. Status of rhinoceros beetle, *Oryctes rhinoceros* (Coleptera: Scarabaeidae) as a pest of young oil palm in Malaysia. *Planter* 73: 5-21.

Norman, H.K., Mohd Basri, W., Zaidi, M.I., and Maimon, A. 1999a. Factors affecting population density of *Oryctes rhinoceros* in a zero-burn replanting environment. *Proceedings of the Third Entomology Association (ENTOMA) Seminar 1999,* pp. 98-100.

Norman, H.K., Mohd Basri, W., Zaidi, M.I., and Maimon, A. 1999b. Population studies of *Oryctes rhinoceros* in an oil palm replant using pheromone traps. *Proceedings of the International Palm Oil Confernce (PIPOC),* Palm Oil Research Institute of Malaysia, Kuala Lumpur, pp. 467-475.

Obasola, C.O., Menendez, T., and Obisesan, J.O. 1978. Inflorescence and bunch abnormalities in the oil palm (*Elaeis guineensis* Jacq.) *J. Nig. Inst. Oil Palm Research* 5(20): 49-57.

Ohlrogge, J.B. 1994. Design of new plant products: Engineering of fatty acid metabolism. *Plant Physiol.* 104: 821-826.

Okwuagwu, C.O. 1985. The genetic base of the NIFOR oil palm program. *Proceedings of the International Workshop on Oil Palm Germplasm,* pp. 228-237.

Oo, K.C. 1988. Biochemistry and biotechnology in research and development of oil palm. *PORIM Bulletin* 17: 22-30.

Oo, K.C., Lee, K.B., and Ong, A.S.H. 1986. Changes in fatty acid composition of the lipid classes in developing oil palm mesocarp. *Phytochemistry* 25: 405-407.

Pantzaris, T.P. and Yusof, B. 1990. Techno-economic aspects of palm oil and palm kernel oil in oleochemicals. *Palm Oil Developments* 14: 1-11.

Paramananthan, S. 2000. Soil requirements of oil palm for high yields. *Proceedings of the Seminar on Managing Oil Palm for High Yields: Agronomic Principles.* Malaysian Society of Soil Science.

Paranjothy, K. 1989. Research strategies and advances in oil palm cell and tissue culture. *Elaeis* 192: 119-125.

Parveez, G.K.A. 1998. Optimization of parameters involved in the transformation of oil palm using the biolistic method. PhD thesis submitted to Universiti Putra Malaysia.

Pichel, R. 1956. L'ameriolation du palmier à huile du Congo Belge Conf. Franco-Brittanique sur le palmier a huile. *Bull Agron. Min. Fr. d'outre Mer.* 14: 59-66.

Piggott, C.J. 1990. *Growing Oil Palm: An Illustrated Guide.* Incorporated Society of Planters, Kuala Lumpur.

Pursglove, J.W. 1972. *Tropical Crop Monocotyledons.* John Wiley & Sons, New York.

Rabechault, H.E.T. and Martin, J.P. 1976. Multiplication vegetative dupalmier à huile (*Elaeis guineensis* Jacq.) à l'aidede cultures de tissue foliares. *Compte Rendu Acad. Sci. Ser.* D 283: 1735-1737.

Rajanaidu, N. 1985. The oil palm (*Elaeis guineensis*) collections in Africa. *Proceedings of the International Workshop on Oil Palm Germplasm and Utilization.* Palm Oil Research Institute of Malaysia, Kuala Lumpur, pp. 59-83.

Rajanaidu, N. 1986. *Elaeis oleifera* collection in Central and South America. *Proceedings of the International Workshop on Oil Palm Germplasm and Utilization.* Palm Oil Research Institute of Malaysia, Kuala Lumpur, pp. 84-94.

Rajanaidu, N. 1987. Collection of oil palm *(Elaeis guineensis)* genetic material in Tanzania and Madagascar. *PORIM Bulletin* 15: 1-6.

Rajanaidu, N. 1994. PORIM oil palm genebank, collection, evaluation, utilization, and conservation of oil palm genetic resources. In conjunction with the Release of Elite Oil Palm Planting Materials and Launching of Oil Palm Genebank. PORIM, Malaysia, September 13.

Rajanaidu, N. and Jalani, B.S. 1994a. Germplasm collection in Gambia. *ISOPB Newsletter* 10(2): 1-16.

Rajanaidu, N. and Jalani, B.S. 1994b. Oil palm germplasm collection in Senegal. *ISOPB Newsletter* 10(1): 1-18.

Rajanaidu, N. and Jalani, B.S. 1994c. Oil palm genetic resources—Collection, evaluation, utilization, and conservation. A paper presented at the PORIM Colloquium on Oil Palm Genetic Resources. Kuala Lumpur, Malaysia, September 13.

Rajanaidu, N. and Jalani, B.S. 1994d. Potential sources of lauric oil for the oleochemical industry. In Applewhite, T.H., Ed., *Proceedings of the World Conference and Exhibition on Lauric Oil: Sources, Processing, and Applications.* AOCS, Manila, Philippines, pp. 47-50.

Rajanaidu, N., Jalani, B.S., Cheah, S.C., and Ahmad Kushairi, D. 1993. Oil palm breeding: Current issues and future developments. *Proceedings of the 1993 PORIM International Palm Oil Congress,* pp. 9-23.

Rajanaidu, N., Jalani, B.S., Kushairi, A., Rafii, Y., and Mohd Din, A. 1996. Breeding for high kernel planting materials PORIM Series 3 (PS3). *PORIM Information Series* 59.

Rajanaidu, N., Jalani, B.S., Rao, V., and Kushairi, A. 1991. New exotic palms for plantations. *Proceedings of the International Palm Oil Conference, Kuala Lumpur.* Palm Oil Research Institute of Malaysia, pp. 19-27.

Rajanaidu, N., Kushairi, A., Jalani, B.S., and Tang, S.C. 1994. Novel oil palm from exotic palms. *Malaysian Oil Sci. and Technol. J.* 3(2): 22-28.

Rajanaidu, N., Kushairi, A., Rafii, M., Mohd Din, A., Maizura, I., and Jalani, B.S. 2000. Oil palm breeding and genetic resources In Yusof, B., Jalani, B.S., and Chan, K.W., Eds., *Advances in Oil Palm Research.* Malaysian Palm Oil Board, Kuala Lumpur, pp. 171-237.

Rajanaidu, N. and Rao, V. 1987. Oil palm genetic collections: Their performance and use to the industry. *Proceedings of the 1987 International Oil Palm/ Palm Oil Conference in Agriculture.* Palm Oil Research Institute of Malaysia, pp. 59-85.

Rajanaidu, N., Rao, V., and Tan, B.B. 1984. Analysis of fatty acid composition in *Elaeis guineensis, Elaeis oleifera,* and their hybrids and its implication in breeding. Palm Oil Research Institute of Malaysia.

Rajanaidu, N., Rohani, O., and Jalani, S. 1997. Oil palm clones: Current status and prospects for commercial production. *The Planter* 73(853): 163-184.

Ramle, M., Mohd Basri, W., Norman, K., Sharma, M., and Siti Ramlah, A.A. 1999. Impact of *Metarhizium anisopliae* (Deuteromycotina: Hyphomycetes) applied by wet and dry inoculum on oil palm rhinoceros beetles, *Oryctes rhinoceros* (Coleoptera: Scarabaeidae). *Journal of Oil Palm Research* 11(2): 25-40.

Ramle, M., Siti Ramlah, A.A., and Basri, M.W. 1995. Histopathology of *Metisa plana* infected with *Beauveria bassiana. Elaeis* 8(1): 10-19.

Rasid, O.A., Cheah, S.C., and Mohd Arif, A.M. 1999. Isolation and sequencing of cDNA clones coding for oil palm *(Elaeis guineesis)* acyl carrier protein (ACP). *Journal of Oil Palm Research* 11: 88-95.

Rohani, O., Sharifah, S.A., Mohd Rafii, Y., Ong, M., Tarmizi, A.H., and Zamzuri, I. 2000. Tissue culture of oil palm. In Yusof, B., Jalani, S., and Chan, K.W., Eds., *Advances in Oil Palm Research.* Malaysian Palm Oil Board, Kuala Lumpur, pp. 238-331.

Sambanthamurthi, R.O., Abrizah, O., and Umi Salamah, R. 1996. Toward understanding the biochemical factors that affect oil composition and quality in the oil palm. *PIPOC ISOPB International Conference on Oil and Kernel Production in Oil Palm,* September 27-28, Kuala Lumpur.

Sambanthamurthi, R., Abrizah, O., and Umi Salamah, R. 1999. Biochemical factors that control oil composition in the oil palm. *Journal of Oil Palm Research* 11: 24-33.

Sambanthamurthi, R. and Oo, K.C. 1991. Thioesterase activity in the oil palm *(Elaeis guineensis)* mesocarp. In Quinn, P.J. and Harwood, J.L., Eds., *Plant Lipid Biochemistry, Structure, and Utilization: The Proceedings of the Ninth International Symposium on Plant Lipids.* Portland Press Ltd, London, pp. 166-168.

Sambanthamurthi, R., Parveez, G.K.A., and Cheah, S.C. 2000. Genetic engineering of the oil palm. In Yusof, B., Jalani. S., and Chan, K.W., Eds., *Advances in Oil Palm Research.* Malaysian Palm Oil Board, Kuala Lumpur, pp. 24-331.

Samsudin, A., Chew, P.S., and Mohd, M.M. 1993. *Oryctes rhinoceros:* Breeding and damage on oil palms in an oil palm to oil palm replanting situation. *Planter* 69: 583-591.

Siti Nor Akmar, A. 1999. Structure and regulation of stearoyl-ACP desaturase and metallothionein-like genes in developing fruits of the oil palm *(Elaeis guineensis).* PhD thesis, University of East Anglia, United Kingdom.

Siti Nor Akmar, A., Cheah, S.C., Aminah, S., Leslie, C.L.O., Sambanthamurthi, R., and Murthy, D.J. 1999. Characterization and regulation of the oil palm *(Elaeis guineensis)* stearoyl-ACP desaturase genes. *Journal of Oil Palm Research* 11: 1-17.

Siti Nor Akmar, A., Farida, H.S., and Cheah, S.C. 1994. Detection of differentially expressed genes in the development of oil palm mesocarp. *Asia Pacific Journal of Molecular Biology and Biotechnology* 2(2): 113-118.

Siti Nor Akmar, A., Farida, H.S., and Cheah, S.C. 1995. Construction of oil palm mesocarp cDNA library and the isolation of mesocarp-specific cDNA clones. *Asia Pacific Journal of Molecular Biology and Biotechnology* 3(2): 106-111.

Siti Ramlah, A.A., Basri, M.W., and Ramle, M. 1996. Isolation and amplification of baculovirus as a biocontrol agent for bagworm and nettle catepillar of oil palm. *Elaeis* 8(1): 1-9.

Smith N.J.H., Williams, J.T., Plucknett, D.L., and Talbot, J.P. 1992. Rubber, oils, and resins. In *Tropical Forest and Other Crops*. Cornell University Press, New York, pp. 207-263.

Sogeke, A.K. 1998. Stages in the vegetative propagation of oil palm (*Elaeis guineensis* Jacq.) through tissue culture. *Journal of Oil Palm Research* 10(2): 1-9.

Soh, A.C. 1986. Expected yield increase with selected oil palm clones from current D×P seedling material. *Oleagineux* 41(2): 51-56.

Stefansson, B.R. 1983. The development of improved rapeseed cultivars. In Kramer, J.K.G., Sauer, F.D., and Pigden, W.J., Eds., *High and Low Erucic Acid Rapeseed Oils*. Academic Press, New York, pp. 143-160.

Stobart, A.K., Stymne, S., and Shewry, P.R. 1993. Manipulation of cereal protein and oilseed quality. In Verma, D.P.S., Ed., *Control of Plant Gene Expression*, pp. 499-533.

Stumpf, P.K. 1987. In Stumpf, P.K. and Conn, E.E., Eds., *Biochemistry of Plants: A Comprehensive Treatise*. Academic Press, New York, pp. 121-136.

Syed, R.A. 1980. Pollinating insects of oil palm. CIBC 1977-1980 Report. Commonwealth Institute of Biological Control, Bangalow, India.

Tarmizi, A.M. 2000. Nutritional requirements and efficiency of fertilizer use in Malaysian oil palm cultivation. In Yusof, B., Jalani, B.S., and Chan, K.W., Eds., *Advances in Oil Palm Research*, Volume 1. Malaysian Palm Oil Board, Kuala Lumpur.

Thomas, R.L., Watson, L., and Hardon, J.J. 1969. Inheritance of some component of yield in the Deli *dura* variety of oil palm. *Euphytica* 18: 92-100.

Toh, P.Y. and Brown, T.P. 1978. Evaluation of carbofuran as a chemical prophylactic control measure for *Oryctes rhinoceros* in young oil palms. *The Planter* 54: 3-11.

Turner, P.D. and Gillbanks, R.A. 1974. *Oil Palm Cultivation and Management*. Incorporated Society of Planters, Kuala Lumpur.

Uhl, N.W. and Dransfield, J. 1987. *Genera Plantarum: A Classification of Palms Based on the Work of Harold E. Moore*. Allen Press, Lawrence, Kansas.

Umi, Salamah R. and Sambanthamurthi, R. 1996. Keto acyl-ACP synthase II in the oil palm (*Elaeis guineensis*) mesocarp. In Williams, J.P., Khan, M.U., and Lem, L.W., Eds., *Physiology, Biochemistry, and Molecular Biology of Plant Lipids*. Kluwer Academic Publishers, Dordrecht, p. 570.

Vanderweyen, R. 1952. La prospection des palmeraies congolaises et ses premier results. *Bull. Inf. INEAC* 1: 357-382.

Varghese, G. 1981. Pests and diseases of oil palm (*Elaeis guineensis* Jacq.): Current status of the problems and strategies for their control. *Planter* 57: 203-209.

Vasil, V., Castillo, A.M., Fromm, M.E., and Vasil, I.K. 1992. Herbicide resistant fertile transgenic wheat plants obtained by microprojectile bombardment of regenerable embryogenic callus. *Bio. Technology* 10: 667-674.

Voelker, T.A., Worrell, A.C., Anderson, L., Bleibaum, J., Fan, C., Hawkins, D.J., Radke, S.E., and Davies, H.M. 1992. Fatty acid biosynthesis redirected to medium chains in transgenic oilseed plants. *Science* 257: 72-73.

Wakker, E. 2000. Funding forest destruction. Available at <www.forest-trends.org>.

Wong, G., Tan, C.C., and Soh, A.C. 1996. Large scale propagation of oil palm clones—Experience to date. Paper presented at the International Symposium on In Vitro Culture and Horticultural Breeding, Jerusalem, Israel.

Wong, S., Goh, T.H., Tan, L.C., and Kerk, P.P. 1991. Recent developments in fractionation of palm oil. *Proceedings of the 1991 PORIM International Palm Oil Conference, Module I*, pp. 144-150.

Wood, B.J. 1968. Pest of oil palms in Malaysia and their control. *Incorporated Society of Planters*, pp. 204.

Wood, B.J. 1976. Insect pests in South-East Asia. In Corley, R.H.V., Hardon, J.J., and Wood, B.J., Eds., *Oil Palm Research*. Elsevier, Amsterdam, pp. 347-367.

Wood, B.J. 1984. *A Brief Guide to Oil Palm Science*. The Incorporated Society of Planters, Kuala Lumpur, Malaysia.

Wooi, K.C. 1995. Oil palm tissue culture: Current practice and constraints. In Rao, V., Henson, I.E., and Rajanaidu, N., Eds., *Proceedings of the 1993 ISOPB International Symposium on Recent Developments in Oil Palm Tissue Culture and Biotechnology*. Palm Oil Research Institute of Malaysia, Bangi, pp. 21-32.

Yong, Y.Y., and Chan, K.W. 1996. Breeding oil palm for competitiveness and sustainability in the 21st century. *Proceedings of PORIM International Palm Oil Congress*, pp. 19-31.

Zamzuri, I. 1998. Efficient rooting of oil palm in vitro plantlets using double-layer technique. *PORIM Bulletin* 36: 23-36.

Zamzuri, I., Mohd Arif, S., Rajanaidu, N., and Rohani, O. 1999. Commercial feasibility of clonol oil palm planting material production. *PORIM Occasional Paper* 40: 1-52.

Zeven, A.C. 1964. On the origin of oil palm. *Grana Palynol* 5:50.

Zeven, A.C. 1967. *The Semi-Wild Oil Palm and Its Industry in Africa*. Pudoc, Wageningen, Netherlands.

Zin Zawawi, Z. 2000. Agronomic utilization of wastes and environmental management. In Yusof, B., Jalani, B.S., and Chan, K.W., Eds., *Advances in Oil Palm Research*, pp. 1413-1438.

Chapter 9

Palmyra

H. Hameed Khan

A. Sankaralingam

ORIGIN AND DISTRIBUTION

The center of origin of palmyra has been traced to Africa. The palm came to prominence only after its introduction into India (Kovoor, 1983). The use of palm leaves as writing material has been in vogue in India since 6000 B.C. Although the leaves of palmyra and talipot were used for writing, the former were said to be of inferior in quality (Davis and Johnson, 1987). The palmyra palm was introduced into North India from the West, across the India Ocean, from where it spread to southern parts of India and established itself well (Losty, 1982). The Tamilians of South India honored the palm by designating it as the state tree in 1978. The close association of the palm with of Tamilian culture can be judged by the way the palm is lauded for its 801 uses in the poem "Talavilasam"' composed by Thiru Arunachalam of Tamil Nadu.

Kovoor (1983) noted two striking features regarding the spread of palmyra from India to Southeast Asian countries. As he pointed out, one could observe the distribution of the palm from India up to Myanmar in the north, down to Indonesia in the south following the trade route across the Indian Ocean. India could be the focus of the spread of *Borassus flabellifer* to countries bordering the Indian Ocean. Another feature is that the dispersal of the palm followed the spread of Buddhism, which originated in India, as palm leaf manuscripts were invariably used by the monks. These may also be the reasons for the palmyra palm being unpopular in Latin and South American countries, although conditions favorable for its growth do occur in these nations.

The palm is distributed in Mauritiana, Senegal, Mali, Gambia, Guinea-Bissau, Guinea, Ivory Coast, Burkina Faso, Nigeria, Gabon, Democratic Republic of the Congo, Republic of the Congo, Sudan, Tanzania, Madagascar, Saudi Arabia, Iraq, Iran, Pakistan, India, Sri Lanka, Bangladesh, Myanmar, Thailand, Cambodia, Vietnam, Malaysia, and Indonesia. However, no reliable current statistics are available on its area and production from many of these countries, since the palm is exploited mainly in the wild.

India has an estimated population of 102 million palms and tops the list in the world. The Indian state of Tamil Nadu has the major share of 51.9 million trees with the district of Thoothukudi ranking first (Harichandran and Bright Junis, 1991). Besides Tamil Nadu., the palm is concentrated in Bihar, Andhra Pradesh, Karnataka, Madhya Pradesh, Kerala, Orissa, and West Bengal (Anonymous, 1988).

Sri Lanka has nearly 10 million palms, most of which are concentrated in the Jaffna and Mannar areas, which are closer to the eastern coastal belt of Tamil Nadu (Nikawela et al., 1998). Palmyra is the national emblem of Cambodia, which has 3 million palms concentrated in ten densely populated provinces. Myanmar has a population of 2.35 million trees, which exist in the wild on the banks of Irawadhy. In Indonesia, the palm is concentrated in the Sulawesi, Timor, Maluku, West Irian, Jawa, and Madura islands (Kovoor, 1983).

If one looked into the extraterritorial distribution of the palm beyond Africa and Asia, one would observe that *B. aethiopum* and *B. flabellifer* was introduced by immigrant laborers and explorers into the cultivable lands, experiment stations, and botanic gardens of Brazil, Virgin Islands, Bahamas, Trinidad, Hawaii, and Florida (Morton, 1988).

David Fairchild, a plant explorer, made several expeditions to African countries and Sri Lanka in his quest to collect palms, and his renowned palmetum at Fairchild Tropical Botanic Garden in Miami, Florida, has the largest and most diverse palm collections in the world, presently having 1,500 accessions with 128 genera and 500 species, including *B. aethiopum* and *B. flabellifer*.

TAXONOMY

Kingdom	Plantae
Division	Angiosperms
Class	Liliopsida
Subclass	Arecidae
Super order	Areciflorae
Order	Arecales
Family	Arecaceae
Subfamily	Coryphoideae
Tribe	Borasseae
Subtribe	Lataniinae
Genus	*Borassus*

In the subtribe Lataniinae, among the four genera, *Borassodendron, Latania, Borassus,* and *Lodoicea, Borassus* is the widely distributed genus from Africa to New Guinea with the following distinct features (Uhl and Dransfield, 1987):

> Dioecious palms; leaves palmate; dissimilar staminate and pistillate inflorescences; staminate flowers in cincinni of 30-60, concealed in pits; bracteoles substending staminate flowers do not bear tuft of hairs; staminate rachillae several, wide; stamens six; pistillate flowers sessile in the axils of large coriaceous bracts; fruits sessile, symmetrical, endocarp forming pyrenes, pyrenes not deeply bilobed, seeds usually three, wider than thick, shallow to deeply bilobed; stigmatic remains apical.

A simple key proposed by Mathew (1991) will enable the readers to distinguish *Borassus* from other genera in Arecaceae (Box 9.1).

Although Beccari proposed seven species of *Borassus,* Uhl and Dransfield (1987) recognized only three to four species. The three major species and their distribution are as follows:

B. aethiopum C. Mart.	Africa
B. flabellifer L.	India, Sri Lanka, Southeast Asia
B. sundaica Becc.	Indonesia

A distinct feature of *B. aethiopum* is the swelling of its stem about 10 m from the base when the palm is around 30 years of age. After the bulge, the trunk narrows, and one or two additional bulges may appear between 90 to 120 years of age (Kovoor, 1983). However, we have noticed single swellings on the trunk at 10-15 m height from the ground level among the populations of *B. flabellifer* also. Another feature, Kovoor (1983) reported was that the drupe of *B. flabellifer* is mostly three seeded, whereas *B. aethiopum* often produces an abnormal number of seeds ranging from one to four. After comparing the chloroplast genome of African and Indian palmyra, Kovoor (1983) concluded that *B. aethiopum* and *B. flabellifer* have to be maintained distinct from each other. Whether the palm of Indonesia is to be kept as a separate species based on morphological features is of concern, and definite conclusions can be drawn by comparing the genomes of the populations from Indonesia with those of India and Africa.

BOX 9.1. A dichotomous key for classification.

1. Leaves simple, palmate
    2. Infl. terminal > 6 m long, monoecious, carpel 1, free, seed without a stony coat; petioles massively spinous
    2. Infl. interfoliar, < 1 m long, dioecious, carpels fused, seeds 3 each with a thick stony wall; petioles shortly spinous
1. Leaves compound, pinnate
    3. Stragglers, spadices as long as leaves with an apical flagellum, unisexual
    3. Shrubs or trees, spadices shorter than leaves without apical flagellum
        4. Leaves 2 pinnate, leaflets obliquely cuneate, apical margin caudate, monoecious
        4. Leaves 1 pinnate, leflets linear, oblong, margin entire
            5. Lowest leaflets modified into spines, dioecious, ovary of 3 free carpels, wild
            5. Lowest leaflets non spinous, monoecious, ovary of 1 carpel, cultivated
                6. Drupes 3.5 cm long, leaflets apically praemorse, leaf
sheath tubular, stem < 15 cm in dia
                6. Drupes 10-30 cm long, leaflets apically acuminate, leaf sheath 0, stem > 40 cm across

Source: Mathew, 1991.

BOTANY

Type Species: **Borassus flabellifer *L.***

Root

The root system is adventitious in nature. Roots originate from the internodes at the base of the stem. New roots are formed and added continuously at the higher levels of the stem as it enlarges. The main roots are string-like, giving rise to many lateral branches that branch off further. Mature roots have a sclerotic exodermis.

Stem

The stem is tall, unbranched with rare exceptions, and grows to a height of 20-30 m. Stem diameter ranges from 60-90 cm. The stem cortex is narrow, bearing vascular bundles that have massive phloem sheaths. Parenchyma cells close to the vascular bundles are isodiametric in shape and found in groups. Vascular bundles are 1,200-3,300 μ in length and 150-250 μ in width (Henry Louis et al., 1991).

Leaf

A crown of 20 to 30 large leaves tops the stem. Each leaf has a stout long petiole of 0.9 to 1.5 m long and a rachis with lamina. The petiole base is broad, having a vertical split, and it is persistent. The petiole base clasps the stem almost half the circumference and is surrounded by a fibrous network of stipules. The leaf petiole is adaxially grooved with sclerotic serrated margins. The leaf lamina is palmate, large, 1.0 to 1.5 m long with 60 to 80 compound segments. The lamina is plicate with ribs, extending along adaxial folds. Leaf segments are in duplicate. Foliar spirals are either left or right handed with alternate phyllotaxy, which helps in efficient utilization of sunlight. The stem of the young palm is ensheathed with a petiole base and dried leaves, whereas the stem of adult palms is marked with scars of the petiole base. The floral biology has been studied in detail by Nambiar (1954).

Inflorescence

Sex in palms can be differentiated only during flowering, which is seasonal. The inflorescence is interfoliar and occurs as a branched spadix, sheathed by many imbricated, fibrous, coriaceous spathes. The outermost spathe is the smallest, whereas the innermost spathe is the longest. A palm produces five to eight spadices annually.

Male Spadix

The male spadix has five to ten branches, and each branch is ensheathed by a spathe. Each branch has two to three branchlets or spikes. Each spike is stout, cylindrical, 30-40 cm long, and 2.5-4.00 cm wide. The width of the spike decreases gradually from base to apex. Numerous bracts imbricate the spikes. The bracts are wedge shaped, cuneate, retuse, and adhere by their lateral margins to the keel or back next to one bract above to form a cavity enclosing small scorpoid spikelets.

The number of spikelets in each spike ranges from 800-1,000, and each spikelet has 12-20 small sessile florets. Each flower in the spikelet is subtended by small, semitransparent, cuneate bracteoles. The flowers in spikelets are arranged in two vertically opposite rows that are serrated into each other, with each spikelet forming an arch with its convex side under most and the common receptacle of florets forming the other. The spikelets are arranged in parallel and nearly straight rows running clockwise or counterclockwise around the spike.

In total, a single spadix may contain 200,000 to 250,000 florets. Flowers are sessile. Sepals are three, imbricate and cuneate with truncate tips. Petals are three, short, ovate, and imbricate. Stamens are six; filaments connate into a stalk with corolla. Anthers are large, sessile, oblong, bilocular, and split longitudinally. Filaments are dark and dorsifixed.

It takes 16-25 days for the flowers to open after the opening of the spathes. Flowers open from the lower half of the spike and extend to both ends. Rarely does more than one flower open at a time. Pollen shedding is simultaneous with flower opening and pollen grains are ellipsoidal.

Female Spadix

The female spadix has only two to four branches or spikes, sheathed by spathes. Bracts imbricate the upper half of the spike, whereas the lower end is a smooth peduncle. A barren bract ensheaths the spike from where the flowers rise, and the terminal of the spike extending to 5-8 cm beyond the flowers is also ensheathed by barren bracts.

The number of female flowers in a spadix ranges 30 to 75. The female flowers are large and globose. Perianth is six lobed, fleshy, imbricate, reniform, and accrescent. Staminodes are six. Ovary is globose, three to four celled, pistils three to four, syncarpous, stigma three, sessile and recurved. Ovules are basal and erect.

Pollination

Although palymra flowers bloom throughout the day, most open between 8 and 11 a.m. Pollination is through insects (bees, wasps, beetles) and wind. It takes 120-130 days for the fertilized female flowers to mature into ripe fruits.

Fruit

The fruit is a fleshy drupe and weighs 1-3 kg. It is nearly spherical with a flat bottom. At the stalk end, six reniform perianth lobes are arranged in two whorls of three each. The epicarp is thin, fibrous, and brittle. The epicarp is creamy when young and turns black or pinkish yellow when it matures. The mesocarp is fleshy and fibrous when fully ripe.

Although the fruit develops from three fused carpels, the number of pyrenes within the mesocarp varies one to three. The endocarp is hard and covers each seed, which has a brown testa. The endosperm is gelatinous when the seed is young, filling the entire cavity after 60 to 70 days of fertilization. As the fruit matures, the endosperm hardens, forming a cavity at its center. The endosperm has fats, oils, and protein but no starch. The embryo is positioned below the germ pore and embedded within the endosperm. The germ pore is situated at the stigmatic end of the fruit.

Germination

Germination of palmyra seed nut is quite distinct from the other related genera of the family *Arecaceae*. Both show extension of the cotyledonary sheath.

Apocolon

Seeds of palmyra germinate within 15 to 20 days of sowing. During germination of the seed, the single cotyledon enlarges and emerges as a germ tube breaking the endocarp. The germtube elongates as a pale yellow sheath enclosing the embryo and carries it down. This hypocotyl is referred as the "apocolon," a storage organ of the first rudimentary leaf, which gets initiated at the center of the apocolon. The apocolon, although referred to as a "tuber," is neither a modified stem nor a root. It is a modified hypocotyl enclosing the coleoptile and coleorhiza. During early stages of apocolon formation, the spongy haustorium that is formed in the center of the seed stores the hydrolyzed sugars of the endosperm.

As days advance during apocolon maturity, the stored metabolites and nutrients from the endosperm of the seed are mobilized, leading to an increase in the weight of the apocolon. This results in a decrease in weight of the pyrene, leading to an empty endocarp, which detaches from the apocolon. As the apocolon matures, the sheath, which initially carried the apocolon down, looses its weight gradually and becomes thin and leathery. When the apocolon reaches optimum maturity, the radicle comes out of the apocolon base leading to the formation of roots. Finally, the plumule comes out of the apocolon and grows from the soil, forming the spindle-shaped first leaf. It takes 150-160 days for the first leaf of the seedling to come out of the soil.

CYTOGENETICS

In palms the gametic chromosome number ranges from 13 to 18, which is referred to as a dysploid series. Members of the subfamily Coryphoideae are considered to be the most primitive group, having $n = 18$ (Uhl and Dransfield, 1987). The genus *Borassus* is not monotypic—Sree Rangaswamy and Devasahayam (1972) examined the sex chromosomes in *B. flabellifer* during meiosis in pollen mother cells and found that the smallest of the 18 bivalents was heteromorphic related to sex determination. An estimation of the sex distribution in field population of the palms indicated a sex ratio of 1:1. Determination of sex at the flowering stage has no practical use if one has the preference to plant either male or female seedlings.

The eukaryotic genome consists of several highly repeated fractions in tandom, which are scattered throughout the DNA. Such sequences are distinguished as satellite (5-100 base pairs long), minisatellite (12-100 base pairs long), and microsatellite (1-5 base pairs long) DNA. A considerable variation occurs between chromosomes in DNA sequences due to a change in the number of copies of such highly repeated sequences. Hence mitotic determination of sex is possible in a dioecious palm such as *Borassus* by differentiating the sex chromosomes by the DNA banding patterns of the highly repeated sequences.

Localization of DNA sequences (genes) in chromosomes can also be performed by fluorescence in situ hybridization (FISH) without using radioactive isotopes. The probe DNA is linked to biotin and allowed to hybridize with the DNA to be detected in the chromosome. Later, the location of the bound, biotin-labelled DNA is detected by fluorescently labeled avidin, a protein that has a high affinity to bind with biotin (Karp, 1999).

Genome Characterization

Genetic improvement in palmyra is relatively difficult to achieve compared to coconut since the former has a long prebearing period. Hence approaches can be made by adopting molecular techniques for genome characterization, and DNA markers can be used to determine genetic diversity.

The chloroplastic DNA of palmyra from Sri Lanka and Senegal were digested with *Eco*RI and *Atug* I, and restriction-banding profiles of both were found to differ, indicating that both pertain to different species (Kovoor, 1983). DNA markers could be used as potential tools to characterize the germplasm and clone genes, as well as to construct linkage maps.

Polymerase chain reaction (PCR) is a highly sensitive technique. A single DNA molecule can be amplified from a mixture of total genomic sequences. The separation of nuclear and organelle genome is not required, and primers could be designed to amplify specific fractions of genomic or organellar DNA. PCR-based techniques are simple compared to other molecular methods, a few of which require radioactive labeling.

Randomly amplified polymorphic DNA (RAPD) is a PCR-based technique wherein oligonucleotides are used as primers to randomly amplify the DNA. Any polymorphism between the DNA sequences could be easily identified. This technique could be adopted to classify genotypes in palmyra (Wadt et al., 1997).

The genes that code for 5.8 S, 18 S and 26 S, ribosomal RNA in eukaryotes are present in tandemly repeated units with many copies per genome. The functional region of the rDNA sequences is highly conserved within a genus. Intervening between the rDNA repeats are the internal transcribed spacer (ITS) sequences; most variation occur in ITS regions within species and between genera. This ITS region could be exploited as a potential tool to study the genetic diversity in *Borassus*.

Germplasm

Gene banks are the sources for any crop improvement program. Breeding work can be contemplated after pooling all the available germplasm. Although efforts are being made to conserve and utilize the gene pool in other plantation crops, little has been done by the countries which exploit palmyra for its multifarious uses.

The known palmyra germplasm collections are in India. The collections for palmyra are at the Palmyrah Research Center at Srivilliputhur of the Tamil Nadu Agricultural University and at the two centers under All India Coordinated Research Project on Palms (of the Indian Council on Agricultural Research [ICAR]), viz. Agricultural College and Research Institute, Killikulam (Tamil Nadu Agricultural University [TNAU]) and Horticultural Research Station, Pandirimamidi (Acharya N.G. Ranga Agricultural University [ANGRAU]).

If the gene pool of *Borassus* palm is not preserved and enriched, a great deal of material may become extinct, as happened in Sri Lanka, where 2.5

million palmyra palms in its arid northeastern province have been devastated out of that country's total wealth of 10 million palms, due to ethnic conflicts. A massive planting program has been taken up by Sri Lanka's Palmyrah Development Board to revitalize the biodiversity available on the island (Prasad, 2000).

In addition to India, genetic materials can be pooled from Africa, Madagascar, Sri Lanka, Myanmar, Thailand, Cambodia, and Indonesia where the *Borassus* palms grow in the wild.

Genetic Improvement

The major economic product of palmyra is sap, and climbing is essential to tap it. Dwarfness and precocity in bearing are the two traits that are looked for while making germplasm collections, since climbing and working on the palm crown to collect sap are arduous. Palms that are precocious and bear fruits at less than 4 m height were identified by Suthanthira Pandian and Doraipandian (1991a) during their survey of the Thoothukudi district of Tamil Nadu. Such genotypes could be exploited in breeding program for selection and hybridization.

Selection of elite genotypes in *Borassus flabellifer* is mainly for sap and fruit yield. Sap yield generally ranges from 100 to 200 liters in a tapping season, although tapping response is a function of genetic and environmental factors and tappers' skill. Observing the sap yield over three to four seasons can help one to easily assess the true genetic potential of a palm.

Female palms are better sap yielders than males. At the Agricultural College and Research Institute at Killikulam, India, female palms yielding more sap than males were identified. One such accession, BF34, yielded 287 liters of sap in 85 days during 1999. This palm has yielded more than 150 liters of sap consistently between 1999 and 2003. Palms capable of giving more than 150 liters can be considered elite types. A shorter tapping duration of 60 to 70 days may reduce the drudgery of the climber. Economic yields can be obtained from certain genotypes even beyond 90 days.

Another parameter for selection is fruit yield. A female palm normally produces five to six spadices with nearly 50-60 fruits. Palms capable of yielding 200-300 fruits have been spotted and selected by the scientists of All India Coordinated Research Project on Palms. The gene bank at Killikulam includes progenies from such mother palms.

Besides sap and fruit yield, components of drupe also exhibit a wide range of variation that can be exploited during selection. Among the four parameters studied, flesh with epicarp had the highest heritability (97.85 percent) and genetic advance, suggesting the role of additive gene effects, whereas the other three traits, fruit weight, seed weight per fruit, and weight per seed, exhibited high heritability and low genetic advance indicating the role of nonadditive gene effects (Suthanthira Pandian and Doraipandian, 1991a). Hence the previously mentioned parameters could be used as yard-sticks for selecting high yielders for tender fruits, drupes, and apocolons.

The color of the epicarp in most females is brown to black and the authors have observed palms bearing drupes with red epicarps. Palms with reddish epicarps are said to be more advantageous in nutritional value and productivity than those bearing drupes with black epicarps (Kovoor, 1983).

The mesocarp of the drupe yields fiber (7-10 percent). If fiber is of major importance, palms with high ratios of fiber to mesocarp are to be selected; when the preference is for edible purposes, palms yielding fruits with low fiber content are to be looked for.

Besides fruit components, variations are also observed in seedling characters. Among traits, viz., seedling height, apocolon length, leaf length, and number of leaves, high estimate of heritability together with genetic advance was observed for seedling height and leaf length (Suthanthira Pandian and Doraipandian, 1991b). The selection of mother palms could also be based on these characters.

GROWING CONDITIONS

Soil, Water, Nutrient Requirements, and Crop Management

Although palmyra has adapted to semiarid regions receiving an annual rainfall below 750 mm, it flourishes well in the fertile soils of the river basins. The palm can grow in elevation up to 80 m above mean sea level (MSL). The palm is versatile, capable of growing in a wide range of soils including *Theri* and wastelands. It thrives well in sandy, red and black soils and river alluvium soils.

Seed sowing generally coincides with the monsoon season. Before sowing, ripe fruits gathered from elite trees are heaped in shade for four weeks. Seeds are loosened from the flesh of the drupe and soaked in carbendazim (0.1 percent) for 24 hrs to reduce the incidence of apocolon rot and increase germination (Sankaralingam, 1999).

A spacing of 3×3 m is recommended since the length of the petiole and lamina together will not exceed 3 m. By adopting this space, one can have 1,110 palms per ha[-1]. Pits are opened to a size of $30 \times 30 \times 60$ cm^3. The pit is half filled with a mixture of farmyard manure (10 kg) and topsoil. The seed is positioned with its germ pore (narrow conical end) facing down or sideways at 5 cm deep, and 100 g of malathion (4 percent) is sprinkled around it. The seed is then covered with soil.

Planting is done during the monsoon season. If there is failure of rain, pot watering immediately after planting and at alternate days up to a month is needed. Follow-up watering is done once a week during nonrainy periods for one year. If rainfall is scarce, pitcher irrigation can be given twice a month during tapping season for increased sap and fruit yield.

Young seedlings require shade with one or two dried palm leaves to protect them against sun and desiccating wind. To prevent stray cattle from pulling out the apocolons and feeding in young foliage, fencing is essential. If not, the tree-bearing period may extend up to 25 years.

Application of farmyard manure (FYM) (10 kg/pit) before planting the seeds and subsequently increasing the dosage (10 kg) for every two years, from 10 kg in the first year until a dose of 60 kg is reached on the eleventh year, is recommended. The same quantity can be continued thereafter. Experiments conducted at Srivilliputhur between 1982 and 1986 revealed that the highest sap yield was obtained when 60 kg FYM was applied before the onset of the northeast monsoon (Velu, 1989). Similarly, when 25-year-old palms at Killikulam were supplied with organic and or inorganic fertilizers, the sap yield was the highest with 50 kg FYM (Arumugam et al., 1994).

Gap filling and ploughing the interspaces before the onset of the monsoon and rectification of the basin (45 cm) around young seedlings are essential. The basin must be widened to 2 m for adult palms. Young leaves should not be removed from juvenile palms. When the palm attains a height of 2 m, one or two leaves may be removed. Adult palms can be defoliated up to 50 percent, leaving 16-22 leaves at the crown. Pruning all the leaves, leaving the bud and two to three leaves in the crown, is totally harmful to the palm. During the summer, leaf bases and old senescenced leaves should not be removed. However, they are to be removed and cleaned before tapping, and the stem should not be injured while cleaning.

During the rainy season, groundnut, gingelly, cowpea, and green gram can be grown as intercrops. *Moringa oleifera* fares well as a border crop. As mixed crops, ber, custard apple, and West Indian cherry can be planted.

Under ideal growth conditions, palmyra flowers at 10-12 years. The palm produces five to eight spadices a year. Sap is tapped from both male and female palms. Flowering is seasonal and it varies between tracts—from March-April to October-November for a period of 90-130 days. Tapping refers to inducing the phloem tissues of the peduncle to exude sap. A mud pot coated with slaked lime to avoid fermentation is tied to the spadix for collecting the sap. In a tapping duration of three to four months, a palm yields 100-200 liters of sap commencing from 12-15 years of planting. Fruit yield from female palms, if left untapped, ranges from 70 to 200 fruits.

Adaptation As Influenced by Biotic and Abiotic Stress

Palmyra palm is so hardy that it suffers little due to diseases, though it harbors a few pests. Since it is not domesticated like other palms, it might have evolved naturally over the years to withstand all adverse conditions, including drought.

The insects that infest the coconut palm, viz., *Oryctes rhinoceros, Rhynchophorus ferrugineus,* and *Opisina arenosella* are found to be associated with palmyra, and suitable management practices have been suggested (Sankaralingam et al., 1999). Palmyra has also been reported as a preferred host, compared to coconut, for *Opisina arenosella* (Murthy et al., 1995). Palmyra harbors the pests throughout the year irrespective of the season. A mealybug, *Palmicultor palmarum* recorded by Ali (1987) in Bangladesh was also been found by the authors to infest the peduncle of the palms.

No major diseases on palmyra have been reported thus far except for the bud rot caused by *Phytophthora palmivora.* The fungus caused serious damage to palmyra in the Krishna and Godavari districts of Andhra Pradesh during the beginning of nineteenth century. A nursery disease leading to rotting of the edible apocolon, caused by *Rhizoctonia solani,* is of concern to farmers. Several foliar diseases of minor importance have been reported (Sankaralingam et al., 1999).

However, in southern Florida, palmyra palms exhibiting symptoms similar to lethal yellowing in coconut were found to contain phytoplasma-like bodies in phloem tissues (Thomas, 1975).

PRODUCT USE IN HOME AND EXPORT MARKETS

A detailed account of the edible palmyra products, viz., sap, tender kernel, fruit, haustorium, and apocolon, their nutritive value, and the value-added products that can be made from this plant, has been given by Sankaralingam et al. (1999).

The Khadi and Village Industries Commission, India, through its state units has been successful in preparing and marketing prepacked sap, jaggery, candy, sugar, syrup, jam, and other value-added products. Palmyra sap has nutritional and medicinal values. However, it is a seasonal produce and can be preserved only for a short while. Research efforts have to be focused for its long-term preservation so that it is available to the consumer throughout the year. Entrepreneurs in Thailand have been successful in preserving, canning, and exporting the tender kernels (Morton, 1988), and such canned foods are available in ethnic stores in the United States. The private entrepreneurs of our country can make similar attempts.

Traditionally, palmyra climbers and their families make either jaggery and palm candy from sap. Governmental, nongovernmental, and state agricultural universities can train them in preparing value-added products, such as palm honey from sap; rava and related products from apocolon; peda and halva from haustorium; squash and leather from fruit; and candy, kheer, sarbat, and kova from tender kernels.

Regarding nonedible produce, all parts of the palm from the root to the crown are useful. Among them fiber from the petiole base and leaves are of major importance.

Members of the Manapad Women Workers' Palm leaf Industrial Cooperative Society Ltd., which was established during 1957 at Manapad in the Thoothukudi district of Tamil Nadu, use palmyra leaves and make 93 types of handicrafts and household articles that are not only marketed outside India but also exported to the United States and the United Kingdom through the Small Industries Producers' Association (SIPA) at Chennai. Recently, scarcity of leaves due to indiscriminate felling of the palms has been viewed as a problem for the society's activities.

In addition to training women in making handicrafts from the palm leaves, the Palmyra Workers' Development Association at Marthandam in the Kanyakumari district of Tamil Nadu collects palm candy produced by the members and trades these for better remuneration to the members.

The village of Kolachal in the Kanyakumari district is known for palmyra fiber processing. The unit, under Tamil Nadu State Palm Gur and Fibre Marketing Federation, pools raw petiole base fibers (kora fiber) collected from palmyra workers throughout the state. The unit combs, trims, sorts, dyes, blends the fibers, and exports them to the United States and the United Kingdom. Several private entrepreneurs are also involved in such trade around Kolachal.

There is wide scope for marketing the products of the palm in the international arena. Researchers at the Indian Plywood Industries Research and Training Institute, Bangalore, have concluded that palmyra wood has a potential to be used as block boards, joinery, and glued laminated wood using the whole stem (Jagadeesh et al., 1996).

RESEARCH AND DEVELOPMENT ORGANIZATIONS

All India Coordinated Research Project on Palms
Central Plantation Crops Research Institute

Indian Council of Agricultural Research
Kasaragod-671 124, Kerala, India
Phone: 91-499-2425533
Fax: 91-499-2425533 / 2430322
E-mail: khanhh–in@yahoo.com / cpcri@yahoo.com

All India Coordinated Research Project on Palms
Tamil Nadu Agricultural University
Agricultural College and Research Institute
Killikulam, Vallanadu (PO)-628 252, Thoothukudi Dist.
Tamil Nadu, India
Phone: 91-4630-261226
Fax: 91-4630-261268
E-mail: rasankar@usa.net
E-mail: library@md3.vsnl.net.in

All India Coordinated Research Project on Palms
Acharya N.G. Ranga Agricultural University
Horticultural Research Station
Pandirimamidi, Ramapachodavaram (PO)-533 288
East Godavari Dist. (AP), India
Phone: 91-8864-43577

Tamil Nadu State Palm Gur and Fibre Marketing Federation
Neera Mansion, 188, Ganga Reddy Road
Egmore, Chennai—600 008, PB No. 783
Tamil Nadu, India
Phone: 91-44-8266490

Small Industries Producers Association (SIPA)
F5, HD Raya Street
Eldoms Road
Teynampet, Chennai-600 018
Tamil Nadu, India
E-mail: sipa@vsnl.com

The Manapad Women Workers' Palmleaf Industrial
Co-operative Society Ltd.
Manapad-628 209
Thoothukudi Dist., Tamil Nadu, India
Phone: 91-4639-26223

Palmyra Workers' Development Association
Marthandam–629 165
Kanyakumari Dist.
Tamil Nadu, India

Andhra Pradesh State palmyra Cooperative Fed. Ltd.
Nidadavole (PO)
West Godavari Dist.
Andhra Pradesh, India

West Bengal State Palm Gur Federation
4, Bipin Pal Road
Calcutta-26, West Bengal, India

Indian Plywood Industries Research and Training Institute
Post Bag–2273, Tumkur Road
Bangalore-560 022
Karnataka, India

Palmyra Development Board (PDB)
244, Galle Road
Bambalapitiya
Colombo 04
Sri Lanka

University of Sri Jayewardenepura
Gangodawila
Nugegoda, Sri Lanka

University of Jaffna
Kokuvil (N.P.)
Sri Lanka

Food and Agricultural Organization of the United Nations
Viale delle Terme di Caracalla
00100 Rome, Italy
Web site: www.fao.org

Dept. of Animal Health and Production
Ministry of Agriculture, Forestry, and Fisheries
P.O. Box 177, Phnom Penn City
Cambodia

Project Manager
Indonesia, Australia Eastern University Project (IAEUP)
P.O. Box 3704, Denpassar 80001, Dall, Indonesia
Fax: 62-391-33482

The International Palm Society, Inc.
P.O. Box 1897, 810 East 10th Street
Lawrence, Kansas 66044-8897
E-mail: j.dransfield@rbgkew.org.uk

The Palm and Cycad Societies of Australia (PACSOA)
P.O. Box 1134
Milton, Queensland, Qld 4064, Australia
E-mail: enquiries@pacsoa.org.au

FUTURE OUTLOOK

Germplasm collection and gene-pool enrichment have to be made systematically and guidelines are to be framed for choosing traits to select elite genotypes.

Since *Borassus* has a long prebearing period compared to other palms, efforts are to be taken to conserve and evaluate the germplasm material properly over the years, which underscores the need to develop a descriptor.

A national-level nodal agency has to be earmarked, which should work in collaboration with the National Bureau of Plant Genetic Resources, New Delhi, in maintaining, cataloging, and exchanging germplasm.

Establishing a world palmyra germplasm repository in Tamil Nadu, India, can be contemplated with the guidelines and assistance of IPGRI and FAO.

The possibility of zygotic embryo collection and retrieval as done in coconut can be explored, especially while introducing exotic planting materials.

Since female palms are higher yielders of sap, fruits can be collected from such elite genotypes and used in crop improvement programs as well as to establish new plantations.

Products obtained from palmyra are natural and environmentally safe since no pesticide is involved. Due to an increase in demand for organic products, as well as the higher prices offered for such items, palm jaggery and candy have good markets.

Fibers from palmyra are eco-friendly and exported to more than 20 countries, with Japan, the United States, and the United Kingdom accounting for nearly 80 percent of exports (Anonymous, 1988). Due to increased consumer preference for natural products, export of palmyra fiber has high prospects.

Palmyra yields 13-20 tons of sugar per ha, compared to 5-8 tonnes per ha by sugarcane, which drains the groundwater potential of India. Raising palmyra, which needs minimum input, can be promoted in wastelands.

Flabelliferin FB, a steroid saponin isolated from palmyra fruit pulp, has antibacterial properties (Nikawela et al., 1998). The possibility of the fruit for pharmaceutical uses can also be explored.

The palm population is dwindling as a result of indiscriminate felling of the trees for fuel. Adult palms of age 80-100 are felled for timber. Even in such cases, replanting is essential.

Creating awareness among those who own and utilize the palm not to overexploit it is the need of the hour. This can be achieved with the cooperation of the state governments, Khadi and Village Industries Commission, Palm Gur and Fibre Marketing Federation, the nongovernmental organizations, and the scientists of agricultural universities, in collaboration with the Indian Council of Agricultural Research, New Delhi.

REFERENCES

Ali, M. 1987. Palm mealybug, *Palmicultor palmarum* (Homoptera: Pseudococcidae), new to the Indian sub-continent. *Annals of the Entomological Society of America* 80(4): 501.

Anonymous. 1988. *The Wealth of India: A Dictionary of Indian Raw Materials and Industrial Products,* Volume 2B. Council of Scientific and Industrial Research, New Delhi.

Arumugam, T., Suthanthira Pandian, I.R., and Doraipandian, R. 1994. Effect of organic and inorganic manuring on neera and yield of palmyrah (*Borassus flabellifer* L.). *South Indian Horticulture* 42(4): 236-238.

Davis, T.A. and Johnson, D.V. 1987. Current utilization and further development of the palmyrah palm in Tamil Nadu State, India. *Economic Botany* 41: 247-266.

Harichandran, C. and Bright Junis, D. 1991. Economics of palmyra industry. In *Proceedings of the Seminar on Modernising the Palmyrah Industry,* February 18 to 19, Haldane Research Center, Nagercoil, Tamil Nadu, pp. 111-117.

Henry Louis, T., Williams, B.C., and Mathai, A.M. 1991. palmyra pal: The state tree of Tamil Nadu. In *Proceedings of the Seminar on Modernising the Palmyrah Industry,* February 18 to 19, Haldane Research center, Nagercoil, Tamil Nadu, pp. 1-6.

Jagadeesh, H.N., Damodaran, K., and Aswathanarayana, B.S. 1996. Palmyrah wood: A potential source of wood raw material. *Wood News* 6(2): 20-23.

Karp, G. 1999. *Cell and Molecular Biology, Concepts and Experiments.* John Wiley & Sons Inc., New York.

Kovoor, A. 1983. The palmyrah palm: Potential and perspectives. FAO Plant Production and Protection paper, Food and Agriculture Organization, Rome.

Losty, 1982. *The Art of Book in India.* The British Library.

Mathew, K.M. 1991. *An Excursion Flora of Central Tamil Nadu, India.* Oxford and IBH Publishing Co., New Delhi.

Morton, J.F. 1988. Notes on distribution, propagation, and products of *Borassus* palms (Arecaceae). *Economic Botany* 42: 420-441.

Murthy, K.S., Gour, T.B., Reddy, D.D., Babu, T.R., and Zeheruddeen, S. 1995. Host preference of coconut black-headed caterpillar *Opisina arenosella* Walker for oviposition and feeding. *Journal of Plantation Crops* 23(2): 105-108.

Nambiar, M.C. 1954. A note on the floral biology of palmyrah palm *(Borassus flabellifer).* *Indian Coconut Journal* 7: 61-70.

Nikawela, J.K., Abeysekara, A.M., and Jansz, E.R. 1998. Flabelliferins: Steroidal saponins from palmyrah (*Borassus flabellifer* L.) fruit pulp. 1. Isolation by flash chromatography, quantification, and saponin-related activity. *Journal of National Science Council, Sri Lanka* 26(1): 9-18.

Prasad, R. 2000. Years of war leave Sri Lankan ecology damaged. Environmental News Service, March 17, 2000 (http://www.forests.org).

Sankaralingam, A. 1999. Management of tuber rot in palmyrah palm. *Journal of Mycology and Plant Pathology* 29: 98-99.

Sankaralingam, A., Hemalatha, G., and Mohamed Ali, A. 1999. A treatise on almyrah. In Hameed Khan, H., Ed., *All India Co-Ordinated Research Project on Palms.* Agricultural College and Research Institute, Killikulam-628 252, Tamil Nadu, India.

Sankaralingam, A. and Sunthanthirapandian, I.R. 1995. Rhinoceros beetle. A serious threat to palmyrah palm. *TNAU News Letter* 24: 1.

Sree Rangaswamy, S.R. and Devasahayam, P. 1972. Cytology and sex determination in palmyrah palm (*Borassus flabellifer* L.). *Cellule* 69(2): 129-134.

Suthanthira Pandian, I.R. and Doraipandian, A. 1991a. Genetic variability for fruit characters in palmyrah. In *Proceedings of the Seminar on Modernising the Palmyrah Industry,* February 18 to 19, Haldane Research Center, Nagercoil, Tamil Nadu, pp. 14-17.

Suthanthira Pandian, I.R. and Doraipandian, A. 1991b. Tuber development and variability for seedling characters. In *Proceedings of the Seminar on Modernising the Palmyrah Industry,* February 18 to 19, Haldane Research Center, Nagercoil, Tamil Nadu, pp. 20-24.

Thomas, D.L. 1975. Possible link between palm species and lethal yellowing of coconut palms. *Proceedings of the Florida Horticultural Society* 87: 502-504.

Uhl, N.W. and Dransfield, J. 1987. *Genera Palmarum: A Classification of Palms Based on the Work of Harold E. Moore Jr.* The International Palm Society, Lawrence, Kansas.

Velu, G. 1989. Impact of organic manuring on palmyrah. *Madras Agricultural Journal* 76(10): 592-598.

Wadt, L.H., Sakyama, N.S., Pereira, M.G., Tupinambo, E.A., Riberio, F.E., and Aragao, W.M. 1997. In *International Symposium on Coconut Biotechnology,* Merida, Yuc. Mexico, December 1-5, 1997, p. 14 (abstract).

Chapter 10

Rubber

Y. Annamma Varghese
Saji T. Abraham

INTRODUCTION

The Para rubber tree of commerce, *Hevea brasiliensis,* produces around 99 percent of the world's natural rubber (NR). The English name "rubber," attributed to the great English scientist Joseph Priestly, was derived from its ability to rub out pencil marks. In all the Indo-European languages except English the name is derived from the Amerindian name for rubber trees: *cachuchu* (weeping wood). The Spanish name *caucho* indicates the ecological origin of the majority of rubber-bearing plants, because Spain was the principal colonial power in South America at the time when rubber started to become known in Europe. French interest continued throughout the eighteenth century, a particularly important contribution being made by the botanist Jean Baptiste Fusée Aublet who published the first taxonomic description of *Hevea* in 1775. The taxonomy of the genus has undergone considerable changes since then. The name itself is a Latinized version of the Ecuadorian Indian name, *Hheve,* and there was some earlier competition with other possible names such as Siphonia and Caoutchoua (Jones and Allen, 1992).

The nineteenth century saw the first vulcanization of rubber, the first development of specialized machinery and techniques for manufacturing rubber goods, the rise of commercial trade in rubber, and the first efforts to cultivate rubber. The rubber boom in Southeast Asia (about 1910) led to the first great surge of planting, after which the crop developed remarkably

The authors are grateful to Dr. N. M. Mathew, Director of Research, RRII, Kottayam for the encouragement, support, and critical review of the manuscript. Acknowledgments are also due to Sri. K. P. Sreeranganathan Sr., artist/photographer, for providing relevant photographs.

403

from a wild jungle tree to a major domesticated crop in about 40 years. The growth of plantations in Southeast Asia was favored by rapid developments in the transportation sector, such as railways and steamships, and the opening up of the Suez Canal. By the end of nineteenth century, NR became one of the major plantation crops. The evolution and adaptation of natural rubber, an important ingredient of many modern facilities, thus illustrates the evolution of modern civilization. Natural rubber is a microcosm of world history, embodying the change from colonialism to independence and from parochialism to internationalism. The industry has withstood the many shocks that it has been subjected to by the vicissitudes of the world economy (Jones and Allen, 1992).

NATURAL RUBBER: ITS VALUE TO MANKIND

The cultivation of rubber is the chief means of livelihood for millions of people in many of the rubber-growing countries of the Far East, who depend directly or indirectly on wages or profits received from the production of plantation rubber. Natural rubber forms the raw material for a very large number of articles useful to humans. Rubber is used to fabricate hard and strong structural materials; soft, yielding, comfortable materials; resilient elastic materials; conductors and nonconductors of electricity; shock absorbers; mountings for motors and other machinery; transmission belts and gaskets; hoses for transporting gases and liquids; transparent materials; translucent materials; articles of clothing to keep out rain or to control the figure; sports goods; cements; paints; plastics; pharmaceuticals; and, above all, tires, the chief outlet for rubber.

ORIGIN AND DISTRIBUTION

Natural rubber, an industrial raw material of strategic importance, is considered to be one of the most versatile agricultural products, finding its use in about 50,000 products the world over. In India, around 35,000 products are made out of NR and it supports an industry with a turn over of about 140,000 million annually. From an agroecological point of view, rubber is an eco-friendly tree species.

The genus *Hevea* occupies the whole of the Amazon River basin in Brazil also extending south and north to parts of Brazil, Bolivia, Peru, Colombia, Ecuador, Venezuela, French Guiana, Suriname, and Guyana. All ten species grow in Brazil, whereas seven are found in Colombia, five in

Venezuela, and four in Peru (Wycherley, 1992). *H. brasiliensis* extends about half of the range of the genus, mainly occupying the region south of the Amazon, extending to the states of Acre, Mato Grosso, and Parana in Brazil, parts of Bolivia, Peru, north of the Amazon to the west of Manaus as far as the extreme south of Colombia.

The species is now grown mainly in the tropical regions of Asia, Africa, and South America, in countries such as Malaysia, Indonesia, India, Sri Lanka, Thailand, China, Philippines, Vietnam, Cambodia, Myanmar, Bangladesh, Singapore, Nigeria, Cameroon, Democratic Republic of Congo, Ivory Coast, Ghana, Zaire, Liberia, Brazil, and Mexico. However, the major share of the total production comes from tropical Asia.

Historic Development of the World Rubber Industry

Pre-Columbian Era of Rubber

The uses of rubber had evidently been well established before the time of Columbus as methods of tapping the trees and processes for making crude articles from the latex were well-known. The earliest European visitors to the new continent recorded the use of articles made of rubber, such as protective garments, balls for playing games, and syringes. This was an illustration of the advanced civilization that European explorers encountered in the Western hemisphere. Schurer (1957) traced the use of rubber in Peru, in the Mayan civilization in Yucatan, and in the ancient Mexican civilization, where it was an important element of religious rites. In Yucatan and Mexico City, rubber was preserved and used in the liquid form and in ceremonies; it symbolized the blood in human sacrifices (Baulkwill, 1989). However, the literature on rubber dates only from the first visit of Columbus to America, when the use of rubber was already well established in the Western hemisphere.

Evolution of Rubber As an Important Commodity (1700-1870)

The sixteenth and seventeenth centuries constituted a period of dullness in the history of development of rubber in Europe. In the mid-eighteenth century, rubber was rediscovered by French scientists. The astronomer-geographer Charles-Marie de La Condamine in 1745 in his book on Amazon described how Amerindians in Ecuador and Brazil made torches, boots, bottles, and syringes employing an elastic resin or gum called *caoutchoua,* "weeping wood," from the sap of a tree called *Hh'eve.* In the period 1750-1800 French scientists discovered solvents for rubber, after which rubber solutions were used to make flexible tubes—catheters for medical purposes. Rubber solutions were applied to the silk fabric of the hydrogen balloons that made successful ascents in France in 1783. By 1790, European scientists reinvented the rubber syringe, a device long known to the Amazonian Indians.

The nineteenth century saw rubber transformed from a curiosity—used chiefly as an eraser—into an important commercial and industrial product. Thomas Hancock in 1821 developed the revolutionary masticator, which shredded and compressed solid rubbers and scraps to a warm, homogenous, moldable, or rapidly soluble mass, which was a remarkable technical advance of that period that made rubber processing possible. Charles Mackintosh in 1823 in Scotland made waterproofed fabrics using his unique fabric-rubber-fabric sandwich and coal tar-derived naphtha as the solvent.

The introduction of vulcanization methods during 1839-1844 by the two giants of the rubber industry, Thomas Hancock and Charles Goodyear, created a new demand for rubber by virtue of the properties, viz., resistance to hot and cold air and melting. The process of cold vulcanization by dipping thin rubber articles in sulphur chloride solution was first patented by Alexander Parkes in England in 1846.

Fast growth in the automobile industry in the nineteenth century resulted in an ever-growing demand for more and more rubber. Annual consumption expanded from a negligible amount to thousands of tons. This century saw tremendous advancement in scientific knowledge and technical development. The chemical structure of the rubber molecule was revealed, and the first "rubberlike" material was synthesized artificially. The three important contributions to the new rubber plantation industry include (1) the discovery of vulcanization, (2) the transfer of *Hevea* from the West to the East in 1876, and (3) the introduction of new methods of tapping and latex coagulation.

The Domestication of Rubber: Early History

The successful transfer of *H. brasiliensis* to the East and the subsequent establishment of rubber plantations there, in response to the rising demand for the raw material, were the results of the political forethought and practical abilities of a few people combined with favorable circumstances. The domestication of *Hevea* in the East is the most spectacular story in the history of rubber, as between 1870 and 1914 a novel plantation industry of some 900,000 ha was established to meet the new industrial demand (Baulkwill, 1989).

The real initiative of the historic domestication of NR in the Far East was taken by Sir Clements Robert Markham of the India office in London (Williams, 1968). Similarly leading roles were played by Joseph D. Hooker, Director of Kew Gardens, London; Henry Wickham, planter, rubber trader, and naturalist; and Henry Nicolas Ridley, protégé of Hooker and from 1888 Director of the Singapore Royal Botanic Gardens (Baulkwill, 1989).

The entire development of rubber plantations in the East is attributed to the famous Wickham collection (Lane, 1953; Wycherly, 1968; Dean, 1987). In 1876, Wickham collected 70,000 seeds of *H. brasiliensis* from Brazil— near Boim on the Rio Tapajoz and from the well-drained undulating country toward the Rio Madeira. This area produced excellent wild rubber. About 2,700 of these seedlings were raised at Kew, England, and 1,919 of them were dispatched during 1876, mainly to Sri Lanka; a few went to Malaysia, Singapore, and Indonesia (Baulkwill, 1989; Dean, 1987). In India, rubber was first received in 1878 from Sri Lanka.

A few initial unsuccessful attempts were made to establish *Hevea* in Calcutta Botanic Gardens, after which Ceylon was selected as the most suitable place for the acclimatization, and thus rubber cultivation moved outside the Indian mainland and was established in Ceylon (now known as Sri Lanka) (Thomas and Panikkar, 2000). Sri Lanka was thus the center of early activity, with the Heneratgoda Botanic Gardens in Colombo becoming a major source of rubber seeds for domestic use and for exports (Jones and Allen, 1992). Early development works on diseases (Petch, 1911), latex flow, and the use of acetic acid to coagulate latex (Parkin, 1910) were also carried out in Sri Lanka.

In India, cultivation of rubber started in 1878 in the town of Nilambur, Kerala, south India, as a forest crop, using the planting materials brought from the Royal Botanic Gardens, Heneratgoda, Ceylon (Petch, 1914; Dean, 1987). With the help of state administrators, British planters initiated commercial rubber plantations in India, and the product was exported to London. Gradually, a rubber-planting community was developed in Kerala, and by 1947, 73 percent of the cultivated area (115,304 ha) under rubber was controlled by Indian companies (Sarma, 1947; Thomas and Panikkar, 2000).

In the twentieth century, rubber experienced a tremendous boost in demand because of the development of automobile industry and its need for tires. This intensified the search for new sources of rubber, encouraged efforts to cultivate rubber-yielding plants, and motivated chemical research into the phenomenon of elasticity of natural rubber. The response of rubber growers to the high increase in demand was quick, and rubber succeeded equally on highly capitalized estates and tiny family holdings. In the two world wars of the twentieth century, rubber was found to be a necessity for national survival. In peace, it became essential to the enjoyment of the conveniences and amenities of modern life.

AREA, PRODUCTION, AND PRODUCTIVITY

Trends in Planted Area and Production

The extent of planted area and production from the 1970s to late 1990s in NR-producing countries reveals the fluctuations in area and production during this period (Table 10.1). Declines, notably in Malaysia and Sri Lanka in Asia and Nigeria and Zaire in Africa, were more than offset by the expansion of more than one million ha in Indonesia, with an increase of 500,000 ha in Thailand, China, and India. In relative terms, many smaller-producing countries, such as Brazil, Vietnam, Guatemala, Ivory Coast, and Cameroon also performed well in terms of area expansion. A few newcomers to the industry include Bangladesh and the most recent entrants to the industry, Gabon and Guinea in Africa, where production started in the 1990s. A number of producers have reentered the industry since the early 1970s, e.g., GREL in Ghana and Mandala in Malawi. Cambodia and Vietnam, whose industries were virtually destroyed by war, have recovered their rubber area, very slowly in the case of Cambodia, where political conflict continues, and quite spectacularly in Vietnam, where most of the present 300,000 ha were planted since reunification in 1975. There has been a rapid expansion of planted area in China, which now has the world's fourth largest rubber base with 610,000 ha.

Area Under Rubber

In the late 1990s, Asian countries held the maximum area under rubber cultivation, constituting about 93 percent of the total cultivated area, with African and Latin American countries contributing 5 percent and 2 percent respectively. Indonesia is the largest cultivator of rubber with an area of 3,329,000 ha, followed by Malaysia (1,315,000 ha), Thailand (1,968,000 ha), China (618,000 ha), and India (573,000 ha) (Table 10.2).

Production and Productivity

Natural rubber is produced largely by small farmers in the developing countries of Asia, Africa, and Latin America. Despite increasing output

TABLE 10.1. Estimates of NR planted area and NR production in the mid-1970s and mid-1990s.

Country	NR Planted area (in thousand ha)		NR production (in thousand tonnes)	
	1970s	1990s	1970s	1990s
Bangladesh	0	47	—	2
Cambodia	45	60	19	40
China	300	610	65	441
India	235	533	137	566
Indonesia	2,322	2,322	842	1,594
Malaysia	1,981	1,635	1,532	980
Myanmar	63	133	15	26
Papua New Guinea	13	18	6	6
Philippines	45	92	29	64
Sri Lanka	228	161	144	105
Thailand	1,400	1,960	382	2,067
Vietnam	42	300	22	203
Asia Total	6,674	7,871	3,193	6,094
Cameroon	20	44	16	58
Ivory Coast	20	72	16	102
Gabon	0	12	—	9
Ghana	6	17	3	10
Guinea	0	6	—	3
Liberia	118	120	84	59
Malawi	0	2	—	2
Nigeria	240	200	66	58
Congo	90	40	30	11
Africa Total	494	513	215	312
Brazil	20	180	19	56
Guatemala	12	36	7	35
Mexico	6	15	—	—
Other L. American countries	8	15	7	23
Latin America Total	46	246	33	114
World	7,214	8,630	3,441	6,520

Source: IRSG, 2000.

from other countries, the share of production of the top three rubber-producing countries, namely Thailand, Indonesia, and Malaysia, was still quite high but falling from its peak of 80.5 percent in 1969 to 74.0 percent in 2003. The influence of Malaysia, the leading producer for many years with

TABLE 10.2. Area under NR in the different rubber-producing countries.

Country	Year	Area (in thousand hectares)
Thailand	1998	1,972.0
Indonesia	1998	3,344.0
Malaysia	1998	1,568.0
India	1998	553.0
China	1996	618.0
Sri Lanka	1998	158.0
Brazil	1998	180.0
Nigeria	1999	150.0
Vietnam	1997	275.0
Ivory Coast	1998	96.0
Cameroon	1997	42.0
Philippines	1999	92.0
Others	1999	658.5
World	1999	9,706.5

Source: IRSG, 2001.

an output of 1.6 million tonnes in 1986, has been declining as economic development has progressed. Malaysia currently produces less than 1 million tonnes annually, whereas Thailand and Indonesia have annual outputs of about 2.9 million tonnes and 1.8 million tonnes, respectively.

Most of the rubber-producing countries have increased their production and productivity substantially since 1980. The increase in Indian yields has been very impressive given the dependence on smallholdings and size of the industry. The production increased from a very low level of 23,730 tonnes in 1955-1956 to a high of 707,100 tonnes in 2003. The average yield per hectare increased from 353 kg/ha in 1955-1956 to 1,592 kg/ha in 2002-2003, the highest productivity among NR-producing countries. Thailand had a remarkable increase in NR production where the productivity as low as 350 kg/ha in 1975 had a sharp increase to 1,130 kg/ha in 1990s. As per the recent trends in production, Thailand contributes 36.0 percent, followed by Indonesia with 22.4 percent, Malaysia with 12.3 percent, India with 8.9 percent, and other countries contributing the balance 20.4 percent in the world NR production scenario (IRSG, 2004) (Table 10.3).

TABLE 10.3. Leading NR producing and consuming countries as in 2000.

	NR production				NR consumption		
Sl no	Country	Tonnes	World Share %	Sl no	Country	Tonnes	World Share %
1	Thailand	2,873,100	36.0	1	USA	1,078,500	13.6
2	Indonesia	1,792,200	22.4	2	China	1,485,000	18.8
3	Malaysia	985,600	12.3	3	Japan	784,200	9.9
4	India	707,100	8.9	4	India	717,100	9.1
5	China	480,000	6.0	5	Malaysia	420,700	5.3
6	Vietnam	384,000	4.8	6	Korea	332,800	4.2
7	Ivory Coast	127,000	1.6	7	Germany	251,000	3.2
8	Sri Lanka	92,100	1.2	8	France	300,200	3.8
9	Liberia	110,000	1.4	9	Thailand	298,600	3.8
10	Philippines	84,000	1.1	10	Brazil	256,000	3.2
11	Others	344,900	4.3	11	Others	1,985,900	25.1
12	World total	7,980,000	100.0	12	World total	7,910,000	100.0

Source: IRSG, 2004.

WORLD TRADE IN NATURAL RUBBER

The total consumption of Natural Rubber in the world was to the tune of about 7,910,000 tonnes in 2003. The United States has been the largest consumer of natural rubber among the world rubber-consuming countries for the past two to three decades with a steep increase in consumption from 666,000 tonnes in 1975 to a high of 1,078,500 tonnes (13.6 percent) in 2003. China now ranks first with an annual consumption of 1,485,000 tonnes (18.8 percent), followed by Japan with 784,200 tonnes (9.9 percent) and India consuming 717,100 tonnes (9.1 percent) annually as per 2003 statistics (Table 10.3).

BOTANY

Cytotaxonomic Background

Hevea brasiliensis belongs to the family Euphorbiaceae. At present, a total of ten species are recognized in the genus *Hevea,* viz., *H. benthamiana,*

H. brasiliensis, H. camergoana, H. camporum, H. guianensis, H. micro-phylla, H. nitida, H. pauciflora, H. rigidifolia, and *H. spruceana. Hevea brasiliensis,* along with three other species originally described under the genus *Siphonia,* was brought under the genus *Hevea* by J. Mueller Argo-viensis in 1865 (Wycherly, 1992). Murca Pires reported the last species, *H. camergoana,* in 1981. All of the species except *H. microphylla* were placed in the subgenus *Heveae,* whereas *H. microphylla* was placed under the sub-genus *Microphyllae* (Wycherly, 1992).

Morphology

The rubber tree is a sturdy, quick growing, erect perennial, growing to a height of about 30 m with an economic life span of more than 30 years in plantations. It has a straight trunk with light gray bark, and the branches de-velop to form an open leafy crown. The young plants show characteristic growth patterns of alternating periods of rapid elongation and consolidated development. The leaves are arranged in groups or storeys, with each storey having a cluster of spirally arranged trifoliate glabrous leaves and extra flo-ral nectaries present in the region of insertion of the leaflets.

Hevea is a deciduous tree that sheds its leaves during December through February (wintering—partial or complete) followed by refoliation and flowering. The plant is monoecious with unisexual flowers produced in pyr-amid-shaped panicles in the axils of leaves having numerous small male flowers and fewer female flowers of bigger size. The female flowers are confined to the tip of the panicles and their branchlets. The ovary is tricar-pellary syncarpous, which on pollination develops in to a three-lobed de-hiscent capsule (regma) with three large, mottled seeds. Pollination is by in-sects, and fruits ripen in five to six months after fertilization. Seeds contain an oily endosperm.

Cytogenetics

Hevea brasiliensis and its congeners are uniformly diploids with $2n = 2x = 36$ (Majumder, 1964). The 36 chromosome genera are therefore probably old tetraploids based on $x = 9$. A few putative hybrids between several dif-ferent species of *Hevea* have been collected in the wild, and interspecific hybrids have also been produced in breeding programs. Indian scientists have reported development of an experimental tetraploid (Saraswathy Am-ma et al., 1984) and synthesis of a triploid (Saraswathy Amma et al., 1980) in the clone RRII 105. A spontaneous triploid (Nazeer and Saraswathy

Amma, 1987) and a genetic variant with dwarf stature (Markose et al., 1981) have also been identified.

Commercial Adoption of Vegetative Propagation

Credit for the widespread adoption of vegetative propagation by budding goes to the horticulturist Van Helten in 1916 in collaboration with two planters, Bodde and Tass, in Indonesia (Dijkman, 1951). This was one of the contributing factors for the rapid growth of the rubber plantations with genetically improved clones. Depending on the color and age of the buds (scion) and the age of the stock seedlings, there are two types of budding, viz., brown and green budding. In brown budding, which was traditionally adopted in *Hevea*, vigorously growing seedlings of about one year of age are used as stock plants with brown buds collected from mature shoots of the same age. In the green budding technique developed in N. Borneo during 1958-1960 (Hurov, 1960), three- to five-month-old, healthy, vigorous seedlings were used as stock, and tender green-colored buds were used as scion. Stock-scion union takes place in about three weeks. A budding success of 90 to 100 percent is possible if quality stock and scions are used and budding is done in favorable seasons.

Brown-budded stumps are hardier than the green-budded ones. In general, green budding results in higher budding success than brown budding during summer (Marattukalam and Premakumari, 1982), which is of advantage since the green-budded plants can be raised in polyethylene bags in summer and can be planted in the field during the ensuing planting season itself. Another type of propagation of young budded plants, viz., young budding, is produced by budding very young stock plants of seven to eight weeks of age (Seneviratne, 1995) with green buds from shoots of comparable girth.

The possibility of forming a three-part tree with improved rootstock, trunk, and crown components originally envisaged by J. S. Crammer at Java in 1926 (De Vries, 1926) was further developed into the crown-budding technique. The crown-budding technique was first adapted for developing trees resistant to South American leaf blight (SALB) (Yoon, 1973). However, due to the practical difficulties of developing crown-budded plants on a large scale, this technique is adapted only on a limited scale. The feasibility of using the bench grafting technique, i.e., indoor grafting of stock plants (Marattukalam and Varghese, 1998) was also established in rubber. This type of budding can be adapted under adverse climatic conditions, such as severe cold, extreme heat, and heavy rains when outdoor grafting will not be successful.

Different types of advanced planting materials were developed to reduce the immaturity period. Among the advanced planting materials, such as mini- or maxistumps, core stumps, and polybag plants, the polybag plants are being used on a commercial scale. Uniform growth, fewer casualties, early establishment, cost reduction, less weed growth, etc., are some of the advantages possible through the use of appropriate advanced planting materials.

Bark Anatomy

The economic product from rubber trees, latex, a specialized cytoplasm, is contained in a laticiferous system in the bark differentiated by the activity of the vascular cambium. The early researchers during the first quarter of the twentieth century realized the importance of studies on the anatomy of the laticiferous system. Our basic understanding of all aspects of the anatomy of laticifers in *Hevea* originated from the contributions of Bobili- off (1919, 1920, 1923) and Gomez (1974), and Premakumari and Panikkar (1992) later provided further detailed information of *Hevea* laticifers.

Hevea bark has two distinct zones: (1) the outer hard bast, which is more protective in function with mostly discontinuous and nonfunctional latici- fers, and (2) the inner soft bast, containing productive and continuos latex vessels differentiated from the vascular cambium. The latex vessels are of the articulated anastomizing type with vessels of the same ring intercon- nected tangentially. Latex vessels are more concentrated in the region near the cambium in the soft bast. The laticifer differentiation from the cambial derivatives is a rhythmic process, and a ring of laticifers is produced each time, forming concentric rings alternating with layers of other phloem tis- sues. The laticifers are generally oriented 3-5° to the vertical, in a counter- clockwise direction.

The quantity of laticiferous tissue in a tree is determined by factors such as the number of latex vessel rows, density of vessels within rows, distribu- tion pattern of and distance between the vessel rings, as well as bark thick- ness and girth of the tree. Highly significant clonal difference has been re- ported for these characters (Premakumari and Panikkar, 1992). During tapping, only a thin slice of bark of 1-1.5 mm thickness is shaved off to cut open the latex vessels, leaving the cambium intact for bark regeneration. Since the renewed bark is exploited after the consumption of virgin bark, bark regeneration is of great significance.

Although the studies on the structural aspects of laticifer anatomy have helped the conceptual development of scientifically correct tapping sys- tems, studies at the ultracytological level (Dickenson, 1965; Gomez, 1974) have contributed significantly in furthering our understanding of the physi- ology and biochemistry of latex.

PHYSIOLOGY OF LATEX FLOW

In an untapped tree, no movement of latex occurs inside the laticiferous systems once it is synthesized. Latex flow from a tapped tree is an abnormal physiological phenomenon inducted by tapping. The cytoplasmic nature of latex, both in terms of structure and function, was elucidated first by high-speed centrifugation (Moir, 1959) and later by electron microscopy (Dickenson, 1965; Gomez and Moir, 1979).

On tapping the trees, the latex flows out mainly because of the very high turgor pressure in the latex vessels and elastic contraction of walls after a sudden release of turgor resultant of cutting the vessels open. After a few hours, capillary forces regulate the flow until it ceases as the latex coagulates and plugs the vessels. An inherent clotting mechanism is present within the latex vessels, which is responsible for the cessation of latex flow (Southern and Yip, 1968) consequent to the plugging of the opened ends of laticifers. Milford et al. (1969) proposed an index—the plugging index, which is a time flow constant expressed as the ratio of the flow for the initial five minutes to the total volume yield—for measuring the extent of plugging.

Sethuraj (1981), through a theoretical analysis of yield components of *H. brasiliensis,* represented the effect of the major yield components by the formula $Y = FlCr/p$ where Y = yield; F = initial flow rate per unit length of the tapping; l = length of the tapping cut; Cr = percentage rubber content; and p = plugging index. Of these four major components, the length of tapping cut is determined by the girth, and thus the growth vigor, and is influenced by the total biomass production and partitioning between growth and latex production. The inherent characteristics, exploitation systems, and the environment also influence the other three components. Yield is a manifestation of the major yield components as well as a number of anatomical, physiological, and biochemical subcomponents influencing the main components (Sethuraj, 1992).

TAPPING OF THE RUBBER TREE

The tree is exploited by tapping the bark—a process of controlled wounding of bark to cut open the latex vessels by regularly removing a thin shaving of bark from the surface of the tapping cut at specified intervals.

The native method of crop harvest from wild rubber trees involved a crude method of latex extraction, which caused irregular secondary growth, poor yield, and early abandonment of trees (Cook, 1928). Henry Ridley, Director of the Singapore Royal Botanic Gardens during 1888-1911, a very prominent figure during this period, made significant contributions to evolve the basic method of the present-day tapping of rubber trees. This excision method of tapping known as Ridley's method involves the removal of a thin layer of bark from the cut end at each tapping, thus permitting a smooth flow of latex and allowing the bark to regenerate (Ridley, 1897). Ridley made other significant contributions such as establishing the importance of tapping in the early morning, the effects of daily and alternate daily tapping, and the best age of starting tapping.

A budded tree is opened for tapping when it attains a girth of 50 cm at a height of 125 cm from the bud union, whereas trees of seedling origin are tapped on attaining 55 cm girth at 50 cm height. One tapping panel covers half the circumference of the tree. The tapping cut of a budded tree should have a slope of about 30° to the horizontal. Since the latex vessels in the bark run at an angle of 3-5° to the right, a cut from high left to low right will open maximum number of vessels. Tapping is a highly skilled work where the depth of tapping should be within 0.5 mm of the cambium to get optimum yield without injuring the cambium. Tapping should commence in the early morning, as late tapping will reduce the latex yield due to increased transpiration leading to lower turgor pressure.

Ethephon As a Chemical Yield Stimulant

The use of yield stimulants such as ethephon (2-chloro-ethyl-phosphonic acid) (Abraham et al., 1968; De'Auzac and Ribaillier, 1969) revolutionized our ability to regulate commercial latex yield at will, based on physiological and economic requirements. Judicious use of ethephon-like stimulants helps to sustain rubber plantations during times of low rubber prices. Detailed field studies in different institutes have resulted in specific commercial recommendations based on different situations. The different methods of application include employment in the bark, groove, lace, and also in multiple bands, tape, and soil.

The action of all yield stimulants is mediated through ethylene production, either endogenously or through hydrolytic decomposition (Abraham et al., 1968; Gomez, 1983). Stimulation delays latex vessel plugging and prolongs duration of latex flow (Boatman, 1966). Tupy and Primot (1976) observed that the increase in pH as a result of stimulation enhanced the rate of latex regeneration through increased invertase activity.

Stimulation with ethephon results in a 20 to 100 percent increase in latex yield. The stimulation effect has been reported to last for three months (Chapman, 1951). However, the yield stimulation will result in a 2 to 7 percent reduction in the dry rubber content, even though it has no effect on rubber properties (Sumarno-Kertowardjono et al., 1976).

Tapping Notations

Tapping notations are a set of internationally approved signs and symbols describing the method of tapping and its frequency. The description of notations for tapping systems was first introduced by Guest (1939) and later revised by Lukman (1983). The notations for tapping method include symbols for length of the tapping cut, direction, frequency, etc. The type of cut given is denoted by a capital letter. "S" indicates spiral cut, "V" indicates a V-cut, "C" indicates circumference (used for two or more unspecified cuts on a tree tapped on the same day), and "Mc" indicates minicut (5 cm or less in length). Length of the cut denotes the proportion of the circumference of the trunk that is cut for tapping, and is represented by a fraction preceding the symbol of cut, except in minicut where the actual length is given in cm. Examples are S for full spiral cut, V for one full V-cut, C for one full circumference (unspecified cut), $\frac{1}{2}$ S for one-half spiral cut, $\frac{1}{4}$ S for one-fourth spiral cut, $\frac{1}{3}$ V for one-third V-cut, $\frac{3}{4}$ S for three-fourths spiral cut, $\frac{1}{2}$ C for one-half circumference cut, and Mc2 denotes a minicut 2 cm in length. Frequency of tapping is denoted as a fraction, which explains the interval between tappings. The numerator of the fraction is d, which denotes the tapping period (day), whereas the denominator denotes the actual interval between tappings in days or in a fraction of a day. Examples are $\frac{d}{1}$ for daily tapping, $\frac{d}{2}$ for tapping once in two days, $\frac{d}{3}$ for tapping once in three days, etc.

The notation for yield stimulation using chemical stimulants consists of three parts. The first part indicates the stimulant and its concentration, the second part indicates the place of application, quantity of stimulant, and method of application, and the third part indicates the number of applications and its periodicity. The notation ET2.5%tPa2(1).16/y(2w) means the tree is stimulated with ethephon (ET) at 2.5 percent concentration applied on panel (Pa) with 2 g of stimulant per application on 1 cm band in 16 applications per year at fortnightly intervals.

Tapping Systems

In rubber, the response to different tapping systems varies from clone to clone. Budded trees are tapped on half-spiral alternate days ($\frac{1}{2}$ S $\frac{d}{2}$) and seedlings are tapped on half-spiral third days ($\frac{1}{2}$ S $\frac{d}{3}$). However, in high-yielding budded clones prone to tapping-panel dryness (TPD), reduced tapping intensity of once in three days is recommended, though, in general small growers prefer higher frequencies. The latex yield from trees varies with clones, age of the tree, fertility of the soil, climatic conditions, tapping system followed, as well as skill of the tapper. Intensive tapping is done for maximum exploitation of the tree before felling it for replanting. The methods adopted include increased tapping frequency, extension of tapping cut, opening of double cuts, and use of yield stimulants. Upper-level panel tapping, such as controlled upward tapping (CUT), is now gaining popularity for longer exploitation of the virgin bark above the lower panel (Vijayakumar et al., 2000).

LATEX BIOCHEMISTRY AND RUBBER PRODUCTION

Latex is a specialized form of cytoplasm containing a suspension of rubber and nonrubber particles in an aqueous serum. Besides rubber and water, the other components of latex are carbohydrates, proteins, lipids, inorganic salts, and other minor substances (Archer et al., 1963). Rubber particles are the most important constituent in latex, making up 30 to 45 percent of the volume. Rubber particles are usually spherical, but are sometimes oval or pear shaped (Dickenson, 1965), and are strongly protected in suspension by a film of adsorbed protein and phospholipids (Archer, 1964). This protein-phospholipid layer imparts a net negative charge to the rubber particle contributing to colloidal stability (Bowler, 1953).

Lutoid particles are next in abundance making up 10 to 20 percent of the volume of latex. Lutoids are subcellular membrane-bound bodies (Southern and Yip, 1968) that enclose a fluid serum known as lutoid serum or B serum, which is a destabilizer of rubber hydrocarbon. One to 3 percent of the latex volume is occupied by Frey-Wyssling particles, which are yellow and spherical (Dickenson, 1965). Quebrachitol is the most concentrated soluble carbohydrate in the latex along with sucrose and glucose (Low, 1978). The total protein content in the fresh latex is about 1 percent, of which about 20 percent is adsorbed on rubber particles, an equal quantity found in the bottom fraction, and the remainder in the serum phase (Archer et al., 1963). Lipids constituting about 1.6 percent of latex, play a vital role in the stability and colloidal behavior of latex. Other minor components are thiols (main reducing agents), various amino acids, organic acids, and inorganic ions, such as potassium, magnesium, copper, iron, sodium, calcium, and phosphate (Archer et al., 1963).

Rubber biosynthesis is initiated with the generation of acetyl coenzyme A, which is converted to isopentanyl pyrophosphate (IPP), and then is polymerized into rubber. The sustained production of rubber is largely dependent on the latex biosynthetic capacity of a clone for in situ latex regeneration between two successive tappings. Biochemical components, viz., total solid content (TSC), thiols, inorganic phosphorus (Pi), magnesium (Mg), and sucrose play an important role in latex regeneration (Jacob et al., 1986). The quantity of rubber per unit of sucrose present in the latex, termed the sucrose conversion efficiency, is an indirect estimate of the overall regulation of the metabolic conversion of sucrose in to rubber (Dey et al., 1995).

GENETIC IMPROVEMENT

Genetic improvement of *Hevea*, despite being a very elaborate and time-consuming procedure, has paid rich dividends in increasing yield by several folds and making available several high-yielding clones for commercial planting.

Breeding and Selection

Rubber breeding in Southeast Asia is one of the outstanding success stories of plant breeding. The first steps in breeding were taken by Dutch workers in Java and Sumatra. They showed that yields per tree were variable, that clonal propagation was possible, and that individual clones differed from one another. With ordinary seeds as planting material in the initial years of rubber research, productivity was around only 300 kg ha per year. Subsequent selection of high-yielding mother trees and multiplication by budding resulted in early primary clones with improved yield potential. From the 1920s onward, budded clones and superior seedling populations progressively supplemented the random open pollinated seedlings. Since World War II, clones have been dominant and rubber breeding has long been a matter of selecting and testing clones from genetically variable populations. Subsequently, breeding new clones through hybridization resulted in a series of high-yielding hybrid clones.

Breeding Objectives

Hevea breeding is aimed at the synthesis of ideal clones with high production potential combined with desirable secondary attributes, such as initial vigor, smooth thick bark with a good latex vessel system, good bark renewal, high growth rate after opening, tolerance to major diseases, wind, TPD, good response to stimulation, and low frequency tapping. Clones with early attainability of tapping girth and high initial yields are preferred over clones with higher yields in later phases of exploitation.

Specific objectives, however, vary depending on agroclimatic and socio-economic requirements. Marginal and nontraditional areas demand priority for development of clones resistant to prolonged drought, high summer and low winter temperatures, strong winds, altitudes, etc. Such situations also demand genotypes responding well under high-density planting, poor soil fertility, and low-input agriculture based on sustainable farming systems. In places where labor is relatively cheap, clones suitable for high-intensity tapping are economic, whereas under labor shortage low-intensity tapping is preferred. Since rubber planters are predominantly small holders, breeding objectives are to be streamlined to take care of their specific needs also.

Breeding Methods

The conventional breeding methods in *Hevea* are introduction, ortet selection, and hybridization and clonal selection.

Introduction of Clones

Introduction or exchange of available clones among the rubber-growing countries in the early years constituted the original breeding pool in each country. Recent introductions under bilateral and multilateral clone exchange programs organized by the International Rubber Research and Development Board (IRRDB) and Association of Natural Rubber Producing Countries (ANRPC) are confined to potential clones of good performance. Thus popular clones evolved in different countries are being introduced to member countries and evaluated under the local agroclimatic conditions, and promising selections are being recommended for large-scale planting. In India, so far, 127 Wickham clones evolved in countries such as Malaysia, Indonesia, Sri Lanka, Thailand, China, and Ivory Coast formed the exotic component of the gene pool, in addition to the recent introduction of around 7,000 genotypes of the wild Brazilian germplasm.

Ortet Selection

Ortet selection or mother tree selection is the oldest breeding method aimed at systematic screening for outstanding seedling genotypes resultant of natural genetic recombination. Clones developed through ortet selection are called primary clones. Screening of extensive seedling plantations in Indonesia, Malaysia, and Sri Lanka resulted in a good number of early primary clones of importance such as Tjir1, PR 107, GT 1, BD 10, AVROS 255, Gl 1, PB 28/59, Mil 3/2, and Hil 28, of which clones such as GT 1 and PB 28/59 are still widely planted. In India, earlier mother tree selections include 46 clones, of which RRII 1, RRII 4, RRII 5, RRII 6, RRII 33, RRII 43, RRII 44, etc., are among the potential clones (Marattukalam et al., 1980).

Hybridization

Hybridization and clonal selection, the most important conventional method in *Hevea* offers scope for exploitation of heterosis in hybrid progeny of potential parent clones. Desirable recombinants once selected can be fixed easily through vegetative multiplication. As a result of hybridization and selection, a good number of hybrid clones of commercial significance have been evolved. The early primary clones as parents of the first hybridization series resulted in early hybrid clones of commercial significance, such as RRIM 500 and RRIM 600 series, the yield levels of which were much superior to those of the parent clones. The best clones in each series were further used as parents in subsequent series. Rubber breeders followed this sort of cyclical generation-wise assortative mating (GAM) over the past 50 years. In Malaysia, the Rubber Research Institute developed clones of RRIM 500 to 1000 series, whereas the Prang Basar Institute in the private sector selected certain PB clones of commercial significance. The Indonesian Research Institute for Estate Crops in Java and Sumatra (Balai Penelitian Purkeburan Medan [BPPM]) evolved Profestation voor Rubber (PR), Algemene Verneiging Rubber Planters Cost Kust Sumatra (AVROS), BPM, S'Lands Caoutchoue Bedrijven (LCB), PPN, and Rubber Research (RR) clones. The Rubber Research Institute of Ceylon (RRIC) clones originate from Sri Lanka, Kohong Rubber Estates (KRS) clones from Thailand and Haiken, and YRITC and South China Academy of Tropical Crops (SCATC) clones from China.

In India crop improvement programs were initiated in 1954 and a large number of hybrid progeny was developed and evaluated for potential recombinants. The early hybrid clones developed by the RRII include RRII 100, 200, and 300 series (Annamma et al., 1990). Among the clones of RRII 100 series, RRII 105 is a highly successful and popular clone (Nair and George, 1969; George et al., 1980). The clones RRII 116 and RRII 118 are outstanding for growth vigor, but with only medium yield. Clones RRII 203 and 208 (Saraswathy Amma et al., 1987) and RRII 300 and 308 (Premakumari et al., 1984) are the best selections in the 200 and 300 series respectively. In another set of hybrid series, nine clones revealed marked heterotic increases for yield over the first three years of tapping (Licy et al., 1998).

Out of this, five clones designated as RRII 400 Series, viz., RRII 414, RRII 417, RRII 422, RRII 429, and RRII 430, have been included in Category III of planting recommendations of the Rubber Board, India, since 2001. These clones have shown significant improvement in yield over the clone RRII 105, to the tune of 23 to 46 percent during the first eight years of tapping in the small-scale evaluation trial (Saraswathyamma, 2003). Another set of 50 hybrid clones were identified as having better potential for genetic advance based on early growth and yield (Varghese and Mydin, 2000).

Overcoming the Constraints in Breeding

The major limitations in rubber that hamper breeding and quick release of cultivars include the narrow genetic base, heterozygous nature of the crop, seasonal and nonsynchronous flowering pattern, low fruit set, long breeding and selection cycle, and lack of fully reliable early selection parameters (Varghese and Mydin, 2000).

Pollen storage and induction of off-season flowering have been suggested to overcome the limitation of nonsynchronous flowering. Low average fruit set of less than 5 percent limits the size of legitimate families available for selection. Treatments such as applying boric acid sucrose solution to the stigma and enclosing the panicles in butter paper (instead of sealing individual flowers) yielded relatively higher fruit set (Mydin et al., 1989), but the extent of fruit set was still low and inadequate demanding further investigations.

The lack of fully reliable early selection parameters hampers quick release of clones. The widely adopted method for juvenile selection, viz., a modified Hamaker-Morri-Mann method (Tan and Subramaniam, 1976), where two- to three-year-old plants are tapped successively for quantifying the latex yield, identifies a fair proportion of high yielders. Annamma, Licy, et al. (1989) suggested an incision method at an age of one year with a view to determining latex yield at a still younger age. A study on early evaluation incorporating clones of high-, medium-, and low-yield potential, revealed performance index at an age of two years to be good enough for selection of a fair proportion of high-yielding clones at an early age (Varghese et al., 1993). Biochemical components such as sucrose, total solid content, thiols, and inorganic phosphorus at immature phase, were suggested to provide more precision for early selection (Licy et al., 1998). However, with the available early prediction methods, nursery yield can be considered as only a fair indicator of mature yield. Detailed investigations at major yield components and stable subcomponent levels are required for development of fully reliable parameters for early prediction of yield.

GENETIC STUDIES

In *Hevea* the characters of economic importance are, in general, controlled polygenically. For effective selection and utilization of genotypes for breeding programs, a clear understanding of the magnitude of genetic variability for the desired characters as well as estimates of genetic parameters, such as heretability and genetic advance, is essential. Successful attempts have been made at RRII in this direction (Premakumari and Panikkar, 1992; Mydin et al., 1992; Licy et al., 1992). Similar studies revealed the preponderance of additive gene action in the statement of yield and yield components suggesting better scope for selection based on specific characters.

An understanding of the genetic divergence of the parent clones used in crossbreeding as well as component clones of polyclonal seed gardens is of paramount importance for proper exploitation of gene recombinants. Studies indicate that in spite of the narrow genetic base, sufficient genetic diversity existed in the original Wickham material (Markose, 1984; Varghese et al., 1997). In studies on genetic diversity of Wickham clones, based on yield and various yield components, two sets of 40 and 35 Wickham clones were grouped into eight (Mydin et al., 1992) and nine (Abraham et al., 1997) genetically divergent clusters. In another recent study, 80 wild accessions were grouped into nine different clusters (Abraham, 2000). Selection of genetically unrelated parent clones offers greater chances of heterosis and superiority of resultant progeny. Use of molecular markers has simplified such studies considerably.

GENETIC BASE AND GERMPLASM RESOURCES

Need for Broadening the Genetic Base

The genetic base of *Hevea* in the East is very narrow, limited to a few seedlings originally collected from a minuscule of the genetic range in Brazil referred to as the "Wickham base" (Simmonds, 1989). These few seedlings were collected from a minuscule sample of the genetic range of the species in Boim, near the Tapajos River in Brazil (Wycherley, 1968; Schultes, 1977). Using this gene pool substantial improvement in productivity has already been achieved over the past five decades. The original narrow base has further narrowed down through the unidirectional selection for yield, a cyclical generation-wise assortative mating pattern and a wider adaptation of clonal propagation by budding. A cyclical breeding pattern with the best genotypes in one breeding cycle used as parents for the next has led to the selection and release of clones that are more or less related. The parentage of popular clones bred in various rubber-growing countries can be traced back to a handful of parent clones (Tan, 1987; Varghese, 1992). The main emphasis given on productivity alone has resulted in a certain amount of genetic erosion with respect to characters such as disease resistance.

Rubber plantations in the east are under the potential threat of the highly damaging disease South American leaf blight, which is caused by *Microcyclus ulei,* a fungus that is prevalent in the American hemisphere (Chee and Holliday, 1986; Edathil, 1986). None of the Wickham clones have been reported to have resistance to SALB. There are also indications of erosion of genes controlling resistance to *Oidium* and *Gloesporium* in the original Wickham material (Wycherley, 1977) and several minor diseases assuming epidemic proportions. Apart from the problem of disease susceptibility, variability for resistance/tolerance to various abiotic stress situations, such as drought, cold, high elevations, and high-velocity winds, assumes significance in the present-day context of extension of rubber cultivation to marginal and nontraditional areas. In order to select clones adaptable to such specific locations, the base material should contain ample genetic variability and a broad genetic base (Varghese and Abraham, 1999).

Fresh Germplasm: The 1981 Collection

Considering the urgent need for broadening the genetic base, fresh, wild germplasm was collected from the center of diversity in Brazil as a result of the joint expedition of the International Rubber Research and Development Board and the Brazilian government in 1981. The team collected 64,736 seeds (Ong et al., 1983; Mohd Noor and Ibrahim, 1986) and budwood from 194 presumably high-yielding mother trees from the states of Acre, Mato Grosso, and Rondonia (Ong et al., 1983). The Malaysian and Ivory Coast germplasm centers established this germplasm in nurseries. Varying proportions of the wild germplasm were established in different rubber research institutes. In India, a total of 7,562 genotypes, including 126 ortet clones, were received from the Malaysian center and have been established in traditional and nontraditional areas (George, 1989). Their conservation, evaluation, and utilization in breeding programs are in progress.

Characterization, Evaluation, and Utilization

Studies of the 1981 collection conserved in nurseries or field evaluation gardens are underway in different institutes across the world, where the genotypes have been introduced and established. The wild genotypes are first characterized using a set of descriptors and subjected to a preliminary evaluation in source bush nurseries. Selections for desirable characters are then planted in field evaluation trials laid out in an augmented design. Wild germplasm, in a phased manner, is subjected to characterization, where it is documented based on a set of descriptors that consists of passport data, plant-type data at the juvenile phase, followed by data on relevant qualitative and quantitative characters of interest in the later stages of growth (Varghese and Abraham, 1999). Data on morphological characterization of a set of wild germplasm at the juvenile stage revealed interesting variability (Abraham et al., 1994). Interesting morphological variants for growth form, flower color, and fruit shape were observed among the wild genotypes (Madhavan et al., 1997).

Preliminary evaluation of the Brazilian genotypes established in the traditional areas in India revealed wide variability in the early growth phase with respect to morphological parameters and juvenile yield (Annamma, Marattu Kulam, et al., 1989; Abraham et al., 1992). Several individual accessions with superiority over the control clone, RRII 105, for certain characters were identified in the immature phase for yield and other characters. Genotype MT 999 had a higher number of latex vessel rows, with higher diameter and cross-sectional area for the latex vessels (Abraham et al., 1992; Reghu et al., 1996). A few genotypes with higher test tap yield in the immature phase than that of the popular clone RRII 105 were identified (Abraham et al., 1992, 2000; Madhavan et al., 1993). Provenance-wise comparison of genotypes for various morphological and anatomical characters and test-tap yield showed the genotypes from Mato Grosso to be superior to those from Acre and Rondonia (Abraham et al., 1992, 2000; Reghu et al., 1996, in India; Lam et al., 1997, in Vietnam). In Malaysia, however, Rondonian genotypes recorded superior yield compared to those from Acre and Mato Grosso (IRRDB, 1996).

Studies for screening of wild germplasm for resistance to major diseases and to drought and cold conditions have been initiated. Field observations have identified a general field tolerance of the wild germplasm belonging to Mato Grosso to shoot rot disease (Mercy et al., 1995). In general, the wild genotypes recorded poor yield, compared to the domesticated control clones. Similar results have also been reported from Malaysia, (IRRDB, 1996), Ivory Coast (Clement-Demange et al., 1997), and Vietnam (Lam et al., 1997). Searches for superior timber characters in wild germplasm have also been initiated in India and other countries.

Incorporation of superior wild genotypes as one of the parents in Wickham × Amazonian hybridization programs has been initiated in various rubber research institutes world wide including those in India. In India, among

12 cross combinations involving wild genotypes and popular Wickham clones, hybrids of the cross RRII 105 and RO/JP/3/6 had superior test-tap yield at two years (RRII, 1994). Another set of nine superior genotypes has been used in crosses with popular Wickham clones, such as RRII 105 and RRIM 600, and the progenies are under evaluation. However, there is a long way to go for realizing the expected benefits from such a large collection of wild germplasm. Utilizing powerful molecular tools, such useful genes should be identified and utilized in the production of transgenic rubber plants for specific stress situations.

BIOTECHNOLOGICAL INTERVENTIONS

In *Hevea,* advances in biotechnology offer various possibilities, such as propagation of elite planting materials, eliminating stock-scion interaction, conservation and characterization of genetic resources, construction of linkage maps, development of transgenic plants incorporated with agronomically important genes, and marker-assisted selection (MAS) (Varghese et al., 2001).

In Vitro Approaches

Micropropagation

The first successful studies on micropropagation by Carron and Enjarlic (1982) were followed by subsequent work on the in vitro multiplication of *H. brasiliensis* using seedling explants by Gunatilleke and Samaranayake (1988), Seneviratne and Flegmann (1996), Seneviratne and Wigesekare (1997), Sobhana et al. (1986), Asokan et al. (1988), and Lardet et al. (1998), reported micropropagation using explants derived from mature clonal trees. In the RRII, protocols were developed for the in vitro generation of *Hevea* plants using shoot tips derived from mature field-grown trees (Sobhana et al., 1986; Asokan et al., 1988).

Protoplast Culture

Only very limited studies related to protoplast culture in *Hevea* are known. Nurhaimi et al. (1993) successfully isolated protoplasts from calli and cell suspensions from anther tissues. During *Hevea* protoplast culture, a rapid decrease in viability was associated with an increase in ethylene production; these reactions are common to stress conditions. A number of physiological phenomena associated with lack of mitotic division of stem protoplasts of rubber were observed. However, after optimizing different parameters Sushamakumari et al. (1998) developed an efficient path way for callus induction from protoplasts.

Somatic Embryogenesis

Attempts to develop somatic embryogenesis as an in vitro propagation technique were first made in the 1970s. The first plantlet from somatic embryogenesis through anther culture was reported from China by Wang et al. (1980). This was followed by reports by Wan et al. (1982) and Asokan et al. (1992). Carron and Enjarlic (1982) reported somatic embryogenesis and plantlet regeneration from inner integumental tissues of fruits in France. Asokan et al. (1992) reported somatic embryogenesis and plantlet formation from some commercial clones using inner integumental tissues. Carron et al. (1998) observed clonal variation in the micropropagation efficiency through somatic embryogenesis. In their study, the in vitro plantlets of clone PB 260 recorded better growth in the field than those of PR 107, and were also superior to other controls. Kumari Jayasree et al. (1999) reported high-frequency somatic embryogenesis and plantlet regeneration from immature anthers of commercial clones.

In Vitro Conservation

In vitro conservation of *Hevea* accessions through cryopreservation is a viable option for the long-term storage of germplasm in field gene banks, although attempts in this direction have been rather scanty. In the late 1990s, two efficient cryopreservation protocols, one using a classical freezing process and the other using a simplified freezing process, were developed for embryogenic calli of a commercial clone of *Hevea*, PB 260 (Englemann et al. 1997). High survival and rapid regrowth, as well as production of somatic embryos, were obtained with calli cryopreserved using both protocols. The simple freezing protocol was used successfully with a second commercial clone, viz., PR 107. However, long-term studies are required to establish the feasibility of this in vitro technique as a convenient and secure alternative to the present ex situ method of conservation.

Genetic Transformation

Gene transfer in *Hevea* has been successfully established using the genes particle gun and by *Agrobacterium* mediation (Arokiaraj and Wan, 1991;

Arokiaraj et al., 1994; Arokiaraj, Jones, et al., 1996). Anther-derived calli were transformed with vectors carrying the beta-glucuronidase (GUS) and the neomycin phosphotransferase (npt II) genes (Arokiaraj, Jaafar, et al., 1996). Arokiaraj et al. (1998) reported the establishment of a gene transfer system for *Hevea* with *Agrobacterium tumefaciens* GV2260 (p35SGUSINT) and LBA4404 (pAL4404/pMON9793), and that the CaMV 35S promoter directs -glucuronidase expression in the laticiferous systems of transgenic plants.

Attempts toward incorporation of tolerance to drought and tapping-panel dryness have been initiated at the Rubber Research Institute of India. Calli derived from immature anthers of the high-yielding clone RRII 105 were transformed with the gene coding for sorbitol 6-phosphate dehydrogenase, isopentanyl transferase, superoxide dismutase, and 1-aminocyclopropane 1-carboxylic acid (ACC) synthase genes in the antisense orientation under the control of the CaMV 35S promoter. The vectors containing the npt II and GUS genes were used for selection and scoring of the transformants.

Biochemical and Molecular Markers

Advances in biochemical markers, viz., isozymes, and DNA markers such as restriction fragment length polymorphism (RFLP), random amplified polymorphic DNA (RAPD), and simple sequence repeats (SSR) or microsatellites, offer fast and attractive adjuncts to the elaborate conventional genetic analysis in perennial species such as *Hevea* (Varghese et al., 2001). These molecular markers are potential tools in characterizing diversity and analyzing the *Hevea* genome. Seguin et al. (1996) were able to group *Hevea* germplasm in to six genetically divergent clusters utilizing data on isozyme and RFLP markers. A comparison of genetic variability in the wild and Wickham germplasm using isozymes (Chevallier, 1988) and RFLPs by Besse et al. (1994) revealed significantly higher variability in the former.

The applicability of RAPD markers for genetic analysis in *Hevea* was evaluated using 42 informative primers in a set of 24 clones from the breeding pool of the Rubber Research Institute of India. Estimation of genetic distance among the tested clones facilitated identification of genetically divergent clusters (Varghese et al., 1997). Polymerase chain reaction (PCR)-based markers, such as RAPDs, have wide application for routine analysis in a breeder's laboratory in comparison to RFLPs since they are easier and faster for screening large numbers of genotypes and assays of genetic variability at the whole genome level (Varghese, 1998). More recently, Lespinasse et al. (2000) reported development of a saturated genetic linkage map of *Hevea* spp. based on RFLP, amplified fragment length polymorphism (AFLP), microsatellites, and isozyme markers.

Molecular aspects of resistance to diseases and drought are areas deserving attention. Rozeboom et al. (1990) isolated and purified hevamine, an enzyme (molecular weight [MW] 29,000) with both lysozyme and chitinase activity, from *H. brasiliensis* latex, which was homologous to certain pathogenesis-related proteins from plants. Variations in RAPD profiles between *Phytophthora*-resistant and susceptible genotypes were reported (Jacob, 1996). Thulaseedharan et al. (1994) observed variation in the RAPD profile of TPD-resistant and susceptible seedling trees and identified two random polymorphic DNA markers tolerant to tapping-panel dryness. Molecular approaches for identification and incorporation of resistance to TPD assume much importance since many high yielders are susceptible to this syndrome, which has not yet proven to be a disease.

CROP MANAGEMENT UNDER DIFFERENT CROPPING SYSTEMS

Rubber is grown under a wide range of management conditions. Management practices in rubber has been modified and improved over the years to suit the different cropping systems adopted. In the Amazonian forests of South America, where wild rubber was exploited for latex, field management was limited to maintaining access paths to the highly scattered trees in the jungle. Here the trees after they reach tappable size are exploited for 20 to 30 years until the bark reserves are exhausted. Intercropping in smallholdings raised in newly cleared forest areas was a very old practice. Food crops, such as rice, maize, cassava, bananas, and other crops, were raised for the first two to three years. The management practices were kept at a minimum, with hand slashing to control the weed growth (Watson, 1989).

In most of the private sector plantings and in some government-funded smallholder plantings and large development schemes, rubber is usually raised as a monocrop. Here intensive management practices are used, as the main objective is to get maximum economic returns from the plantations by reducing the immaturity phase and maximizing the yield during the mature phase. Over the years, the cropping cycle has been reduced with the existing plantations being replanted with more high-yielding clones as they become available.

Crop Management Practices

Rubber trees face more competition in the immature phase than in the mature phase. As rubber is usually grown in plantations in association with ground cover, and with intercropping becoming a common practice in the modern cropping system, proper field upkeep is essential during this phase so as to provide the best environment for the young rubber plants to survive the competition. Important management practices during this phase include establishment of ground cover, intercropping, weed management, mulching, induction of branching, pruning, and thinning. These practices are meant to reduce the immaturity period and maintain optimum plant density (Punnoose et al., 2000).

Cover Crop Establishment

Leguminous creepers were found to be the best cover crops as they were superior in terms of high nitrogen-fixing capacity, fast growth, tolerance to shade, drought, pests, and diseases, and provided minimum competition for nutrients with rubber plants. Leguminous cover crops reduce the cost of cultivation by substantially reducing the amount of costly nitrogenous fertilizers, which are supplemented by the cover crop. They reduce soil erosion to a great extent from hilly rubber-growing tracts in addition to reducing expenses on weeding, increasing soil moisture, and organic matter. Some of the important leguminous cover crops grown in rubber plantations include *Pueraria phaseoloides, Mucuna bracteata, Calapogonium muconoides,* and *Centrosema pubescens.*

Weed Management

Traditionally, weed control in rubber plantations has been a hand labor by slashing the planting strips in the first four years until the canopy closes. Application of herbicides for weed control in rubber plantations is limited to mostly dry seasons. The two main types of herbicides are preemergent (applied in soil) and postemergent (applied on the weeds). Some of the common herbicides used in rubber plantations are paraquat (Gramoxone), 2,4-D (Fernoxone), which controls broad-leaved weeds, and Glyphosate (Glycel, Roundup, or Weedoff), which controls grass weeds. To achieve weed management most economically and efficiently, an integrated approach is practiced in which a combination of manual and chemical controls are used along with the establishment of cover crops.

Intercropping

In general, rubber will grow best in monoculture, with interrow areas protected by leguminous cover crops during the immaturity period. On the other hand, the land in such a system is underutilized with regard to the interrow space. In the first two to three years after planting, before the tree canopy closes, it is possible to cultivate a variety of suitable crops in the interspaces available in the young plantations. This brings income to the grower during rubber's immaturity period of six to seven years. When properly done, intercropping has been reported to enhance the growth of rubber (Jessy et al., 1996). Legume cover can also be incorporated in the intercropping system by growing legumes in the alternate interrows or available interspaces.

Intercrops should not be planted too close to the young rubber plants in order to minimize competition for nutrients. Since intercropping necessitates soil tilling at different degrees, it is desirable to restrict the practice to level lands and gentle slopes so as to avoid soil erosion (Punnoose et al., 2000). Commonly cultivated intercrops include pineapple, banana, ginger, turmeric, vegetables, coffee, some medicinal plants, and fodder crops. Yogaratnam et al. (1995) observed intercropping of tea with rubber to be successful in Sri Lanka.

Irrigation

Irrigation in rubber is usually restricted to nontraditional rubber-growing areas, where drought is more pronounced and the plants are under severe water stress. Irrigation during the summer can enhance growth and reduce the immaturity period (Pushparajah and Haridas, 1977; Omont, 1982; Jessy et al., 1994). Irrigation in rubber nurseries is also practiced in traditional areas for good growth of seedlings and polybag plants.

In the traditional areas, summer irrigation during the young phase can reduce the immaturity period by six months to one year (Jessy et al., 1994). Philip (1997) reported the beneficial effect of irrigation in reducing the effect of cold weather in northeastern parts of India where the winter temperature can drop to –10°C. In some drought-prone, nontraditional rubber-growing areas in India, irrigation up to 50 percent of the estimated crop water requirement could reduce the immature period from ten years to six years (Vijayakumar et al., 1998).

Soils and Manuring

Rubber grows on a vast majority of the acidic soils of the humid tropics. Deep, well-drained soils of pH below 6.5 and free from underlying sheet rocks are well suited for rubber. In the traditional tract in India, the soils are mostly of laterite and lateritic types; however, red and alluvial soils are also seen in some areas. In general, the laterite and lateritic soils are very porous, well drained, moderately to highly acidic, deficient in available phosphorous, and variable with regard to available potassium and magnesium. However, adverse physical, chemical, and physiographic features of soils influence growth and productivity of rubber to a considerable extent (Krishnakumar and Potty, 1992).

Rubber plants have been found to respond well to systematic manuring. Proper manuring during the immature stage accelerates growth, thereby reducing the unproductive phase and optimizing productivity in the mature phase. The major nutrients N, P, K, and Mg have positive effects on growth and yield, and application of NPK increases the yield substantially (George, 1962). Fertilizer recommendations have been formulated for different regions based on soil types. A system for routine discriminatory fertilizer recommendation based on soil and leaf analysis has been set up in the different rubber-growing regions of India. This facility has the advantage of avoiding unnecessary use of fertilizers resulting in monetary benefit and reduced environmental pollution. A diagnosis and recommendation integrated system can be used to improve DFR.

DISEASES AND PESTS

Rubber plants are susceptible to various diseases but their economic importance and severity vary with the climatic conditions and cultural practices of a region. Important leaf diseases include abnormal leaf fall caused by *Phytophthora* spp., or secondary leaf fall by *Oidium heveae*, *Colletotrichum gloeosporioides*, and *Corynespora cassicola*.

Until recently, abnormal leaf fall disease (*Phytophthora* spp.) and powdery mildew (*Oidium* spp.) were the two major leaf diseases in India. However, the threat of minor diseases caused by such fungi as *Corynespora* has increased significantly. As a result of the severe incidence of *Corynespora* leaf spot disease in Sri Lanka, RRIC 103, one of the most popular high yielders planted extensively, had to be withdrawn from the planting recommendation, necessitating replanting of vast areas under this clone (Liyanage et al., 1991). In India, reports (RRII, 1998) on the incidence of *Coryenespora* leaf spot observed in the popular clones RRII 105 and RRIM 600 in the rubber growing tracts of Karnataka and certain parts of North Kerala have caused serious concern over the possibility of the disease attaining epidemic proportions.

South American leaf blight caused by *Microcyclus ulei,* fortunately still confined to the tropical Americas, is the most devastating disease of rubber. However, rubber plantations in the East are also under the potential threat of SALB due to favorable climatic conditions. Hence breeding and selection of SALB-resistant clones assumes significance for the Asian countries as well. Available reports indicate that disease resistance in *H. brasiliensis* is inherited polygenically, implying that the possibility exists of obtaining horizontal resistance, which is more stable and durable (Varghese, 1992).

Pink disease caused by *Corticium salmonicolor* is an important stem disease in rubber causing drying of the main stems and branches of three- to seven-year-old immature plants, particularly in the fork region. Dry rot disease by *Ustilina deusta* and patch canker by *Phytophthora palmivora* cause occasional damage. Among the panel diseases, black stripe by *Phytophthora* spp. cause damage to the tapping panels. Three root diseases observed are white root disease *(Rigidoporus lignosus)* brown root disease *(Phellinus noxius),* and red root disease *(Ganoderma philippii).*

Damage by pests is fairly insignificant in rubber. Termites, cockchafers, mites, and thrips are some of the pests in rubber plantations. Plant protection schedules for most diseases and pests have been developed.

Tapping-Panel Dryness (Brown Bast)

Brown bast or tapping-panel dryness is a syndrome characterized by prolonged dripping of latex with a gradual decline in volume yield, precoagulation of latex, and partial or complete drying of the tapping area. In some instances browning and thickening as well as cracking and deformation of the bark take place. High-yielding clones are most susceptible to the syndrome, resulting in drying up of the tapping panels in 10 to 25 percent of the trees. Reduced tapping intensity or tapping rest for three to twelve months are considered curative measures for this syndrome. It is considered to be a stress-related physiological disorder.

TECHNOLOGICAL DEVELOPMENTS

Primary Processing

NR harvested from the tree as latex accounts for 70 to 80 percent of the crop; the rest is the field coagulum. Latex can be processed into ribbed smoked sheets (RSS), pale latex crepe (PLC), technically specified rubber (TSR), speciality rubbers, preserved field latex, and latex concentrate, whereas field coagulum is processed and marketed as either block or crepe rubber. Conversion of fresh latex into ribbed smoked sheets is the oldest method of processing and has been widely adopted, especially by small growers, due to its simplicity and low cost. At present, sheet rubber accounts for around 75 percent in the Indian NR market (Kuriakose and Thomas, 2000).

The need for production of technically specified rubber (TSR) arose when the traditional visually graded sheet and crepe rubbers were not considered adequate enough to face stiff competition from the synthetic rubbers. Consequently, new methods of processing were developed to produce and market TSR in compact medium sized blocks, wrapped in polyethylene, and graded adopting technical standards. The implementation of new manufacturing practices and international standards has resulted in an increased use of TSR compared to other marketable forms, which on a global scale accounts for around 50 percent of the NR processed (George, Alex, and Mathew, 2000). In block rubber processing, the coagulum is made into crepe by passing it through macerators, granulating it in hammer mills, and pressing it into blocks.

Concentrated latex is produced by various processes, viz., evaporation, electrodecantation, creaming, and centrifugation, of which centrifugation has been developed on a large scale and is currently the most widely accepted method accounting for more than 90 percent of the concentrated latex produced (Mathew and Claramma, 2000). Since a number of products, such as foam, elastic threads, carpet backings, adhesives, gloves, balloons, rubber bands, and other dipped goods, are made directly from latex, latex preservation with appropriate chemicals for long-term storage and concentrations are essential. Although ammonia is being used conventionally as an ideal preservative, low ammonia preservation systems have been developed to rectify some of its inherent defects.

Modified Forms of Rubber

Latex harvested from the tree contains around 94 percent hydrocarbon and nonrubber substances, such as proteins, fats, fatty acids, and carbohydrates, which influence the chemical and physical properties of the hydrocarbon polymer. Studies on physical and chemical modifications of NR have resulted in several modified forms suitable for specific processes and applications. These include viscosity-stabilized rubber, thermoplastic NR, oil-extended natural rubber (OENR), epoxidized NR, graft copolymers, chlorinated rubber, liquid natural rubber (LNR), etc.

An increase in viscosity of NR during primary processing and subsequent storage under ambient conditions leads to storage hardening. Therefore, constant viscosity (CV) rubbers have been developed by adding small amounts of chemicals, such as hydroxylamine neutral sulfate, which help to preserve the original Mooney viscosity for a long time. The controlled and stable Mooney viscosity provides easy and uniform processing and also minimizes the premastication time, resulting in energy savings. In comparison with NR vulcanizates, thermoplastic natural rubber (TPNR) blends are remarkably resistant to heat aging and ozone (Elliot, 1982). Chlorinated rubber with 65 percent chlorine is highly resistant to chemicals, and is used in anticorrosive and heat-resistant paints, coatings, adhesives, printing inks, textile finishers, etc. Epoxidized natural rubber, a chemically modified form, has qualities such as improved resistance to hydrocarbon oils and low air permeability, while retaining the high strength properties of NR (Baker et al., 1985).

Oil-extended natural rubber has reduced tensile strength and resilience, but has a good resistance to tear and wear when blended with polybutadiene rubber. OENR has definite advantages when used in the manufacture of tire treads, which have better grip on snow and ice and have gained considerable acceptance in European countries. Deproteinized natural rubber (DPNR), a highly purified form of NR, is suitable for electrical and engineering applications (for reduced creep and stress relaxations, superior dynamic properties, and consistency to stiffness).

Natural rubber can be modified chemically to graft copolymers by polymerizing vinyl monomers such as methyl methacrylate, styrene, and acrylonitrile (Claramma et al., 1984, 1989). Methyl methacrylate (MMA)-grafted NR and styrene-grafted NR are the two graft copolymers developed. The major use of the former is in adhesives for the shoe and tire industries, whereas the latter finds application in micocellular solings in place of the high styrene resin grade of solid black rubber (SBR).

Extensive size reduction of molecular chains of NR by depolymerization leads to the formation of liquid natural rubber. This serves as a reactive plasticizer in compounding and also as a substitute to synthetic liquid elastomers, such as fluid silicones used in elastic molds for various industrial and art work (George, Reghu, and Nehru, 2000). A wide range of engineering applications of rubber have been developed in almost every industrial sector (e.g., engine mounts, suspension mounts, bridge bearings, earth quarter protection of buildings, etc.).

Discovery of Vulcanization

Credit of the historic discovery of vulcanization, a process of heating rubber with sulfur at a high temperature to improve its strength properties, goes to Charles Goodyear (1800-1860) an American rubber manufacturer. All articles of rubber were observed to become sticky in hot weather and brittle in cold weather. Goodyear worked on the problem for making rubber more stable and less susceptible to heat, cold, and light. The fundamental changes in the properties of NR through vulcanization removed most of its susceptibilities to climatic conditions and its limitations as a raw material for mechanical applications.

Natural rubber is basically a high molecular weight polymer, cis-1, 4-polyisoprene with viscoelastic properties. The elasticity of rubber depends on the predominant cis configuration of the polymer. In vulcanization, the cis-isoprene units form intermolecular cross-links with sulfur, making it hard, resistant to abrasion, heat, light, and oils while retaining its unique elasticity, which makes rubber useful in manufacturing a variety of products.

Product Development

The different forms of natural rubber and its chemical modifications are being used in the manufacture of thousands of products. The major share of the dry forms of natural rubber finds applications in the tire sector. Other products include belts and hoses, footwear, and molded items, such as mountings, bushes, seals, mats, and pharmaceutical closures. Concentrated latex is the raw material used in the production of foam, condoms, gloves, latex threads, and adhesives.

Technology has been developed for the rubberization of roads, which is becoming popular. Rubberized roads are more durable compared to asphalt roads, and the service life increases by 50 percent or more. Moreover, the surface bears heavy traffic, provides skid resistance, and withstands both extreme cold and hot weather conditions. The additional cost for seal coats using 2 percent rubber in asphalt is 12 to 15 percent, whereas the additional cost for rolled asphalt with 4 percent rubber is 16 percent (Haridasan and Gopalakrishnan 1980; Gopalakrishnan, 1994).

Ancillary Products

The major ancillary products from a rubber plantation, especially in the mature phase, are rubber honey, seeds, and wood. Commercial exploitation of these ancillary products adds to the efforts to maximize returns from this crop, especially to the small farmers.

Rubber Honey

Rubber plantations have been identified as an important source of honey. During the refoliation and flowering period, honey bees collect large quantities of nectar from the extrafloral nectaries in the leaflets. Lack of honey flow in the rest of the year in rubber plantations is managed by alternate bee flora.

In India, rubber honey production faced a drastic reduction from a peak of 2,750 tonnes in 1990-1991 to an all-time low of 550 tonnes in 1993-1994 due to the outbreak of sacbrood disease of bees (George and Joseph, 1994). The rehabilitation measures by introduction of *Apis mellifera* have resulted in the rise of honey production to 1,500 tonnes in 1997-1998 (RRII, 1998). *Apis mellifera* had a reported average yield of 60 kg per hive per year compared to 19.46 kg per hive per year for the popular Indian honey bee, *Apis indica* (Haridasan et al., 1987). On an average, 15 to 20 *A. indica* hives can be placed per ha, and the results of a recent survey showed an average yield of 12.1 kg per hive per year for the Indian honey bee (Chandy et al., 1998). Mature plantations in India thus have the potential to produce 67,886 tonnes of rubber honey annually, though only 2 percent of this potential was exploited in 1996-1997 (George, Reghu, and Nehru, 2000).

Rubber Seed Oil

A rubber seed has an average weight of 3 to 5 g, of which about 40 percent is kernel, 35 percent is shell, and 25 percent is moisture. Oil content in the kernel ranges from 35 to 38 percent, and the recovery rate of seed cake is in the range of 57 to 62 percent. The two major products processed from rubber seeds are rubber seed oil and rubber seed cake.

In India, the seeds mature between July and September. The estimated production potential is about 150 kg per ha. Production of rubber seed oil and cake in India for 1997-1998 was 2,890 and 4,710 tonnes respectively (RRII, 1998). Among the three methods of extracting rubber seed oil, viz., solvent, expeller, and rotary extraction, the rotary method is commonly used. The recovery of oil and cake depends on the quality of the kernel, the extent of drying, and quantity of molasses used for processing (George, Reghu, and Nehru, 2000).

Rubber seed oil is used in the soap-manufacturing industry (Hardjosuwito and Hoesnan, 1976). Epoxidized rubber seed oil is used in the formulation of anticorrosive coatings and polymer additives, and in alkyl resin casting (Vijayagopalan and Gopalakrishnanan, 1971). It also serves as a substitute to linseed oil in the paint industry. Rubber seed cake is rich in protein and is a source for cattle and poultry feeds (Amritkumar et al., 1985).

Rubber Wood

Rubber plantations have become a major source of industrial timber in the rubber-growing countries of South and Southeast Asia. Since the 1980s rubber wood has been internationally accepted as an eco-friendly source of timber. Because of its excellent physical properties, rubber wood has become an important source of raw material for the manufacture of panel products, such as particleboard, block board, medium-density fiberboard, etc. (Yusoff, 1994). The current size of the world market for rubber wood-based furniture and other products is more than U.S. $1.5 billion (George, Reghu, and Nehru, 2000).

The current estimated average production of rubber wood worldwide is 150 and 180 m^3 per ha in the smallholding and estate sectors respectively. The annual gross availability of rubber wood was 1.27 million m^3 during 1997-1998, and the projected estimate for 2010 is 4.24 million m^3 (George and Joseph, 1999).

Rubber wood is a light to moderately heavy timber with medium density (515 kg/m^3), low shrinkage, straight grains and attractive color, which makes it suitable for furniture manufacture (Reghu, 1998). The structural and anatomical features of rubber wood resemble those of other hardwood species with many physical and mechanical properties comparable with that of teak (George, Reghu, and Nehru, 2000). However, tension wood formation is a natural defect causing a variety of drying, woodworking, and finishing problems.

The distribution pattern of tension wood was reported by Reghu et al. (1989).

Since rubber wood is sensitive to biodeterioration by various biological agents, processing has to be done very meticulously in two stages: preservative impregnation and drying. Preservative treatments are aimed at either short-term/temporary protection or long-term/permanent protection. For temporary protection, the timber is dipped in wood preservatives, such as insecticides and fungicides. For long-term protection, the wood preservatives are allowed to penetrate deep into the timber by providing maximum dry salt retention and complete preservative penetration by the dip diffusion and pressure impregnation processes.

In dip diffusion, freshly sawed timber is immersed for an adequate period in a mixture of boric acid and borax in water (Gnanaharan and Mathew, 1982). In pressure impregnation, preservatives are impregnated into the wood by creating a vaccum under pressure and dried to optimum moisture levels in the timber to avoid dimensional variations in the end product.

Development of latex timber clones (LTC) aimed at higher latex and timber yield for ensuring economic viability of NR cultivation is gaining significance. Identification of genotypes with superior qualitative and quantitative timber characters is a priority, and work has already been initiated in this line in various countries, including India (Reghu, 1998; Lam et al., 1997; IRRDB, 1996). In Malaysia, 20 fast-growing genotypes were identified as potential timber clones with a total wood volume ranging from 1.4 to 2.52 m^3 per tree (RRIM, 1996).

ECO-FRIENDLINESS OF RUBBER

Rubber, domesticated as a plantation tree crop more than 100 years ago has already proved its worth as an ecofriendly tree species and an ecologically viable cash crop. Natural rubber, a deciduous forest tree, helps for restoration of denuded and depleted forestlands. With a stand of 450 plants per ha, and a canopy that closes in less than five years, *Hevea* is a good candidate for afforestation of marginal and denuded soils. Rubber as an ecofriendly crop contributes mainly to soil conservation, nutrient recycling, soil fertility, and biomass generation, and serves as an alternate source to traditional timber.

Rubber cultivation involves measures to protect soil from erosion and preserve its fertility. The planting of rubber in hilly terrain in inwardly slopping contoured terraces and digging of silt pits in the terraces reduce soil erosion. Cultivation of leguminous crops helps to prevent soil erosion. Because rubber is a surface feeder, its root system helps to bind soil.

The recommended agromanagement practices in rubber are aimed at improving soil properties. Soil fertility is sustained through regular application of N, P, K, and Mg during the immature and mature phases. The fertilizer inputs are optimized by discriminatory fertilizer recommendations to supply only the required quantity to the soil. Nitrogen-fixing cover crops improve soil fertility. Rubber plantations adopting proper agromanagement practices help enrich soil organic matter, which consequently improves physical properties, such as bulk density, soil porosity, moisture retention, and infiltration (Krishnakumar et al., 1990).

Mature rubber requires comparatively less fertilizer to sustain high productivity in comparison to annual crops. The amount of nutrients removed from the ecosystem by a latex crop is relatively small when compared to other crops, such as coconut and tea (Samarappuli, 1996).

The rubber tree is a potential solar energy harvester with an annual biomass production potential of around 35 tonnes of dry matter per ha per year, which is comparable with any fast-growing tropical forest tree species (Sethuraj, 1996). The biomass production potential of a plant species is related to its photosynthetic capacity per unit leaf area and the total leaf area produced per individual plant (Jacob, 2000). The high biomass production potential of *Hevea* makes it an excellent candidate for energy plantation, and as a source of timber and fuel wood.

The growth of biomass, microflora, and understory vegetation in rubber plantations is comparable to that in teak plantations (Krishnakumar et al., 1991), indicating the ecological desirability of rubber in terms of habitat diversity, soil physical properties, and nutrient recycling. Considering the energy requirements, energy for the production of NR is derived mostly from sunlight through photosynthesis, whereas nonrenewable oil sources are used for the cost of production of synthetic rubber, which is about 11 times higher than that for NR (Mathew, 1996). Thus natural rubber has enormous potential in contributing to the cause of solving energy crises.

Similarly, as an alternative timber source, wood from natural rubber has become the main non-forest timber resource, decreasing the logging pressure on natural forests and teak plantations in rubber-growing countries.

RESEARCH AND DEVELOPMENT ORGANIZATIONS

Secretary General
Association of Natural Rubber Producing Countries
7th Floor Bangunan Getah Asli (Menara)
148 Jalan Ampang, Kuala Lumpur 50450, Malaysia
Phone: +60-3-2611-900
Fax: +60-3-2613-014

International Rubber Research and Development Board
P.O. Box 10150
Kuala Lumpur 50908, Malaysia

Executive Director
International Natural Rubber Organisation

P.O. Box 10374, Kuala Lumpur 50712, Malaysia
Phone: +60-3-248-6466
Fax: +60-3-248-6485
E-mail: inro@po.jaring.my

Director General
Malaysian Rubber Board
P.O. Box 10150, Kuala Lumpur 50908, Malaysia
Phone: +60-3-456-7033
Fax: +60-3-457-4512
E-mail: igm.gov.my

Secretary General
International Rubber Study Group
8th Floor, York House, Empire Way
Wembley, Middlesex HA9 OPA, UK
Phone: +44-181-903-7727
Fax: +44-181-903-2848
E-mail: irsg@compuserve.com

Director of Research
Tun Abdul Razak Research Centre
MRPRA, Brickendonbury
Hertford, SG 13 8 NL, UK
Phone: +44-1992-584966
Fax: +44-1992-554837
E-mail: general@tarrc.tcom.co.uk

Responsible du Programme hévéa
Centre de Cooperation International
en Recherche Agronomique pour le Development
Department des plantes perennes, Programme hévéa
2477, Avenue d'Agropolis, BP 5035
34032 Montpellier Cedex I, France
Phone: +33-4-67-61-71-55
Fax: +33-4-67-61-71-20
E-mail: banchi@cirad.fr

Professor President
Chinese Academy of Tropical Agricultural Sciences
Baodao Xincun, Danzhou,
Hainan 571737, China

Phone: +86-890-330-0157
Fax: +86-890-330-0157 / 0776
E-mail: rcri@public.dzptt.ha.cn

Director
Shanghai Rubber Products Research Institute
381 Fanyau Road
Shanghai 200052, China
Phone: +86-21-62523050
Fax: +86-21-62512948

Director
Rubber Research Institute, Department of Agriculture
Chatuchak, Bangkok 109900, Thailand
Phone: +66-2-579-7557/8
Fax: +66-2-561-4744
E-mail: rk000051@netaccess.loxinfo.co.th

Director
Rubber Research Institute of Sri Lanka
Dartonfield, Agalawatta, Sri Lanka
Phone: +94-34-47426 / 47383 / 23078
Fax: +94-34-47427
E-mail: director@rri.ac.lk

Rubber Research Institute of Vietnam
177, Hai Bai Trung Street, Ward 6, District 3
Ho Chi Minh City, Vietnam
Phone: +848294139
Fax: +848298599

Executive Head
Institut de Recherches sur la Caoutchouc au Cambodge
Numero 48, Rue Neary Khlahan
Pnomh Penh, Cambodia
Phone: +855-23-27498
Fax: +855-23-27498

Executive Head
Centre de Cooperation Internationale en
Recherche Agronomique pour le Development
Department des cultures perennes

2477, Avenue du Val de Montferrand, BP 5035
34032 Montpellier Cedex I, France
Phone: +33-4-67-61-58-00
Fax: +33-4-67-61-56-59
E-mail: devernou@cirad.fr

Director
Indonesian Rubber Research Institute
Sungei Putih, P.O. Box 1415, Medan
Sumatera Utara 20001, Indonesia
Phone: +62-61-958-045
Fax: +62-61-958-046
E-mail: rubbersp@indosat.net.id

Head of Station
Bogor Research Station for Rubber Technology
BRTK Bogor, Jalan Salak No. 1, Bogor
Jawa Barat 16151, Indonesia
Phone: +62-251-319-817 / 324-048
Fax: +62-251-324-047
E-mail: bptkbgr@indonet.id

Executive Head
Indonesian Planter's Association for Research and Development
Jalan Tanjung Karang No.5
Jakarta 10230, Indonesia
Phone: +62-21-324-473 / 325-478
Fax: +62-21-325-517

Director
Rubber Research Institute of India, Rubber Board
Kottayam-686009, Kerala, India
Phone: 91-481-353311
Fax: 91-481-353327
E-mail: rrii@vsnl.com

Chairman
Rubber Board
Kottayam-686002, Kerala, India
Phone: 91-481-571231
Fax: 91-481-571380
E-mail: rbktm@ker.nic.in

Executive Head
Rubber Technology Centre
Indian Institute of Technology
Kharagpur W.B. 721 302, India
Phone: +91-3222-55221
Fax: +91-3222-55303

Director
Rubber Research Institute of Nigeria
PLMP 1049, Benin City, Bendel City
Bendel State, Nigeria
Phone: +234-52-60-18-85
Fax: +234-52-60-18-85

Chief of Programme
National Rubber Research Programme
Institute of Agricultural Research for Development
PMP 25, Buea, Cameroon
Phone: +237-32-2221
Fax: +237-32-22-44

Amazon Agroforestry Research Centre
Km 28 da Rodovia AM-010
Manaus, Amazonas CP 31969011970, Brazil
Phone: +55926222012
Fax: +55926221100

Director General
National Research Center for Genetic Resources and Biotechnology
SAIN - Parque Rural - CP 02372, Brasilia DF
CEP 70.770-849, Brazil
Phone: +55-61-340-3500 / 3501
Fax: +55-61-340-3502 / 3624
E-mail: valois@cenargen.embrapa.br

Director
Center of Research and Technological Development
for the Rubber Industry
Av. General Paz entre Albarellos y Constituyentes
San Martin, Pcia. De Buenos Aires, C.P. 1650, Argentina
Phone: +54-1-754-4141 / 45 or 5151 / 55

Fax: +54-1-753-5781
E-mail: lrehak@inti.gov.ar

Executive Head
Rubber Institute of Rosario
Argentina
Phone: +54-41-587110
Fax: +54-41-264020

FUTURE PROSPECTS

Research and development in NR production in various rubber-growing countries has enabled the plantation industry to become increasingly science based and economically viable. Conventional breeding has brought about a tenfold increase in the productivity of rubber trees within a short period. Now that fresh germplasm has been introduced to broaden the original narrow gene pool, incorporation of desired variability for yield and yield components is expected to bring in further genetic improvement. With pressure on expanding rubber cultivation to nontraditional rubber-growing areas in all major rubber-producing countries, development of location-specific clones capable of withstanding various stress situations is a major priority. In this context, identification of molecular markers linked to specific characters, such as resistance to diseases, drought, cold, etc., are important. Marker-assisted selection will play a significant role in the selection of the desired variability in the future thereby enhancing the yield stability of high-yielding cultivars.

Because tapping-panel dryness syndrome is an elusive physiological disorder, the search for its causes continues to be an important area of research. Basic physiobiochemical investigations on latex flow, environmental and stress physiology, rubber biosynthesis and biochemistry, and molecular biology of isoprene production are priority areas. Systems such as controlled upward tapping have been perfected for long-term exploitation of high panels. Low-frequency tapping systems are also in demand in view of increasing labor shortage worldwide. Optimization of exploitation by latex diagnosis is being perfected. Nutritional aspects and fertilizer recommendation systems based on soil nutrient analysis have been well established. Research on cropping systems, soil and water conservation, integrated weed management, etc., are in progress. Effective control measures have been achieved for most of the serious diseases in the rubber-growing countries. However, identification of sources of durable resistance is the current priority in this area, and the wild germplasm will be a rich source of many valuable genes for disease resistance. Although SALB, the devastating disease in *Hevea,* is still confined to South America, sources of resistance must be located due to the potential threat of the disease in the Asian rubber-producing countries.

In addition to the primary rubber product, NR latex, optimum exploitation of the ancillary products is very important. Rubber wood is assuming importance for various industrial applications. Considering the inherent natural defects of rubber wood, technology has to be perfected to increase the utilization of rubber timber in a more cost-effective manner. In this context, selection and development of timber latex clones is gaining importance as a breeding objective. Desirable qualitative and quantitative traits of timber should be identified from the available germplasm. The selections could be used directly for raising rubber forests or incorporated in breeding programs for developing latex timber clones. Being a deciduous forest tree, *Hevea* is basically an environment-friendly tree. However, more data need to be generated on the environmental impact of natural rubber cultivation on the physical, chemical, and biological properties of rubber soils as well as water consumption by rubber plantations.

Many milestones have been reached in rubber processing and technology. Viscosity-stabilized rubber, thermoplastic NR, chlorinated rubber, liquid rubber, etc., are some of the modified forms of rubber that have already been produced commercially, the demand of which depends on the emerging market force. In latex technology the priorities include preservation of latex, prevulcanization of latex, radiation vulcanization, deproteinization, etc. Since the global production of NR has almost reached a plateau due to a shortage of land and other reasons, it is quite likely that the global demand for NR in the next century will be higher than global production. Technologies to recycle rubber from used waste products (reclamation) and attempts to enhance the service life of conventional rubber products could lead to conservation of the polymer in the future.

As with other perennial plantation crops, the market for natural rubber has been subjected to frequent price fluctuations over the years mainly due to its relatively inelastic supply in the short run. A major issue debated in the world NR market is the steep decrease in rubber prices that has occurred since 1996, following a record peak in 1995. According to projections by the World Bank, the NR price is expected to increase at a declining rate until 2010. The forecast is for a rise of 5.7 percent in 2002 and then an annual growth rate of 3.6 percent during 2003-2005, followed by a rate of 2.4 percent during 2006-2010. A recent forecast by the Economist Intelligence Unit of the International Rubber Study Group (2000) indicates that NR consumption will outpace production during the next three years. The growth rate of NR consumption is forecast to be 4.0 percent, compared to 2.7 percent NR production for the year 2002. In absolute figures, the deficit increased from 80,000 tonnes to 355,000 tonnes in 2002. However, IRSG forecasts a significant and steady increase in NR production until 2005 with an increase in total world production from 6,750,000 tonnes in 1999 to 8,331,000 tonnes in 2005 (IRSG, 2000). In general, the projections indicate that the world elastomer industry can look forward to a relatively healthy growth rate in supply and demand for the rest of this decade.

In the context of globalization, product diversification and development of products meeting customer quality requirements and import substitution are of utmost significance. The R&D institutions and rubber-based industries across the world are gearing up to meet the challenges ahead.

REFERENCES

Abraham, P.A., Wycherley, P.R., and Pakianathan, S.W. 1968. Stimulation of latex flow in *Hevea brasiliensis* by 4-amino-3, 5, 6-trichloropicolinic acid and 2-chloroethane phosphonic acid. *Journal of the Rubber Research Institute of Malaya* 20: 291-305.

Abraham, S.T. 2000. Genetic parameters and divergence in certain wild genotypes of *Hevea brasiliensis* (Willd. ex Adr. de Juss.) Muell. Arg. PhD thesis submitted to Mahatma Gandhi University, Kottayam, India.

Abraham, S.T., Panikkar, A.O.N., George, P.J., Reghu, C.P., and Nair, R.B. 2000. Genetic evaluation of wild *Hevea* germplasm: Early performance. Paper presented at the Fourthteenth PLACROSYM International Conference on Plantation Crops, December 12-15, Hyderabad, India.

Abraham, S.T., Reghu, C.P., George, P.J., and Nair, R.B. 1997. Studies on genetic divergence in *Hevea brasiliensis*. Paper presented at the Symposium on Tropical Crop Research and Development. September 9-12, Tichur, India.

Abraham, S.T., Reghu, C.P., Madhavan, J., George, P.J., Potty, S.N., Panikkar, A.O.N., and Saraswathy, P. 1992. Evaluation of *Hevea* germplasm 1: Variability in early growth phase. *Indian Journal of Natural Rubber Research* 5(1 and 2): 195-198.

Abraham, S.T., Reghu, C.P., Panikkar A.O.N., George, P.J., and Potty, S.N. 1994. Juvenile characterisation of wild *Hevea* germplasm. *Indian Journal of Plant Genetics Resources* 7(2): 157-164.

Amritkumar, M.N., Sundaresan, K., and Sampath, S.R. 1985. Effect of replacing cottonseed cake by rubber seed cake in concentrate of cows on yield and composition of milk. *Indian Journal of Animal Sciences* 55(12): 1064-1070.

Annamma, Y., Licy, J., John, A., and Panikkar, A.O.N. 1989. An incision method for early selection of *Hevea* seedlings. *Indian Journal of Natural Rubber Research* 2(2): 112-117.

Annamma, Y., Marattukalam, J.G., George, P.J., and Panikkar, A.O.N. 1989. Nursery evaluation of some exotic genotypes of *Hevea brasiliensis* Muell. Arg. *Journal of Plantation Crops* 16(Suppl): 335-342.

Annamma, Y., Marattukalam, J.G., Premakumari, D., Saraswathy Amma, C.K., Licy, J., and Panikkar, A.O.N. 1990. Promising rubber planting materials with special reference to Indian clones. *Proceedings, Planters Conference,* Kottayam, Rubber Research Institute of India.

Archer, B.L. 1964. Site and mechanism of rubber biosynthesis from mevalonate. *Proceedings of the Natural Rubber Producers Research Association Jubilee Conference, 1964,* Cambridge, London, pp. 101-112.

Archer, B.L., Barnard, D., Cockbain, E.G., Dickenson, P.B., and McMullen, A.I. 1963. Structure composition and biochemistry of *Hevea* latex. In Bateman, L., Ed., *The Chemistry and Physics of Rubber-like Substances.* Maclaren & Sons, Ltd., London, pp. 41-72.

Arokiaraj, P., Jaafar, H., Hamzah, S., Yeang, H.Y., and Wan Abdul Rahaman, W.Y. 1996. Enhancement of *Hevea* crop potential by genetic transformation: HMGR activity in transformed tissue. In *Proceedings of the Symposium on Physiological and Molecular Aspects of the Breeding of Hevea* brasiliensis, Penang, Malaysia, November 6-7, 1995, International Rubber Research and Development Board, Hertford, England, pp. 74-82.

Arokiaraj, P., Jones, H., Cheong, K.F., Coomber, S., and Charlwood, B.V. 1994. Gene insertion into *Hevea brasiliensis. Plant Cell Reports* 13: 425-431.

Arokiaraj, P., Jones, H., Jaafar, H., Coomber, S., and Charlwood, B.V. 1996. *Agrobacterium* mediated transformation of *Hevea* anther calli and their regeneration into plantlets. *Journal of Natural Rubber Research* 11(2): 77-87.

Arokiaraj, P. and Wan, A.R.W.Y. 1991. Agrobacterium-mediated transformation of *Hevea* cells derived from in vitro and in vivo seedling cultures. *Journal of Natural Rubber Research* 6(1): 55-61.

Arokiaraj, P., Yeang, H.Y., Cheong, K.F., Hamzah, S., Jones, H. Coomber, S., and Charlwood, B.V. 1998. A gene transfer system for *H. brasiliensis* was established with A. tumefaciens GV2260 (p35SGUSINT) and LBA4404 (pAL4404/pMON9793): CaMV 35S promoter directs beta-glucuronidase expression in the laticiferous system of transgenic *Hevea brasiliensis* (rubber tree). *Plant Cell Reports* 17(8): 621-625.

Asokan, M.P., Kumari Jayasree, P., Sushamakumari, S., and Sobhana, P. 1992. Plant regeneration from anther culture of rubber tree *(Hevea brasiliensis). Abstract of International Natural Rubber Conference, Bangalore,* February, Rubber Research Institute of India, p. 49.

Asokan, M.P., Sobhana, P., Sushama Kumari, S., and Sethuraj, M.R. 1988. Tissue culture propagation of rubber *(Hevea brasiliensis)* Willd. ex Adr. de Juss. Muell. Arg. Clone GT (Gondang Tapen) 1. *Indian Journal of Natural Rubber Research* 1(2): 1-9.

Baker, C.S.L., Gelling, I.R., and Newell, R. 1985. Epoxidised natural rubber. *Rubber Chemistry and Technology* 58(1): 67-85.

Baulkwill, W.J. 1989. The history of natural rubber production. In Webster, C.C. and Baulkwill, W.J., Eds., *Rubber*. Longman Scientific & Technical, New York, pp. 1-56.

Besse, P., Seguin, M., Lebrun, P., and Chevallier, M.H. 1994. Genetic diversity among wild and cultivated populations of *Hevea brasiliensis* assessed by nuclear RFLP analysis. *Theoretical and Applied Genetics* 88: 199.

Boatman, S.G. 1966. Preliminary physiological studies on the promotion of latex flow by plant growth regulators. *Journal of the Rubber Research Institute of Malaya* 19(5): 243-258.

Bobilioff, W. 1919. Onderzoekingen over het onstan van latexvaten enlatex *bij Hevea brasiliensis. Archive voor de Rubber Cultur* 3: 43.

Bobilioff, W. 1920. Correlations between yield and number of latex vessel rows of *Hevea brasiliensis. Archive voor de Rubber Cultur* 4: 391.

Bobilioff, W. 1923. *Anatomy and physiology of* Hevea brasiliensis. Anatomy of Art Institute, Orall Fusse.

Bowler, W.W. 1953. Electrophoretic mobility of fresh *Hevea* latex. *Industrial Engineering Chemistry* 45: 1790.

Carron, M.P. and Enjarlic, F. 1982. Studies on vegetative micropropagation of *Hevea brasiliensis* by somatic embryogenesis and in vitro microcutting. *Proceedings of the Fifth International Congress on Plant Tissue and Cell Culture, Tokyo,* pp. 751-752.

Carron, M.P., Lardet, L., and Dea, B.G. 1998. *Hevea* micropropagation by somatic embryogenesis. *Plantation Research Development* 5: 187-192.

Chandy, B., Joseph, T., and Mohanakumar, S. 1998. Commercial exploitation of rubber honey in India: Report of a sample survey. *Rubber Board Bulletin* 27(3): 9-15.

Chapman, G.W. 1951. Plant hormones and yield in *Hevea brasiliensis. Journal of the Rubber Research Institute of Malaya* 13: 167-176.

Chee, K.H. and Holliday, P. 1986. South American leaf blight of *Hevea* rubber. *Malaysian Rubber Research and Development Board Monograph,* No. 13.

Chevallier, M.H. 1988. Genetic variability of *Hevea brasiliensis* germplasm using isozyme markers. *Journal of Natural Rubber Research* 3: 42-53.

Claramma, N.M., Mathew, N.M., and Thomas, E.V. 1989. Radiation induced graft copolymerization of acrylonitrile in natural rubber. *Radiation Physics and Chemistry* 33(2): 87-89.

Claramma, N.M., Rajammal, G., and Thomas, E.V. 1984. Graft co-polymerisation of acrylonitrile with natural rubber. *Proceedings of the International Rubber Conference* 2(1): 13-18.

Clement-Demange, A., Legnate, H., Chapuset, T., Pinard, F., and Seguin, M. 1997. Characterization and use of the IRRDB germplasm in Ivory Coast and French Guyana: Status in 1997. In *Proceedings of the IRRDB Symposium on Natural Rubber,* Volume 1. *General, Soils and Fertilizations and Breeding and Selection Session.* Ho Chi Minh City, October 14-15, 1997, pp. 71-88.

Cook, O.F. 1928. Beginnings of rubber culture. *Journal of Heredity* 19: 204-215.

De'Auzac, J. and Ribaillier, D. 1969. L'ethylene, nouvel agent stimulant de la production de latex chez l' *Hevea brasiliensis*. *Revue Generale du Caoutchouc et des Plastiques* 46: 857-858.

De Vries, O. 1926. Superieur plantmateriaal (zaailingen en oculaties). *De Bergcultures* 1: 404.

Dean, W. 1987. *Brazil and the Struggle for Rubber: A Study in Environment History*. Cambridge University Press, Cambridge, UK.

Dey, S.K., Thomas, M., Sathik, M.B.M., Vijayakumar, K.R., Jacob, J., and Sethuraj, M.R. 1995. The physiological and biochemical regulation of yield in clones of *Hevea brasiliensis*. *Proceedings of the IRRDB Symposium on Physiological and Molecular Aspects of Breeding of* Hevea brasiliensis, November, Malaysia, pp. 168-174.

Dickenson, P.B. 1965. The ultrastructure of the latex vessel of *Hevea brasiliensis*. *Proceedings of the Natural Rubber Producers Research Association Jubilee Conference,* 1964 (Ed. L. Mullins), Cambridge, London, MaClaren and Sons, pp. 52-66.

Dijkman, M.J. 1951. *Hevea: Thirty Years of Research in the Far East*. University of Miami Press.

Edathil, T.T. 1986. South American leaf blight: A potential thread to the natural rubber industry in Asia and Africa. *Tropical Pest Management* 32(4): 296-303.

Elliot, D.J. 1982. Natural rubber systems. In Whelan, A. and Lee, K.S., Eds., *Developments in Rubber Technology*. Applied Science Publishers, London, pp. 221-223.

Englemann, F., Lartaud, M., Chabrillange, N., Carron, M.P., and Etienne, H. 1997. Cryopreservation of embryogenic calluses of two commercial clones *Hevea brasiliensis*. *Cryoletters* 18: 107-116.

George, B., Alex, R., and Mathew, N.M. 2000. Modified forms of natural rubber. In George, P.J. and Kuruvilla Jacob, C., Eds., *Natural Rubber: Agromanagement and Crop Processing*. Rubber Research Institute of India, Kottayam, pp. 453-470.

George, C.M. 1962. Mature rubber manuring: Effect of fertilizers on yield. *Rubber Board Bulletin* 5: 202.

George, K.T. and Joseph, T. 1994. Commercial exploitation of rubber honey in India: A preliminary assessment. Unpublished monograph, Kottayam, Rubber Research Institute of India.

George, P.J. 1989. Need for conservation of crop genetic resources with special emphasis on rubber. *Rubber Board Bulletin* 24(4): 13-17.

George, P.J., Nair, V.K.B., and Panikkar, A.O.N. 1980. Yield and secondary characters of the clone RRII 105 in trial plantings. Abstract of paper, *International Rubber Conference,* November, Kottayam, Rubber Search Institute of India.

George, T.K., Reghu, C.P., and Nehru, C.R. 2000. Byproducts and ancillary sources of income. In George, P.J. and Kuruvilla Jacob, C., Eds., *Natural Rubber: Agromanagement and Crop Processing*. Rubber Research Institute of India, Kottayam, pp. 507-520.

Gnanaharan, R. and Mathew, G. 1982. Preservative treatment of rubber wood. *Research Report, Kerala Forest Research Institute* 15: 1-16.

Gomez, J.B. 1974. Ultrastructure of mature latex vessels in *Hevea brasiliensis. Proceedings of the Eighth International Congress on Electron Microscopy.* Australian Academic of Science, Canberra.

Gomez, J.B. 1983. *Physiology of Latex (Rubber) Production.* Malaysian Rubber Research and Development Board, Kuala Lumpur, pp. 71-98.

Gomez, J.B. and Moir, G.F.J. 1979. The ultracytology of latex vessels in *Hevea brasilensis. Malaysian Rubber Research and Development Board Monograph,* No 4.

Gopalakrishnan, K.S. 1994. An insight into rubberised roads. *Indian Road Congress, Seminar on Bituminous Roads: Design and Construction Aspects,* August 25-26, 1994, New Delhi, India, pp. I.116-I.125.

Guest, E. 1939. The international notation for tapping systems and its use in the tabulation of yield records. *Journal of the Rubber Research Institute of Malaya* 9(239): 142-163.

Gunatilleke, D.I. and Samaranayake, C. 1988. Shoot tip culture as a method of micropropagation of *Hevea. Journal of Rubber Research Institute of Sri Lanka* 68: 33-44.

Hardjosuwito, B. and Hoesnan, A. 1976. Rubber seed oil analysis and its possible use. *Merara Perkebunan* 44: 225-229.

Haridasan, V. and Gopalakrishnan, K.S. 1980. Economics of rubberised road. *Rubber News* May: 37-38.

Haridasan, V., Jayaratnam, K., and Nehru, C.R. 1987. Honey from rubber plantation: A study of its potential. *Rubber Board Bulletin* 23(1): 18-21.

Hurov, H.R. 1960. Green bud strip budding of two to eight month old rubber seedlings. *Proceedings of the Natural Rubber Research Conference,* Part II, Natural Rubber Production, Kuala Lumpur, Malaya, pp. 419-428.

International Rubber Research and Development Board (IRRDB). 1996. *Quarterly Report, International Rubber Research and Development Board,* Hertford, UK.

International Rubber Study Group (IRSG). 2000. *IRSG Rubber Economics Year Book 2000.* UK.

International Rubber Study Group (IRSG). 2001. *IRSG Rubber Statistical Bulletin* 55(8): 44.

Jacob, C.K. 1996. *Phytophthora* diagnosis. *IRRDB Information Quarterly* 5:14.

Jacob, C.K., Edathil, T.T., Idicula, S.P., Jayarathnam, K., and Sethuraj, M.R. 1989. Effect of abnormal leaf fall disease caused by *Phytophthora* spp. on the yield of rubber tree. *Indian Journal of Natural Rubber Research* 2(2): 77-80.

Jacob, J. 2000. Rubber tree, man, and environment. In George, P.J. and Kuruvilla Jacob, C., Eds., *Natural Rubber: Agromanagement and Crop Processing.* Rubber Research Institute of India, Kottayam, pp. 599-610.

Jacob, J.L., Eschbach, J.M., Prevot, J.C., Roussel, D., Lacrotte, R., Chrestin, H., and d'Auzac, J. 1986. Physiological basis for latex diagnosis of the functioning of the laticiferous system in rubber trees. *Proceedings, International Rubber Conference* 3: 43-65.

Jessy, M.D., Mathew, M., Jacob, S., and Punnoose, K.I. 1994. Comparative evaluation of basin and drip systems of irrigation in rubber. *Indian Journal of Natural Rubber Research* 7(1): 51-56.

Jessy, M.D., Philip, V., Punnoose, K.I., and Sethuraj, M.R. 1996. Multispecies cropping system with rubber: A preliminary report. *IRRDB Symposium on Farming System Aspects of the Cultivation of Natural Rubber* (Hevea brasiliensis), 1996, Beruwela, Sri Lanka, International Rubber Research and Development Board, Hertford, UK, pp. 81-89.

Jones, K.P. and Allen, P.W. 1992. Historical development of the world rubber industry. In Sethuraj, M.R. and Mathew, N.M., Eds., *Natural Rubber: Biology, Cultivation, and Technology*. Elsevier, Amsterdam, pp. 1-25.

Krishnakumar, A.K., Eappen, T., Rao, N., Potty, S.N., and Sethuraj, M.R. 1990. Ecological impact of rubber *(Hevea brasiliensis)* plantations in North East India: 1. Influence on soil physical properties with special reference to moisture retention. *Indian Journal Natural Rubber Research* 3(1): 53-63.

Krishnakumar, A.K., Gupta, C., Sinha, R.R., Sethuraj, M.R., Potty, S.N., Eappen, T., and Das, K. 1991. Ecological impact of rubber *(Hevea brasiliensis)* plantations in North East India: 2. Soil properties and biomass recycling. *Indian Journal of Natural Rubber Research* 4(2): 134-141.

Krishnakumar, A.K. and Potty, S.N. 1992. Nutrition of *Hevea*. In Sethuraj, M.R. and Mathew, N.M., Eds., *Natural Rubber: Biology, Cultivation, and Technology*. Elsevier, Amsterdam, pp. 239-262.

Kumari Jayasree, P., Asokan, M.P., Sobha, S., Sankari Ammal, L., Rekha, K., Kala, R.G., Jayasree, R., and Thulaseedharan, A. 1999. Somatic embryogenesis and plant regeneration from immature anthers of *Hevea brasiliensis*. *Current Science* 76(9): 1242-1245.

Kuriakose, B. and Thomas, K.T. 2000. Ribbed sheets. In George, P.J. and Kuruvilla Jacob, C., Eds., *Natural Rubber: Agromanagement and Crop Processing*. Rubber Research Institute of India, Kottayam, pp. 386-398.

Lam, L.V., Ha, T.T.T., Ha, V.T.T., and Hong, T. 1997. Studies of *Hevea* genetic resources in Vietnam: Results of evaluation and utilization. In *Proceedings of the IRRDB Symposium on Natural Rubber, Volume 1, General, Soils and Fertilization and Breeding and Selection Sessions, 3. Breeding and Selection. Ho Chi Minh City, October 14-15*, pp. 89-100.

Lane, E.V. 1953. The life and work of Sir Henry Wickham. *The Indian Rubber Journal* 126: 25-27, 65-68, 95-98, 139-142, 177-180.

Lardet, L., Aguilar, M.E., Michaux-Ferriere, N., and Berthouly, M. 1998. Effect of strictly plant-related factors in the response of *Hevea brasiliensis* and *Theobromo cacao* nodal explants cultured in vitro. *In Vitro Cellular Development and Biology* 34: 34-40.

Lespinasse, D., Rodier-Goud, M., Grivet, L., Leconte, A., Legnate, H., and Seguin, M. 2000. A saturated genetic linkage map of rubber tree (*Hevea* spp.) based on RFLP, AFLP, microsatellites, and isozyme markers. *Theoretical and Applied Genetics* 100: 121-138.

Licy, J., Panikkar, A.O.N., Premakumari, D., Saraswathy Amma, C.K., Nazeer, M., and Sethuraj, M.R. 1998. Genetic parameters and heterosis in rubber *(Hevea brasiliensis)* Muell. Arg.: 4. Early versus mature performance of hybrid clones. In Mathew, N.M. and Kuruvilla Jacob, C., Eds., *Developments in Plantation Crops Research*. Allied Publishers Limited, New Delhi, pp. 9-15.

Licy, J., Panikkar, A.O.N., Premakumari, D., Varghese. Y.A., and Nazeer, M.A. 1992. Genetic parameters and heterosis in *Hevea brasiliensis:* 1. Hybrid clones of RRII 105 x RRIC 100. *Indian Journal of Natural Rubber Research* 5(1 and 2): 51-56.

Liyanage, A. de S., Jayasinghe, C.K., and Liyanage, N.I.S. 1991. Losses due to *Corynespora* leaf fall disease and its eradication. In *Proceedings of the Rubber Research Institute of Malaysia Rubber Growers Conference,* 1991, Malacca, pp. 401-410.

Low, F.C. 1978. Distribution and concentration of major soluble carbohydrates in *Hevea* latex, the effects of ethephon stimulation, and the possible role of these carbohydrates in latex flow. *Journal of the Rubber Research Institute of Malaysia* 26(1): 21-32.

Lukman, R. 1983. Revised international tapping notation for exploitation systems. *Journal of the Rubber Research Institute of Malaysia* 31(2): 130-140.

Madhavan, J., Abraham, S.T., Reghu, C.P., and George, P.J 1997. A preliminary report on two floral variants in the 1981 wild *Hevea* germplasm collection. *Indian Journal of Natural Rubber Research* 10(1 and 2): 1-5.

Madhavan, J.M., Reghu, C.P., Abraham, S.T., George, P.J., and Potty, S.N. 1993. Resources of *Hevea*. Paper presented at the Indian Society of Plant Genetic Resources Dialogue in Plant Genetic Resources: Developing National Policy Conference held at National Bureau of Plant Genetic Resources, New Delhi.

Majumder, S.K. 1964. Chromosome studies of some species of *Hevea*. *Journal of the Rubber Research Institute of Malaya* 18: 253-260.

Marattukalam, J.G. and Premakumari, D. 1981. Seasonal variation in budding success under Indian conditions in *Hevea brasiliensis* Muell. Arg. *Proceedings of the Fourth Annual Symposium on Plantation Crops,* 1981, Mysore, India, Indian Society of Plantation Crops, Kasaragod, Kerula, pp. 81-86.

Marattukalam, J.G., Saraswathy Amma, C.K., and George, P.J. 1980. Crop improvement through ortet selection in India. *Abstract of Papers: International Rubber Conference,* November, Kottayam, Rubber Research Institute of India.

Marattukalam, J.G. and Varghese, Y.A. 1998. Bench grafting with green buds in *Hevea brasiliensis*. *Thirteenth Symposium on Plantation Crops,* 1998, Coimbatore, India, United Planters Association of South India, Nalparai.

Markose, V.C. 1984. Biometric analysis of yield and certain yield attributes in the Para rubber tree: *Hevea brasiliensis* Muell. Arg. PhD thesis, Kerala Agricultural University, India.

Markose, V.C., Saraswathy Amma, C.K., Licy, J., and George, P.J. 1981. Studies on the progenies of a *Hevea* mutant. *Proceedings of Placrosym IV,* Mysore, India, pp. 58-61, Indian Coffee Research Institute, Chickamangalore.

Mathew, N.M. 1996. Natural rubber and the synthetics: Energy, environment, and performance considerations. In *Natural Rubber: An Ecofriendly Material*. Rubber Board, Kottayam, pp. 77-99.

Mathew, N.M. and Claramma, N.M. 2000. Latex preservation and concentration. In George, P.J. and Kuruvilla Jacob, C., Eds., *Natural Rubber: Agromanagement and Crop Processing*. Rubber Research Institute of India, Kottayam, pp. 414-433.

Mercy, M.A., Abraham, S.T., George, P.J., and Potty, S.N. 1995. Evaluation of *Hevea* germplasm: Observations on certain prominent traits in a conservatory. *Indian Journal of Plant Genetic Resources* 8(1): 35-39.

Milford, G.F.J., Paardekooper, E.C., and Ho, C.Y. 1969. Latex vessel plugging, its importance to yield and clonal behavior. *Journal of the Rubber Research Institute of Malaya* 21: 274-282.

Mohd Noor, A.G. and Ibrahim, N. 1986. Characterisation and evaluation of *Hevea* germplasm. *IRRDB Symposium*, November 11-13, 1986, Malaysia.

Moir, G.F.J. 1959. Ultracentrifugation and staining of *Hevea* latex, *Nature* 161: 177.

Mydin, K.K., Nair, V.G., Sethuraj, M.R., Saraswathy, P., and Panikkar, A.O.N. 1992. Genetic divergence in *Hevea brasiliensis*. *Indian Journal of Natural Rubber Research* 5(1 and 2): 120-124.

Mydin, K.K., Nazeer, M.A., Licy, J., Annamma, Y., and Panikkar, A.O.N. 1989. Studies on improving fruit set following hand pollination in *Hevea brasiliensis* (Willd. ex Adr. de Juss.) Muell. Arg. *Indian Journal of Natural Rubber Research* 2(1): 61-67.

Nair, V.K.B. and George, P.J. 1969. The Indian clones: RRII 100 series. *Rubber Board Bulletin* 10(3 and 4): 115-140.

Nazeer, M.A. and Saraswathy Amma, C.K. 1987. Spontaneous triploidy in *Hevea brasiliensis* (Willd. Ex. Adr. de Juss.) Muell. Arg. *Journal of Plantation Crops* 15: 69-71.

Nurhaimi, Haris, Darussamin, A., and Dodd, W.A. 1993. Isolation of rubber tree (*Hevea brasiliensis* Muell. Arg.) protoplasts from callus and cell suspension. *Menara Perkebunan* 61(2): 25-31.

Omont, H. 1982. Plantations d'heveas en zone climatique marginale. *Revue Generale des Caoutchoucs et du Plastiques* 625: 75-79.

Ong, S.H., Ghani, M.N.A., Tan, A.M., and Tan, H. 1983. New *Hevea* germplasm: Its introduction and potential. *Proceedings of the Rubber Research Institute of Malaysia, Planters Conference* 3: 3-17.

Parkin, J. 1910. The right use of acetic acid in the coagulation of *Hevea* latex. *Indian Rubber Journal* 40: 752.

Petch, T. 1911. *The Physiology and Diseases of* Hevea brasiliensis. Dulau, London.

Petch, T. 1914. Notes on the history of the plantation rubber industry in the east. *Annals of the Royal Botanic Gardens, Peradeniya* 5: 440-487.

Philip, V. 1997. Soil moisture and nutrient influence on growth and yield of *Hevea brasiliensis* in Tripura. PhD thesis, Indian Institute of Technology, Kharagpur, India.

Premakumari, D., George, P.J., Panikkar, A.O.N., Nair, V.K.B., and Sulochanamma, S. 1984. Performance of RRII 300 series clones in the small scale trial. *Proceedings of the Fifth Annual Symposium on Plantation Crops*, 1982, Kasaragod, India, pp. 148-157, Indian Society of Plantation Crops, Central Plantation Crops Research Institute, Kasaragod, Kerala.

Premakumari, D. and Panikkar, A.O.N. 1992. Anatomy and ultracytology of latex vessels. In Sethuraj, M.R. and Mathew, N.M., Eds., *Natural Rubber: Biology, Cultivation, and Technology*. Elsevier, Amsterdam, pp. 67-87.

Punnoose, K.I., Kothandaraman, R., Philip, V., and Jessy, M.D. 2000. Field upkeep and intercropping. In George, P.J. and Kuruvilla Jacob, C., Eds., *Natural Rubber: Agromanagement and Crop Processing*. Rubber Research Institute of India, Kottayam, pp. 149-169.

Pushparajah, E. and Haridas, G. 1977. Developments in reduction in immaturity period of *Hevea* in peninsular Malaysia. *Journal of the Rubber Research Institute of Sri Lanka* 54: 93-105.

Reghu, C.P. 1998. Indian rubber wood: A versatile timber for future. In Kuriakose, Kumbalakuzhy, Ed., *Smrithimanjari*. Rev. Fr. Paul Pazhampally Memorial Charitable Trust., Kottayam, pp. 259-264.

Reghu, C.P., Abraham, S.T., Madhavan, J., George, P.J., Potty, S.N., and Leelamma, K.P. 1996. Evaluation of *Hevea* germplasm: Variation in bark structure of wild Brazilian germplasm. *Indian Journal of Natural Rubber Research* 9(1): 28-31.

Reghu, C.P., Premakumari, D., and Panikkar, A.O.N. 1989. Wood anatomy of *Hevea brasiliensis* (Willd. ex Adr. de Juss.) Muell. Arg.: 1. Distribution pattern of tension wood and dimensional variation of wood fibres. *Indian Journal of Natural Rubber Research* 2(1): 27-37.

Ridley, H.N. 1897. Rubber cultivation. *Agricultural Bulletin of Malay Peninsula* 7: 136-138.

Rozeboom, H.J., Budiani, A., Beintema, J.J., and Dijkstra, B.W. 1990. Crystallization of hevamine, an enzyme with lysozyme/chitinase activity from *Hevea brasiliensis* latex. *Journal of Molecular Biology* 212(3): 441-443.

Rubber Research Institute of India (RRII). 1994. *Annual Report 1992-93*. RRII, Kottayam, India.

Rubber Research Institute of India (RRII). 1998. *Annual Report 1997-98*. RRII, Kottayam, India.

Rubber Research Institute of Malaysia (RRIM). 1996. *Annual Report 1995*. RRIM, Kuala Lumpur, Malaysia.

Samarappuli, L. 1996. The contribution of rubber plantations toward a better environment. *Bulletin of the Rubber Research Institute of Sri Lanka* 33: 45-54.

Saraswathyamma, C.K. 2003. Advances in crop imporvement in *Hevea* in the traditional rubber-growing tract of India. In *Global Competiveness of Indian Rubber Industry-Rubber Planters' Conference, India 2003*, Eds. C. Kuruvilla Jacob and Kurian K. Thomas. Rubber Research Institute of India, Kottayam, Kerala, India, pp. 101-116.

Saraswathyamma, C.K., George, P.J., Nair, V.K.B., and Panikkar, A.O.N. 1980. RRII 200 series clones. *International Rubber Conference*, Rubber Board, Kottayam, India.

Saraswathyamma, C.K., George, P.J., and Panikkar, A.O.N. 1987. Performance of a few RRII clones in the estate trials. *Rubber Board Bulletin* 23(2): 5-9.

Saraswathyamma, C.K., Markose, V.C., Licy, J., and Panikkar, A.O.N. 1984. Cytomorpholicial studies in an induced polyploid of *Hevea brasiliensis* Muell. Arg. *Cytologia* 49: 725-729.

Sarma, P.V.S. 1947. A short note on the rubber plantation industry in India. India Rubber Board, Kottayam.

Schultes, R.E. 1977. Wild *Hevea*: An untapped source of germplasm. *Journal of the Rubber Research Institute of Sri Lanka* 54: 1-17.

Schurer, H. 1957. Rubber, a magic substance of ancient America. *Rubber Journal* 7(April): 543-549.

Seguin, M., Besse, P., Lespinasse, D., Lebrun, P., Rodier-Goud, M., and Nicholas, D. 1996. *Hevea* molecular genetics. *Plantations, recherche, development* 3(2): 85-87.

Seneviratne, P. 1995. Young budding: A better planting material. *Bulletin of the Rubber Research Institute of Sri Lanka* 32(1): 35-38.

Seneviratne, P. and Flegmann, A. 1996. The effect of thidiazuron on axillary shoot proliferation of *Hevea brasiliensis* in vitro. *Journal of the Rubber Research Institute of Sri Lanka* 77: 1-14.

Seneviratne, P. and Wigesekare, G.A. 1997. Effect of episodic growth pattern on in vitro growth of *Hevea brasiliensis*. *Journal of Plantation Crops* 25: 52-56.

Sethuraj, M.R. 1981. Yield components in *Hevea brasiliensis:* Theoretical considerations. *Plant Cell Environment* 4: 81.

Sethuraj, M.R. 1992. Yield components in *Hevea brasiliensis*. In Sethuraj, M.R. and Mathew, N.M., Eds., *Natural Rubber: Biology, Cultivation, and Technology*. Elsevier, Amsterdam, pp. 137-163.

Sethuraj, M.R. 1996. Impact of natural rubber plantations on environment. In *Natural Rubber: An Ecofriendly Material*. Rubber Board, Kottayam, Kerala, pp. 55-59.

Simmonds, N.W. 1989. Rubber breeding. In Webster, C.C. and Baulkwill, W.J., Eds., *Rubber*. Longman Scientific & Technical, England, pp. 86-124.

Sobhana, P., Sinha, R.R., and Sethuraj, M.R. 1986. Micropropagation on *Hevea brasiliensis:* Retrospect and Prospect. In *International Congress of Plant Tissue and Cell Culture*, Minnesota, p. 106.

Southorn, W.A. and Yip, E. 1968. Latex flow studies: 3. Electrostatic considerations in the colloidal stability of fresh *Hevea* latex. *Journal of the Rubber Research Institute of Malaya* 20(4): 201-215.

Sumarno-Kertowardjono, S., Taha Surida, A., and Ain-Tjasadihardja, N.M. 1976. Some influence of ethrel stimulation on properties of natural rubber and latex. *Menara Perkebunan* 44:75-81.

Sushamakumari, S., Asokan, M.P., Sobha, S., Kumari Jayasree, P., Kala, R.G., Jayashree, R., Power, J.B., Davey, M.R., and Cochking, E.C. 1998. Protoplast culture of *Hevea brasiliensis:* Optimization of parameters affecting isolation and culture. *International Tree Science Conference*, New Delhi, p. 15.

Tan, H. 1987. Strategies in rubber tree breeding. In Abbot, A.J. and Atkin, R.K., Eds., *Improving Vegetatively Propagated Crops*. Academic Press, London, pp. 27-62.

Tan, H. and Subramaniam, S. 1976. A five-parent diallel cross analysis for certain characters of young *Hevea* seedlings. *Proceedings, International Rubber Conference*, International Rubber Research Organization, Kuala Lumpur, Malaysia, 2: 13-26.

Thomas, K.T. and Panikkar, A.O.N. 2000. Indian rubber industry: Genesis and development. In George, P.J. and Kuruvilla Jacob, C., Eds., *Natural Rubber:*

Agromanagement and Crop Processing. Rubber Research Institute of India, Kottayam, pp. 1-19.

Thulaseedharan, A., Chattoo, B.B., Asoken, M.P., and Sethuraj, M.R. 1994. Preliminary investigations on variations in RAPD profiles between TPD affected and normal plants from a seed propagated rubber plantation. *Proceedings of the Workshop on Tapping Panel Dryness of Hevea. Chinese Academy of Tropical Agricultural Sciences,* Hainan, China, pp. 26-28.

Tupy, J. and Primot, L. 1976. Control of carbohydrate metabolism by ethylene in latex vessels of *Hevea brasiliensis* in relation to rubber production. *Biologia Plantarum* 18: 373-382.

Varghese, Y.A. 1992. Germplasm resources and genetic improvement. In Sethuraj, M.R. and Mathew, N.M., Eds., *Natural Rubber: Biology, Cultivation, and Technology.* Elsevier, Amsterdam, pp. 88-115.

Varghese, Y.A. 1998. Random amplified polymorphic DNA (RAPD) technique and its applications. In Varghese, J.P., Ed., *Molecular Approaches to Crop Improvement.* Church Mission Society (CMS) College, Kottayam, India, pp. 53-69.

Varghese, Y.A. and Abraham, S.T. 1999. Germplasm conservation, utilisation and improvement in rubber. In Ratnambal, M.J., Kumaran, P.M., Muralidharan, K., Niral, V., and Arunachalam, V., Eds., *Improvement of Plantation Crops.* Central Plantation Crops Research Institute, Kasaragod, India, pp. 124-133.

Varghese, Y.A., John, A., Premakumari, D., Panikkar, A.O.N., and Sethuraj, M.R. 1993. Early evaluation in *Hevea:* Growth and yield at the juvenile phase. *Indian Journal of Natural Rubber Research* 6(1 and 2): 19-23.

Varghese, Y.A., Knaak, C., Sethuraj, M.R., and Ecke, W. 1997. Evaluation of random amplified polymorphic DNA (RAPD) markers in *Hevea brasiliensis. Plant Breeding* 116: 47-52.

Varghese, Y.A. and Mydin, K.K. 2000. Genetic improvement. In George, P.J. and Kuruvilla Jacob, C., Eds., *Natural Rubber: Agromanagement and Crop Processing.* Rubber Research Institute of India, Kottayam, pp. 36-46.

Varghese, Y.A., Thulaseedharan, A., and Kumari Jayasree, P. 2001. Rubber. In Parthasarathy, V.A., Bose, T.K., Deka, P.A., Das, P., Mitra, S.K., and Mohandas, S., Eds., *Biotechnology of Horticultural Crops,* Volume I. Naya Prokash, Calcutta, pp. 630-635.

Vijayagopalan, P.K. and Gopalakrishnan, K.S. 1971. Epoxitation of rubber seed oil. *Rubber Board Bulletin* 11: 52-64.

Vijayakumar, K.R., Dey, S.K., Chandrasekhar, T.R., Devakumar, A.S., Mohanakrishna, T., Rao, P.S., and Sethuraj, M.R. 1998. Irrigation requirement of rubber in the subhumid tropics. *Agricultural Water Management* 35: 245-259.

Vijayakumar, K.R., Thomas, K.U., and Rajagopal, R. 2000. Tapping. In George, P.J. and Kuruvilla Jacob, C., Eds., *Natural Rubber: Agromanagement and Crop Processing.* Rubber Research Institute of India, Kottayam, pp. 215-238.

Wan, A.R., Ghandimathi, A., Rohani, O., and Paran Jothy, K. 1982. Tissue culture of economically important plants. (Ed. A.N. Rao). pp. 152-158. COSTED, Singapore.

Wang, Z., Zeng, X., Chen, C., Wu, H., Li, Q., Fan, G., and Lu, W. 1980. Induction of rubber plantlets from anther of *Hevea brasiliensis* Muell. Arg. in vitro. *Chinese Journal of Tropical Crops* 1: 25-26.

Watson, G.A. 1989. Field maintenance. In Webster, C.C. and Baulkwill, W.J., Eds, *Rubber*. Longman Scientific & Technical, New York, pp. 245-290.

Wycherly, P.R. 1968. Introduction of *Hevea* to the Orient. *The Planter* 44: 127-137.

Wycherley, P.R. 1977. Motivation of *Hevea* germplasm collection and conservation. Workshop on inter-collaboration in *Hevea* breeding and the collection and establishment of materials from the neo-tropics, International Rubber Research and Development Board, Kuala Lumpur.

Wycherley, P.R. 1992. The genus *Hevea*: Botanical aspects. In Sethuraj, M.R. and Mathew, N.M., Eds., *Natural Rubber: Biology, Cultivation, and Technology*. Elsevier, Amsterdam, pp. 50-66.

Yogaratnam, N., Iqbal, S.M.M., and Samarappuli, I.N. 1995. Intercropping of rubber with tea feasible. *Rubber Asia* 9(5): 75-79.

Yusoff, M.N.M. 1994. Panel products from rubber wood: Particle board, block board, and medium density fibre board. In Hong, L.T. and Sim, H.C., Eds., *Rubber Wood: Processing and Utilisation*. Forest Research Institute of Malaysia, Kuala Lumpur, pp. 185-200.

Chapter 11

Tea

W. W. D. Modder

Although tea is grown mostly as a monocrop, its story is a tale of diversification and adaptation to different growing conditions and varying customer requirements. It is an account of diversity all the way from its biotypes and varietal forms, through planting and processing techniques for making a bewildering array of teas and tea products, through strategies for trading in a range of international markets, to the varying preferences of consumers' palates.

From the earliest days of its cultivation, the requirements and convenience of trade and palate have driven the form and style of the tea crop, and that of the tea made from it. The old marketing cliche, "the customer is king," was literally true when only princes and rich men could afford tea, and even now when it is commonly called the cheapest drink after water, the export tea trade is still a buyers' market.

ORIGIN AND HISTORY

Legend credits the origin of the tea infusion to leaves from a tea tree falling by accident into drinking water intended for the Chinese emperor, Shen Nung, in 2737 B.C. It is the very stuff of romance for this serendipitous discovery to have led to the monumental consumption today, nearly 5,000 years later, of an estimated 3.5 billion cups of tea every day worldwide.

The plantation crop cultivated as tea (a name derived from the Chinese ideograph *ch'a,* first used about 725 A.D.) belongs to the genus *Camellia* (family Theaceae), which also contains nearly 90 species of wild and ornamental nontea forms. The genus appears to be native to the Southeast Asian landmass, and the "true" tea species, *Camellia sinensis* (L.) O. Kuntze, is native to southwestern China from where it spread to central China and southern Japan 1,000-2,000 years ago.

Many so-called herbal "teas" are retailed around the world, but only the terminal shoots of *C. sinensis* are used for manufacturing the ubiquitous and singular product, tea sensu strictu, that feeds a huge global demand. (World tea production was 2.8 billion kilograms [kg] in 1999.)

In the 1830s, tea plants for commercial cultivation were imported into Assam in northeast India and into Sri Lanka (then Ceylon) from China. Although indigenous, wild tea was already present in Assam, being "discovered" in upper Assam in 1823. Assamese tea seeds were introduced to nurseries on the island of Ceylon by the British in the 1840s, and the first viable tea plantation was established there in 1876. In the 1820s, tea from Japan was planted in Java (now a part of Indonesia) by the Dutch.

After the discovery of Assam wild tea, bioprospecting in the adjacent areas of Assam and Burma (now Myanmar) led to the discovery of other germplasm, which gave rise to the tea stocks now used for commercial cultivation in the older and newer tea-producing countries of Asia, Africa, South America, and Australia. Now that a plateau is being, or has been, reached in the yield levels of existing tea cultivars in major producing countries, it seems timely to try to resume exploration in likely geographical locations, for example in China, Myanmar, and elsewhere, for new tea germplasm, such as natural polyploids, which may be more commercially productive and have better resistance to biotic and abiotic stresses. The official exchange of established, commercially grown germplasm between tea-growing countries is now difficult or impossible owing to governmental or industry restrictions.

THE PRACTICES OF TEA CONSUMPTION

The earliest record of tea cultivation and consumption is found in ancient Chinese literature dating back to 1100 B.C. As it is still consumed today, tea was taken as an infusion after the leaves had been brewed in hot water. Tea consumption evolved into a part of Chinese religious symbolism and culture (indeed the first monograph on tea, written about 780 A.D., was called *Ch'a Ching,* "Tea Scripture"), and continues to this day in the elegant and ornate tea ceremonies of far-eastern countries. Green tea was brought to Japan in the thirteenth century by Zen Buddhist monks returning from China, and *Chado,* the "Way of Tea" (the discipline of tea preparation and consumption), arose in the fifteenth century. These rituals, still extant, have distinct reverential and aesthetic connotations.

The British Empire established formalized social and domestic tea rituals in the mother country and the outposts. Not the least of these is the "tea break" in the stately flow of the British game of cricket, now continued with aplomb in the former colonies.

These far-eastern and British rituals have survived and transmuted into the widespread, modern habit of drinking tea, either hot or cold, with or without milk and sugar, as a pleasant, relaxing, and beneficial beverage.

THE BOTANY OF TEA

In nature, the tea plant grows into a small- to medium-sized tree. The China-type tree is small or shrublike with small, erect, dark-green leaves and single flowers. The Assam-type tree is taller with broad, light-green to yellowish leaves that droop at their ends, and which have clustered flowers. The Cambod type has oval, semierect leaves.

The natural plants derived from seed ("seedling tea") have a main stem, and branches that grow from the leaf axils. Four to seven foliage leaves grow alternately, above two scale leaves, after which they become dormant. When the plant is under stress (low ambient temperature or lack of water), dormancy may ensue after the production of only two or three foliage leaves. The small first scale leaf is shed; the second larger scale leaf (the "fish leaf") persists. The bud in the axil of the first (topmost) foliage leaf either expands or becomes dormant (a so-called "banji" shoot). The crop of plucked shoots (the "flush") consists of both actively growing and banji shoots. Desirable flush for the manufacture of tea comprises the bud, the intervening portion of stem, and either two or three of the uppermost foliage leaves.

Seedling tea has deep, anchoring tap roots and lateral roots which produce a surface mat of feeder roots. The roots store starch as a food reserve.

When cultivated, the plant is pruned to a low bush for obtaining continuous vegetative growth, and its leaf and shoot morphology and growth habit, although not completely reliable for the purpose, have been used for categorization of bush types and varieties. The identification of tea bushes by their vegetative characteristics, which are plastic and show continuous variation, is therefore more or less subjective, and not as reliable as biochemical or molecular characterization, although broad phenotypic characterizations useful for breeding programs have been achieved.

The reproductive characteristics in the flowers are less variable than the vegetative characteristics, and are therefore taxonomically reliable. The flowers have a persistent calyx and usually five sepals and petals, 100-300 stamens, and an ovary with three carpels. The styles and stigma are used for classification. Being almost self-sterile, tea flowers are usually cross-pollinated by insects, and the flowers then develop into three-lobed capsules, each of which when mature produces one or two hard, spherical seeds. Their cotyledons are rich in oil.

TAXONOMY AND GENETICS

Research into the taxonomy and genetics of tea has been driven by pragmatic, commercial imperatives for acquiring the best cultivars for yield and quality. Since the tender shoots are harvested for processing, only the biotypes and hybrids that possess the chemical compounds desirable in made tea are of economic interest. For this reason, hardly any extensive knowledge is available, based on modern techniques, on the evolutionary interrelationships of the taxa. Banding patterns that reveal isozyme polymorphisms and, more reliably, DNA isolated from leaves for restriction fragment length polymorphism (RFLP) and random amplified polymorphic DNA (RAPD) analysis have however been used, in recent years, for fixing interrelationships and constructing dendrograms.

Camellia sinensis is diploid ($2n = 30$ chromosomes), as are most other *Camellia* species. This suggests a monophyletic evolutionary origin for the genus. On the other hand, it has been suggested that the wild species are also the result of widespread hybridization (with no real pure lines), in which case the genus could have a polyphyletic origin.

Although mostly diploid, polyploidy may occur naturally in tea (up to $2n = 90$) or be induced by the alkaloid colchicine ("colchiploidization") and by mutagens. Increased ploidy in plants; above the usual mono- and diploid levels, impart vigor and hardiness, and, in tea, would also give higher productivity. Rooting, leaf size, and dry weight, are better in tri- and tetraploid tea plants.

For practical purposes, three types (or varieties) of cultivated teas are recognized: the China type (the original *C. sinensis*), the Assam type (considered a distinct species, *C. assamica*), and the Cambod type (considered a subspecies, *C. assamica* ssp. *lasiocalyx*). Tea has a high degree of self-sterility and cross-pollination gives vigorous, better quality progeny than the parents. Biotypes cross freely with each other. In commercial seedling plantations, the crop consists of extremely mixed, hybridized populations of the China, Assam, and Cambod types, and perhaps even of *Camellia* species other than *C. sinensis,* particularly *C. irrawadiensis* from upper Myanmar. Even species (such as *C. irrawadiensis*), which do not contain the compounds essential in made tea, may be used in hybridization.

Seedling teas therefore have high genetic variability, and heterogeneous anatomical, morphological, physiological, and biochemical characteristics, ranging in a wide spectrum from the typical Assam to the typical China, with merging and overlapping in between. As a result, many localized varieties of seedling tea have been recognized, and given a plethora of ornate, vernacular names, in different tea-growing countries.

GROWING CONDITIONS

Although primarily tropical and subtropical, the tea plant adapts to wide extremes: in latitude, from 42°N (Georgia in the Commonwealth of Independent States) to 42°S (Tasmania, the most southern state of Australia); in elevation, from sea level to 2,600 meters; in soil types (although well-drained, acidic soils, pH 4.5-5.5, are optimal); in temperature, from -8°C to 35°C; and in annual rainfall, from < 700 to > 5,000 mm. However, for optimum growth, data from different tea-growing regions indicate that 23°-30°C and an annual rainfall of 2,500-3,000 mm are necessary. Uniform rainfall over the year (minimum: 100-150 mm per month) gives better growth than seasonal rains, although seasonal stress conditions (bright days, cold nights, and dry desiccating winds) produce chemicals in the plant, such as geraniol, linalool oxides and methyl salicylate, that give a desirable flavor and aroma (called "seasonal quality") to manufactured tea.

Tea is not an efficient photosynthesizer, having only the Calvin-Benson or "C-3" cycle. C-3 plants are temperate in origin, and their photosynthetic rates increase when incident light is optimized by shading. Thus shade trees are grown in tea fields for maximum productivity.

NUTRIENT REQUIREMENTS

The soils in which tea is cultivated are, in general, poor in nutrients, particularly in the tropics. Fertilization is therefore crucial and expensive (8 to 12 percent of the total cost of tea production).

Primary Nutrients

Of the major or primary plant nutrients (those required in relatively large amounts), namely nitrogen (N), potassium (K), and phosphorus (P), N and

K leach quickly out of typical tea soils, the process being assisted by high rainfall. However, the low soil pH of tea soils serves to fix applied N and K and increase their uptake by the plant.

Relatively large quantities of N are lost in the harvested shoots and need to be replaced frequently by high rates of application of N fertilizer, such as urea and sulfate of ammonia. The same high-rate replacement of K is not necessary for maintaining yield, although it is controversial whether high K applied to the soil is needed for withstanding drought and for the development of a sturdy bush frame.

In contrast to N and K, applied inorganic P is held effectively in the soil, mainly as iron- and aluminium-bound P, which, though insoluble, are now known to be readily taken up, owing to the presence of acid-producing organic salts (citrate and malate) released from the roots. Applied rock phosphate (calcium-bound P) is a good P source because it is solubilized both by the acidic conditions near the roots, and by the application of N fertilizer in the acid-promoting ammonium form, rather than in the alkaline nitrate form. Notwithstanding the importance of inorganic NPK fertilizer, crop responses are optimized when NPK use is balanced by that of organic manure.

Secondary Nutrients

Calcium (Ca), magnesium (Mg), and sulfur (S) are called secondary nutrients because they are required in smaller amounts than NPK. In tea soils, Ca and Mg are deficient owing to leaching by rain and their displacement by applied K. Liming is therefore an essential operation in tea cultivation. Application of dolomitic limestone is particularly useful because it supplies both Ca and Mg in addition to raising soil pH. S is also low in tea soils (even if the soils have a high organic matter content), and has to be applied.

Micronutrients

Micronutrients (that is, nutrients needed only in trace amounts), namely iron, zinc (Zn), manganese, copper, boron, molybdenum, and chlorine, are adequate in most tea soils, with the exception of Zn. Translocation of Zn from the roots is poor, and it is therefore applied to the foliage, rather than to the soil, and most conveniently as zinc sulfate which is water-soluble and cheap.

NURSERY PRACTICES

Nursery tea plants are raised from both seeds and stem cuttings under shade.

The seeds after germination in sand beds or boxes, or the cuttings, are transferred to polyethylene bags filled with nematode-free soil (loamy soil in which "rehabilitation" grass has been grown for about two years, or jungle soil) or soil substitutes. Growth in the nursery is faster under shade, or in sealed polyethylene tunnels, and when specially formulated nursery fertilizer mixtures are used. Following rooting (in 6-18 months), the bags are placed in planting holes in the field. Under rain-fed conditions, the resulting bushes are ready for plucking in about five years. With irrigation, the period is considerably reduced.

Although cuttings for vegetative propagation may be obtained from bushes at any phase of cultivation, mother bushes that are dedicated for the purpose and are spaced, pruned, and fertilized appropriately are the best source of cuttings.

Composite plants may be produced by grafting scions with desirable attributes, such as high-yielding potential, onto hardy (usually drought-resistant or nematode-tolerant) stocks.

CROP MANAGEMENT

Spacing

Cultivated tea is easily tailored to the convenience and requirements of growers. Tea bushes are grown at different planting densities: for the older seedling tea from 3,000 to 14,000 plants per hectare; for VP tea 12,500 to 18,000. Spacing of plants has to take into account such variables as soil, climate, habit, and growth pattern of the cultivars, the requirement for a continuous ground cover, access for agricultural operations, and bush architecture suited to mechanical harvesting.

Pruning

Periodic pruning or removal of mature foliage is one of the major practices adopted for preventing the tea plant from becoming a tree, and turning it rather into a low-spreading bush that is vegetatively vigorous and able to generate a continuous crop of fresh shoots at a convenient height for plucking. Pruning also ensures optimal utilization and productivity of the land available, as well as allowing space for agricultural operations.

Pruning is done either manually or by machines. The main pruning styles are low (clean or hard), high (cut across or light), and lung. In lung pruning, one or two leafy branches, or "lungs," are left for photosynthetic support of the bush during its post-pruning recovery. Bush debilitation, for example from wood rot and cankers during dry weather, may occur after a low prune, whereas during a high prune, dead or diseased branches, or those affected by insect pests, live-wood termites, or shot-hole borer, cannot be satisfactorily removed. Lung pruning is a useful compromise. Apart from supplying carbohydrates, lung branches maintain the flow of nitrogenous beads and other material from the feeder roots to the developing shoots, and function as a sink for toxic root metabolites.

Pruning just before the onset of dry weather is avoided because wound healing and new growth are quicker during sap flow. A cessation of plucking ("resting"), two to three months before pruning, is necessary for increasing starch reserves.

The frequency of pruning, or the length of the pruning cycle, is determined by the particular type of tea, the cropping pattern, and elevation. A cycle length of two years is suitable for tea near sea level, with the length increasing with elevation (six years at 2,000 meters). In Sri Lanka, pruning cycles are two to three years in the low country, and four to five years in the high country.

Harvesting or Plucking

The harvesting, or plucking, interval of the tender shoots or flush is adjusted according to environmental conditions and the need to maintain bush vigor, for obtaining the highest yield at the lowest cost, and for the desired quality.

The shoots' growth rate depends on conditions such as general climate and weather, ambient temperature, day length, and bush vigor and nutrient status, and it therefore varies at different latitudes and altitudes. The average time period for a bud to develop into a shoot ready for plucking varies from 40 days in southern Africa to 55 days and 80 days in Sri Lanka's low country and high country, respectively. In the tropics, although shoot growth fluctuates with annual variations in moisture conditions and temperature, plucking is done uniformly over the year at intervals of four to ten days. In Sri Lanka, plucking rounds are four days for fast-growing, vegetatively propagated teas in the warm, low country, and ten days for seedling teas under cold, high country conditions. In Malawi (southern Africa), rounds are 7 to 14 days. Shorter rounds give increased crop yield, but at higher plucking costs.

In the tropics, dormant buds increase in number during unfavorable dry conditions, but when conditions become favorable, as with the onset of the rains, "bud break" or a simultaneous growth of dormant buds occurs. This results in a peak or "rush" crop during the wet months of the year.

In temperate regions (such as in Japan), a growing season takes place in the spring, when most of the shoots become harvestable, and cropping for the year is done over a continuous period of one or two months only.

In east Africa (exemplified by Kenya), with a high altitude and the annual rainfall evenly spaced, crop is taken every month although the yields fluctuate. In contrast, in southern Africa, with lower altitudes, higher temperatures, and seasonal rain, 90 percent of the annual crop is taken in four months.

The top of the bushes can be trained by the plucking process to a level "plucking table" (for manual plucking), or to a dome shape (usually for mechanized plucking).

The removal of the terminal bud with the plucked shoot causes the axillary bud just below it to produce a new shoot having at least four new leaves, after which leaf production on the shoot ceases temporarily. Flush consists of both actively growing and temporarily dormant (or *banji*), shoots.

The quality of the made tea is better under high-altitude (low temperature), slow-growing conditions. It is determined by the "leaf standard" (whether only one, two, or three of the topmost immature leaves and the bud are taken, or whether more mature leaves are included as well), by the care taken in handling and conveying the flush to the factory, and by the chemical constituents and fiber content of the shoots.

For the manufacture of tea with good flavor or aroma (as in Sri Lankan "Ceylon" teas with Western High Grown or Uva quality), the leaf standard should not be less than 75 to 80 percent of shoots comprising bud and two leaves. This finer, or lighter, plucking, however, reduces yield, and quality and yield are therefore mutually exclusive. Coarser, or harder, plucking gives more yield and is appropriate wherever teas with more color and strength, or "cuppage" (as in Sri Lankan low-growns), are required by the trade. Crop management is therefore adapted to needs of the trade.

PROPAGATION AND GENETIC IMPROVEMENT OF TEA STOCKS

Seed and stem cuttings were the only source of new tea plants until the 1920s and 1930s, when the technique for the vegetative propagation of tea from single-node cuttings was discovered and became established. However, for accelerated progress in improving plants genetically, modern biotechnological methods have a much greater potential. Although conventional screening, selection, and breeding of tea clones will continue, biotechnology is now being used to incorporate a range of desirable attributes (including high yields, quality, tolerance to drought, pests, and diseases, and suitability for mechanized harvesting) into vegetatively propagated material.

Propagation from Seed

Since, in plucking fields, only bushes with decreased vigor will flower and produce seeds, plants that give good yield and quality can only be produced from seeds developed in special "seed gardens." In these gardens, tea plants are allowed to develop out of the vegetative phase by growing into trees and becoming reproductive. The practice of the commercial seed garden or *bari* was innovated in Assam.

Selection of plants for seed gardens was based mainly on morphology and growth characteristics. Crosses between Assam- and China-type "seedbearers" gave hybrids with improved yields and quality characteristics that were tolerant of poor soil and suited to diverse climatic conditions. The hybrids came to be referred to as Assam or China hybrids depending on which parent they resembled more. A negative feature was that the large variation in morphology, yield potential, etc., of hybrid plants, resulting from an unpredictable admixture of parental attributes, was not conducive to good crop management.

Vegetative Propagation

Starting from the 1920s and 1930s, new genotypes have been generated mainly by vegetative propagation in tea-growing countries, such as Japan, India, and Sri Lanka. Plants are raised by direct planting of vegetative material from carefully selected mother bushes. Since they have a genotype identical to that of the single-parent mother bush, these vegetatively propagated (VP) plants, or "clones," have attributes that are at least theoretically predictable and modified only by environmental variations and agricultural practices.

The development of an economical method of producing desirable planting material, by vegetative propagation from single-node leaf cuttings, was a giant step for the tea industry. VP tea has a much greater yield potential than seedling tea and greater tolerance against, or resistance to, drought, pests, and diseases. In addition, VPs reduce erosion because the spreading canopies of contiguous bushes gives a much better cover to the soil. Although considerable genetic improvements to tea stocks have been made by VP production (selection from good seedling teas and hybridization of the selected material), the process is empirical, slow, and laborious, with an improved VP taking about 15 to 20 years to produce. VP production is frequently hampered by characteristics inherent in the tea plant such as self-incompatibility, the high level of heterozygosity, and a long generative cycle.

Grafted plants are also used in commercial cultivation. The stock is selected for vigor and drought resistance, and the scion for quality and yield potential.

Characteristics of Seedling and VP Tea

Seedlings are heterozygous and have high genetic variability, and can therefore adapt to widely differing growing conditions and management and cropping systems much more efficiently than VP planting material. The productive life of VPs is 30 to 50 years, although senescence at 15 to 20 years or even earlier seems to occur. Bushes raised from seed last much longer—75 to 100 years or more.

VP cultivars comprise about 18 percent of the tea extent in China, although as much as 90 percent of the tea is VP in the Fujian Province. In Japan, about 80 percent of the tea is VP. In Sri Lanka, about 70 percent of the tea is VP in smallholder fields, compared to 45 percent on the estates.

Micropropagation

As a more rapid alternative to conventional vegetative propagation, tissue culture is used for the mass production of tea plantlets ("micropropagation") and plant regeneration. However, tissue culture is expensive (roughly five times the cost of producing plants by the conventional method), and is therefore usually restricted to the multiplication of elite clones.

Virtually any part of the plant (seed cotyledons, shoot meristems and tips, axillary buds, and nodal "explants"), in a suitable medium containing growth hormones, can give embryos in vitro, which develop into plantlets ("somatic embryogenesis"). To give an example of the order of the scale of production, one explant may give 500 genotypically identical plantlets in a relatively short period, say eight months.

Following the formation of root hairs in another so-called rooting medium, and hardening or acclimatization, the plantlets are transferred to soil and placed in a nursery. After one year or so, the young plants can be transferred to the field in the normal way, rooted and grown, either for evaluating breeding lines or for developing bushes in bearing.

Biotechnological Screening for Desirable Traits

Cells or protoplasts in culture can be exposed to biotic and abiotic stresses, and those surviving because they are tolerant or resistant to these stresses may be regenerated into whole plants, hopefully with the same tolerance or resistance. However, although this technique has been successful in some plant species, this has not been so with tea.

Molecular marker technology involving DNA fingerprinting of clones using the techniques of RFLP and RAPD can be used for making significant improvements to the existing genetic pool. Marker-assisted selection allows earlier evaluation of breeding lines and reduces the number of plants needing to be screened.

Embryo Rescue

In tea breeding, hybridizing species and genera usually fail because the hybrid embryos abort at an early stage of development. However, by excising the immature embryos and culturing them in vitro, it is possible to raise hybrid plants. This so-called "embryo rescue" technique is used for interspecific and intergeneric hybridization.

Anther and Microspore Culture

The production of homozygous diploids by inbreeding using conventional methods is, in practice, impossible in tea. Anther and microspore culture have therefore been tried for the production of haploid plants, which could be used for developing material possessing hybrid vigor. There has only been limited success of anther culture in tea wherever it has been tried, and haploids have apparently not been used successfully in tea breeding programs.

Somatic Hybridization Using Protoplast Fusion

Somatic hybridization between divergent tea species could give improvements in qualitative attributes of the resulting hybrids, such as growth rates and resistance to pests and environmental stresses. Protoplast fusion and plant regeneration have not been reported so far with tea. However, protoplasts have been isolated from leaf mesophylls and embryogenic cell suspensions, and attempts are being made to achieve plant regeneration from these isolated protoplasts.

Cryopreservation

Although seedling and VP tea bushes having genotypes valuable for future hybridization are preserved in germplasm- or gene banks set up in field conditions, the extended storage of germplasm (as shoot tips and axillary buds) under conditions of reduced growth in vitro, at temperatures below freezing, is employed as necessary.

THE EVOLUTION OF RATIONAL PEST AND DISEASE MANAGEMENT

"Naturally" Grown Tea

In the earlier periods of tea cultivation, seedling tea was grown as a monocrop mostly in jungle clearings of southern Asia. The tea agroecosystem was therefore little more than an extension of the natural jungle ecosystem, and subject to the same abiotic and biotic influences.

Since 1920 in Sri Lanka, and for longer periods of time in the older tea-growing countries (China and India), management practices based on organic manuring and mulching, with some minimal use of the inorganic compounds then available, must have kept pests and diseases below economic injury levels, at least for the most part, by keeping their natural enemies in a state of ecological balance with them. Availability (yields of well under 500 kg/ha were the norm) and demand for the crop were small in modern terms, inputs were financially modest, and therefore tea cultivation was a profitable enterprise.

The Downside of VP Tea Cultivation

With the development and introduction of VP planting material, yield potentials increased enormously, provided high levels of expensive, inorganic fertilizers were applied to meet the characteristically greater nutrient demands of the VPs. Typically, annual average yields can vary from 1,000-1,500 kg/ha for seedling tea fields, 2,500 kg/ha for rain-fed VP fields at high elevations, considerably more at low elevations (Sri Lanka), and 9,000 kg/ha for irrigated VP fields (Malawi).

Although specific VPs could be bred for pest and disease resistance and tolerance (as well as for other desirable attributes), VPs in general, because of their lowered variability, were much more susceptible to pests and diseases than the hardier, more genetically variable seedling teas. The widespread planting of VPs has led, therefore, to an increase in the complex of pests and pathogens attacking tea, and their absolute numbers.

Coincidentally, together with the planting of VPs, a profusion of synthetic chemical pesticides became available after World War II for use in agriculture and for the control of human and animal disease vectors, and naturally these began to be used for controlling tea pests as well. However, the synthetic pesticides decimated beneficial organisms and natural enemies (parasitoids, predators, and pathogens) as well. As a result, pests that were of minor importance because they had been kept in check by natural enemies began to take on a new significance. On a worst case basis, losses in tea crop of 15 to 25 percent have been estimated from insect pests alone.

The Adaptation of Tea to IPM

Just as the sustained use of inorganic, particularly nitrogen-based, fertilizers has led to increasing pollution of soil and water, the use of synthetic, persistent pesticides and herbicides has led to the contamination of food, including made tea, with pesticide residues. These unwelcome spin-offs from modern agriculture have grown to such proportions that there are now national and international movements against the unregulated use of agrochemicals. The threat posed by agrochemicals to human and animal health and genetics, to soil and water, and to the whole natural order of food webs and interrelationships that sustains the planet has been realized.

Happily, the indiscriminate use of persistent and hazardous synthetic pesticides has now given way in many agricultural systems, including that of tea, to integrated pest management (IPM). In Sri Lanka, for example, stemming from the work of its Tea Research Institute on IPM, tea growers employ minimal quantities of nonpersistent and less hazardous pesticides in combination with resistant VPs, as well as cultural methods that include agronomic operations timed to avoid or reduce pest damage. As a result, Sri Lankan tea was described by the Technical Sub Committee on Tea of the International Organisation for Standardization (ISO), in February 1997, as "the cleanest tea in the world" as far as pesticide residues are concerned. This description is based on routine analyses by ISO of made tea samples from all the major tea-producing countries in the world. The description was repeated at the succeeding ISO conference in November 1999.

Although the use of synthetic pesticides would still be necessary to keep pest outbreaks in check, there are indications that IPM and de-emphasis of synthetic chemicals will be a guiding principle in tea agriculture worldwide. The primary aim (apart from savings in foreign exchange arising from curtailment of their imports), is to move away from reliance on conventional, synthetic pesticides which leave residues in made tea, thereby lowering its global market value. International tea drinkers, particularly in developed countries, are becoming increasingly sensitive to additives and chemical residues, and demand is growing for tea as a natural health drink free of all extraneous chemicals.

THE VARIATIONS IN TEA MANUFACTURE

There are three main types of tea manufactured from plucked shoots: black, green, and oolong tea. In each case, the processing methods employed are different. On delivery to the factory, shoots are either withered (for making black tea), steamed, or subjected to dry heat (for green tea).

Black Tea: Orthodox and CTC

For withering, batches of shoots are spread out on tats or netting, or in troughs on mesh, in layers less than 30 cm thick. Fans are used to force air through the shoots in order to prevent their being heated by internal respiration, and to reduce their moisture content to an appropriate level, if necessary by preheating the air stream depending on the relative humidity. Withering for 6-20 hours releases "flavor compounds," caffeine, and amino acids.

In black tea manufacture, the cells of the withered leaf (moisture content: 55 percent) are either broken up in machines called rollers (the "orthodox" process), or the leaf is cut into particles of less than 1 mm (the cut, tear, and curl [CTC] process). (In southern Africa, the Lawrie Tea Processor is used instead of the CTC machine.)

These processes cause the mixing of polyphenols and the polyphenol oxidase enzyme system from the leaf's palisade cells and epidermal cells, respectively. Polyphenol oxidase catalyzes the oxidation of the polyphenols by atmospheric oxygen, and more polyphenol units (or monomers) are added to the oxidized polyphenols to form dimers (two monomers) or polymers (more than two monomers). Although the chemical process is polymerization; it is, incorrectly referred to in tea parlance as "fermentation."

Fermentation results in dimeric, orange-red theaflavins (TF) and predominantly polymeric, dark-brown thearubigins (TR). Black tea is 80 to 100 percent fermented, and its mixture of dimeric and polymeric pigments, TF and TR, may be referred to as oligomeric polyphenols.

The tea maker adjusts fermentation time (45 minutes to more than 2 hours) and temperature, almost intuitively, in order to make black tea with the desired attributes. The proper control of temperature during fermentation can result in a TF:TR ratio of 1:10, which is known to give optimal color and brightness to the tea liquor.

The reaction of oxidized polyphenols with other leaf compounds, such as carotenes and amino acids, early in fermentation results in "flavor compounds." The periods of fermentation are shortened during the "flavor seasons" in order to maximize flavor and minimize liquor strength.

"Firing," usually by using heated air in a machine called a drier, deactivates the enzymes and stops fermentation, and removes most of the moisture.

Green Tea

In green tea manufacture, the initial heating or steaming of the shoots denatures the polyphenol oxidase at the very outset, and no fermentation can occur. Green tea is thus rich in monomeric polyphenols (or catechins).

Oolong Tea

Oolong tea is semi- or 50 percent fermented and contains polyphenol dimers. It is produced only in China (mainly in Fujian, Guangdong, and Taiwan).

Antioxidants in the Main Types of Tea

The percentage of catechins against total polyphenols in black tea is less than 20, in oolong teas (as well as in Sri Lankan uvas) 50 to 60, and in green teas over 65. It used to be thought, until relatively recently, that the antioxidant activity, which makes tea beneficial to health, was due primarily to catechins. However, it is now known that the TFs, found exclusively in black teas, are equally effective antioxidants, and even more beneficial as a counter to the onset of certain health conditions.

TEA PRODUCTION AND THE GLOBAL TRADE

The Early Tea Trade

The first international trade in tea began out of China, more than 1,000 years ago, exports going overland through Asia as far as western Europe. During the periods of the European sea voyages of discovery and commerce, from the 1500s onward, trading in tea and tea drinking were reintroduced by the Portuguese, the Dutch, and the British to Europe, by way of the sea route from China. China was the largest producer of tea, and supplied 90 percent of the global requirements in the latter part of the nineteenth century and for some years after, when fast sailing ships from Europe (the "tea clippers") would race with the precious commodity to their home markets.

Tea Production, Exports, and Consumption

World and Country Production

World tea production has been stable over the first half of the last decade at about 2.5 billion kg (1990-1995), but since 1996 there have been more or less regular increases in production. In 1999, production stood at 2.8 billion kg, up by 17 percent over the 1989 figure of 2.4 billion.

The major four tea-producing countries during the last decade in volume terms have been India, China, Sri Lanka, and Kenya. The fifth place was taken either by Turkey or Indonesia. Among the minor tea producers, Japan, Iran, Argentina, Bangladesh, and Vietnam were prominent, with a distinct geographical group, the eastern and southern African countries, becoming significant also, together with Taiwan.

Production by Type

Of the black teas, more CTC is produced worldwide than orthodox tea. Because it consists of uniform particles, CTC tea is considered better for value addition, particularly for tea bags.

The main CTC producers are India and Kenya, with much smaller though sizeable volumes being produced in Bangladesh and Malawi. As with Malawi, the newer eastern and southern African tea-growing countries produce CTC tea only, in keeping with demand.

The main orthodox producers are Sri Lanka, Turkey, and Indonesia. India (one-tenth of its total production), Iran, Argentina, and the other South

American tea-growing countries, and China also produce orthodox tea. India and China are apparently looking to increase their orthodox production, in order to have a larger share in the world market.

Sri Lanka produces 93 percent orthodox and 6 percent CTC; 93 to 95 percent of its production is exported. Kenya produces about 99 percent CTC; more than 85 percent of its production is exported.

Approximately one fifth of world tea production is green tea, most of it produced and consumed, as in the earliest days of tea, in far-eastern countries. Japan's production is 100 percent green tea, China 71, Vietnam 75, and Indonesia 20; these countries together produce 98 percent of the world's green tea.

Green tea from China is less expensive than that from Japan, and most of the green tea consumed in the United States, for instance, is from China. In 2000, 62.3 percent of China's exports was green tea with only 11.8 percent being black tea. Most of Japan's green tea production is for internal consumption, but it is also a net importer of orthodox tea.

Country Exports

The major tea-exporting countries are Sri Lanka, Kenya, China, India, and Indonesia. The world's largest exporter of tea, Sri Lanka, keeps marginally ahead of Kenya. Sri Lankan export volumes have been climbing steadily, with all-time records being set every year since 1993. In the year 2000, exports reached an unprecedented 306 million kg.

China and India export less than the two top rankers, despite their larger production, because they meet a large internal demand with their huge populations. Approximately 80 percent and 65 percent of India's and China's production, respectively, are retained for internal consumption. It has been estimated that if every Indian drank an extra cup of tea every day, India would become a net importer of tea.

World and Country Consumption

The highest consumers are the CIS (or the USSR before it was disbanded), the United Kingdom, Pakistan, the United States, and Egypt. The total imports for consumption in 1999 were 1.2 billion kg, up from 1.1 billion kg in 1989, an increase of 10 percent.

The countries with the highest annual per capita consumption rates are Ireland, Turkey, Libya, the United Kingdom, Qatar, and Kuwait (above about 2 kg). Many of the Arabic countries and Sri Lanka rank next. Most of the highest consumers of tea in absolute volumes are India, China, the CIS, Turkey, the United Kingdom, and Japan.

VALUE ADDITION AND PRODUCT DIVERSIFICATION

Unlike in Asia, where tea, unadorned, is widely drunk as it always has been (as a cheap beverage, particularly by a nonaffluent underclass), in affluent, western societies and in the higher social strata of developing countries, variety is the key and diversification into innovative forms and presentations is an urgent requirement in tea marketing and promotion, particularly among the young. Value addition has myriad faces, and is proving to be highly profitable.

The tea-growing countries have traditionally exported their produce as a commodity, that is, as loose or bulk tea, with value being added by commercial conglomerates in tea-importing countries. However, value addition is on the increase in the tea trade of producing countries as well, to different extents and forms, by different companies and governmental agencies. Of the producing countries, Sri Lanka is ahead of India, China, Kenya, Indonesia, and Bangladesh at present in the field of value addition.

Packaging

Value addition to made tea involves packaging and imparting reliable information on quality and origin to international consumers. Packaging alone can increase the profit margin of made tea by as much as ten times: for example, Sri Lankan tea was exported to the United States in 1997 at an average price of U.S. $2.6 per kg, and sold in supermarkets there, in 1998, at an average price of U.S. $23 per kg.

The brand name on the package is a guarantee of quality, and the label can be used to give information on origin, blending details, and natural health benefits. These are all essential for the discriminating consumer to make an informed choice between a bewildering shelf display of various teas and other beverages. The packaging may be cardboard or foil packets, or tins of loose tea.

Tea Bags

The introduction of the tea bag revolutionized tea-drinking habits in the United Kingdom, in Western Europe, in the United States, and in Japan, owing to their being hygienic and economical, and easy to use and dispose of;

that is, to make and drink hot tea on the run, as it were, in keeping with the pace of modern lifestyles. Almost half the weight of tea sold in the developed countries is sold as tea bags. They are of various shapes and forms. A bag is simply and conveniently dropped into hot water or suspended in it by a string.

CTC teas, nearly 60 percent of global black tea production, are suited for tea bags because of their uniform particle size, which makes infusion easy. However, orthodox and oolong teas are also adaptable to tea bags. Tea bagging uses expensive machinery and is a large industry in itself.

Instant Tea

An alternative to the tea bag from the point of view of convenience is instant tea as powder or granules, manufactured from green leaf or fermented tea waste, which gives it a pronounced tea character. The tea solids are extracted with hot water, concentrated, and freeze-dried. It is available in a hot- or cold-water soluble form, plain or with milk and/or sugar incorporated into it, and is suitable for vending machines.

RTD Teas

Instant tea can also be used as a base for ready-to-drink (RTD) teas, for sale in cans, bottles, or in tetrapacks. RTDs do not in general have those tea characteristics which the connoisseurs of tea look for. RTDs serve only the temporary thirst of casual drinkers. However, a tea concentrate, appropriate for making RTDs, developed by the Tea Research Institute of Sri Lanka in the late 1980s, with only 4 to 5 percent of soluble tea solids, has now been improved to 10 percent of solids.

In the United States, 24 to 30 percent of the RTDs are consumed by 20- to 39-year-olds, with those over forty consuming only 14 to 18 percent. Teenagers consume only 11 percent.

On the other hand, a phenomenal growth in the production and consumption of RTDs (canned and bottled, green, Ooloong and orthodox) by all age groups has been observed in Japan, in the face of overwhelming competition from multinational soft drink and beverage giants.

Iced Tea Drinks

Soft drinks based on tea are consumed cold or iced, except in Japan, where they face strong competition from expensively promoted, branded, nontea soft drinks.

At the present time, iced tea, instant or brewed, comprises about 80 percent of the tea drunk in the United States, most of it in the southern states. Presumably it is being perceived as a component of a healthy, low-calorie diet and a pleasant, relaxing drink of convenience even for a more mature age group. Twenty-four to 27 percent of the iced tea consumed in the United States is by the thirty- to over-fifty-years age group, with 9 to 10 percent in the below thirty group.

Scented and Flavored Teas

Teas containing essential oils to give flower scents are popular in Eastern countries, the Middle East and in some Western countries. Teas of all kinds are easily treated with liquid flavors, either natural or artificial. Flavors in granular form are also incorporated into tea bags. The granules slowly release the flavor through the tea bag during infusion.

Alcoholic Tea Drinks

Tea wines and sherries are available in world markets. A tea-based alcoholic beverage with an alcohol content of 20 percent was developed by the Tea Research Institute of Sri Lanka.

Non-Beverage Tea Products

In Japan, a wide range of products based on green tea and its catechin components are available. Freshly picked leaves are used as a vegetable. As powder, green tea is incorporated into common items of food, such as noodles, rice, and bread, and into toothpaste. Tea extracts and extracted catechins are used to enrich drinks, bagged tea, chewing gum, laxatives, mouthwashes, deodorants, air-purifiers, soaps and cosmetics, sun-blocks, and molluscicidal sprays.

Tea is reprocessed and used for making sweets, jams, jellies, and even massage oils. New tea-based products emerge all the time. For instance, in Taiwan and elsewhere in Asia, and now in the United States, a new drink, glutinous tapioca balls (from cassava) shaken up with green or black tea, is becoming popular.

Nonfood Tea Products

Tea is used for stuffng pillows and bathing sponges, and its brew has been used even for making "tea-stain" paintings.

RESEARCH AND DEVELOPMENT INSTITUTIONS FOR TEA

In most of the tea-producing countries, institutions are dedicated to tea research and development (R & D), or having a mandate for tea, as well as for vegetable crops. Much of the research is concerned with studies on field production, but there is not as much on the chemical and engineering aspects of tea processing. Some details of these institutions are given here.

Bangladesh

Bangladesh Tea Research Institute
Address: Srimanga-3210, Moulvibazar, Bangladesh
Telephone: 0861-3328
Activities: Development of high-yielding clones; soil rehabilitation; fertilizer use; pest management; drainage
Publications: *Tea Journal of Bangladesh* (half-yearly), annual report, quarterly bulletin

Central Africa

Tea Research Foundation (Central Africa)
Address: PO Box S 1, Mulanje, Malawi
Telephone: 462 261; 462 271
Activities: Promotion of tea research in the central and southern Africa regions. Field trials in Zimbabwe and South Africa. Breeding for high quality clones; efficient fertilizer use; response of bush to mechanization; pesticide residue monitoring; modernizing manufacture
Publications: Annual report, newsletter (quarterly)

China

A large number of R & D organizations operate within the 19 provinces in which tea is grown. They include:

The International Institute of Tea Science and Culture
Tea Industry R & D Centre, Chinese Academy of Social Sciences
Tea Research Institute, Chinese Academy of Agricultural Sciences
Agricultural University of North-Western China
Agricultural University of North China
Agricultural University of Central China
Zhejiang Agricultural University

Hunan Agricultural University
Sichuan Agricultural University
Anhui Agricultural University
Fujian Agricultural University
Yunnan Agricultural University
Hangzhou Tea Research Institute, All-China Supply
    and Marketing Cooperative
Tea Research Institute, Agricultural Scientific Academy, Hunan Province
Tea Research Institute, Agricultural Scientific Academy, Fujian Province
Tea and Silkworm Research Institute, Agricultural Scientific Academy,
    Jiangxi Province
Tea Research Institute, Guizhou Province
Tea Research Institute, Sichuan Province
Tea Research Institute, Guangdong Province
Fruit and Tea Research Institute, Hubei Province
Tea Research Institute, Agricultural Scientific Academy, Yunnan Province
Tea Research Institute, Agricultural Scientific Academy, Anhui Province
Tea Research Institute, Agricultural Scientific Academy, Wuxi, Jiangsu
    Province

Zhejiang. Agricultural University
Address: Hangzhou 310029, Zhejiang Province, China
Telephone: 42605
Activities: Soil science, seed science, plant breeding, crop husbandry; rice,
wheat, cotton, citrus, peach, pear, tea, and vegetables
Publications: *Zhejiang Agricultural University Journal* (quarterly)

Department of Tea Science, Zhejiang Agricultural University
Activities: Tea manufacture; tea plant protection; tea tasting and grading;
cultivation of the tea plant; tea plant breeding
Publications: *Tea Bulletin* (bimonthly)

India

United Planters' Association of Southern India (UPASI)
Tea Research Institute
Address: Nirar Dam B. O., Valparari 642127, Coimbatore District
Tamil Nadu, India
Telephones: 4253 71038, 4253 71038
Fax: 4253 7102, 44253 71419

Activities: Soil and water conservation; plant production (improving tea yield and quality, development of new varieties and clones); husbandry; plant protection; management systems for economic viability of tea cultivation; food science; agricultural engineering
Publications: Annual Report of UPASI Scientific Department; bulletin

Tea Research Association (Tocklai)
Address: Tocklai Experimental Station, 785008, Assam, India
Telephone: 0376 320054 Fax: 0376 325589, 0376 327253
Activities: Plant breeding; physiology; soil and water management; field technology; agricultural practices; plant protection; tea biochemistry and processing; tea machinery development; agronomy; statistics; tissue culture; biotechnology
Publications: Annual scientific report; advisory leaflets; bulletin

Institute of Himalayan Bioresource Technology
Address: P.B. No. 6, Palampur (H.P.), 176 061 India
Telephone: 01894-3042-26, 30741 (Office); Fax 91-1894-30433
Activities: Citrus, pecan, tea, stone fruit

Indonesia

Research Institute for Tea and Cinchona
Address: Gambung, P.O. Box 148, Bandung 40010, Indonesia
Activities: Plant production by genetic manipulation; plant protection; product diversification, quality improvement, modification of processing equipment; value addition, waste utilization; studies on government tea smallholder program

Iran

State Tea Organisation, Ministry of Agriculture
Address: 202, Sornmaye Avenue, Teheran, Iran
Activities: Funding and training tea farmers and factory operatives; buying green leaf; supplying planting material and agricultural inputs; supervising manufacture; importing tea; research for increasing crop and quality

Japan

National Research Institute of Vegetables, Ornamental Plants, and Tea
Address: 2769 Kanaya, Shizuoka 428, Japan
Telephone: (05474) 5 4101
Activities: Tea agronomy; tea processing and technology
Publications: Bulletin, series B (Kanaya)

Kenya

Tea Research Foundation of Kenya (TRFK)
Address: Box 820, Kericho, Kenya
Telephone: 20598
Activities: R & D on all problems related to tea production, including better planting material, productivity, quality, and land suitability, pests and diseases, and costs of production
Publications: Annual report; *Tea* (biannual); *Tea Growers Handbook*

Mauritius

Activities: Identifying high-yielding clones with quality characteristics, and importing such clones for testing; improving productivity from mechanical plucking; appropriate fertilizer usage; identifying food crops and fruit trees for abandoned tea lands for export

Nepal

National Tea and Coffee Development Board
Address: Singhadurbar, Kathmandu, Nepal
Telephone: 01-228638
Activities: Research and development on all problems related to production, processing, and marketing of tea and coffee

Nigeria

Cocoa Research Institute of Nigeria
Address: PMB 5244, lbadan, Oyo State, Nigeria
Telephone: (022) 412430
Activities: Research into factors affecting cocoa, kola, coffee, cashew, and tea
Publications: Annual reports; technical bulletins; *Farmers' Field Days* (three yearly)

Pakistan

National Tea Research Institute
Address: Shinkiari, Mansehra, NWFP, Pakistan
Telephone: 05922-2553, 2125
Activities: Nutritional and propagational studies; extension work

Russia

Russian Research Institute in Floriculture and Subtropical Crops
Address: Fabritsius str., 2/28, Sochi-354002, Russia
Activities: Breeding and production of planting material; land preparation; soil, fertilization, and nutrition; water conservation and irrigation; mechanized plucking; weed control

South Africa

Burgershall Research Station
Address: Private Bag X501, Kiepersol 1241, 86 Hazyview, South Africa
Telephone: (0131 242)
Activities: Research on bananas, macadamia nuts, avocados, granadillas, coffee, tea, pepper, ginger

Sri Lanka

Tea Research Institute of Sri Lanka
Address: St. Coombs, Talawakelle, Sri Lanka
Telephone: 94 51 22601
Fax: 94 52 58229
Activities: Breeding and production of improved planting material through hybridization and biotechnology; productivity of tea lands through research on cultural practices, soil reconditioning, erosion control, weed control, soil-plant-water relationships; harvesting and pruning methods including mechanization; intercropping; the tea plant in relation to environment and cultural practices; characterization of tea soils, responses of the tea plant to inorganic and organic fertilization; rational control and integrated management of insect, mite, and nematode pests and tea diseases; processing and packaging for quality improvement and cost reduction; computerization and automation of manufacture; value addition and product development; health benefits of tea consumption; economic evaluation of technical recommendations, worker productivity, and sociology

Publications: Annual reports in English (E), Sinhala (S) and Tamil (T); *Sri Lanka Journal of Tea Science* (semiannually in E); *Tea Bulletin* (semiannually in E, S, T); monographs and pamphlets (in E, S, T); advisory circulars (as necessary, in E); *The Handbook on Tea* (in E); *The Field Guide* (in E, S, T); *TRI Update* (biannual newsletter in E)

Taiwan

Taiwan Tea
Address: Pushin, Yangmei, Taoyuan 32613, Taiwan
Telephone: (03) 478 2059
Activities: Breeding and cultivation, soils and fertilizers, tea-plant physiology and protection, farm management; manufacture and new methods of tea processing, packaging, and storage; tea chemistry and taste evaluation, instant and flavored teas; farming and irrigation machinery
Publications: *Taiwan Tea Research Bulletin;* annual report

Tanzania

Tanzania Agricultural Research Organization (TARO)
Address: PO Box 9761, Dar es Salaam, Tanzania
Telephone: (OS 1) 74247
Activities: Interdisciplinary research into all aspects of crops, soil, and farming systems; coordination of all crop research in Tanzania: maize, sorghum and millet, grain legumes, *Phaseolus* beans, oilseeds; cotton, sisal, coffee, tea, tobacco, banana, cashew, sugar cane, root and tuber crops, wheat, rice; soils and farming systems, resource-efficient farming
Publications: Annual report; research reports; *TARO Newsletter* (quarterly); booklets

TARO Marikitanda Agricultural Research Institute
Address: PO Box 8, Amani, Tanga, Tanzania
Activities: Tea, legumes
Publication: Annual report

TARO Maruku Agricultural Research Institute
Address: PO Box 127, Bukoba, Kafera Region, Tanzania
Telephone: 533
Activities: Banana, coffee, tea, cereals, legumes; soils; farm management
Publications: Annual report, newsletter (monthly)

Vietnam

Tea Research Institute
Address: Vinh Phu Province, Vietnam
Activities: Collection and importation of tea varieties for breeding high-quality teas; soil and farming surveys; improvements to processing equipment including semi-mechanized, small capacity units for rural households; training

THE FUTURE OUTLOOK FOR TEA

The future of tea as a profitable crop in a variety of agricultural environments around the world, and the adaptability of the types and forms of tea to changing consumer requirements, can be assessed by an acronymal SWOT (strength, weakness, opportunity, threat) analysis, wherein the internal strengths and weaknesses of the global tea-producing and tea-marketing segments can be looked at in conjunction with the external opportunities and threats to which these segments are subjected.

Strengths

The Traditions of Tea Drinking

Tea consumption is a part of the traditional culture of Far-Eastern countries such as China and Japan, and in the diaspora of these countries in all parts of the world. Apart from ritualistic tea consumption enshrined in traditional tea ceremonies and banquets, the continuous sipping of green tea at ordinary meals, and the sheer sizes of the populations involved in this practice, ensure the future of green tea. It is unlikely that Far-Eastern cultures will ever dispense with a centuries-old habit whatever the onslaughts of modernity.

In other parts of Asia, notably in the Indian subcontinent, where black tea is the traditional, time-honored drink of millions, a huge market has always been assured.

Large populations in the west Asian countries and Russia are inveterate consumers of strong black teas (such as Indian teas and the Sri Lankan low-growns). In the Arabic countries of west Asia, small cups of black tea taken strong are consumed at all hours of the day, and is the customary offering denoting hospitality and friendship.

The ex-colonies of the British Empire and those countries influenced by British mores, such as the United Kingdom, the United States, Canada, Australia, and a host of others in all the continents, many of them now economically sound in their own right, have never abandoned tea drinking completely. The habit, ingrained in their collective psyche as it were, continues in their domestic and social settings.

In addition, there has always been, particularly in the West and Japan, a coterie of connoisseurs of good teas, such as the Indian Darjeelings and Sri Lankan speciality teas (those well-known as NUwara Eliya, Dimbula, Uva, Uda Pussellawa, Kandy, and Ruhuna), grown under conditions nonreproducible anywhere else. These conditions may obtain in unique agroclimatic zones and subzones, or even in individual plantations which only the ultimate connoisseur can distinguish.

One of tea's main strengths therefore is its diversity and adaptability as a product for all manner and conditions of humankind (there are an estimated 3,000 different types of tea), its long history as a traditional beverage, its relative cheapness and availability, and an upmarket image among connoisseurs.

The Tea Research Culture

There are a large number of research institutes in tea-growing countries as well as other institutions, including universities, both in the developed and developing worlds, where studies on all aspects of tea cultivation, processing, and marketing are going on apace. Some of these, as in India, Sri Lanka, and China, have been in existence for decades. This continuous examination of tea and the tea trade from all angles, and the findings continuously being disseminated, must be seen as another strength of the worldwide tea industry.

The Favorable Environmental Impact of Tea

The adaptation of tea into a relatively environmentally benign, nonpolluting crop is being done within many tea-growing systems, based on the use of VPs, integrated nutrient and pest management, organic farming, precision agriculture, microirrigation, and intercropping. As in certain areas of Sri Lanka, tea may be intercropped with rubber, coconut, coffee, spices, fruit crops, and legumes; this serves to increase profitability as well as biodiversity.

A well-managed, modern tea plantation is a prime exemplar of environmentally rational agriculture. It approximates to a natural forest in that it is multicanopied: the thick tea canopy over which are the two canopies of the medium- and high-shade trees. Tea plantations with their large biomass cools the environment, encourages an animal biodiversity, and acts as a buffer against environmental pollution by absorbing carbon dioxide.

Agricultural practices and factory processes connected with tea have evolved in such a manner that they now have the smallest impact on soil, water, and air when compared to some other agricultural production systems.

Being completely worker-dependent, field operations conventionally use a minimum of machinery. Motorized harvesters and brush cutters are used but, with good management, the pollution effects arising from them are negligible.

Soil disturbance and erosion, particularly in the uplands and on slopes where tea is grown, are ever-present hazards, having both negative on- and off-site effects. However, researchers and tea growers have evolved and can now adopt soil conservation measures that keep soil erosion well below the tolerance limit (9 tonnes/ha/year in Sri Lanka). These measures include bush spacing for maximum canopy cover, mulching and planting of cover crops during land preparation and the first two years after planting, construction of leader and contour drains and contour terraces, and the establishment of hedgerows in the sloping agricultural land technology system. A combination of manual, cultural, and chemical methods for managing weeds, rather than the use of scrapers, also conserves the soil.

The use of soil fumigants against pests and diseases is kept to a minimum, and a particularly damaging fumigant, methyl bromide, has been withdrawn internationally.

Tea manufacture makes use of fuel wood, oil, or solar radiation as energy sources. Trees are renewable, and if their removal for energy generation is controlled, this will not have a chronic environmental impact. The usual location of tea factories within large plantations and close to forests allows for the sequestration of greenhouse gases (mainly carbon dioxide). The small amounts of heat given out by scattered tea factories in these locations are not significant in warming the environment.

No chemicals are used in tea manufacture. Tea factory waste is merely refuse tea and wood ashes. Both are valuable in that they are composted for use as manure in the tea fields. Refuse tea can be used for making instant tea.

Weaknesses

The Oversupply Situation

Global tea consumption has been lagging behind production, with the result that oversupply is growing. The increase in overall production between

1989 and 1999 was 17 percent, although imports by consuming countries increased only by 10 percent).

A Requirements Mismatch

Another current weakness in the global tea industry is a mismatch between the types of beverages consumers seem to want, and what are available in the form of tea and tea-based drinks. Even in traditional tea-drinking, Western countries such as the United Kingdom, a discernible reduction in tea imports has been observed. There is pessimism, in sections of the tea trade in the United Kingdom, that tea has an outdated, old-fashioned image in the minds of the younger generation, and that sales will inevitably decrease as a result.

Tea being a perennial crop, with a production momentum that is difficult to slow down or redirect, in order to cut back on supply, or in order to adjust product specifications, short- to medium-term responses to market requirements are virtually impossible. As a result, prices tend to move from boom to bust over relatively short periods of time, whatever the improvements in production and marketing efficiency.

The Marginalization of Producers

Unless a seamless vertical integration occurs between production, marketing, and retail purchasing, with a "bush-to-cup" marketing strategy based on well-accepted national brand images, tea producers, both in the corporate sector and in smallholders, will continue to lose out on profit margins to middlemen. They will continue to be, in the main, suppliers of a primary commodity to sophisticated, often multinational, buying and export enterprises with their own well-financed branding and distribution arrangements.

Agrochemical Residues in Made Tea

Consumers and the international trade are becoming increasingly sensitive to additives and agrochemical residues in tea, and the marketing of tea as a pleasant and beneficial health drink, free of contaminants, is being jeopardized.

At the Technical Sub Committee Meeting on Tea of the International Organization for Standardization, held in Calcutta in November 1999, the levels of pesticide residues in tea exports, as revealed by European surveillance during 1994-1997 and 1997-1998, were reported on. Although most of the exporting countries did not exceed the maximum residue limits set by the European Union, it was indicated that certain Asian countries, both in 1994-1997 and in 1997-1998, had incidence of residues in their teas which were at actual or potential violation of EU limits. The Technical Sub Committee advised that the use of all waste-soluble pesticides (whose residues find their way into the tea brew) should be stopped.

Opportunities

Optimizing on the Health Benefits of Tea

People in general are now aware that tea (of all types, both black and green) is rich in polyphenols, which are highly effective antioxidants, and that tea's health-giving and beneficial effects have been scientifically and medically proven by independent research in different countries.

Even a relatively slight increase in per capita consumption worldwide, under the impetus of renewed generic promotion efforts based on the recent health findings, should serve to correct the present oversupply situation. For instance, an increase in the current daily consumption of charateristically well-informed U.K. consumers from the present 3.2 cups per person to four cups would help in rectifying the production-consumption imbalance.

The Emergence of New Markets

The opening up of new markets in countries where tea had not been commonly consumed hitherto is a new growth opportunity for the tea trade. Growing populations and higher incomes in developing countries entering the free-market nexus have resulted in higher demands for inexpensive beverages. Thus, consumer preferences for tea in the countries of central and eastern Europe, such as Poland, the Czech Republic, Slovakia, and Hungary, and the CIS are growing.

With the emergence of the new states of the CIS derived from the former USSR, and their opening to free trade, large new Muslim populations in the CIS countries have become contiguous with the habitual tea drinkers of west Asia. The tea-drinking habit is therefore being reinforced in the CIS, and has enormously expanded the outlook for tea.

Adapting to a New Generation of Consumers

There is guarded optimism in the global tea trade that, with value addition and product development moving away from traditional forms and styles of tea consumption, the present need–availability mismatch will reverse, and that tea will come to be accepted, particularly in Western countries in newer, more upbeat forms, by young and old, for example as a part of a healthy diet.

The age-old adaptation of tea, this time to present-day consumerist trends by a new generation, will determine its future. There is a real opportunity for the world's tea industry to grow and remain viable, with value addition and new forms of the product adapting quickly and in the short term to the convenience of consumers and their evolving tastes and lifestyles.

The decreasing consumption by hitherto dedicated tea drinkers, as in Western countries, could perhaps be reversed by creating new national branding images. Sophisticated and informed consumers would be willing to pay higher prices for a brand that gives assurance of quality and maintenance of product standards.

Cleaning Up the Product

Happily, there is now a clearly perceptible and growing determination on the part of all tea producers and exporters, particularly in Asia, that unacceptable levels of residues shall not be allowed to stigmatize their tea abroad or be detrimental to their local consumers. This determination extends to calls for growing tea under stringent, organic farming conditions according to internationally accepted norms.

Growing Tea Alongside Other Profitable Crops

The practice of growing tea with other tree crops, or intercropping, has emerged from field research efforts as being viable and profitable. Intercropping serves to maximize land productivity and minimize the economic and environmental risks involved in growing a monocrop. Added advantages are favorable cost benefits and increased employment.

In Sri Lanka, for example, commencing in the early 1980s, studies on intercropping tea with rubber proved successful at the lower elevations where rubber is widely grown. Tea is also intercropped successfully with coconut in the low country.

For many years prior to this, it was the practice for smallholders in the Sri Lankan mid-country to cultivate pepper, coffee, and cloves mixed with tea. Intercropping tea with fruit crops and legumes has also proved profitable.

Threats

Competition from Other Beverages

The external threats to the tea trade come from the globally widespread and hugely expensive promotional campaigns of multinational beverage and soft-drink marketers, which tea promotional efforts cannot match owing to the costs involved. Thus, the soft-drink market out competes tea in most countries, with the exception of Japan. Even though coffee consumption in the United States has fallen precipitously over the last 25 years or so, it is now picking up because of the introduction of new forms and images of coffee. The typical U.S. consumer takes more coffee than tea, health effects notwithstanding.

The Flight of Workers

Many of the tea economies, based as they are in developing countries characterized by increasing levels of education, industrialization, and urbanization, are facing a serious and growing threat of worker shortages in field and factory. This situation is due to "pull" factors (better prospects and wages outside the plantations) and "push" factors (the stigma attached to plantation labor, and unpleasant working and living conditions).

SWOT: The Last Word

Although tea producers cannot afford to be sanguine at the present time, it is clear that the strengths of tea, and the fresh opportunities opening up in its favor, outweigh its weaknesses and the threats it faces both nationally and on the global scene. The key factor that will swing the balance teaward, and make it the beverage of choice, is that it is clearly an inexpensive and, most of all, health-beneficial component of a modern diet. Tea will surely adapt now to becoming an item of food.

Chapter 12

Wattle

N. Kumar

Wattle belongs to *Acacia* (Family: Fabaceae), a large genus comprised of more than 500 species found in the warmer and drier parts of the world, chiefly in Australia and Africa. Species with pinnately compound leaves are found throughout the tropics, and the phyllodineous ones are natives of Australia. In India, there are about 22 indigenous species, distributed throughout the plains. However, all species which yield wattle are exotic and are introduced from Australia.

TYPES OF WATTLE

Acacia mearnsii *(Syn.* A. mollissima*)—Black Wattle*

It closely resembles *A. decurrens* in its habit and is sometimes considered a mere variety of the latter. In the Nilgiris of India alone, the area under this crop would be around 6,000 ha (Gupta, 1993). The main difference from *A. decurrens* lies in the shape of the leaflets, which in *A. mearnsii* are short, obtuse, 5.0-7.5 cm long and closely spaced; in *A. decurrens* they are narrow, 7.5-10 cm long, and widely spaced. The leaves of the former are dark green, whereas those of the latter are yellowish green. *A. mearnsii* is also a native of South Australia and Tasmania. It is the principal species grown in South Africa and is the chief source of the wattle bark of commerce. It was introduced into India around 1840, along with *A. dealbata*, mainly to provide fuel, of which a shortage existed at that time. It thrives at elevations of 1,665-2,330 m above mean sea level (MSL), where there is a well-distributed rainfall of about 150 cm per annum. It is now grown in the Nilgiris, the Pulneys, and in central Kerala on a small scale. It also grows in the highlands of Sri Lanka, but does not appear to be suited to Myanmar. A few trees are also found in tea estates as windbreaks and shade trees.

Acacia decurrens—*Green Wattle*

It is an evergreen tree, reaching a height of 12 m. In general, the bark is olive green, but is dark grey on older trees. The pale yellow flowers are less plentiful and less scented than those of *A. dealbata*. The bark is rich in tannin (36-41 percent at 11 percent moisture content).

A. dealbata—*Silver Wattle*

Acacia dealbata is an evergreen tree native to Tasmania and South Australia. It is now regarded as a variety of *A. decurrens*. It was introduced to the Nilgiris in 1840, where it became naturalized and is now a characteristic feature of the vegetation from 1,250 m upwards. It has also been planted in the Himalayas (Shimla, Nainital, and Almora hills) chiefly between 1,800 and 2,500 m. It suffers from snowbreak considerably, but regenerates through its numerous root suckers.

On the Nilgiris, *A. dealbata* reaches a height of more than 12 m and a girth of more than 1.2 m. Its bark is thinner than that of *A. decurrens,* and is silvery grey in appearance. The young shoots and foliage are also of the same color. It blooms profusely, producing large quantities of yellow flowers. The pods are broader and less constricted between the seeds than those of *A. decurrens.*

The tree has extraordinary powers of reproduction through its spreading root suckers and is of great value for clothing unstable hill slopes. The bark has only one-third the tannin content of that of *A. decurrens.* The tannin content of Nilgiri bark varies from 1 to 15 percent at a moisture content of 10 percent.

A. pycnantha—*Golden Wattle*

Acacia pycnantha is a native of South Australia, where, along with *A. mearnsii*, it constitutes the chief source of tan bark. In size it is a smaller tree than the latter and its bark is thinner. It is also said to be a less hardy species and has not found favor in South Africa. Experimental plantations are being raised on the Nilgiris. The species is very rich in tannin. Analyses showed up to 50 percent of tannin in the air-dried material. The best commercial bark has an average of more than 38 percent tannin in the Nilgiris.

The trees yield Australian gum, which is principally an arabogalactan. The extract of the bark is said to be equal to the best Indian catechu. In view of the keen demand for wattle and wattle extracts from the leather industry, large-scale plantations of black wattle were taken up in the Nilgiri and

Pulney hills during the 1950s. Nearly 18,000 ha have been brought under wattle plantation, and the annual production of bark at present is 8,000 to 10,000 tonnes (Sherry, 1971). Currently the area under wattle in Tamil Nadu is estimated to be 36,660 ha (Anonymous, 1999). A private factory in Mettupalayam, India, produces wattle extracts with an annual production capacity of 3,750 tonnes. Wattle bark contains tannins, nontannins, insolubles, and fibers. The tannins in wattle bark belong to the catechol group and consist of a complex mixture of polyphenols, of which catechin and its derivatives are the most important.

GROWING CONDITIONS

Climate and Soil

Black wattle comes up well at elevations between 1,950 to 2,500 m above MSL. A well-distributed rainfall of 150 cm with 100 rainy days is essential for its growth. In regions of high rainfall, the bark may be infested with lichen, resulting in deterioration of quality. Winds and frosts during the early part of the rainy season (June-July) adversely affect growth of the trees. In East Africa, it grows well where annual rainfall is 1,041-1,321 mm, (about 75 percent between April and September). On the equator, where black wattle is grown in South America, the rain pattern is nearly opposite: the mean annual temperature range is 17-23°C, with little seasonal variation but considerable diurnal variation. At higher altitudes in South America, frost is a risk, and heavy snows may break tree limbs. Tannin content varies inversely with precipitation. Ranging from Warm Temperate Dry through Tropical Thorn to Tropical Moist Forest life zones, black wattle is reported to tolerate annual precipitation of 660-2,280 mm and annual mean temperatures of 14.7-27.8°C. In South India, 8° to 11° North latitude, black wattle grows well at elevations above 1,500 meters. At higher elevations, due to incidence of heavy frost and strong cold winds, the initial growth is very slow. Thereafter, growth (height and diameter) is rapid for five or six years and then falls off gradually. Except in the Nilgiris, wattle is planted in regions of heavy rainfall of around 2,500 mm and more (Rajagopalan, 1973). Black wattle plantations are confined to forest areas of the eastern highlands of Zimbabwe, where the mean annual rainfall exceeds 1,000 mm. Within this zone, the altitude varies from 700 m to more than 2,500 m, but the higher collar regions are the most suitable. The mean temperature of the hottest month is 20-24°C in Zimbabwe and mountain mists are frequent.

Wattle grows on poor dry soils but favors deeper, moist, and more fertile soils. In Australia, black wattle occurs on soils derived from shells, mudstones, sandstones, conglomerates, and alluvial deposits. In South America, it grows on red clay or sandy soils that have suffered from severe erosion and depletion (ferruginous clay loams with little or no free silica).

However, deep, well-drained soils are preferred for good growth. Sandy loams are deemed optimum. Clay pans and hard laterites are not suitable and should be avoided. In Zimbabwe, the soils of the main wattle-growing areas are derived from granite, quartzite, and dolerite with textures varying from sand to clay loam. Higher yields of wattle bark are obtained from deep, well-drained sites irrespective of the parent rock.

In the Nilgiri and Pulney hills of south India, growth of wattle in areas having adequate soil depth, protection from cold winds, and receiving an annual rainfall of around 1,500 mm is comparable with other wattle-growing countries. However, in areas receiving high rainfall (2,500 mm and more), wattle plantations are poorly grown and stocked thinly.

Pretreatment of Seeds

The common method of propagation of wattle is by seeds. Before sowing, the seeds are subjected to hot water treatment. After treatment, the seeds are washed four or five times with cold water to remove the gummy mucilage. They are then dried under shade and stored in lead-lined boxes. These seeds may retain viability for six months to one year. The pretreatment involves boiling in a drum or a pot with watering about five times the volume of the seeds. When the water begins to boil, the drum is removed from the fire and the seeds are put into the boiling water and allowed to cool for 24 hours. Care is taken not to boil the water after putting in the seed. After cooling, the seeds are washed in cold water to remove the gummy mucilage and dried in the shade. Gupta and Thapliyal (1974) recommended that pre-germination treatments for *A. mearnsii* involve soaking seeds in hot water for 12-24 hours or stirring in concentrated H_2SO_4 for 1 hour. This is sufficient to break dormancy.

Nursery Practices

Treated seeds are sown in raised standard beds of 10 m × 1 m; 300 g of seeds are sown per bed. The beds are covered with bracken ferns and watered. The object of providing bracken cover is to protect the beds from excessive evaporation and also to facilitate quick germination. Germination occurs in 10 to 15 days when the bracken cover is removed from the beds.

Seeds are sown during October-November and pricked out when they are 5 cm tall. Two-month-old seedlings should be pricked out from the bed and planted in small polyethylene bags (size 15 × 10 cm) filled with good soil, as done for any potting of nursery seedlings. The seedlings in the polyethylene containers should be watered twice daily for the first two months and then once daily on nonrainy days. In this condition, the seedlings are kept for three to four months. By this time, they reach a height of 25 to 30 cm and are ready for planting in the field. One m² of bed provides 400 to 500 seedlings. Nine- to twelve-month-old seedlings are used for planting. About 450 g of seeds will give 3,500 to 4,000 seedlings, which can be planted in an area of 3 to 6 ha. The seedlings in the nursery have to be protected against frost during October to January and against rabbits by providing rabbit-proof fencing.

In Zimbabwe, root pruning is done during three to four months of the nursery period. It is essential to confine the roots either by drawing taut steel wires below the polyethylene bags or by lifting the plants individually and trimming the roots. Seedlings may also be shoot-pruned to the ideal planting size of 15-20 cm if planting is delayed.

Planting Practices

Planting of the container seedlings should be done at the onset of the monsoon season during May-June in pits of 30 cm³ at a spacing of 2 × 2 m. The plantations can accomodate 2,500 seedlings per hectare. Wattle can also be propagated by stump cutting for which 7.5 cm of shoot and 22.5 cm of root have been found optimum. While planting polyethylene-container seedlings, care should be taken to ensure that the polyethylene bags are removed without disturbing the ball of earth.

Preparation of Land

The pits may be filled with composted humus. A half kilogram of potato fertilizer may be applied to each pit as an alternative practice. This helps to boost growth of the plants at the initial stages.

Planting and Aftercare

Nine- to twelve-month-old seedlings are planted in the main field during June-July depending upon the monsoon. While planting, mossing of roots is done. A spacing of 3 x 3 or 4 × 4 m is adopted. At the Wattle Research Institute, Pietermaritzburg, South Africa (Anonymous, 1960), it was found that the highest yield of tannin per acre was obtained from 11-year-old trees at a density of 600 per acre and from 9-year-old trees at 800 per acre. Adequate windbreaks have to be provided for the plantations. Weeds have to be removed from around the young wattle seedlings. At higher elevations, seedlings may be provided with frost covers to guard against frost in the first year of planting.

Growth of seedlings raised by direct sowing is extremely slow during the first year of planting. Thereafter growth increases both in diameter and height. In well-spaced plantations, the growth rate is more rapid and continues up to ten years. The growth rate of wattle trees is higher in the Nilgiris than at Kodaikanal due to better soil and climatic conditions. In South America, fields are usually ploughed and harrowed in April or May. Seedlings are set out from May to November, but usually in winter, from June to August, after a rain. Plants are spaced 2 × 2 m, at the rate of 2,500 per ha. Propagation by cuttings is almost impossible without mist. Air layering is more promising. Two types of people grow acacia. The tanner or business owner plants 200 ha or so entirely to black wattle, usually one section at a time so that he or she can plant and harvest within the same year and continue year after year. The farmer plants half or less of his or her land to black wattle and the rest to crops such as corn, beans, cassava, sugarcane, other vegetables, or pasture. The farmer plants 2 to 6 hectares of acacias each year and thus evenly distributes work and production. Oxen may be useful for ploughing, but most work is done by hand. Usually only ploughs and hoes are used in cultivation. Intercrops may be grown the first year during which trees grow about 4-5 m in height, and about 2.5 cm in diameter (Duke, 1981).

Thinning

Wattle trees exhibit strong affinity to light and do not tolerate shade. Thinning is necessary in the third year of planting if plantations are closely spaced. However, in wider spacing, thinning is not necessary.

Manuring

Samraj and Chinnamani (1978) recommended that nitrogen, phosphorous, and potassium (NPK) be applied to each plant of black wattle in the first year to boost growth. Soil working should be done, leaving 15 cm around the plant, first in July-August and again in November-December. In the second year, only one working is carried out following the rains. They further suggested that 200 g of NPK fertilizer (wattle mixture) be applied to each plant in the Nilgiris to boost initial growth. The NPK fertilizer should be made up of ammonium sulfate, superphosphate, and muriate of potash respectively. Recent experiments conducted in the Nilgiris reveal that black wattle responds better to the application of 45 g of ammonium sulfate, 90 g of superphosphate, and 45 g of muriate of potash compared to other combinations of NPK fertilizers for boosting initial growth. Hussain et al. (1980) found that application of NPK [45 g $(NH_4)_2SO_4$, 90g superphosphate, 45 g KCl] at establishment gave the best results.

Aftercare

Wattle is susceptible to root competition from grasses, hence scrape weeding to a diameter of one meter around the plant after establishment (one month) is essential. Subsequently, soil working is done leaving 15 cm around the plant, first in July-August and again in November-December. Casualties are replaced periodically in the first year along with weeding and soil working. In the second year, one soil working is given to plants after the rains. No more care is necessary in subsequent years except protection against trampling by cattle, illicit cutting, and damage by fire and wind. For protection against damage by fire, clearing of fire lines 3 m wide around the plots during December-January in the first year of raising the plantation is the recommended practice. Rescraping of fire lines and maintenance by fire patrols should be done in subsequent years during December-January and every year thereafter until rotation age.

Management and Harvest

Throughout its growth period, black wattle sheds its twigs, leaves, flowers, and fruit. Being a leguminous tree, in addition to adding nutrients to the soil by way of litter accumulation, it also enriches soil fertility. The total quantity of this leaf litter, recycled by trees, is estimated about one and a half tonnes per hectare per year under protected conditions. The average depth of litter accumulated was also found to be about 5 cm for a period of ten years under black wattle cover. Venkataramanan et al. (1983) found that in black wattle, the annual dry weight of litter was 960 kg/ha; made up of 22.1 kg N, 0.5 kg P, 3.4 kg K, 3.3 kg Ca, 0.9 kg Mg, and 662 kg organic matter. The maximum leaf fall is during June and July. Recycling of these nutrients keeps the land highly fertile. A comparison of data from *shola* forests in the Nilgiris showed that *shola* leaf litter contained more nutrients. The undergrowth consists mainly of wattle regenerating from the mother trees. Wherever the canopy is not too close, grasses and other herbs and shrubs also make their appearance. The leaf litter should be protected to safeguard the soil from erosion, whereas the twigs and lower branches can be pruned for use as fuel. This can provide periodical additional returns.

The crop is ready for harvest in ten years. The bark is first cut around the tree at the base and peeled off during the rainy season. The debarked tree is then cut 10 cm above ground level by means of a sharp axe or preferably a saw and then billeted into 1 meter length pieces and stacked. The bark is dried and disposed of depending upon the quality of the bark and prevailing market rate. The debarked wood is estimated five times the quantity of bark.

Natural Regeneration and Rotation

The second rotation crop is obtained mainly by profuse natural regeneration from seeds and also by root suckers. This renders black wattle particularly suitable for a silvicultural system of clear felling, followed by natural regeneration. Burning is an effective means of controlling grass and inducing germination of wattle seeds. After extraction of bark and wood, the remaining slash is spread evenly and burned during February-March. Seeds are also sown in blanks before burning. By this method of control burning, profuse and adequate regeneration is achieved.

On receipt of premonsoon showers, seeds germinate, and if any blanks are noticed, nursery-raised container seedlings are planted. The resulting stand obtained by natural regeneration is usually very dense. Therefore a mechanical thinning is done in the third year to reduce the stand to 2,500 trees per hectare. Subsequently, a silvicultural thinning is done wherever necessary, taking into consideration growth and form of trees.

Recent observations indicate that the second rotation crop is more vigorous and healthier than the original planting. This may be due to accumulation of leaf litter, ash, etc., resulting from burning, and absence of grass and other weeds.

Rotation

Rotation of wattle depends on site and care given to plantations. A 12-year rotation is adopted in good quality sites in South Africa. In poor quality sites, an eight- to ten-year rotation is followed. Length of duration depends on conditions and growth of trees. In the Nilgiris, a ten-year rotation has been adopted. In the Nilgiris, Samaraj et al. (1977) compared runoff and soil loss during 1958-1970 in the plots of (1) natural "shola" (wet, mountain, temperate, evergreen) forest; (2) a blue-gum *(Eucalyptus globulus)* plantation; (3) a black wattle *(Acacia mearnsii)* plantation; and (4) a mixed plantation of (2) and (3), on the Nilgiri plateau, Tamil Nadu. Little difference existed in runoff between the vegetation types; soil loss (slight) was observed only from (2). Annual litter accumulation was about 2.9, 2.3, and 1.3 tonnes per ha for (1), (2), and (3) respectively. The yield of timber (at the rotation age of ten years) from both *E. globulus* and *A. mearnsii* was greater from mixed plantations of these species than from pure plantations of either. Such mixed plantations are therefore recommended for both timber production and soil and water conservation in the region.

UTILITY OF BARK AND WOOD AND SOIL FERTILITY

The bulk of the bark goes as industrial raw material for the manufacture of wattle extract. The average percentage of tannins from black wattle is 35. The debarked wood is used for the manufacture of rayon-grade pulp and paper pulp. It is a very good mine crop. It adds more than 1,600 to 2,000 kg of leaf litter per hectare per year and enriches the soil with nitrogen-fixing bacteria through root nodules.

The per hectare production of bark and wood in the Nilgiris is perhaps lower than it ought to be, though in good pockets the production can be as high as 40 tonnes of bark and 200 tonnes of wood. Though it may not be possible to bring all the areas to this production level, it may be possible by 2010 or so to bring the bulk of the areas to produce at least 30 tonnes of bark and 150 tonnes of wood per hectare on average. Intensive soil preparation, addition of optimum dosage of fertilizers, and strict fire protection measures seem to be the solution to bring the poor quality wattle areas to a reasonably productive level. This is more important as the Nilgiris have a considerable area of land that is too steep for terracing and unfit for agriculture, but which is ideal for black wattle cultivation. Both in terms of profit and the utilization of natural resources, the cultivation of black wattle is evidently an excellent investment in the Nilgiris.

YIELD

Black wattle stands are felled about ten years of age. Sometimes, it may be necessary to carry out a prefelling thinning on the clearing about 18 months before the end of rotation. The objective is to remove diseased and nonstrippable trees to facilitate the main felling operation. Felling is done by axe or bow saw during the time of the year when the bark is most easily stripped from the trees. Bark strippability is the best when the sap is rising and this occurs during warm, wet summer months.

At clear felling, the trees will have the breast height diameter of 15-20 cm with height up to 18-22 m. The felled trees are debranched and the bark is stripped as far up the stem as possible, cut into lengths of 1.2 m, and tied in 50 kg bundles for transport to the tannin-extraction factory. Thoroughly dried bark is arranged in bales of 75 to 80 kg when ready for transportation. Tanning power improves by 10 to 15 percent in bark carefully stored for a season. Percent tannin does not differ between bark harvested in dry and wet seasons. The amount of bark on trees may be less on poor soils than on rich soils. Tannin runs about 25 to 35 percent per kilo of dried bark, on either poor or rich soil. Prior to felling the trees, horizontal cuts are given about a meter from the ground level. Then the bark is stripped down. This makes the complete removal of bark on the stump. Then the trees are felled and the bark is stripped upward to the point at which immature bark is reached. The bark stripped from the tips of the trees is generally green in color, whereas that obtained from below is gray in color. The difference in quality between the two extracts is relatively slight (Grey and Jones, 1960).

The green bark is sent to the factory. During drying, the bark is spread with its outer surface exposed. The inner surface is not exposed to the sun because it may lead to discoloration. Drying is done until the bark becomes brittle and brown in color. Instead of sun drying, bark may be dried in hot-air ovens. The dried bark is cut into small pieces of 15-25 cm length and pressed into bales and sent to tanneries. Bark should be as fresh as possible for delivery to the factory, any delay may result in oxidation and excessively red-colored extract or the formation of fungal mold, which is an undesirable characteristic. The yield of bark varies from 8 to 30 tonnes per ha. In the Nilgiris, it varies from 8 to 17 tonnes of bark per hectare of plantation. With respect to wood for fuel, it ranges from 75 to 100 tonnes per hectare.

An average plantation yields about 40 tonnes of bark and 200 tonnes of wood per hectare. Taking the 2001 market prices of bark and wood (bark at the rate of Rs. 500 per tonne and wood at Rs. 125 per tonne) and also taking into account the expenditure on conversion, supervision, transportation, and overhead charges (Rs. 100 for one tonne of wood and Rs. 100 for peeling of one tonne of bark), a hectare of wattle plantation fetches about Rs. 21,000 on average for a rotation of ten years. This is Rs. 2,100 per year. In the second rotation, the returns will be much more than Rs. 21,000.

The percentage of tannin content is correlated directly with bark thickness. The diameter of the tree is correlated positively with bark thickness and tannin content. The larger the mean diameter of the tree, the thicker its bark will be (Rajagopalan, 1973).

At the Wattle Research Station, Ootacamund, Tamil Nadu, India, it was found that there was not much difference between the age group and tannin content, though the maximum percentage of tannin was observed in the ninth year (41.12 percent) and the minimum in the sixth year (37.19 percent) (Rajagopalan, 1973) (see Table 12.1).

Although the increase in tannin content is small, the increase in bark weight is considerable.

Experiments conducted in South Africa on quality and yield of black wattle bark in regard to different densities, i.e., 1,000, 1,250, 1,500, 1,750, and 2,000 per hectare and managed on rotation ranging from 8, 10, and 12 years revealed that for the rotation lengths the percentage of stripping decreased more or less regularly with decreasing stand density (see Table 12.2).

The mortality percentage, however, though following a similar trend on short rotation, behaved quite differently on rotations of 10 and 12 years. Mortality was invariably the lowest at density of 1,500 trees per ha and showed a more or less regular size with successive density levels above and below this point. It was presumed to be due to the fact that mortality resulting from suppression rises as stand density increases above 1,500 trees per

TABLE 12.1. Tannin content of the bark from trees of different ages.

	Age in years				
	4	5	6	7	8
Tannin (%)	36.10	37.70	38.90	39.30	39.40
Moisture-free bark (tonnes/ha)	4.09	5.93	6.50	7.76	9.64
Tannin/ha (tonnes)	1.48	2.23	2.51	3.04	3.76

Source: Rajagopalan, 1973.

TABLE 12.2. Yield of bark from different stand of densities per hectare.

No. of stems	Yield of bark (tonnes)
2,000	23.98
1,750	23.38
1,500	23.15
1,250	22.32
1,000	20.39

Source: Rajagopalan, 1973.

ha, and mortality from disease increases as densities fall below this level (Sherry, 1966).

The average yield of bark for the South African wattle industry is estimated at 18.83 tonnes per hectare. Experiments conducted at the Wattle Research Institute, Pietermaritzburg, to determine the optimum time of felling the wattle stand in tonnes of tannin content revealed that the tannin content of the bark increased significantly with age (Anonymous, 1964).

Except for some mangrove species, black wattle in pure stand produces more tannin per hectare than most tanniniferous plants. In South Africa, well-managed plantations produced the equivalent of 3 tonnes per ha tannin, about twice the average, when grown in rotations in excess of 12 years. Twelve trees produce 1 m^3 of firewood. The wood of debarked trees is dried and used for pulpwood, fuel, and mine timbers. Moisture loss is rapid during the first four weeks after felling, then much slower. Wood weighs 708.7 kg per m^3. One tree produces up to 10 cwt of bark or about 5 cwt stripped. One tonne of black wattle bark is sufficient to tan 2,530 hides, best adapted for sole leather and other heavy goods; the leather is as durable as that tanned with oak bark. One tonne of bark yields 4 cwt of extract tar. Destructive distillation of the wood yields 33.2 percent charcoal, 9.5 percent lime acetate, and 0.81 percent methyl alcohol. As a source of vegetable tannin, black wattle shares a large portion of the world market for vegetable tannins with quebracho and chestnut. According to Sherry (1971), plantation-grown wattle in South Africa, Rhodesia, Tanzania, Kenya, and Brazil supplied about 38 percent of world demand for tannin. *Eucalyptus grandis* produces more wood than wattle, but is inferior for fuel and charcoal. At one time in South Africa, 56 percent of the proceeds from wattle were from bark, the balance from timber (Duke, 1981).

PROCESSING OF BARK

Wattle bark as a tanning material appears to have entered the European market in 1908, and its exceptional merits were immediately recognized by the tanning industries in England and continental Europe. It contains astringent catechol tannin and lends itself particularly to sole-leather manufacture, but it can also be used very successfully for light leather. Wattle leather is firm and durable.

Although classified as a rapid tanning material, the leather is much less red than that obtained from many other catechol tans. The solubility of wattle tannin compares very favorably with that of other commercial vegetable tannins, and the temperature and concentration of extraction are not such important factors as in quebracho. The tan liquors produce very little acid on fermentation, and in consequence do not plump well. Wattle therefore makes a good blend with acid-producing tanning materials, such as myrobalans.

Gupta et al. (1981) reported that condensed tannin (used in adhesives) from *A. mearnsii* bark was extracted in boiling water and purified using hide powder. Crude yield was 39 percent of bark dry weight, and purified yield was 52 percent of this. MCL weight was 1,025 kg and the contents of total and phenolic hydroxyl groups were 26 and 14 percent respectively.

The tannin content of dried wattle bark is 30 to 40 percent. Wattle bark extract is manufactured by a countercurrent extraction process (Williams, 1954) whereby shredded bark is leached with hot water in a series of vessels. The resulting liquor is then clarified and concentrated by evaporation in multistage evaporators and vacuum pans until a point is reached where it solidifies on cooling. It is then run to cooling racks where the product is left to harden. Any contact with iron must be carefully eliminated from the process because of the discoloration that occurs when tannin comes in contact with this material.

PESTS AND DISEASES

One factor that limits the yield of bark is the damage caused by wattle bagworm. The wattle bagworm (*Acathopsyche junodi* Hely) causes serious damage. This pest causes complete defoliation of trees of all ages with consequent permanent stunting of growth. The other principal insects attacking Brazilian wattle are *Molippa sabina, Achryson surinamum, Placosternus cycleme, Eburodacrys dubitata, Neoclytus pusillus, Oncideres impluviata, Oncideres saga,* and *Trachyderes thoracica.* Ants, termites, and borers also cause damage. The dauva ant, which attacks the leaves, is fought constantly with arsenicals and carbon disulfide. Nematodes reported on this species include *Meloidogyne arenaria, M. incognita* ssp. *acrita, and M. javanica.*

The most serious disease is dieback, caused by *Phoma herbarum.* Other fungi attacking black wattle include *Chaetomium cochliodes, Daldinia* spp., and *Trichoderma viride.*

Ribeiro et al. (1988) found that several *Acacia decurrens* plants of the Capao Bonito region of Brazil were affected with symptoms of wilting, wood splitting, and gum exudation. Transversely cut wood showed ashy-colored pith, which liberated numerous perithecia when incubated in a humid chamber. The pathogen was identified as *Chaetominu fimbriata.* The four-month-old plants inoculated with isolate died after 14 days. In cross-inoculation tests, isolates from *Acacia* and mangoes were pathogenic to both hosts.

Morris et al. (1998) recorded a rust fungus on *A. mearnsii* for the first time in South Africa. It is a uredial rust, and comparison with rusts on this host in Australia suggests that it is probably the uredial state of *Uromycladium alpinum.*

REFERENCES

Anonymous. 1960. Report of Wattle Research Institute, the University of Natal, Pietermaritzburg, S. Africa.

Anonymous. 1964. Report of Wattle Research Institute, the University of Natal, Pietermaritzburg, S. Africa.

Anonymous. 1999. Forest survey of India. Indian Council for Forest Research and Education, Dehradun.

Duke, J.A. 1981. *Handbook of Legumes of World Economic Importance.* Plenum Press, New York.

Grey, G.G. and Jones, R.A. 1960. Report of the Wattle Research Institute, University of Natal, Pietermaritzburg, S. Africa.

Gupta, B.N. and Thapliyal, R.C. 1974. Presowing treatment of black wattle (*Acacia mearnsii* De Wild) and Australian black wood (*Acacia melanoxylon* R. Br.) seed. *Indian Forester* 100(12): 733-735.

Gupta, R.K. 1993. *Multipurpose Trees for Agroforestry and Wasteland Utilisation.* Oxford and IBH publications, New Delhi.

Gupta, S., Singh, S.P., and Gupta, R.C. 1981. Studies on tannin from Indian wattle (*Acacia mearnsii*) bark. *Indian Journal of Foresty* 4(1): 18-21.

Hussain, A.M.M., Ponnuswamy, P.K., and Viswanathan, M. 1980. Nutrition studies on black wattle I: Effect of different NPK fertilizer combinations up to the 6th year. *Indian Forester* 106(6): 397-402.

Morris, M.J., Wingfield, M.J., and Walker, J. 1988. First record of the rust on *Acacia mearnsii* in southern Africa. *Transactions of British Mycological Society* 90(2): 324-327.

Rajagopalan, K.P. 1973. Seminar on Forest Institutes, Tamil Nadu Forest Department, Government of Tamil Nadu, Madras.

Ribeiro, I.J.A., Ito, M.F., Paradela Filho, O. and De Castro, J.L. 1988. Gummosis of *Acacia decurrens* Wild. caused by *Ceratocystis fimbriata* Ell. and Halst. *Bragantia* 47(1): 71-74.

Samraj, P. and Chinnamani, S. 1978. Black wattle cultivation in Nilgiris. *Indian Farming* 28(1): 15-18.

Samraj, P., Chinnamani, S., and Haldorai, B. 1977. Natural versus man-made forest in Nilgiris with special reference to runoff, soil loss, and productivity. *Indian Forester* 103(7): 460-465.

Sherry, S.P. 1966. Report of the Wattle Research Institute for 1965-66. Wattle Research Institute, University of Natal, Pietermaritzburg, South Africa.

Sherry, S.P. 1971. *The Black Wattle* (Acacia mearnsii De Wild.). University of Natal Press, Pietermaritzburg.

Venkataramanan, C., Haldorai, B., Samraj, P., Nalatwadmath, S.K., and Henry, C. 1983. Wattle cultivation. *Indian Forester* 109(6): 370-378.

Williams, W. 1954. The Natal wattle industry. *World Crops* 6: 417-419.

Index

Page numbers followed by the letter (t) indicate tables; those followed by the letter (f) indicate figures.